Entropy and Free Energy in Structural Biology

Foundations of Biochemistry and Biophysics

This textbook series focuses on foundational principles and experimental approaches across all areas of biological physics, covering core subjects in a modern biophysics curriculum. Individual titles address such topics as molecular biophysics, statistical biophysics, molecular modeling, single-molecule biophysics, and chemical biophysics. It is aimed at advanced undergraduate- and graduate-level curricula at the intersection of biological and physical sciences. The goal of the series is to facilitate interdisciplinary research by training biologists and biochemists in quantitative aspects of modern biomedical research and to teach key biological principles to students in physical sciences and engineering.

New books in the series are commissioned by invitation. Authors are also welcome to contact the publisher (Lou Chosen, Executive Editor: lou.chosen@taylorandfrancis.com) to discuss new title ideas.

System Immunology: An Introduction to Modelling Methods for Scientists
Jayajit Das, Ciriyam Jayaprakash

An Introduction to Single Molecule Biophysics
Yuri L. Lyubchenko

Light Harvesting in Photosynthesis
Roberta Croce, Rienk van Grondelle, Herbert van Amerongen, Ivo van Stokkum (Eds.)

An Introduction to Single Molecule Biophysics
Yuri L. Lyubchenko (Ed.)

Biomolecular Kinetics: A Step-by-Step Guide
Clive R. Bagshaw

Biomolecular Thermodynamics: From Theory to Application
Douglas E. Barrick

Quantitative Understanding of Biosystems: An Introduction to Biophysics
Thomas M. Nordlund

Quantitative Understanding of Biosystems: An Introduction to Biophysics, Second Edition
Thomas M. Nordlund, Peter M. Hoffmann

Entropy and Free Energy in Structural Biology: From Thermodynamics to Statistical Mechanics to Computer Simulation
Hagai Meirovitch

https://www.crcpress.com/Foundations-of-Biochemistry-and-Biophysics/book-series/CRCFOUBIOPHY

Entropy and Free Energy in Structural Biology

From Thermodynamics to Statistical Mechanics to Computer Simulation

Hagai Meirovitch

CRC Press
Taylor & Francis Group
Boca Raton London New York

CRC Press is an imprint of the
Taylor & Francis Group, an **informa** business

To Eva, Avner, Ofra

Alon, Amir, Haim, and Noga

Contents

Section I Probability Theory

Section II   Equilibrium Thermodynamics and Statistical Mechanics

Section III   Topics in Non-Equilibrium Thermodynamics and Statistical Mechanics

Section IV   Advanced Simulation Methods: Polymers and Biological Macromolecules

Preface

Statistical mechanics is a well established basic discipline encompassing all the exact sciences; its theoretical foundations, laid down at the end of the nineteenth century, have been summarized in the early classical monographs of Tolman, as well as Landau & Lifshitz. In many later books, the scope of the theory has been extended to new systems and phenomena. Statistical mechanics received a big boost in the 1950s of the twentieth century with the advent of Monte Carlo (MC) and molecular dynamics (MD) simulation techniques. Thus, computer simulation, not only has become an organic part of statistical mechanics, but it is currently the main engine of development in this field, reflected by a vast literature including the early books of Binder, Allan & Tildesley, Frenkel & Smith, and more recent ones. The constant progress in this field requires timely reviews of new material, which is one aim of this book.

The present book is an extension of a course given first to graduate students in the biophysics program at Bar Ilan University, Israel, and later to students in the graduate program of the Department of Computational and Systems Biology at the University of Pittsburgh School of Medicine. The book, which is structured as a course, consists of four parts: (1) probability theory, (2) equilibrium thermodynamics and statistical mechanics, (3) non-equilibrium problems, and (4) computer simulation of polymers and biological macromolecular systems. The 23 chapters of the book stand to a large extent by themselves. The book requires entry level mathematics—derivatives and integrals—and thus can benefit graduate and undergraduate students, as well as researchers from all the exact sciences (in particular, chemistry, chemical engineering, biophysics, and structural biology), who seek to get acquainted with statistical mechanics theory and computer simulation methodologies for treating polymers and biological macromolecules. We elaborate on the (somewhat vague) notion of entropy, pointing out the difficulties encountered in its computation and in the computation of the related properties—free energy, chemical potential, and the potential of mean force. Sophisticated methodologies (including ours) for the calculation of these properties are presented and classified with respect to performance and functionality.

While the material covered in the book can be learned from the contents, we highlight below, in some detail, unique topics and focuses of the book, such as teaching aspects of probability theory, the non-traditional derivation of statistical mechanics, or our methods for calculation the entropy.

In Chapter 1, which contains standard probability material, we elaborate (with examples) on the related notions of *experimental probability*, *probability space*, and the *experimental probability on a computer*, advocating *probability space* to be used as a framework for solving probability problems in a systematic way. The distinction between these three phases, which is sometimes confusing to students, is essential for devising simulation methods, in particular, for the entropy. We devote longer discussions than usual to the concept of *product space*, which constitutes the theoretical foundation for simulation and estimation theories; we also discuss the related "central limit theorem" (CLT)—a highly used tool throughout the book. Chapter 2 provides a short review of classical thermodynamics. An emphasis is given to the notion of entropy in equilibrium and non-equilibrium conditions, illustrated by solvable problems for an ideal gas, where its entropy, S, is shown to include probabilistic elements, $S \sim \ln P$, already on the thermodynamics level. This leads to our *non-traditional* derivation of statistical mechanics (Chapter 3), where thermodynamic parameters (e.g., energy) become statistical averages based on initially unknown probability density, P, and an entropy function, S'. Based on $\ln P$ above, and experimental properties (e.g., extensivity), an S' function is defined, becoming part of a free energy functional $A(P)$, which is minimized with respect to P, leading to the Boltzmann probability density.

In Chapter 4, we derive the statistical mechanics equations of the ideal gas and the harmonic oscillator on three levels: macroscopic, microscopic, and quantum mechanics. Elaborating on the differences among these models deepens the understanding of the probabilistic nature of statistical mechanics. Fluctuations (standard deviations), σ, are treated in Chapters 5 and 6. For the exact free energy, A, $\sigma(A) = 0$—a completely unrecognized result, which can lead to the correct A by extrapolating to $\sigma(A) = 0$

approximate results of A versus $\sigma(A)$. The behavior of a large system, $\sigma(E)/E \to 0$, provides a way for solving problems in statistical mechanics based on "the most probable energy term," in addition to using thermodynamic derivatives and statistical averages. Likewise, different ensembles lead to the same averages (but not fluctuations). This versatility of alternative treatments (which is not always emphasized) is demonstrated by carrying out multisolutions of specific problems, in particular, for calculating the entropic forces required to stretch a one-dimensional ideal chain (Chapter 8). First- and second-order phase transitions, the Ising model, and the corresponding critical exponents are shortly discussed in Chapter 7, and self-avoiding walks (SAWs) with and without attractions are reviewed in Chapter 9, and the relation of their properties to the behavior of real macromolecules is discussed.

Chapters 11–13 cover several topics in non-equilibrium thermodynamics and statistical mechanics *close to equilibrium*. However, being relaxation-type methods, MC/MD are presented first, and their efficiencies are discussed (Chapter 10) as a precursor to the relaxation phenomena studied in the subsequent chapters. In Chapter 11, Onsager relations, microscopic reversibility, the notion of a steady state, and *the principle of minimum entropy production* are described in detail with solved problems. Chapter 12 presents complete derivations of the two Fick's laws, the diffusion equation (and Einstein's derivation for Brownian motion), Langevin and Fokker-Planck equations, and the stochastic dynamics simulation. Chapter 13 is devoted to the master equation, with a specific example from nuclear magnetic resonance (NMR). Finally, a statistical mechanics example of the principle of minimum entropy production is presented. The effects of the *close to equilibrium* condition and *the local equilibrium hypothesis* are carefully examined.

Chapter 14 deals with methods that generate a SAW step-by-step (from nothing) with the help of transition probabilities (TPs); thus, unlike the case of MC, the construction probability, $P = \Pi_k TP_k$, is known and S is known as well. "*Simple sampling*" is an exact, but inefficient method (due to "sample attrition") since a direction, v, is chosen "democratically" ($p_v = 1/2d$) ($N < 90$ in $d = 2$). With the Rosenbluth & Rosenbluth (RR) method ($N < 160$, $d = 3$), v is chosen out of the *unoccupied* sites, a procedure which introduces a bias that should be removed. More efficient are "*the enrichment method*" ($N \sim 1000$) and "*the dimerization method*" ($N \sim 10^4$). With our "*scanning method*" (SM), the one step ahead ($f = 1$) scanning of RR is replaced by a scanning parameter $f > 1$, and TP_k is obtained by exact enumeration ($f < 20$ for $d = 2$). SM ($N \sim 1500$ $d = 2$) provides a lower and upper bounds for S, and a procedure for estimating accuracy; it is, in particular, efficient for chains under geometrical constraints and for Ising models. Chapter 15 is devoted to the *pivot algorithm*, which is an MC-type dynamic method, where global pieces of the chain are moved. The method is extremely efficient for SAWs in the bulk ($N \sim 10^6$), but very inefficient for chains under geometrical constraints. Combined methods, enrichment/dimerization and enrichment/RR are also discussed.

Chapter 16 is devoted to proteins: We define the protein chain (e.g., by dihedral angles), its energy function (force field), and implicit and explicit models for the solvent. The difficulty to fold a protein is attributed to the rugged potential energy surface. Energy minimization techniques, and the (non-statistical mechanics) methods for conformational search, "MC *minimization*" and "*simulated annealing*" are discussed. We elaborate on the much neglected concepts of a *microstate* (e.g., the localized region occupied by an α-helix) and *intermediate flexibility*, which characterizes a peptide or a loop populating significantly several microstates in equilibrium. Intermediate flexibility should be considered in *solution* structure determination of flexible proteins/peptides based on NMR data. Chapter 17 deals with the calculation of free energy differences (ΔF) by *calorimetric* thermodynamic integration (TI). This is a general robust approach, depending *only* on thermodynamic variables, such as T and E, without the need to consider the microscopic structure of the system studied. Zwanzig's free energy perturbation (FEP) and Kirkwood's TI pertain to this category. To obtain the absolute F, the integration should start from a reference state with known F. Two such integrations for a SAW and a peptide are presented. Thermodynamic cycles are described as a tool for comparing the free energy of binding of two ligands to the same enzyme. This procedure is appealing since the transformation of ligand A to B relies only on the ligand-environment interactions. However, the process suffers from poor convergence of the protein's side chains and difficulties of the energy to lead to the correct structure of the active site/B complex. In Chapter 18, non-integration methods for calculating the *absolute* S and F are discussed: among them are, "*the harmonic approximation*," "*the quasi-harmonic approximation*" (QH), "*the second generation mining minima*," and others. The efficiencies of these methods are discussed; in particular, QH can handle reliably only a single microstate. These methods enable one to calculate the absolute entropy S_a of a microstate a of a peptide/protein and thus the difference $\Delta S = S_b - S_a$ between microstates a and b—a simple alternative to carrying out a complex integration from a to b.

In Chapter 19, we describe our approach for extracting the absolute entropy from a single MC/MD sample. This approach is based on the recognition that two large samples simulated by different *exact* techniques are equivalent in the sense that they lead to the same averages and fluctuations. Therefore, one can treat a given MC sample of SAWs, for example, as generated by the scanning method, which allows reconstructing for each chain, the TPs that *hypothetically* were used by SM to construct it. This leads to an upper bound, $S^A(f)$, and the method is called "*hypothetical scanning*" (HS). One can also define the stochastic HS, where each TP at step k is calculated, not by exact enumeration, but by an additional MC simulation (based on an n-size sample) applied to the f future steps, or to the entire future of $N - k + 1$ steps, leading to *partial* HSMC and *complete* HSMC, respectively. One can also define several lower bounds denoted by S^{Bi} and the corresponding averages, $S^{Mi} = (S^A + S^{Bi})/2$, which are expected to outperform S^A and S^{Bi} individually. This methodology has been applied to peptides, Ising models, Lennard-Jones fluid (argon), and TIP3P water. Combining HSMD with TI creates HSMD-TI, which allows calculating $\Delta F(a,b)$ for both the peptide and the surrounding water. Unlike QH, the HS approach is applicable to any chain flexibility, ranging from harmonic microstates to the random coil state. Another method is the "*local states*" (LS) method, where the TPs are calculated from the frequencies of occurrence of certain local states. LS was applied to Ising models, peptides, and fluid dynamics.

Chapters 20–22 present advanced simulation techniques (based on MC/MD) for different purposes. The first group, which seeks to outperform FEP and TI for calculating ΔF, includes among others "*umbrella sampling*," "*Bennett's acceptance ratio*," and the Jarzynski's approach. The "*weighted histogram analysis method*" (WHAM) (and its extensions) is a central method for the calculation of the potential of mean force. Conformational search techniques are "*temperature and Hamiltonian replica exchange*," the "*multicanonical method*," the "*method of Wang and Landau*," and the "*method of expanded ensembles*." Also, to gain further efficiency and to widen applicability, some methods have been combined to create hybrid techniques. We also discuss methods for calculating the chemical potential, including "*Widom's insertion method*," the "*deletion*" and the "*insertion/deletion*" methods of Shing and Gubbins, and more recent techniques based on Personage's ideas and others. Performance studies based on toy or simplified models have been inconclusive. Therefore, we follow carefully the applications of these techniques in the literature, providing in detail the systems studied. Our criterion of performance is based on the maximal system size and complexity that a method can handle.

In Chapter 23, the statistical mechanics equation of Boresch et al. for the standard free energy of reaction, ΔA^0, is derived. A common way to calculate ΔA^0 is by TI, where the ligand-environment interactions are gradually eliminated in both the solvent and the protein. However, this (rigorous) approach (called DAM) might encounter convergence problems in the final TI stages, where due to weak ligand–protein interactions, the ligand might leave the active site ("the end-point problem"). This problem has been rigorously solved by an appropriate addition of harmonic restraints. HSMD-TI differs from these approaches in two main aspects. (1) TI is carried out on structurally *fixed* structures, therefore, the "end-point problem" is eliminated and the need to apply restraints is avoided. (2) The decrease in the ligand's entropy in going from the solvent to the protein's active site is calculated. HSMD-TI was applied to the ligand, FK506 (126 atoms) complexed with the protein FKBP12, where $\Delta A^0 = -12.8$ kcal/mol is known experimentally. The HSMD-TI result, -13.6 ± 1.1 kcal/mol is one of the best in the literature. The ligand's entropy is reduced by 7.1 ± 1.2 kcal/mol. Finally, the pitfalls encountered in estimation ΔA^0 with respect to the force field and the trajectories' length are critically discussed.

Acknowledgments

This book includes research conducted in my group over the years; thus, I would like to thank my mentors and collaborators. I am indebted to my PhD supervisor Professor Zeev Alexandrowicz (Weizmann Institute), who has planted in me the passion for entropy. I am grateful to my post doctoral supervisor Professor Harold A. Scheraga (Cornell), who acquainted me with the fascinating field of protein structure and folding. I appreciate all my collaborators, mostly graduate students and post docs, for their significant contributions to the research in ideas and hard work. I would like to mention several colleagues who have made a strong impact, in particular, on our structural biology studies: Max Vásquez, Canan Atilgan (Baysal), Iksoo Chang, Eva Meirovitch, Ron P. White, Srinath Cheluvaraja, Mihail Mihailescu, and Ignacio J. General. This book was initiated in the Department of Computational and Systems Biology, University of Pittsburgh School of Medicine; I thank Professor Ivet Bahar for her support. I thank also the Physics Department at Bar Ilan University, Israel for its hospitality. Finally, I would like to thank my wife, Eva, for her help and encouragement.

Author

Hagai Meirovitch is professor Emeritus in the Department of Computational and Systems Biology at the University of Pittsburgh School of Medicine. He earned an MSc degree in nuclear physics from the Hebrew University, a PhD degree in chemical physics from the Weizmann Institute, and conducted postdoctoral training in the laboratory of Professor Harold A. Scheraga at Cornell University. His research focused on developing computer simulation methodologies within the scope of statistical mechanics, as highlighted below. He devised novel methods for extracting the absolute entropy from Monte Carlo samples and techniques for generating polymer chains, which were used to study phase transitions in polymers, magnetic, and lattice gas systems. These methods, together with conformational search techniques for proteins, led to a free energy-based approach for treating molecular flexibility. This approach was used to analyze NMR relaxation data from cyclic peptides and to study structural preferences of surface loops in bound and free enzymes. He developed a new methodology for calculating the free energy of ligand/protein binding, which unlike standard techniques, provides the decrease in the ligand's entropy upon binding. Dr. Meirovitch conducted part of the research indicated here, and other studies, at the Supercomputer Computations Research Institute of the Florida State University, Tallahassee.

Section I

Probability Theory

1

Probability and Its Applications

1.1 Introduction

Probability is a basic scientific concept in all the sciences, in particular, in statistical mechanics, quantum mechanics, and the theory of computer simulation. However, the application of probability theory is not always straightforward, and we thus find it important to provide an extensive review of this topic. We emphasize the essential notions of experimental probability, a *formal* probability space, and a product probability space, which provide a systematic approach for solving probability problems. In addition, a standard review is provided of discrete probability theory, random variables, the central limit theorem, Markov chains, etc. All these elements are imperative for devising new computer simulation methods, in particular, methods for calculating the entropy and the free energy. The chapter is based on several books appearing as the references [1–4].

1.2 Experimental Probability

We first define the notion of *experimental probability*. Assume that a die is rolled n times, and we are interested in the chance to get an odd number, $f(m) = m/n$, where m is the number of times an odd number has been obtained. The ratio, $f(m)$, is the *relative frequency* of an odd number. The results of multiple experiments appear in Table 1.1.

The table suggests that for $n \rightarrow \infty$ $f(m)$ would converge to:

$$f(m) \rightarrow P = 0.5 \tag{1.1}$$

where P is defined as the *experimental probability* (in some cases the time, t replaces n). It should be pointed out that P can be defined only if the same experiment (the same die) is repeated. However, while the exact $P(n \rightarrow \infty)$ can never be obtained experimentally, the larger is n, the better is the estimation of P. Still, under the assumption of ideal conditions, P can be guessed, and instead of carrying out experiments, it is more efficient to describe the experimental situation within the framework of an *idealized* mathematical model called *probability space*, where the probability, in principle, is known *exactly*.

TABLE 1.1

Experimental Probability, $f(m) = m/n$ from Flipping a Coin
Experiment

n	10	50	100	400	1000	10,000
m	7	29	46	207	505	5034
$f(m)$	0.7	0.58	0.46	0.517	0.5050	0.5034

Note: n and m are the number of experiments, and the number of
times an odd number has been obtained, respectively.

1.3 The Sample Space Is Related to the Experiment

We start by defining the notion of a *sample space*, which consists of *events* and relations among them.

The **elementary event** is the possible outcome of a *single* experiment. For a tossed coin with two possible outcomes, A and B, there are two elementary events, A happened or B happened; for a die, with six possible outcomes, the number of elementary events is six: 1, 2, 3, 4, 5, or 6 happened.

The **event** is any combination of elementary events. For a die, one can define the event: "an even number happened," that is, *one* of the following happened: 2, 4, or 6. Another event is: "a number larger than 3 happened," meaning that, 4, 5, or 6 happened.

The **empty event** is the impossible event denoted by $\emptyset$. An empty event for a die is: "a number between 2 and 3 happened."

The **certain event** is denoted by Ω; a certain event for a coin is: A or B happened.

The **complementary event** of an event, A is denoted by $\bar{A}$, where $\bar{A} = \Omega - A$; for example: $(1,2,3,4,5,6) - (2,4,6) = (1,3,5)$.

We also define two operations among events depicted in Figure 1.1:

The **union** (denoted by $\cup$); under union, all elementary events are added only *one* time; for a die: $(1,2) \cup (2,3,5) = (1,2,3,5)$.

The **intersection** (denoted by $\cap$); this operation selects the common elementary events in two groups; for example: $(1,2,4) \cap (2,5,6) = (2)$.

$A\cup B$ = A and B $A\cup B$ = whole
$A\cap B = \emptyset$ $A\cap B$ = grey intersection

FIGURE 1.1   Illustration of union and intersection.

1.4 Elementary Probability Space

An elementary probability space is a sample space, where for each event, a probability is assigned (Figure 1.2), thus,

1. The sample space consists of a finite number, k of elementary events (points) B_i, $i = 1,2, ..., k$
2. Every partial set is an event
3. A probability $P(A)$ is defined for each event A
4. The probability of an elementary event, B_i is $P(B_i)=1/k$
5. $P(A) = l/k$, where l is the number of points defining event, A; $0 \leq P(A) \leq 1$; see Figure 1.2
6. For two events A and C, $P(A \cup C) \leq P(A) + P(C)$
7. $\sum_i P(B_i) = 1$
8. Notice that elementary events can have different probabilities (e.g., a non-symmetric coin)

The probability space is an *ideal model* of reality. In the *experimental world*, rolling a die would never be exactly random, and the (measured) probabilities of the elementary events thus would not be equal. However, in many cases, under the assumption of ideal conditions, the probabilities of the elementary events can be guessed and thus be incorporated within the corresponding probability space. This enables one to calculate the probability of the events of interest without the need to carry out experiments. However, one should always make the distinction between the probabilities in the experimental world and their counterparts in the probability space.

It should be pointed out that the structure of the probability space defined above provides a *systematic way* for treating problems involving probability; this procedure is based on three successive steps: (1) define the elementary events, (2) calculate their probability, and (3) define the events of interest and determine their probability; these steps are applied in the following simple problem and in other examples presented later in this chapter.

Example

A box contains 20 marbles, 9 white and 11 black. A marble is drawn at random. What is the probability that it is white?

Solution

The first step is identifying the basic elementary event (EE), which is the "selection of one of the 20 marbles." Since there are $k = 20$ EEs, $P(EE) = 1/20$, and the event A—"a white marble was chosen" contains $l = 9$ EEs; thus, $P(A) = 9/20$. This consideration involves the ideal probability space; the real world significance is that P is a result of n experiments, $P = f(m)$, $n \rightarrow \infty$, where m is the number of white marbles obtained (see Table 1.1).

The relation between the experimental world and the probability space becomes even more complex in statistical mechanics. In general, one seeks to study a realistic system (say, a

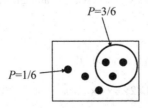

P=3/6

P=1/6

FIGURE 1.2   The number of elementary events, B_i is 6, thus $P(B_i) = 1/6$, and the probability of the circled event, A of 3 points is $P(A) = 3/6 = 1/2$.

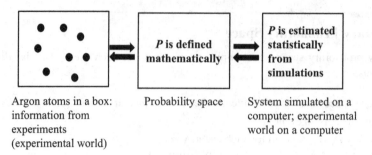

FIGURE 1.3   An illustration of the three phases related to probability, *P*: (1) The experimental system modeled by (2) a probability space, where *P* is well defined mathematically, but typically with an unsolved integral. Therefore, one moves to (3), the experimental world on a computer, where *P* and other parameters are estimated by simulations.

container of liquid argon) by theoretical means. Thus, the experimental system is modeled by a probability space [called the canonical (*NVT*) ensemble], where each spatial arrangement of the *N* molecules within the fixed volume, *V* (called a configuration), becomes an elementary event with (mathematically) well defined probability—the Boltzmann probability. However, unlike the simple case of marbles discussed above, the probability for argon is defined by an integral which cannot be evaluated analytically, and a common option is to resort to computer experiments (simulations). Thus, a system of argon atoms is built within the computer, the molecules are moved according to their intermolecular forces, and the probabilities and other properties of *the probability space* are *estimated* by statistical procedures. In other words, we return to the experimental world on a computer (Figure 1.3). Therefore, when dealing with probability issues (in particular, related to entropy), it is mandatory to make the distinction between the three possible phases—the original experimental world, the corresponding probability space, and the experimental world on the computer. In general, these phases are theoretically separated, however, in some theories they are mixed (e.g., Sections 19.2, 20.2, and 20.4.2). We shall return to this (sometimes confusing) subject later in this book. For a short philosophical discussion about probability, see the end of Chapter 6 of [3].

1.5 Basic Combinatorics

Assume a set of *n* cells and *r* different balls. What is the number of ways to arrange the balls in these cells? Clearly, every cell can contain any number of balls and each ball can be put in any of the *n* cells, therefore:

$$\# \text{ways} = n \times n \times n \times n \times \cdots \times n = n^r \tag{1.2}$$

n^r is also the number of ways to arrange *n* objects in *r* cells with repetitions. Notice that n^r is also the number of words of length *r* (with repetitions) based on *n* letters. For example, for the *n* = 3 letters, A, B, and C, one obtains: AA, BB, AB, BA, CC, CA, AC, BC, and CB, that is $n^r = 3^2 = 9$.

1.5.1 Permutations

We first define the operation *n*-factorial, denoted, *n*!,

$$n! = 1 \times 2 \times 3 \times 4 \times \cdots \times n \qquad\qquad 0! = 1 \tag{1.3}$$

Then we seek to find $(n)_r$—the number of groups of *r* objects which can be defined out of *n* different objects *without* repetitions (while the order is considered). Thus, the first object can be selected in *n* ways, the second in $n - 1$ ways, etc., where the rth object can be selected in $n - r + 1$ ways; therefore, the number of permutations $(n)_r$ is:

$$(n)_r = n(n-1)\cdots(n-r+1) = \frac{1\times 2\cdots(n-r)(n-r+1)\cdots n}{1\times 2\cdots(n-r)} = \frac{n!}{(n-r)!} \tag{1.4}$$

Notice that $(n)_n = n!$, and in the example, $(3)_2 \equiv (1,2), (1,3), (2,1), (2,3), (3,1), (3,2)$, there are no repetitions and the order is considered, that is $(1,3) \neq (3,1)$. $(3)_2$ is also the number of $r = 2$ letter words that can be obtained from $n = 3$ letters: AB, BA, CA, AC, BC, CB.

Example

Find the probability that r people ($r \leq 365$) selected at random will have different birthdays?

Solution

We first define the sample space, that is, all the EEs, which are all the arrangements of r objects (people) in $n = 365$ cells (days). The number of these EEs is 365^r [Equation (1.2)], and the probability of each EE is thus, $P(EE) = 1/365^r$. Now, one has to calculate the number of EEs, which define event A: "not two birthdays fall on the same day." This is obtained by realizing that the birthday of the first person is allowed to fall in any of the 365 days, the birthday of the second person in $361 - 1$ days, where the birthday of the rth person can happen only in $365 - r + 1$ days. Therefore, the number of points in A is:

$$365 \times 364 \times \cdots \times (365 - r + 1) = (n)_r \text{ and}$$

$$P(A) = \frac{(n)_r}{365^r} = \frac{365!}{(365-r)!\,365^r}$$

1.5.2 Combinations

What is the number of ways (combinations) to select r objects out of n objects if (unlike the case of permutations) the order is *not* considered? Clearly, the number of combinations is equal to the number of permutations $(n)_r$ divided by $r!$.

$$\binom{n}{r} = \frac{(n)_r}{r!} = \frac{n!}{(n-r)!\,r!}; \quad \binom{n}{0} = 1 \tag{1.5}$$

For $n = 3$; $r = 2$, the number of combinations, $\binom{n}{r} = 3$ is half the 6 permutations $(1,2), (2,1), (1,3)$, etc. $\binom{n}{r}$ is the binomial coefficient; see below.

Example

In how many ways can n objects be divided into k groups of $r_1, r_2,\ldots,r_k$ objects, where $\Sigma_k\, r_k = n$ without considering the order in each group, but considering the order between the groups (that is, $r_1 = 3, r_2 = 2$ is different from $r_1 = 2, r_2 = 3$)?

Solution

The number of ways is:

$$\binom{n}{r_1}\binom{n-r_1}{r_2}\binom{n-r_1-r_2}{r_3}\cdots = \frac{n!}{r_1!\,r_2!\ldots r_k!} \tag{1.6}$$

The first group (r_1) can be chosen in $n!/r_1!(n - r_1)!$ ways. We remain with $n - r_1$ objects from which we select the second group of r_2: $(n - r_1)!/(n - r_1 - r_2)!r_2!$, etc. We obtain the product:

$$\left[n!/r_1!(n-r_1)!\right]\left[(n-r_1)!/(n-r_1-r_2)!r_2!\right]\left[(n-r_1-r_2)!/(n-r_1-r_2-r_3)!r_3!\right]\ldots$$
$$\left[(n-r_1-r_2\ldots-r_{k-1})!/(n-r_1-r_2\ldots-r_k)!r_k!\right] = n!/\left[r_1!\ r_2!\ \ldots r_k!\right]$$

Notice that $n!/r_1!r_2!\ldots r_k!$ is the multinomial coefficient of $x_1^{r_1}x_2^{r_2}x_3^{r_3}\ldots x_k^{r_k}$ in the expansion of $(x_1 + x_2 + x_3 +\ldots+ x_k)^n$, and for $(1 + x)^n$, one obtains the binomial expansion,

$$(1+x)^n = \binom{n}{0}+\binom{n}{1}x+\cdots+\binom{n}{r}x^r+\cdots+\binom{n}{n}x^n. \tag{1.7}$$

A related problem is, "how many events exist in a sample space of n elementary events?" The answer is given in Equation (1.8) below, where the contributions of the different groups are added up; notice that the empty event (first term) and the certain event (last term) are also included,

$$\binom{n}{0}+\binom{n}{1}+\binom{n}{2}+\cdots+\binom{n}{n} = (1+1)^n = 2^n \tag{1.8}$$

Example

23 chess players are divided into 3 groups of 8, 8, and 7 players. What is the probability that players A, B, and C are in the same group (event A′)?

Solution

We first define the elementary event as a possible arrangement of the 23 players in the three groups; according to Equation (1.6), the total number of such arrangements is $23!/(8!8!7!)$.

If A, B, and C are in the first group, the number of players to be arranged decreases to 20 and the total number of arrangements decreases to $20!/(5!8!7!)$. In the same way, one calculates the number of arrangements as the three players belong to the second and third groups. Therefore,

$$P(\text{A}') = \frac{2\times 20!}{5!8!7!} + \frac{20!}{8!8!4!}\bigg/ \frac{23!}{8!8!7!} = \frac{21}{253}$$

Example

What is the number of ways to arrange n objects that $r_1, r_2, r_3, \ldots, r_k$ of them are identical, and $\Sigma r_i = n$?

Solution

If all the objects were different, the number of permutations would be $n!$. Because $r_1, r_2, r_3, \ldots, r_k$ are the same, one has to divide $n!$ by $r_1!r_2!r_3!\ldots r_k!$. Thus,

$$\#\text{ways} = \frac{n!}{r_1!r_2!\ldots r_k!} \tag{1.9}$$

Example

52 cards are divided among 4 players. What is the probability that every player will have a king?

Solution

The 52 cards are composed of four groups of 13 cards. The elementary event is a possible division of these cards among the four players; according to Equation (1.9), their number is $52!/(13!)^4$.

If every player has a king, only 48 cards are left to be distributed into 4 groups of 12 cards. This event (A) contains $48!/(12!)^4$ EEs, therefore,

$$P(A) = \frac{4!\,48!/(12!)^4}{52!/(13!)^4}$$

where the added factor, 4! is the number of distributions of 4 kings.

Solve the following problems within the probability space, as in the above examples.

PROBLEM 1.1

A die is rolled four times. What is the probability to obtain "6" exactly one time?

ANSWER

$P = 0.3858$.

PROBLEM 1.2

A box contains 6 red balls, 4 white balls, and 5 blue balls. 3 balls are drawn successively. Find the probability that they are drawn in the order red, white, and blue if the ball is (a) replaced, (b) not replaced.

ANSWER

$P(a) = 0.0356$, $P(b) = 0.0440$.

PROBLEM 1.3

Six passengers sit on a bus which visits 9 bus stops. What is the probability that two of the passengers will never get off the bus at the same bus stop?

ANSWER

$P = 0.113$.

1.6 Product Probability Spaces

The probability space defined thus far models a *single* experiment; for example, for a *single* tossing of a coin, we defined the probabilities $P(1) = p$ and $P(0) = q$ for the elementary events 1 or 0 (denoting "success" and "failure"), respectively. One might be interested in the probabilities for *two* subsequent tossing of a coin; here, four elementary events are defined, where each is a possible outcome of the double tossing, that is, (0,0), (0,1), (1,0), and (1,1). The corresponding probabilities can be estimated in the *experimental world* by tossing a coin twice n times ($n \to \infty$), where the relative frequencies $f(i, j)$ of successive outcomes (i, j) are calculated; this would lead to the *exact* probabilities, $P(i, j)$, defined in the probability space. In the case of *independent* tossing, the values of $P(i, j)$ can be obtained from the single tossing probabilities, $P(0)P(0)$, $P(0)P(1)$, $P(1)P(0)$, and $P(1)P(1)$.

In the same way, one can build a product probability space for n experiments, where the sample space can be represented schematically by the following structure, where each column contains all the EEs of the single experiment:

$$\begin{matrix} 1 & & 2 & & 3 & & & n \end{matrix}$$
$$\binom{1}{0} \times \binom{1}{0} \times \binom{1}{0} \times \cdots \binom{1}{0} \tag{1.10}$$

The EEs in the (product) sample space are the 2^n n-dimensional vectors consisting of 0 and 1:

$$(0,1,1,0,0,1,\ldots,1); \quad (1,0,0,1,1,1,\ldots,1); \cdots$$
$$(1,2,3,4,5,6,\ldots,n); \quad (1,2,3,4,5,6,\ldots,n); \cdots$$

If the experiments are statistically independent, and $p = q = \frac{1}{2}$, then the probability of each vector is the same, $P(\text{vector}) = (1/2)^n$.

In the same way, one defines a product space for n experiments of a die, where each column now contains the six EEs, 1,2...,6 of a single experiment,

$$\begin{matrix} 1 & & 2 & & 3 & & & n \end{matrix}$$
$$\begin{pmatrix} 1 \\ 2 \\ 3 \\ 4 \\ 5 \\ 6 \end{pmatrix} \times \begin{pmatrix} 1 \\ 2 \\ 3 \\ 4 \\ 5 \\ 6 \end{pmatrix} \times \begin{pmatrix} 1 \\ 2 \\ 3 \\ 4 \\ 5 \\ 6 \end{pmatrix} \times \cdots \times \begin{pmatrix} 1 \\ 2 \\ 3 \\ 4 \\ 5 \\ 6 \end{pmatrix} \tag{1.11}$$

The EEs in the (product) sample space are the 6^n n-dimensional vectors of the type,

$$(2,1,3,4,6,1,\ldots,5); \quad (3,4,2,1,1,5,\ldots,2); \cdots$$
$$(1,2,3,4,5,6,\ldots n); \quad (1,2,3,4,5,6,\ldots,n); \cdots$$

and if the experiments are statistically independent, and $P(i) = 1/6$ for $i = 1,\ldots,6$, one obtains for any vector, $P(\text{vector}) = (1/6)^n$.

The product probability space constitutes the theoretical basis for the sampling theory, in particular, by computer simulation (e.g., using the Monte Carlo method; see Chapter 10), where typical results are obtained only from k experiments, that is, only a *single* vector (realization) out of the 2^k total vectors is generated. Being the theoretical basis for sampling, the notion of probability product space is central in this book (see Sections 1.14–1.16).

Example

15 dice are rolled. Find the probability to obtain three times the numbers 1, 2, and 3 and twice, 4, 5, and 6?

Solution

The elementary events are all the possible outcomes of 15 experiments, thus $\#(\text{EEs}) = 6^{15}$. According to Equation (1.9), $\#(A) = 15!/[(3!)^3(2!)^3]$, therefore,

$$P(A) = \frac{15!}{6^{15}(3!)^3(2!)^3}$$

Also: "1" can be chosen in $(15 \cdot 14 \cdot 13)/3!$ ways, "2" in $(12 \cdot 11 \cdot 10))/3!$ ways, etc.

1.6.1 The Binomial Distribution

The product space discussed above for a coin can actually describe other phenomena as well. In the literature, statistically independent experiments of this type are called Bernoulli experiments, where $p + q = 1$, but p and q are typically not equal. In this case the probability of a specific vector is:

$$P(\text{vector}(n)) = p^m q^{n-m} \tag{1.12}$$

where m is the number of times the number 1 appears in the vector and $n - m$ is the number of zeroes. One can ask: "what is the probability to obtain exactly m 'successes' ('1')?" Clearly, there are $\binom{n}{m}$ ways (combinations) to arrange the m digits "1" within the vector, and thus the desired probability, $P_n(m)$ is:

$$P_n(m) = \binom{n}{m} p^m q^{n-m}. \tag{1.13}$$

$P_n(m)$ is called the binomial distribution. Based on the binomial expansion [Equation (1.7)], $P_n(m)$ is a normalized distribution,

$$\sum_{m=0}^{m=n} \binom{n}{m} p^m q^{n-m} = (p+q)^n = 1 \tag{1.14}$$

1.6.2 Poisson Theorem

Assume a sequence of n Bernoulli experiments, where the probability for a success is $p = \lambda/n$ ($\lambda > 0$ is a constant); in this case, the probability for m successes, $P_n(m)$ is:

$$P_n(m)_{n \to \infty} \to \frac{\lambda^m}{m!} e^{-\lambda} \tag{1.15}$$

Proof:

$$P_n(m) = \binom{n}{m} p^m (1-p)^{n-m} = \binom{n}{m} \left(\frac{\lambda}{n}\right)^m \left(1-\frac{\lambda}{n}\right)^{n-m} = \frac{n(n-1)\ldots(n-m+1)}{m!} \frac{\lambda^m}{n^m} \left(1-\frac{\lambda}{n}\right)^n \left(1-\frac{\lambda}{n}\right)^{-m}$$

For constant m and $n \to \infty$, the limiting values of the following expressions lead to Equation (1.15).

$$\frac{n(n-1)\ldots(n-m+1)}{n^m} \to 1; \quad \left(1-\frac{\lambda}{n}\right)^n \to 1; \quad \left(1-\frac{\lambda}{n}\right)^{-m} \to e^{-\lambda}$$

The Poisson theorem enables one to estimate $P_n(m)$ by Equation (1.15), which constitutes a good approximation to the binomial distribution when p is small and n is large. Equation (1.15) is called the Poisson distribution, which satisfies the normalization condition,

$$\sum_{m=0}^{\infty} \frac{\lambda^m}{m!} e^{-\lambda} = e^{-\lambda} \sum_{m=0}^{\infty} \frac{\lambda^m}{m!} = e^{-\lambda} e^{\lambda} = 1 \tag{1.16}$$

1.7 Dependent and Independent Events

Event A is defined as independent of event B if $P(A)$ is not affected if B occurred; we write,

$$P(A/B) = P(A) \tag{1.17}$$

where $P(A/B)$ is called the *conditional probability*.

Example

The following probabilities are defined for a die: $P(\text{even}) = P(\{2,4,6\}) = 1/2$, $P(\text{square}) = P(\{1,4\}) = 1/3$, and $P(2) = 1/6$. Therefore, "even" is independent of "square" since $P(\text{even/square}) = P(\text{even}) = 1/2$. On the other hand, "2" depends on "even" since $P(2/\text{even}) = 1/3 \neq P(2) = 1/6$. Also, "even" and "odd" are called *disjoint* events because $P(\text{even/odd}) = 0$.

1.7.1 Bayes Formula

$$P(A/B) = \frac{P(A \cap B)}{P(B)}; \qquad P(B) > 0 \tag{1.18}$$

$$P(B/A) = \frac{P(A \cap B)}{P(A)}; \qquad P(A) > 0 \tag{1.19}$$

therefore,

$$P(A/B) = \frac{P(B/A)P(A)}{P(B)} \tag{1.20}$$

Examples for a die: Defining A = (2) and B = even = {2,4,6} $\rightarrow P(A \cap B) = P(2) = 1/6$; using Equations (1.18) and (1.19) lead to:

$$P(A/B) = \frac{1/6}{1/2} = 1/3 \qquad P(B/A) = \frac{1/6}{1/6} = 1$$

Using Equation (1.20), one obtains as above, $P(A/B) = 1/3$ since:

$$P(A/B) = \frac{1 \times 1/6}{1/2} = 1/3$$

An equivalent condition for independency: Event A is independent of B if:

$$P(A \cap B) = P(A)P(B). \tag{1.21}$$

This stems from Equation (1.18), which lead to,

$$P(A/B)P(B) = P(A \cap B) = P(B)P(A)$$

or $P(A/B) = P(A)$.

Thus, in the example above, event A depends on B since: $P(A \cap B) = 1/6 \neq P(A)P(B) = 1/6 \times 1/2 = 1/12$. However, A = (1,2,3) is independent of B = (3,4) since $P(A \cap B) = 1/6$ is equal to $P(A)P(B) = 1/2 \times 1/3 = 1/6$.

Based on the Bayes formula [Equation (1.18)], one can show that if an event A must result in mutually exclusive events, $A_1,..., A_n$, that is, $A = A \cap A_1 + A \cap A_2 + ... A \cap A_n$, then:

$$P(A) = P(A_1)P(A/A_1) + \cdots + P(A_n)P(A/A_n) \qquad (1.22)$$

Example

Two cards are drawn successively from a deck. What is the probability that both are red (r)?

Solution

The elementary events are defined in the product space; they are: (r, r), (r, no), (no, r), and (no, no). The events of interest are: $A \equiv$ ("first card is red," which includes {(r, r); (r, no)}, and $B \equiv$ ("second card is red," which includes {(no, r); (r, r)}, thus:

$$P(A \cap B) = P(r,r) = P(A)P(B/A) = \frac{1}{2}\left(\frac{25}{51}\right)$$

1.8 Discrete Probability—Summary

Thus far, we have dealt with discrete probability theory. We started by defining the experimental probability as a limit of relative frequency, and then described a model for the experimental reality called elementary probability space, where the probability $P(EE)$ of any EE is known *exactly*. This space enables one addressing in a systematic way complicated problems without the need to carry out experiments. In solving such problems, it is crucial to make the distinction between the experimental world, the corresponding (modeled) probability space, and when simulations are performed, the experimental world on a computer.

We have defined the notion of permutations, combinations, and conditional probability and demonstrated how related problems can be solved *systematically* within the framework of the probability space. A special emphasis has been given to product probability spaces, which constitute the basis of sampling theory. Next, we advance to the more complex theory of a *continuous* probability space, starting with a discussion on random variables.

1.9 One-Dimensional Discrete Random Variables

Assume a probability space consisting of a set of elementary events, $\{\omega\} = \Omega$ and the corresponding events. A random variable is a numerical function $X = X(\omega)$ from the elementary events, ω to the real line, $-\infty < X(\omega) < \infty$. Thus, replacing the objects (events) of the probability space by the real line has the advantage that numerical analysis methodologies will become available for treating complex probability problems.

As a simple example, let us consider a coin with "p" and "q" denoting a head and a tail, respectively (p and q were used to denote probabilities in Section 1.6; one should distinguish between the different definitions). One can define, $X(p) = 1$ and $X(q) = 0$, while any other definition is acceptable, e.g., $X(p) = 15$ and $X(q) = 2$. The choice of X is dictated by the problem (Figure 1.4). Tossing a coin n times

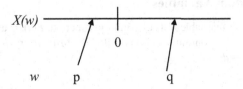

FIGURE 1.4 A random variable function from the elementary events p and q of a coin to the real line. Any two points on the line are acceptable. The points accept the probability of p and q, respectively.

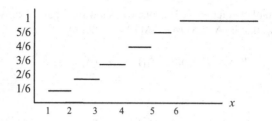

FIGURE 1.5 $F_x(X)$ for a die. The random variable function includes the numbers, 1, 2, 3, 4, 5, and 6 for the six elementary events, respectively. The probability is accumulated from 0 to 1 in going from 0 to 6, respectively.

is described by a product space where {ω} is the set of vectors of the type, (1,0,0,1…) with a probability, $P(1,0,0,1…)$. In this case, one can also define $X = m$, where m is the number of successes (heads) in a vector. This example shows that a random variable is not necessarily a one-to-one function!

1.9.1 The Cumulative Distribution Function

For a random variable $X(\omega)$, $(-\infty < X(\omega) < \infty)$, the distribution function, $F_x(X)$, is defined by:

$$F_x(X) = P[X(\omega)] \leq x. \tag{1.23}$$

Thus, $F_x(X)$ accumulates the probability defined along the x-axis from $-\infty$ to the value x of interest, and therefore it is a monotonically increasing function, which is also called the *cumulative distribution function* (Figure 1.5).

1.9.2 The Random Variable of the Poisson Distribution

The random variable of the Poisson distribution is an integer, $x = m \geq 0$, defined by the distribution itself [Equation (1.15)], and unlike a coin, it is not originated on a different probability space, thus:

$$P(x = m) = \frac{\lambda^m}{m!}\exp(-\lambda) \qquad m \geq 0 \tag{1.24}$$

The corresponding cumulative distribution function (denoted for simplicity by $F(x)$) is:

$$F(x) = \sum_{m \leq x} \frac{\lambda^m}{m!}\exp(-\lambda) \tag{1.25}$$

1.10 Continuous Random Variables

X is called a continuous random variable if its distribution function, $F(x)$, is a continuous function with a derivative, $f(x) = F'(x)$ that is also continuous. $f(x)$ is called the *probability density function*; $f(x)dx$ is the probability between x and $x + dx$.

$$F(x) = \int_{-\infty}^{x} f(t)dt; \qquad \int_{-\infty}^{\infty} f(t)dt = F(\infty) = 1 \tag{1.26}$$

1.10.1 The Normal Random Variable

The normal random variable is the entire real line $(X = x)$, and the corresponding probability density is:

$$f(x) = \frac{1}{\sigma\sqrt{2\pi}} \exp{-\left(\frac{x^2}{2\sigma^2}\right)}, \tag{1.27}$$

where the significance of the parameter σ will be discussed in Section 1.12.2. The cumulative distribution function is,

$$F(x) = \frac{1}{\sigma\sqrt{2\pi}} \int_{-\infty}^{x} \exp\left(-\frac{t^2}{2\sigma^2}\right) dt. \tag{1.28}$$

The normalization condition is satisfied [see Appendix, Equation (A1.1)],

$$\frac{1}{\sigma\sqrt{2\pi}} \int_{-\infty}^{\infty} \exp\left(-\frac{t^2}{2\sigma^2}\right) dt = 1 \tag{1.29}$$

The normal distribution, also called "the Gaussian distribution," is the most important distribution in probability theory and statistics, as many populations and phenomena are normally or close to normally distributed.

1.10.2 The Uniform Random Variable

The uniform random variable is defined on a segment of the real line, $a \le x \le b$, where the probability density function, $f(x)$, is constant over this segment (Figure 1.6),

$$f(x) = \begin{cases} c & a \le x \le b \\ 0 & \text{otherwise} \end{cases} \tag{1.30}$$

The constant c and $f(x)$ are determined from the normalization condition (Figure 1.6),

$$1 = \int_{-\infty}^{+\infty} f(x) = c \int_{a}^{b} dx = c(b-a) \Rightarrow f(x) = \frac{1}{(b-a)} \tag{1.31}$$

$F(x)$ is:

$$F(x) = \int_{-\infty}^{x} f(t)dt = \begin{cases} 0 & x < a \\ \dfrac{x-a}{b-a} & a \le x \le b \\ 1 & x \ge b \end{cases} \tag{1.32}$$

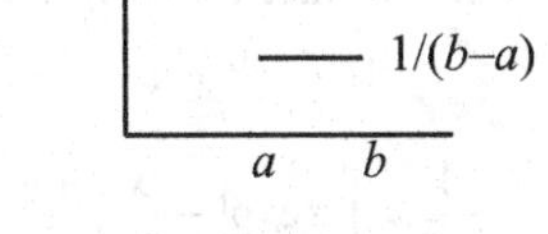

FIGURE 1.6 The uniform random variable.

1.11 The Expectation Value

Assume a discrete random variable X with n values, $x_1, x_2,...,x_n$ and $P(x_1), P(x_2),...,P(x_n)$; the *expectation value*, $E(X)$, is an *operator* applied to X,

$$E(X) = \mu = <X> = \sum_{i=1}^{n} P(x_i) x_i \tag{1.33}$$

$E(X)$ is also called *the statistical average* and *the mean*, and in probability theory, it is also denoted by μ and by the angle brackets, $<>$; we shall use these definitions in the present Chapter 1. Notice, however, that in later chapters, E will also be used to denote the energy and μ will stand for the chemical potential.

1.11.1 Examples

For a coin: $X \equiv 1, 0$ with P and $1 - P \rightarrow$

$$E(X) = P \cdot 1 + (1-P) \cdot 0 = P \tag{1.34}$$

For a die: $X \equiv 1, 2, 3, 4, 5, 6$ with $P = 1/6$ for all X, therefore,

$$E(X) = (1/6) \cdot (1+2+3+4+5+6) = 21/6 = 3.5 \tag{1.35}$$

For a *continuous* random variable, $E(X)$ is defined by the following integral provided that the integral exists:

$$E(X) = \mu = \int_{-\infty}^{\infty} x f(x) dx \tag{1.36}$$

Thus, for a uniform random variable,

$$E(X) = \int_{a}^{b} x \frac{1}{b-a} dx = \frac{x^2}{2(b-a)} \Big|_a^b = \frac{b^2 - a^2}{2(b-a)} = \frac{b+a}{2} \tag{1.37}$$

and for the Poisson random variable, we obtain $E(x) = \lambda$,

$$E(x) = \sum_{m=0}^{\infty} mP(x = m) = \sum_{m=0}^{\infty} m \frac{\lambda^m}{m!} e^{-\lambda} = \lambda \sum_{m=1}^{\infty} \frac{\lambda^{m-1}}{(m-1)!} e^{-\lambda} = \lambda e^{-\lambda} e^{\lambda} = \lambda. \tag{1.38}$$

Notice that for the normal distribution, $E(x) = 0$ since the integrand is anti-symmetrical around zero,

$$E(x) = \frac{1}{\sigma\sqrt{2\pi}} \int_{-\infty}^{\infty} x \exp\left(-\frac{x^2}{2\sigma^2}\right) dt = 0. \tag{1.39}$$

$E(X)$ is a linear operator: If X and Y are two random variables defined on the same probability space and c is a constant:

$$E(X+Y) = E(X) + E(Y); \qquad E(cX) = cE(X) \tag{1.40}$$

Proof:

$$\sum_\omega [X(\omega) + Y(\omega)]P(\omega) = \sum_\omega X(\omega)P(\omega) + \sum_\omega Y(\omega)P(\omega) = E(X) + E(Y); \quad \omega \in \Omega$$

1.12 The Variance

The variance, $V(X)$ (also denoted by $\sigma^2(X)$), is an operator which measures the distribution of the values of X around $E(X)$; for a continuous random variable,

$$V(X) \equiv \sigma^2(X) = E[(X-\mu)^2] = \int_{-\infty}^{\infty} (x-\mu)^2 f(x)dx. \tag{1.41}$$

$V(X)$ can be expressed in terms of $E(X)$,

$$V(X) = E[(X-\mu)^2] = E[X^2 + \mu^2 - 2X\mu] = E(X^2) + \mu^2 - 2\mu E(X) = E(X^2) - \mu^2$$
$$= E(X^2) - E(X)^2 \tag{1.42}$$

Notice that V will also be used to denote the volume, and $\sigma(X) = \sqrt{V(X)}$ is called the *standard deviation*. $V(X)$ is not a linear operator, since for a constant c, $V(cX) \neq cV(X)$,

$$V(cX) = E(c^2X^2) - [E(cX)]^2 = c^2E(X^2) - [cE(X)]^2 = c^2V(X) \tag{1.43}$$

Example

A random variable with an expectation value, but without a variance (c is a constant):

$$P(x=n) = \frac{c}{n^3} \qquad E = \sum_n n\frac{c}{n^3} = c\sum_n \frac{1}{n^2} \leftarrow \text{converges}$$

$$E(X^2) = \sum_n n^2 \frac{c}{n^3} = c\sum_n \frac{1}{n} \leftarrow \text{diverges}$$

Examples:

For the coin, we obtained $E(X) = p$, while $E(X^2) = 1^2p + 0^2q = p$; thus, $V(X) = p^2 - p = p(1-p) = pq$.
    For a die, using $E(X) = 7/2$ [Equation (1.35)], we obtain,

$$V(X) = E(X^2) - E(X)^2 = \frac{1}{6}(1^2 + 2^2 + \cdots + 6^2) - \left(\frac{7}{2}\right)^2 = \frac{91}{6} - \frac{49}{4} = \frac{35}{12} \tag{1.44}$$

and for the uniform distribution, [Equation (1.37)],

$$V(X) = E(X^2) - E(X)^2 = \int_a^b x^2 \frac{1}{b-a} dx - \frac{(b+a)^2}{4} = \frac{(b-a)^2}{12} \tag{1.45}$$

1.12.1 The Variance of the Poisson Distribution

For the Poisson variable, we have found $E(m) = \lambda$, where $m = 0,1,2,\ldots$; we show below that $V(m) = \lambda$ as well,

$$E(m^2) = \sum_{m=0}^{\infty} m^2 \frac{\lambda^m}{m!} e^{-\lambda} = \sum_{m=0}^{\infty} (m^2 - m) \frac{\lambda^m}{m!} e^{-\lambda} + \sum_{m=0}^{\infty} m \frac{\lambda^m}{m!} e^{-\lambda} = \sum_{m=0}^{\infty} m(m-1) \frac{\lambda^m}{m!} e^{-\lambda} + \sum_{m=1}^{\infty} \frac{\lambda^m}{(m-1)!} e^{-\lambda}$$

$$= \lambda^2 \sum_{m=2}^{\infty} \frac{\lambda^{m-2}}{(m-2)!} e^{-\lambda} + \lambda \sum_{m=1}^{\infty} \frac{\lambda^{m-1}}{(m-1)!} e^{-\lambda} = (\lambda^2 + \lambda) e^{\lambda} e^{-\lambda} = \lambda^2 + \lambda \tag{1.46}$$

Thus,

$$V(X) = E(X^2) - E(X)^2 = \lambda^2 + \lambda - \lambda^2 = \lambda \tag{1.47}$$

1.12.2 The Variance of the Normal Distribution

For the normal distribution, we have found, $E(x) = 0$ [Equation (1.39)], hence, the variance is based on $E(x^2)$ alone, which can be obtained from the integral defined in Equation (A1.2),

$$V(X) = E(x^2) = \int_{-\infty}^{\infty} x^2 f(x) dx = \frac{1}{\sigma\sqrt{2\pi}} \int_{-\infty}^{\infty} x^2 \exp - \left[\frac{x^2}{2\sigma^2} \right] dx = \frac{\sqrt{\pi}(2\sigma^2)^{\frac{3}{2}}}{2} \frac{1}{\sigma\sqrt{2\pi}} = \sigma^2 \tag{1.48}$$

Therefore, the normal distribution (Gaussian) has a symmetrical bell-shape around $x = 0$. This distribution is determined by two parameters, the expectation value, $E(x) = 0$, and the standard deviation, σ. One can show that 68% and 95% of the probability is accumulated within one and two standard deviations, (σ), respectively (Figure 1.7).

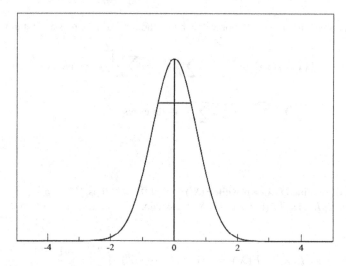

FIGURE 1.7   The normal distribution and its variance.

It should be noted that the most general Gaussian is distributed around $E(x) = x_0$ rather than $x = 0$; the corresponding probability density is:

$$f(x) = \frac{1}{\sigma\sqrt{2\pi}} \exp\left[-\frac{(x-x_0)^2}{2\sigma^2}\right],$$ (1.49)

where σ remains the standard deviation.

1.13 Independent and Uncorrelated Random Variables

Random variables X and Y defined on the same sample space are called *independent* if:

$$P(X,Y) = P(X)P(Y)$$ (1.50)

This should be compared with $P(A \cap B) = P(A)P(B)$ [Equation (1.21)]. In the case of a successive tossing of a coin with the elementary events, (1,1), (0,1), (0,0), and (1,0), the probability of the second toss is independent of the first one. Independence in the product space is defined in the same way,

$$P(X_1, X_2, ..., X_n) = P(X_1)P(X_2)...P(X_n)$$ (1.51)

where $P(X_1, X_2 ..., X_n)$ is called the joint probability.

1.13.1 Correlation

Random variables X and Y defined on the same sample space are *uncorrelated* if:

$$E(XY) = E(X)E(Y).$$ (1.52)

One can show that if X and Y are *independent* random variables, $[P(x_iy_j) = P(x_i)P(y_j)]$, defined on the same sample space, they are also uncorrelated.
  Proof:

$$E(XY) = \sum_{ij} x_i y_j P(x_i y_j) = \sum_{ij} x_i y_j P(x_i)P(y_j) = \sum_i x_i P(x_i) \sum_j y_j P(y_j) = E(X)E(Y)$$ (1.53)

Notice, however, that the opposite is not always true, that is, if X and Y are uncorrelated they are not necessarily independent. If X and Y are uncorrelated random variables defined on the same sample space, then:

$$V(X+Y) = V(X)+V(Y) \text{ [but still } V(cX) = c^2V(X)]$$ (1.54)

PROBLEM 1.4

Let x be a random variable and c an arbitrary constant. Determine the value of c which makes $E[(x-c)^2]$ minimum.

ANSWER

$c = E(x)$

PROBLEM 1.5

For the function $p(x) = a\exp(-2a|x|)$, show that it is normalized, and that $E(x) = 0$ and $V(x) = 1/2a^2$.

PROBLEM 1.6

The random variable y is a sum of N random variables x_n $y = \sum_{n=1,N} x_n$. Derive the expression for the variance of y when: (a) the random variables x_n are uncorrelated and (b) when they are correlated.

1.14 The Arithmetic Average

Let $X_1, X_2, \ldots, X_n$ be n random variables defined on the same sample space with the same expectation value, $\mu = E(X_i)$, $i = 1,2, \ldots,n$, then the arithmetic average $\bar{X}$ is:

$$\bar{X} = \frac{X_1 + X_2 + \cdots X_n}{n}. \tag{1.55}$$

$\bar{X}$ is a random variable of the random variables, X_i, with the same expectation value, $E(\bar{X}) = \mu$; indeed,

$$E(\bar{X}) = E\left(\frac{X_1 + X_2 + \cdots X_n}{n}\right) = \frac{1}{n}E(X_1 + X_2 + \cdots X_n)$$
$$= \frac{1}{n}[E(X_1) + E(X_2) + \cdots + E(X_n)] = \frac{n\mu}{n} = \mu \tag{1.56}$$

Of a special interest is the variance of the arithmetic average. Thus, let $X_1, X_2, \ldots,X_n$ be a set of *uncorrelated* random variables with the same μ and σ^2, then:

$$\sigma^2(\bar{X}) = V(\bar{X}) = V\left(\frac{X_1 + X_2 + \cdots X_n}{n}\right)$$
$$= \frac{1}{n^2}V(X_1 + X_2 + \cdots X_n) = \frac{1}{n^2}nV = \frac{V}{n} = \frac{\sigma^2}{n} \tag{1.57}$$

Therefore, while the expectation values of $\bar{X}$ and every X_i are the same $[\mu = E(X_i)]$, the variance of $\bar{X}$, $\sigma^2(\bar{X})$ is smaller than $\sigma^2(X_i)$ decreasing with increasing n, as $\sigma(\bar{X}) = \frac{\sigma}{\sqrt{n}}$. This result is extremely important, playing a central role in statistics and in the analysis of simulation data, as discussed below.

To understand the significance of the σ^2/n result, it should first be noted that both $E(\bar{X})$ and $V(\bar{X})$ are defined over the product space created by $X_1, X_2, \ldots,X_n$. Now, let us consider a set of n *equal* and *statistically independent* random variables, $X_1, X_2, \ldots,X_n$, that is, all the expectation values and the variances are equal, $\mu_i = \mu$ and $\sigma_i = \sigma$, $i = 1, 2,\ldots,n$. X_i is based on k elementary events, x_j, and probabilities $P(x_j)$, $j = 1, \ldots,k$. The product space can be visualized by the following set of n columns (see also Section 1.6):

$$
\begin{array}{cccccc}
1 & 2 & 3 & & n & \\
\begin{pmatrix} x_1 \\ x_2 \\ x_3 \\ x_4 \\ \vdots \\ x_k \end{pmatrix} & \times \begin{pmatrix} x_1 \\ x_2 \\ x_3 \\ x_4 \\ \vdots \\ x_k \end{pmatrix} & \times \begin{pmatrix} x_1 \\ x_2 \\ x_3 \\ x_4 \\ \vdots \\ x_k \end{pmatrix} & \times \cdots \times & \begin{pmatrix} x_1 \\ x_2 \\ x_3 \\ x_4 \\ \vdots \\ x_k \end{pmatrix} & \\
\mu & \mu & \mu & & \mu & \\
\sigma & \sigma & \sigma & & \sigma &
\end{array}
\tag{1.58}
$$

The elementary events of this product space are the k^n possible n-dimensional vectors, where their ith component is one of the k components of the ith column. For each n-dimensional vector, l, a probability, P_l, is defined as the product of the probabilities of its components. Also, each vector, l, contributes a component, μ_l, to the total arithmetic average μ. These contributions are depicted schematically below, where the entire set of vectors, l, $l = 1, 2, \ldots, k^n$ are shown with their probabilities, P_l, and their contributions, μ_l,

$$
\begin{array}{cccc}
l & \text{vector} & P_l & \mu_l \\
\hline\hline
1 & (x_1, x_2, x_3, \ldots x_n)_1 & P_1 & \mu_1 = 1/n \sum_{i=1}^{n} x_i(1) \\
\\
2 & (x_1, x_2, x_3, \ldots x_n)_2 & P_2 & \mu_2 = 1/n \sum_{i=1}^{n} x_i(2) \\
" & & & \\
m & (\cdots\cdots\cdots)_m & P_m & \mu_m = 1/n \sum_{i=1}^{n} x_i(m) \\
" & & & \\
k^n & (\cdots\cdots\cdots)_{k^n} & P_{k^n} & \mu_{k^n} = 1/n \sum_{i=1}^{n} x_i(k^n) \\
" & & &
\end{array}
\tag{1.59}
$$

$$
\mu = \sum_{l=1}^{k^n} P_l \mu_l
$$

The fact that the variance decreases with increasing n means that the contribution μ_l of the most probable vectors l converges to μ as n is increased; moreover, we shall see that this applies with a very high chance to any generated vector, which is very useful in simulations where one seeks to estimate μ typically by generating *only a single* vector out of the total set of k^n vectors! Thus, the longer the generated vector the better the approximation, that is, the closer is μ_l to μ. This conclusion is explained more quantitatively by the central limit theorem discussed next.

1.15 The Central Limit Theorem

The central limit theorem is fundamental in probability theory and statistics. The theorem is stated here without a proof, which can be found in [1] and standard probability textbooks.

Let $X_1, X_2, \ldots, X_n$ be mutually independent random variables with mean values, $\mu_1, \mu_2, \ldots, \mu_n$ and variances $\sigma_1^2, \sigma_2^2, \ldots, \sigma_n^2$, which define the random variable Y_n:

$$
Y_n = \frac{X_1 + X_2 + \cdots X_n - \mu_1 - \mu_2 \cdots - \mu_n}{\sqrt{\sigma_1^2 + \sigma_2^2 + \cdots + \sigma_n^2}}
\tag{1.60}
$$

Then, under various conditions, Y_n is normally distributed, where for $n \to \infty$:

$$F(x) = P(Y_n \le x) \to \frac{1}{\sqrt{2}} \int_{-\infty}^{x} \exp\left[-\frac{t^2}{2}\right] dt \qquad (1.61)$$

Thus, this Gaussian is defined solely by the two parameters, $\mu = 0$ and $\sigma^2 = 1$.

Now we return to our discussion at the end of Section 1.14 about the ability of a single vector to provide an estimation for the expectation value, μ. To recall, we deal there with a set of n statistically independent random variables X_i, where $\mu_i = \mu$ and $\sigma_i = \sigma$; therefore, Y_n [Equation (1.60)] becomes,

$$Y_n = \frac{X_1 + X_2 + \cdots X_n - n\mu}{\sqrt{n\sigma^2}} = \frac{\bar{\mu}_n - \mu}{\sigma/n^{1/2}} \qquad (1.62)$$

The right hand side of this equation is obtained by dividing both the numerator and denominator by n. $\bar{\mu}_n$ denotes the arithmetic average, $(X_1 + X_2 + \cdots + X_n)/n$ defined over the product space of the X_i [Equation (1.58)]. We are interested in calculating $P[\bar{\mu}_n - \mu] \le x]$, and for that, we define : $x' = \frac{x}{\sigma/n^{1/2}}$; from the central limit theorem [Equation (1.61)] we obtain,

$$P[\bar{\mu}_n - \mu \le x] \to \frac{1}{\sqrt{2\pi}} \int_{-\infty}^{x'} \exp\left[-\frac{t^2}{2}\right] dt = \frac{1}{\sqrt{2\pi\sigma^2/n}} \int_{-\infty}^{x} \exp\left[\frac{t^2}{2\sigma^2/n}\right] dt \qquad (1.63)$$

Thus, $[\bar{\mu}_n - \mu]$ is distributed normally around 0 with the standard deviation $\sigma/\sqrt{n}$, that is, $\bar{\mu}_n \to \mu$, as $n \to \infty$. Since 68% and 95% of the probability of a Gaussian is located within one and two standard deviations, respectively, there is a very high chance that a generated vector will belong to one of these regions and the corresponding μ_n will thus become a better and better approximation for μ as n is increased (Figure 1.8).

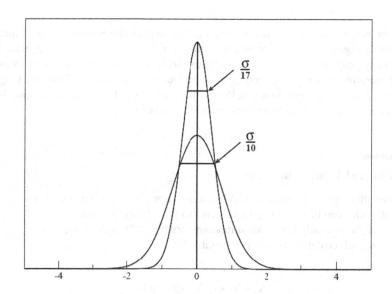

FIGURE 1.8 Two Gaussians with decreasing standard deviations, $\sigma/\sqrt{n}$ as the number of random variables, n is increased.

1.16 Sampling

The fact that the expectation value of the total product space can be obtained from a single long vector is very appealing. We have already mentioned (Section 1.4) that N atoms of liquid argon contained in a volume V at temperature T can be treated in statistical mechanics within a probability space called the canonical (NVT) ensemble. The corresponding elementary events (random variables, X_i) are all the system configurations, which are the $3N$-size vectors of Cartesian coordinates for the N atoms, $\mathbf{X}_i \equiv (\mathbf{x}_1(i), \mathbf{x}_2(i), \ldots, \mathbf{x}_N(i))$ (for simplicity $\{\mathbf{X}_i\}$ are discretized). An interaction energy is defined among the atoms, and the total potential energy of $\mathbf{X}_i$, denoted, $E(\mathbf{X}_i)$, is another random variable ($E(\mathbf{X}_i)$ should be distinguished from the expectation value denoted $E(X)$ in Equation (1.33)) (for simplicity, the effect of the velocities is ignored).

However, the corresponding Boltzmann probability density, $P^B(\mathbf{X}_i)$, cannot be calculated analytically. In this case, it would be tempting to move to the experimental world on a computer, that is, to build a finite system there and simulate a long vector of configurations from which statistical averages, such as the energy can be calculated. However, this simulation would require using the *unavailable* probabilities $P^B(\mathbf{X}_i)$! Luckily, this problem has been solved, as methods exist that enable one to sample argon configurations (for example), according to P^B *without the need to know the value of* P^B! One famous method in this category is Metropolis Monte Carlo, which will be discussed in Section 10.4. Thus, using Metropolis Monte Carlo, one can generate a single long trajectory (sample) of configurations from which averages can be *estimated*, the longer the trajectory, the closer the estimations to the correct values (defined in the probability space).

It is important to emphasize the distinction between four levels of information related to probability, P: (1) its exact *value* is known a priori or (2) unknown, where in case (2) it is still (3) possible to sample according to P *without knowing its value*. Sometimes the value of P (4) can be obtained from the simulation itself. To explain this issue further, we consider first an uneven coin with *a priori unknown* probabilities P(head) and $(1–P)$(tail) (case 2); still, if the coin is tossed randomly, it will fall on its sides proportional to P and $(1–P)$ (cases 3 and 4). For argon (case 2), the situation is more complex, because the number of configurations is huge (as compared to two states in the case of a coin), while only a tiny part of them contribute significantly to the averages (therefore, a random search will fail, as this part will be missed entirely). On the other hand, sampling according to $P^B(X_i)$ (case 3) with MC will direct the simulation towards this tiny region. However, the percentage of visited configurations is typically very small, hence the values of $P^B(X_i)$ cannot be obtained (unlike in the coin example).

Thus, knowing how to sample according to P without knowing the value of P is an essential property of sampling theory—useful in particular in computer simulation. Still, in statistical mechanics, it is important to be able to calculate also the *value* of $P^B(X_i)$, which leads to the absolute entropy, $\sim\ln P^B(X_i)$—a property that is extremely difficult to obtain otherwise. An appreciable part of this book is devoted to methods for the calculation of the entropy and the free energy.

1.17 Stochastic Processes—Markov Chains

In Section 1.6, we have discussed n *independent* experiments of tossing a coin, with two elementary events (outcomes) i and j; the independency means that the corresponding probabilities P_i and $P_j = 1 - P_i$ of the tth experiment do not depend on the event having occurred at experiment $t - 1$ or at previous experiments. We shall consider now a different process, a Markov chain, where $P_i(t)$ depends on the outcome at $t - 1$. Thus, if the event i occurred at $t - 1$, event j will occur at (time) t with the conditional probability $p(j/i)$, and i will occur again with $p(i/i)$, etc.

$$
\begin{array}{ll}
t-1 & t \\
\hline
\end{array}
$$

$$
\begin{aligned}
i \quad &\to p(j/i) \quad \& \, p(i/i) \quad p(j/i) + p(i/i) = 1 \\
j \quad &\to p(i/j) \quad \& \, p(j/j) \quad p(i/j) + p(j/j) = 1
\end{aligned}
\tag{1.64}
$$

For simplicity, we shall change the notation $p(j/i)=p(i \to j)$ to p_{ij}, where p_{ij} is also called transition probability—the probability to obtain j after having i in the previous experiment. As in Section 1.6, the states (which can be converted into random variables) will be denoted by i and j. Assume now that the initial probabilities at time 0 are P_i^0 and P_j^0, where $P_i^0 + P_j^0 = 1$, and we are interested in P_i^1 and P_j^1—the probabilities of the next experiment (at time 1); they are:

$$P_i^1 = P_i^0 p_{ii} + P_j^0 p_{ji}$$
$$P_j^1 = P_i^0 p_{ij} + P_j^0 p_{jj}.$$

(1.65)

or in a matrix form:

$$\left(P_i^1, P_j^1\right) = \left(P_i^0, P_j^0\right)\begin{pmatrix} p_{ii} & p_{ij} \\ p_{ji} & p_{jj} \end{pmatrix} \leftarrow M.$$

(1.66)

M is a *stochastic matrix*, where all of its components are ≥ 0, and the sum of probabilities in each row is 1 ($p_{ii} + p_{ij} = 1$; $p_{ji} + p_{jj} = 1$). One can ask: what is P_i^2 and P_j^2? Applying the same procedure,

$$\left(P_i^2, P_j^2\right) = \left(P_i^1, P_j^1\right)M = \left(P_i^0, P_j^0\right)M^2$$

(1.67)

and after t experiments:

$$\left(P_i^t, P_j^t\right) = \left(P_i^0, P_j^0\right)M^t = \left(P_i^0, P_j^0\right)\begin{pmatrix} p_{ii} & p_{ij} \\ p_{ji} & p_{jj} \end{pmatrix}^t$$

(1.68)

Thus, at each time, t, a probability space is defined with P_i^t and P_j^t. Also, the set of t experiments defines a product space based on vectors of the type $(i(0), j(1), i(2), \dots j(t-1), i(t))$; the probability of a vector is the product of the corresponding transition probabilities. The probability at time t, $P_i^t(P_j^t)$, is the sum of the probabilities of the 2^t vectors that end with i (j), as described in Equation (1.69) below; in this equation, the vectors (counted by k) are divided into two groups, according to their last state, i or j,

$$(1,2,3,\dots,t-1,t)$$

$$k=1 \qquad (i,j,j,\dots,j,i) \qquad P_i^t(1) = P_i^0 p_{ij} p_{jj} \dots p_{ji}$$

$$k=2 \qquad (j,i,i,\dots,i,i) \qquad P_i^t(2) = P_j^0 p_{ji} p_{ii} \dots p_{ii}$$

$$k=3 \qquad (i,i,j,\dots,j,i) \qquad P_i^t(3) = P_i^0 p_{ii} p_{ij} \dots p_{ji}$$

$$\vdots$$

$$P_i^t = \sum_{k=1}^{2^t} P_i^t(k)$$

(1.69)

$$k=1 \qquad (i,j,j,\dots,i,j) \qquad P_j^t(1) = P_i^0 p_{ij} p_{jj} \dots p_{ij}$$

$$k=2 \qquad (j,i,i,\dots,j,j) \qquad P_j^t(2) = P_j^0 p_{ji} p_{ii} \dots p_{jj}$$

$$k=3 \qquad (i,i,j,\cdots,i,j) \qquad P_j^t(3) = P_i^0 p_{ii} p_{ij} \dots p_{ij}$$

$$\vdots$$

$$P_j^t = \sum_{k=1}^{2^t} P_j^t(k)$$

The significance of the Markov chain in the *experimental world* is the following: if one repeats the set of t experiments many times always starting with P_i^0 and P_j^0, the number of times i and j will occur at the tth experiment is proportional to P_i^t and P_j^t, respectively.

One can define Markov chains with any number of states, N; the stochastic matrix (p_{ij}) will be of size $N \times N$.

$$P_j^{t+1} = \sum_i P_i^t p_{ij} \quad i, j = 1, N \tag{1.70}$$

1.17.1 The Stationary Probabilities

The interesting question is whether for a large number of experiments (denoted t_c), the set of probabilities $\{P_i^{t_c}\}$ will converge to values that are not changed for $t > t_c$, and $\{P_i^{t_c}\}$ is independent of the initial set $\{P_i^0\}$. For a "well behaved" Markov chain, where all states are irreducible and aperiodic, the answer is positive. The term "irreducible" means that for every pair of states i and j, there is t for which $(M^t)_{ij} > 0$, meaning that all states are reachable; "aperiodic" means that this condition is satisfied for all sufficiently large t—see [2]. Under these conditions, one obtains a *unique* set of stationary probabilities $\{\pi_i\}$ with the following properties,

$$\lim_{t \to \infty} P_i^t = \pi_i \qquad \{\pi_i\} \text{ are independent of } \{P_i^0\} \tag{1.71}$$

$$\pi_i > 0; \quad \sum_i \pi_i = 1; \quad \pi_j = \sum_i \pi_i p_{ij} \quad \text{as compared to}: \quad P_j^{t+1} = \sum_i P_i^t p_{ij}.$$

It should be pointed out that it is relatively easy to calculate realizations (vectors) of t experiments of a Markov chain using a computer, provided that the matrix M of transition probabilities is known; such (experimental world) realizations are obtained typically with the help of a random number generator, which is a computer program generating numbers uniformly distributed [Equations (1.31)] within the range [0,1]; see Appendix, A3.

The existence of stationary probabilities means for $t \geq t_c$, a (stationary) probability space is defined over the N-states of the Markov chain. We shall see that there is a strong interest in calculating expectation values and variances of random variables defined over this space. An example is the average energy $<E>$,

$$<E> = \sum_{i=1}^N \pi_i E_i \tag{1.72}$$

where E_i is the energy of state i. In principle, one can carry out a large number, n of Markov chain runs of length t_c starting from the same initial set of probabilities $\{P_i^0\}$. From the number of visits n_i to state i, at the last step, (t_c), one will obtain the estimates $\{\bar{\pi}_i = n_i / n\}$ for the correct $\{\pi_i\}$ and thus an estimate, $\bar{E}$ for $<E>$ (estimates are denoted by a bar) is,

$$\bar{E} = \sum_i \bar{\pi}_i E_i \tag{1.73}$$

However, this procedure becomes inefficient for a large t_c, and estimating the values of π_i is impractical for a large number of states, N. Alternatively, one can generate a *single* realization (vector) of the Markov chain for $t > t_c$ estimating $<E>$ based on the ergodic theorem (see next section) in a similar way that it is done for independent sampling (e.g., for a regular coin or a die),

$$\bar{E} = \frac{1}{t - t_c} \sum_{k=1}^{t - t_c} E_{i(k)} \tag{1.74}$$

where $i(k)$ is the state i obtained at time k of the process. Two main points differentiate between independent sampling and dependent sampling based on a Markov chain. With an independent sampling of a coin, for example, all the experiments, starting from the first one, are considered, and due to independence, the variance of a random variable decreases as $\sigma_t = \sigma/t^{1/2}$ [Equation (1.57)]. On the other hand, in the case of a Markov chain, one has to ignore the first t_c experiments, where the probabilities relax to their stationary values. Also, because the experiments are not independent, the random variables are correlated. Considering the method based on the ergodic theorem, if the correlation disappears after l experiments, the variance decreases slower than in an independent sampling, that is, only as $\sigma_t = \sigma/(t/l)^{1/2}$.

So, why use a Markov chain if it is inferior in efficiency to independent sampling? This stems from the fact that a Markov chain can lead to stationary probabilities $\{\pi_i\}$ of interest, in particular, to the Boltzmann probabilities $\{\pi_i = P_i^B\}$ using the Monte Carlo method (mentioned earlier), which is a Markov chain. Therefore, while in general the values of P_i^B are not provided by the Monte Carlo method, systems of interest in statistical mechanics can be simulated correctly *according* to P_i^B and various statistical averages can be obtained using, for example, Equation (1.74) (see also Section 1.16). Indeed, Monte Carlo techniques have become very important tools in statistical mechanics and other branches of science (see Section 10.4).

Mentioning the Monte Carlo method as an example of a Markov chain, we point out that this procedure is based on *microscopically reversible* transition probabilities, which satisfy the *detailed balance* condition,

$$\frac{p_{ij}}{p_{ji}} = \frac{\exp- E_j / k_B T}{\exp- E_i / k_B T} = \exp-(E_j - E_i) / k_B T \qquad (1.75)$$

Thus, the transition from i to j depends on the corresponding energies E_i and E_j of these states. For a more complete discussion, see Sections 10.4 and 11.8.

1.18 The Ergodic Theorem

We have seen that a large set of mutually independent random variables obey the central limit theorem. Applied to the energy of n system configurations, for example, the expectation value of a single realization, j (vector), of the product space, $[E_1(j) + E_2(j) + \dots + E_n(j)]/n$ converges to $<E>$ as $n \to \infty$ [Equation (1.63)]. However, this theorem does not apply to correlated random variables defined, for example, by a Markov chain; in this case, the convergence to $<E>$ relies on the ergodic theorem.

Thus, concentrating on statistical mechanics, we have already defined the canonical ensemble as a probability space with the Boltzmann probability, P_i^B, and the Monte Carlo method as a Markov chain leading to $\{\pi_i = P_i^B\}$. Typically, one is interested in *ensemble* averages, such as:

$$<E> = \sum_i P_i^B E_i = \sum_i \pi_i E_i = <E>_\pi \qquad (1.76)$$

The ergodic theorem states that the *ensemble* average ($<E>_\pi$) can be obtained by *time* averaging of a very long *single* realization of the Markov chain considered for $t \geq t_c$:

$$<E>_\pi = \lim_{t \to \infty} \frac{1}{t - t_c} \sum_{t'=t_c}^{t'=t} E_{i(t')} \qquad (1.77)$$

where $E_{i(t')}$ is the energy obtained at time t'. This stems from the fact that in a very long realization, all states i are expected to appear proportional to their probability π_i.

The ergodic theorem applies not only to the energy, but to other thermodynamic properties and to other stochastic processes, such as molecular dynamics (Section 10.6) and Langevin dynamics (Section 12.4.2). The ergodic theorem can be written in a more general way for a trajectory of a function f defined over a *continuous* probability space, $\mathbf{X}^N$ (see Section 1.16), which is propagated in time according to a stochastic process (trajectory),

$$<f>_{\text{time}} = \lim_{T \to \infty} \frac{1}{T} \int_{t=t_0}^{t=t_0+T} f(\mathbf{X}^N(t))dt = <f>_{\text{ensemble}}. \tag{1.78}$$

This theorem is essential in simulation theory.

1.19 Autocorrelation Functions

As discussed above, for a Markov chain in statistical mechanics, the ensembles (probability spaces) defined for $t > t_c$ are identical and the corresponding ensemble averages (denoted by $<>$) are the same as well. However, for small t', successive values E_t and $E_{t+t'}$ along a trajectory (a vector in the product space) are expected to be similar, where this correlation decreases gradually, with increasing t' becoming zero for large t'. The autocorrelation function, denoted "Corr" (written as an example for the energy), is defined by:

$$\text{Corr}(E_t, E_{t+t'}) = <(E_t - <E>)(E_{t+t'} - <E>) = <E_t E_{t+t'}> - <E>^2 \tag{1.79}$$

Thus, for large t',

$$<E_t E_{t+t'}> = <E_t><E_{t+t'}> = <E>^2 \to \text{Corr} = 0 \tag{1.80}$$

To calculate $<E_t E_{t+t'}>$, we write (for a Markov chain) first the probabilities of the N states at time $t + t'$, provided that at time t, the system resided only at state i, $P_i^t = 1$,

$$\left(P_{1(i)}^{t+t'}, \quad P_{2(i)}^{t+t'}, \quad ..., \quad P_{N(i)}^{t+t'}\right) = \left(P_1^t = 0, \quad 0, \quad , P_i^t = 1, ..., 0\right)M^{t'} \tag{1.81}$$

and $<E_t E_{t+t'}>$ becomes,

$$<E_t E_{t+t'}> = \sum_{i=1}^{N} \pi_i E_i \sum_{j=1}^{N} P_{j(i)}^{t+t'} E_j \tag{1.82}$$

On the computer (experimental world), $<E_t E_{t+t'}>$ is estimated by $\overline{E_t E_{t+t'}}$, that is, by averaging $E_t E_{t+t'}$ along a long trajectory (realization),

$$\overline{E_t E_{t+t'}} = \frac{1}{n} \sum_{t=t_c}^{t=t_c+n} E_t E_{t+t'} \qquad n \to \infty \tag{1.83}$$

Again, these definitions presented for the energy, also hold for other properties and stochastic processes. The normalized autocorrelation function $R(t,t')$ is:

$$R(t,t') = \frac{<E_t E_{t+t'}> - <E>^2}{<E^2> - <E>^2} \tag{1.84}$$

For $t' = 0$, $R = 1$, and for a large t', $R = 0$ stemming from Equation (1.80).

(a)

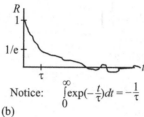

Notice: $\int\limits_{0}^{\infty}\exp(-\frac{t}{\tau})dt=-\frac{1}{\tau}$

(b)

FIGURE 1.9   (a) A correlated time series of the energy. (b) An exponential decay of $R(t)$.

1.19.1 Stationary Stochastic Processes

A well behaved Markov chain in the region $t > t_c$ is an example of a stationary stochastic process for which $\{P_i^t\}$ is equal to $\{P_i^{t+t'}\}$ for any t and t'. Therefore, the autocorrelation function depends only on t' (but not on t):

$$R(t,t') = R(t') \tag{1.85}$$

A typical $R(t')$ decays exponentially as $\sim\exp(-t'/\tau)$, where τ is the correlation time and $R(\tau) = 1/e$ (Figure 1.9).

$$\int\limits_{0}^{\infty}\exp(-\frac{t}{\tau})dt=-\frac{1}{\tau}=\frac{\log R(t)}{t} \tag{1.86}$$

Our Markov chain is discrete in time. One can define a continuous stochastic process where t can assume any real value; the random variables can be continuous as well.

For a stationary stochastic process, $R(t')$ can be estimated from a long realization (time series) by averaging $E_t E_{(t+t')}$ in "windows" of size t' along the trajectory. Otherwise, $R(t')$ is averaged by initiating n different simulations (trajectories). Sometimes one is interested in the cross correlation between two random variables x and y defined on the same process:

$$\rho(t,t')=\frac{<[x(t)-<x>][y(t+t')-<y>]>}{\sigma_x\sigma_y} \tag{1.87}$$

HOMEWORK FOR STUDENTS

Besides the examples and problems presented in this chapter, Feller's book [1] provides many problems with solutions (presented at the end). In Schaum's book by Spiegel et al. [3], probability theory is introduced through large problems/solutions sets.

A COMMENT ABOUT NOTATIONS

Some of the notations used in this probability chapter have different meanings in the chemistry and statistical mechanics literature, and to avoid confusion, they are specified here. Thus, the letter V used for the *variance* in Chapter 1 (also σ^2) will be used for the *volume* in the rest of the book. The *expectation value,* denoted by E and μ, will be replaced by the parentheses, $<>$, where from now on, E and μ will stand for the energy and chemical potential (or the magnetic moment), respectively; the probability density, $f(x)$, will be denoted in most cases by $P(x)$ or $\rho(x)$, where global (homogeneous) density will be denoted by ρ or c, where local density by $c(x)$. W denotes work and the potential of mean force. This discussion shows that even within statistical mechanics there are multiple uses of a given letter in different sub-disciplines. Other examples are the well defined critical exponents ν, β, and γ, where ν also denotes a stoichiometry coefficient, a unit vector on a lattice, and an energy level (E_ν); β also denotes the inverse temperature, $\beta = 1/k_B T$. We have tried to keep the special letters as used in the different disciplines.

REFERENCES

1. W. Feller. *An Introduction to Probability Theory.* (John Wiley & Sons, New York, 1970).
2. J. G. Kemeny and J. L. Snell. *Finite Markov Chains.* (Springer, New York, 1976).
3. M. R. Spiegel, J. Schiller and R. A. Srinivasan. *Schaum's Outline of Probability and Statistics.* (McGraw Hill Education, Europe, 2002).
4. K. A. Dill and S. Bromberg. *Molecular Driving Forces, Statistical Thermodynamics in Biology, Chemistry, Physics, and Nanoscience.* (Garland Science, New York, 2010).

Section II

Equilibrium Thermodynamics and Statistical Mechanics

2

Classical Thermodynamics

2.1 Introduction

The theory of classical thermodynamics was developed in the nineteenth century, and due to its general applicability, it is still used extensively in chemistry, physics, and engineering. Historically, classical thermodynamics preceded the advent of statistical mechanics, and indeed, we show that the theory of the latter can be derived straightforwardly on the basis of the former. Therefore, this chapter is not a complete review of thermodynamics, but (based on [1–3]) contains only the fundamentals of the theory that are needed for the derivation of statistical mechanics. For example, we accept the notion of temperature without elaborating on its origin; also, for simplicity, most examples involve an ideal gas. This chapter is based on the books [1–6]. Since thermodynamics describes the effect of heat, it proves educational to emphasize first the difference between its stochastic character and the deterministic nature of macroscopic mechanical systems, which is thus outlined below.

2.2 Macroscopic Mechanical Systems versus Thermodynamic Systems

Classical mechanics deals with forces, $\mathbf{F}$, which are three-dimensional vectors, described here for simplicity, in one dimension. Examples are: (1) Hook's law, $F = -fx$ where x is the stretching distance of a spring and f is a constant. (2) The gravitation force, $F = kMm/r^2$, where the masses m and M are distant by r, and k is a constant; on earth, R and M are large leading to $g = kM/R^2$. (3) Coulomb's law, $F = k_e q_1 q_2/r^2$; q_1 and q_2 are charges distant by r, and k_e is a constant. The forces satisfy Newton's second law: $F = m\mathbf{a}$, where $\mathbf{a}$ is the acceleration.

Fundamental concepts in mechanics are the mechanical work the energy. Thus, if a constant force is exerted on a mass along a distance x, the mechanical work done is $W = Fx$, or more generally, $W = \int F dx$. If, for example, a mass m is raised to height, h, a negative work is done, $W = -mgh$, and the mass gains potential energy, $E_p = -W = +mgh$, which measures the mass's ability to do mechanical work; W is released when m falls dawn, and E_p is converted into kinetic energy, E_k, which becomes maximal at the floor, $E_k = mv^2/2$, where $v^2/2 = gh$. The total energy in a closed system, $E_t = E_p + E_k$, is constant (conservation of mechanical energy), but the ratio, E_p/E_k, can change, as in the oscillations of a mass hung on a spring. The related conclusions are:

1. Ignoring friction, mechanical work, kinetic and potential energies in a closed system can be converted to each other without a loss of the total mechanical energy.
2. The dynamical state of a mechanical macroscopic system is *deterministic*, that is, if the forces are known, and the positions and velocities of the masses at time $t = 0$ are known as well, their values at time t can *in principle* be determined by solving Newton's equation (second law). Simple examples are the harmonic oscillator (a spring), a trajectory of a projectile, or a movement of a spaceship, while many-body problems can also be treated (e.g., many particles enclosed in a box). In complex cases, Newton's equations can only be solved numerically using strong computers.

Thermodynamics deals with the effect of heat, which is inherently a disordered kind of energy. Therefore, an amount of heat added to a system (e.g., gas at a high temperature enclosed in a cylindrical

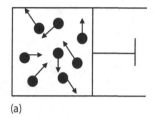 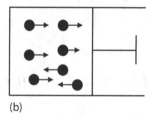

(a)   (b)

FIGURE 2.1   (a) Added heat increases the velocities randomly, (b) horizontally ordered velocities.

container with a moving piston) leads to a *random* increase in the velocities of the molecules, as shown in Figure 2.1a; hence, this heat cannot be converted entirely into mechanical work. A full heat/work conversion would require the piston to be "attacked" by horizontally ordered velocities, as shown in Figure 2.1b. As discussed below, the maximal work can be gained in a reversible process. This energy/work relation is an essential difference between thermodynamic and macroscopic mechanical systems. Another (related) essential difference is the fact that the definition of the (dynamic) state in terms of the coordinates and velocities of the $\sim 10^{23}$ molecules is impossible; therefore, in thermodynamics, a state of the system is defined only by several *macroscopic* variables, such as the volume $-V$, temperature $-T$, and the pressure $-p$. These variables are not independent, but are connected by an equation of state:

$$f(p,V,T) = 0. \tag{2.1}$$

Thus, the general non-specific theoretical aspects of thermodynamics make it applicable to a wide range of systems.

2.3 Equilibrium and Reversible Transformations

For simplicity, we assume an ideal situation, where the system is homogeneous and in *equilibrium*, that is, a system where the external conditions have not changed for a long time and all relaxation processes have ended. It should be pointed out that at equilibrium, fluctuations in macroscopic parameters, such as the energy, do exit, but in most situations they are extremely small, and thus are ignored in thermodynamics. In this context, one also defines a *reversible transformation* from state a to state b as a very slow transformation, composed of many intermediate equilibrium states (such a transformation is of course an idealization).

2.4 Ideal Gas Mechanical Work and Reversibility

We now consider a macroscopic system of a dilute gas of N particles ($N \sim N_A = 10^{23} - N_A$ is Avogadro's number) contained in a volume, V. The system is at equilibrium with a very large (infinite) heat bath (reservoir) with relatively high absolute temperature T, which defines the system's temperature, $T_{\text{system}} = T$. Due to dilution, the particles practically do not "meet" each other and can be considered as non-interacting particles; this system, which is called an "*ideal gas*," can interact mechanically and thermally with its surroundings. It has been established *experimentally* that the equation of state is:

$$pV = Nk_BT = nRT \tag{2.2}$$

where p is the pressure, k_B is the Boltzmann constant, n is the number of moles, and $R = k_BN_A = \sim 0.002$ kcal/mol/deg. Historically, R was determined empirically, V and N are extensive parameters, that is, both grow with the system size, and p and T_{system} are intensive parameters, which need time to equilibrate.

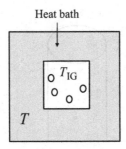

FIGURE 2.2   A system of ideal gas (IG) in contact with a heat bath at temperature, $T = T_{IG}$.

p is related to the average number of "knocks" of the particles on the system's walls, and as shown later, T_{system} is proportional to the kinetic energy per particle (from now on we shall not distinguish between T and T_{system}) (Figure 2.2).

It is of interest to calculate the mechanical work done by an ideal gas. Thus, let us consider first a cylindrical container filled with gas (not necessarily ideal gas), where its volume V, hence, its pressure p, are determined by a moving frictionless piston of area A (Figure 2.3). At equilibrium, the internal pressure on the piston (exerted by the gas) is equal to the external pressure, p_e, that acts in the opposite direction.

To generate a reversible infinitesimal transformation toward the outside, p should become slightly larger than p_e, $p + dp > p_e$. Thus, the gas exerts a force pA on the piston, and if the piston is moved by a distance dh, the amount of *positive* work $d'W$ done *on the surroundings* is: $(p + dp)dV \sim pdV$ or

$$d'W = pAdh = pdV \tag{2.3}$$

and the total work is $W = \int d'W$ (Figure 2.4).

The apostrophe in $d'W$ stems from the fact that W is not a state function, because dW is not an *exact* (total) differential, and thus its integral $\int pdV$ depends on the path (exact differential is defined in the

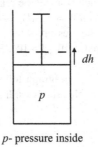

p- pressure inside

FIGURE 2.3   A cylindrical container with gas exerting pressure p on the piston, which is moved by dh.

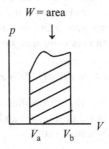

FIGURE 2.4   The work, W, done by increasing the volume from V_a to V_b is the area under the $p(V)$ line.

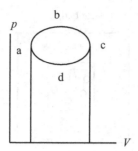

FIGURE 2.5   Two integration paths for the mechanical work W, abcda and adcba are not identical.

Appendix, Section A2). The path related integral is demonstrated by the cyclical integration described in the $p-V$ plot in Figure 2.5.

Thus, if one uses first the path abc, dV is positive, hence, W_{abc} is positive; however, in the integration along cda, dV is negative, and negative is also W_{cda}. Thus, $W_{total} = W_{abc} - W_{cda}$ is the *positive* area of the ellipse abcd. On the other hand, if adc is integrated first, followed by cba, W_{total} is the *negative* area of the ellipse. We emphasize again that this consideration applies to any gas, not necessary an ideal gas.

We calculate now the amount of work done when the volume of an *ideal gas* at constant T is changed from V_1 to V_2. Using Equation (2.2) one obtains,

$$W = \int_{V_1}^{V_2} p(V)dV = \int_{V_1}^{V_2} \frac{Nk_BT}{V} dV = Nk_BT \ln\left(\frac{V_2}{V_1}\right) = Nk_BT \ln\left(\frac{p_1}{p_2}\right) \tag{2.4}$$

Again, this result is correct only if the piston is moved infinitesimally slowly to allow *equilibration* of the system at each step, that is, when the transformation is *reversible*. In an irreversible process, where ΔV is moved fast, there will not be enough time for the pressure to build up at the piston (ΔV will be too dilute), meaning that $p_{fast} < p$ and $d'W_{fast} < d'W$ [Equation (2.3)]; the conclusion is that reversible work done on the surroundings is maximal. A related conclusion can be reached if the volume is compressed; in this case, the reversible work done on the gas is $-d'W = p_eAdh = -pdV$. Now, if an abrupt (irreversible) move of the piston (toward the inside) is applied, the pressure at the piston will increase, that is, $-p_{fast} < -p$, and the work invested in compression will increase, $|W_{fast}| > |W|$. Hence, the reversible work in compression is the minimal, that is, the most efficient.

We shall discuss this subject later (see the paragraph following Equation (2.26) and Section 2.8.3), with respect to entropy in reversible/irreversible processes. It should be emphasized that the theory will be developed at equilibrium.

2.5 The First Law of Thermodynamics

The first law is the conservation of energy for thermodynamic systems. It states that the variation in energy during any transformation (from state a to state b) is equal to the amount of energy that the system receives from the surroundings, and it depends on states a and b and not on the intermediate process. The law implies the existence of *internal energy*, E, which is an *extensive state function* defined up to an additive constant,

$$E_b - E_a = \Delta E = Q - W + Z. \tag{2.5}$$

Q is heat—a disordered energy *received from* the surroundings; W is the mechanical work done *by* the system *on the* surroundings, Z stands for other contributions, and T is the absolute temperature of the surroundings (heat bath). The first law in a differential form is:

$$dE = d'Q - d'W + d'Z \tag{2.6}$$

where *dE is an exact* differential, while $d'Q$, $d'W$, and $d'Z$ are not. In most cases in this book, we use $d'W = pdV$ and $d'Z = \mu dN$, where μ is the chemical potential. A nineteenth-century question: can Q be converted totally into mechanical work? The answer is "no," but the maximum W is obtained in a reversible transformation (see discussions in Sections 2.2, 2.4, and based on the second law of thermodynamics, in Section 2.9). The first law can be expressed by different sets of variables,

$$d'Q = dE + d'W \qquad d'W = pdV \qquad \textbf{Variables}$$

$$dE = \left(\frac{\partial E}{\partial T}\right)_V dT + \left(\frac{\partial E}{\partial V}\right)_T dV \qquad T,V \tag{2.7}$$

$$d'Q = \left(\frac{\partial E}{\partial T}\right)_V dT + \left[\left(\frac{\partial E}{\partial V}\right)_T + p\right] dV \qquad T,V \tag{2.8}$$

$$d'Q = \left[\left(\frac{\partial E}{\partial T}\right)_p + p\left(\frac{\partial V}{\partial T}\right)_p\right] dT + \left[\left(\frac{\partial E}{\partial p}\right)_T + p\left(\frac{\partial V}{\partial p}\right)_T\right] dp \qquad p,T \tag{2.9}$$

$$d'Q = \left(\frac{\partial E}{\partial p}\right)_V dp + \left[\left(\frac{\partial E}{\partial V}\right)_P + p\right] dV \qquad p,V \tag{2.10}$$

$$d'W = pdV(p,T) = p\left(\frac{\partial V}{\partial T}\right)_p dT + p\left(\frac{\partial V}{\partial p}\right)_T dp \qquad p,T \tag{2.11}$$

Equation (2.11) is based on Equation (2.9). Using Equations (2.8) and (2.9), we define the thermal (heat) capacities, C_V and C_p, at constant V and p, respectively,

$$C_V = \left(\frac{\partial'Q}{\partial T}\right)_V = \left(\frac{\partial E}{\partial T}\right)_V \tag{2.12}$$

$$C_p = \left(\frac{\partial'Q}{\partial T}\right)_p = \left(\frac{\partial E}{\partial T}\right)_V + p\left(\frac{\partial V}{\partial T}\right)_p \tag{2.13}$$

Notice that specific heat is the heat capacity per one gram of substance.

2.6 Joule's Experiment

Figure 2.6 describes a calorimeter. Initially, compartment A is filled up with ideal gas (IG), where compartment B is empty. Then, the piston is removed, and the ideal gas expands and occupies the two compartments. While $\Delta V > 0$, the thermometer remains at the same temperature T or $\Delta T = 0$, meaning that $\Delta Q = 0$, thus $\Delta E = 0$. The conclusion of this Joule's experiment is that for an ideal gas, the energy, E, depends on the temperature, but does not depend not on the volume. Experimentally, C_V of an ideal gas does not depend on T, therefore, using Equations (2.7) and (2.12) one obtains:

$$dE = C_V dT \qquad \rightarrow \qquad E_{IG} = C_V T + \text{const.} \tag{2.14}$$

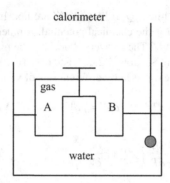

FIGURE 2.6 A calorimeter consisting of two compartments A and B (separated by a removable piston), is immersed in a container filled with water at temperature, T measured by a thermometer. Initially, compartment A contains an ideal gas, while compartment B is empty.

Indeed, from statistical mechanics one derives, $E_{IG} = 3Nk_BT = C_VT$, [see Equations (4.11) and (4.12)], meaning that the constant is equal to 0. Because the energy is an exact differential, and from Joule's experiment $\partial E/\partial V = 0$,

$$\left(\frac{\partial C_V}{\partial V}\right)_{T,N} = \left(\frac{\partial(\partial E / \partial V)_{T,N}}{\partial T}\right)_{V,N} = 0 \tag{2.15}$$

Therefore, C_V does not depend on V, and the *first law* [(Equation (2.8)] becomes:

$$d'Q = C_V dT + p dV \tag{2.16}$$

Differentiating the equation of state [Equation (2.2)]) for one mole, $d(pV) = RdT$ leads to

$$pdV + Vdp = RdT \qquad \rightarrow \qquad pdV = RdT - Vdp \tag{2.17}$$

and substituting this result in Equation (2.16) leads to:

$$d'Q = (C_V + R)dT - Vdp. \tag{2.18}$$

At constant pressure, $dp = 0$ and the specific heat at constant pressure [Equation (2.13)] is:

$$C_p = \left(\frac{\partial'Q}{\partial T}\right)_p = C_V + R \tag{2.19}$$

Example

Show that the relation $pV^\gamma = $ const. holds in a reversible adiabatic process of an ideal gas, where γ is the ratio of the specific heats of constant pressure and constant volume.

Solution

For an ideal gas, C_V is independent of T and adiabatic means $d'Q = 0$. Thus, $d'Q = C_V dT + pdV = 0$, using $pV = RT$, one obtains $C_V dT + RTdV/V = 0$ or $CvdT/T + RdV/V = 0$; since, $C_p = C_V + R$, one obtains $C_V dT/T + (C_p - C_V)dV/V = 0$. Now, integration leads to $C_V \ln T + (C_p - C_V)\ln V = $ const. or $TV^{(Cp/Cv-1)} = $ const. $TV^{(\gamma-1)} = (pV/R)V^{(\gamma-1)} = $ const., leading to $pV^\gamma = $ const.

PROBLEM 2.1

A mole of an ideal gas is expanded reversibly at a constant temperature of 20°C from an initial pressure of 20 atm to 1 atm. What is the work done by the gas in Joule? How much heat in calories must be supplied to the gas?

ANSWER

The work in Joules is 7.29×10^3 joule/mole. The heat in calories is 1.744 kcal/mol.

2.7 Entropy

According to the first law, $dE = d'Q - pdV + \mu N$, where p and μ are intensive parameters, affecting changes in the volume and the number of particles, respectively. Because temperature differences lead to the transfer of heat, one can formally define $d'Q = TdS$, where S is a thermodynamic variable called *entropy*. S is an extensive parameter defined up to an additive constant, and it is a state function, that is, dS is an exact differential (while $d'Q$ is not). A given thermal equilibrium state, a, is chosen as a standard state. The entropy $S(b)$ of the system at state b is:

$$S(b) = \int_a^b \frac{d'Q_{\text{rev}}}{T} \tag{2.20}$$

The integration is carried out over *any* reversible path connecting state a to state b. T is the temperature of the surroundings, and $d'Q$ is the heat absorbed by the system. The physical meaning of entropy will be discussed first with respect to the second law of thermodynamics and later in statistical mechanics.

It is of interest to calculate the entropy of an ideal gas. First, we demonstrate that for an ideal gas, dS is an exact differential, while $d'Q$ is not. Thus, for an ideal gas, we have obtained [Equation (2.16)],

$$d'Q = C_V dT + pdV \tag{2.21}$$

where p can be substituted by its value from the equation of state [Equation (2.2)], $p = RT/V$; this leads to:

$$d'Q = C_V dT + \frac{RT}{V} dV \tag{2.22}$$

It is evident that $d'Q$ is not an exact differential since the derivatives of the two terms are not equal [see Equation (2.15) and Section, A.2]

$$\frac{\partial C_V}{\partial V} = 0 \neq \frac{\partial (RT/V)}{\partial T} = \frac{R}{V} \tag{2.23}$$

On the other hand, dS is an exact differential because for $dS = d'Q/T$, the two derivatives $= 0$; in particular, $d(R/V)/dT = 0$. Correspondingly, one can show that $Q_{\text{I}} \neq Q_{\text{II}}$ (and $W_{\text{I}} \neq W_{\text{II}}$), where Q_{I} and Q_{II} are calculated between states a and b along different reversible paths (see discussion related to Figure 2.5). One can calculate an expression for the entropy of an ideal gas by dividing Equation (2.22) by T, and by replacing R by Nk_{B}; one obtains,

$$dS_{\text{IG}} = C_V \frac{dT}{T} + k_{\text{B}} N \frac{dV}{V} \tag{2.24}$$

and an integration leads to,

$$S_{IG} = C_V \ln T + k_B N \ln V + k_B N g(N) \qquad (2.25a)$$

$$S_{IG} = C_V \ln T + k_B N \ln V + k_B N [-\ln(N) + const.] \qquad (2.25b)$$

$$S_{IG} = C_V \ln T - k_B \ln(\frac{1}{V})^N - N k_B \ln N + N k_B \, const. \qquad (2.25c)$$

The first two terms in Equation (2.25a) result from the direct integration of Equation (2.24), while the third term denoted $g(N)$ is an integration constant, which in Equation (2.25b) is defined as $-\ln(N)+const.$ The *const.* cannot be determined from thermodynamics, and will be obtained by statistical mechanics in Chapter 4. The first term, $-k_B N \ln N$ ($\sim -k_B \ln N!$), has been introduced to keep S_{IG} extensive. Thus, if two ideal gas systems defined by $(N/2, V/2, T)$ are mixed, one would expect, $S_{IG}(N, V, T) = 2S_{IG}(N/2, V/2, T)$. However, this condition is not satisfied by the $\ln V$ term in Equation (2.25a) because $k_B N \ln V \neq 2 k_B N/2 \ln V/2 = k_B N \ln V/2$. However, adding $-k_B N \ln N$, solves this problem since, $k_B N \ln(V/N) = 2 k_B N/2 \ln[(V/2)/(N/2)] = k_B N \ln(V/N)$. Actually, this problem is a manifestation of Gibbs paradox that applies to systems of indistinguishable particles. We shall return to this issue later in Sections 3.3.2 and 4.2. Notice also that the entropy is related to probability, as the term $(1/V)^N$ [Equation (2.25c)] is the probability density of a configuration of N non-interacting particles in a box of volume V. In Chapter 4, we show that $C_V \ln T$ is related to probability (a Gaussian) as well [Equation (4.27)], which, however, cannot be deduced from Equation (2.25c).

To understand the significance of S_{IG}, we calculate the increase in S_{IG}, as the volume is increased from V_1 to V_2, while T and N remain constant. Since C_V is independent of V one obtains:

$$\Delta S_{IG} = N k_B \ln\left(\frac{V_2}{V_1}\right) \quad T \Delta S_{IG} = W \qquad (2.26)$$

Now, according to Equation (2.4), $T\Delta S_{IG} = W$, where W is the work done on the surroundings by an ideal gas enclosed in a cylindrical container as its volume is increased (reversibly) from V_1 to V_2 due to the moving piston. From Joule's experiment, we know that changes in V do not affect the energy. Therefore, the work done is caused by the entropic changes alone. Now, the energy and entropy for the work done on the surroundings are provided by the heat bath, meaning that W/T is the entropy lost by the heat bath, and the total entropy change, ΔS_{total}, in this *reversible* process, as expected, is zero, $\Delta S_{total} = \Delta S_{IG} - W/T = 0$. On the other hand, if the piston moves fast, $W_{fast} < W$ and $\Delta S_{total} = \Delta S_{IG} - W_{fast}/T > 0$, which is a manifestation of the second law of thermodynamics, discussed in the next sub-section, stating that in an irreversible process, $\Delta S_{total} > 0$.

PROBLEM 2.2

Calculate the increase in entropy when an ideal gas is heated from T_1 to T_2: (1) at constant pressure and (2) at constant volume. Show that the entropy in the first case is γ times that in the second case (see the example solved above).

2.8 The Second Law of Thermodynamics

The second law of thermodynamics is based on the realization that while mechanical work can be transferred completely to heat, there is a limitation in transferring heat to mechanical work (Sections 2.2 and 2.4). From a microscopic point of view of an ideal gas, molecular motion is not ordered and the kinetic energy cannot be fully harnessed (Section 2.2). The law was introduced initially in terms of engines, perpetuum mobile, or a Carnot cycle, by Kelvin, Clausius, and others. We shall not cover this topic in detail, but will present a postulate based on the entropy. The second law states,

$$\int_a^b \frac{d'Q}{T} \le S(b) - S(a) \tag{2.27}$$

where equality occurs for a reversible path. For an infinitesimal process,

$$d'Q/T \le dS. \tag{2.28}$$

The significance of this equation is the following: dQ'/T is the entropy accepted by the system from the surroundings (reservoir), which in a reversible process, $d'Q/T$ is equal to the increase in the system's entropy, dS. On the other hand, in an irreversible process, the amount of $d'Q/T$ generated is smaller than dS, which is a state function; an example is discussed in the paragraph following Equation (2.26).

Another way to formulate the second law is by defining two kinds of entropies, an *external* entropy added to the system from the outside, $d_e S$, and an internal entropy created in the system, $d_i S$, also called *entropy production*. One obtains:

$$dS = d_i S + d_e S \tag{2.29}$$

and an equivalent statement of the second law [Equation (2.28)] is,

1. $d_i S \ge 0$; $d_i S = 0$ in a reversible process, while $d_i S > 0$ in an irreversible process

2. $d_e S$ can take any sign and value.

$$\tag{2.30}$$

We shall return to this definition in Chapter 11. To see the identity of the two formulations, one has to define the "system" adequately. Thus, in the example following Equation (2.26), the "system" is the cylindrical container together with the heat bath, therefore, $d_e S = 0$ and $d_i S = dS_{\text{total}} \ge 0$.

2.8.1 Maximal Entropy in an Isolated System

In an isolated system, where no interaction with the surroundings occurs, heat cannot be transferred from the outside, that is, $d'Q = 0$ and $S(b) \ge S(a)$; therefore, S of the final state can never be smaller than S of the initial state. If the process is reversible, no change in the entropy can happen.

Thus, when an isolated system is in the state of maximum S (consistent with its energy), it cannot transform further because S will decrease. In other words, the state of maximum entropy is the most stable for an isolated system (equilibrium state), more specifically:

$$S = \text{maximum}(S[k^*]) \tag{2.31}$$

where $[k^*]$ is the optimal set of system parameters for given E and V. The fact that the entropy is maximized, $\delta(S)_{E,V} \ge 0$, means that a *spontaneous* process always goes in one direction of increasing entropy, as in the example depicted in Figure 2.7.

In Figure 2.7, an *equilibrated* ideal gas occupies initially a small compartment of volume V_1, which is a part of a larger *isolated* system of volume V_2. Then, the partition that holds the gas is removed (no work is done), the gas expands irreversibly, and finally gets equilibrated in V_2. The energy, E_{IG}, remains constant during the process since it is independent of the volume (Joule's experiment). E_{IG} depends only on the temperature, T, which remains constant as well (T should be considered as T_{system}, which was determined when the system was prepared.) However, the entropy increases, and since it is a state function, ΔS_{IG}

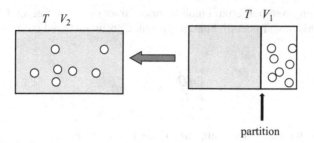

partition

FIGURE 2.7 Spontaneous expansion of an ideal gas from volume V_1 to volume V_2.

depends only on the initial and final states of the system, which are determined by the volumes V_1 and V_2, respectively. Thus, ΔS_{IG} is given by,

$$\Delta S_{IG} = Nk_B \ln\left(\frac{V_2}{V_1}\right). \tag{2.32}$$

Notice that in an isolated system, S is not related to temperature or heat.

2.8.2 Spontaneous Expansion of an Ideal Gas and Probability

ΔS_{IG} [Equation (2.32)] can be written in terms of a ratio between the probability densities of a particle in the two volumes,

$$\Delta S_{IG} = Nk_B \ln\left(\frac{V_2}{V_1}\right) = -k_B \ln\frac{(1/V_2)^N}{(1/V_1)^N} = -k_B \ln\frac{\left(\rho_2^{particle}\right)^N}{\left(\rho_1^{particle}\right)^N} \tag{2.33}$$

This equation demonstrates that the increase in entropy in the above experiment has also a statistical meaning. Thus, for an ideal gas, each system configuration of the N atoms [that is, $(\mathbf{x}_1, \mathbf{x}_2, \mathbf{x}_3, \cdots, \mathbf{x}_N)$, where $\mathbf{x}_i$ is the Cartesian vector of atom i] in the container has the same probability density to occur, $[\sim(1/V_2)^N]$. The number of available configurations within V_2 (the entire container) is tremendously larger than the number in V_1 [a ratio of $(V_2/V_1)^N$]; therefore, in practice, the confined configurations (in V_1) will never occur again after the expansion has been completed. This connection between entropy and probability is a statistical mechanics consideration, which explains the tendency for disorder (maximum entropy) in an isolated system.

2.8.3 Reversible and Irreversible Processes Including Work

Figure 2.8 describes a closed vessel of volume V_2 containing a smaller compartment of volume V_1, which can be increased by a moving piston. The compartment contains an ideal gas, and the

T piston

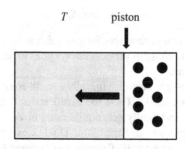

FIGURE 2.8 Ideal gas in a moving volume in contact with a heat bath.

system is at equilibrium with a heat bath with temperature, T. This system has already been depicted in Figure 2.3 and discussed in the paragraphs following Equations (2.4) and (2.26); the present discussion is a summary involving the second law. Thus, the piston is moved *reversibly*, and the volume of the gas increases from V_1 to V_2. The mechanical work done [see Equation (2.4)] is, $W = Nk_BT \ln(V_2/V_1)$. The required energy for moving the piston is provided by the heat bath in the form of heat, Q, which is absorbed by the system, that is, $W = Q$. Being a state function, the increase in the entropy depends only on the final states due to Equation (2.32). In an equilibrium process, the increase in entropy is $\Delta S_{IG} = W/T$, where ΔS_{IG} is equal to the *decrease* in the bath's entropy Q/T, that is, $\Delta S_{IG} = Q/T$, or:

$$\Delta S_{total}(\text{bath} + \text{IG}) = \Delta S_{IG} - \frac{Q}{T} = 0 \qquad (2.34)$$

If the piston is moved fast, ΔS_{IG} is still defined by Equation (2.32). However, the amount of work decreases, $\Delta W_{fast} < \Delta W$ [see the discussion following Equation (2.4)], hence, the total heat, Q_{fast}, absorbed from the bath is smaller, $Q_{fast} < Q$; one obtains,

$$\Delta S_{total} = \Delta S_{IG} - \frac{Q_{fast}}{T} > 0. \qquad (2.35)$$

This equation is equivalent to Equations (2.27) and (2.28) for the second law,

$$\int \frac{d'Q_{fast}}{T} = \frac{Q_{fast}}{T} < \Delta S_{IG}. \qquad (2.36)$$

In summary, in all the above examples involving an ideal gas, the energy is not existent, and the work done is due to entropy changes alone. Also, the maximum amount of work on the surroundings is achieved in a reversible process.

2.9 The Third Law of Thermodynamics

The entropy of a chemically uniform body of finite density approaches a limiting value as the temperature goes to absolute zero regardless of pressure, density, or phase. It is therefore convenient to take the state at 0°K as the standard state by assuming:

$$\lim_{T \to 0} S = S_0 \equiv 0 \qquad (2.37)$$

This is the third law of thermodynamics suggested by Nernst. The conclusions are: (1) one can never attain $T = 0°K$ by any means. (2) The specific heat and similar quantities are bound to approach to 0 as $T \to 0$. As mentioned earlier, for an ideal gas [Equation (2.25)], it is impossible to determine the integration constant by evaluating the equation at $T = 0$, because this model describes a gas at a high temperature.

2.10 Thermodynamic Potentials

In this section, we present the various thermodynamic potentials, in addition to the energy and entropy discussed thus far.

2.10.1 The Gibbs Relation

Combining the first law of thermodynamics and the definition of entropy, one obtains Gibbs relation,

$$dE = d'Q - d'W + d'Z = TdS - pdV + \sum_i \mu_i dN_i - \sum_i X_i dx_i. \qquad (2.38)$$

For completeness, we have added into Equation (2.38) the chemical potentials, μ_i, and the corresponding numbers of molecules, N_i, and other external potentials, X_i, with their conjugate variables x_i (e.g., a magnetic field and the related magnetization). This is a fundamental equation in thermodynamics. Ignoring *for simplicity* the last two terms (which include X_i and μ_i), one obtains the energy as a function of the entropy and volume variables alone, $E = E(S,V)$, therefore,

$$dE = \left(\frac{\partial E}{\partial S} \right)_V dS + \left(\frac{\partial E}{\partial V} \right)_S dV. \qquad (2.39)$$

A comparison with Equation (2.38) leads to:

$$\left(\frac{\partial E}{\partial S} \right)_V = T \qquad \left(\frac{\partial E}{\partial V} \right)_S = -p, \qquad (2.40)$$

and because dE is an exact differential, the second derivatives are equal; thus, one obtains the Maxwell relation,

$$\left(\frac{\partial T}{\partial V} \right)_S = -\left(\frac{\partial p}{\partial S} \right)_V \qquad (2.41)$$

2.10.2 The Entropy as the Main Potential

Sometimes it is convenient to use the entropy $S(E, V, N_i, x_i)$ as the main potential function, instead of the energy $E(S, V, N_i, x_i)$; thus, from Equation (2.38) one obtains,

$$dS = \left(\frac{1}{T} \right) dE + \left(\frac{p}{T} \right) dV - \sum_i \left(\frac{\mu_i}{T} \right) dN_i + \sum_i \left(\frac{X_i}{T} \right) dx_i, \qquad (2.42)$$

with the corresponding equations to those derived above for the energy,

$$dS = \left(\frac{\partial S}{\partial E} \right)_V dE + \left(\frac{\partial S}{\partial V} \right)_E dV \qquad (2.43)$$

$$\left(\frac{\partial S}{\partial E} \right)_V = 1/T \qquad \left(\frac{\partial S}{\partial V} \right)_E = p/T \qquad (2.44)$$

and the Maxwell relation

$$\left(\frac{\partial(1/T)}{\partial V} \right)_E = \left(\frac{\partial(p/T)}{\partial E} \right)_V \qquad (2.45)$$

The energy in the Gibbs equation is a function of the extensive parameters S, V, N_i, and x_i which are not always convenient to handle experimentally. Therefore, other potentials (free energies) have been defined, where extensive parameters are replaced by their conjugate intensive ones.

2.10.3 The Enthalpy

We wish to define a potential which is a function of S and p. For that, we first write pdV as

$$pdV = d(pV) - Vdp \qquad (2.46)$$

and substitute it in dE [Equation (2.38)],

$$dE = TdS - pdV = TdS - d(pV) + Vdp. \qquad (2.47)$$

Thus, a new function called *enthalpy* and denoted by $H = E + pV$ is defined,

$$dH = d(E + pV) = TdS + Vdp. \qquad (2.48)$$

$H = E + pV$ is a function of p rather than V, that is, $H = H(S, p, N_i, x_i)$; H (like E) is a state function, hence, dH is an exact differential, and following Equations (2.39) through (2.41) above one obtains:

$$\left(\frac{\partial H}{\partial S} \right)_p = T \qquad \left(\frac{\partial H}{\partial p} \right)_S = V \qquad (2.49)$$

and the Maxwell relation is:

$$\left(\frac{\partial T}{\partial p} \right)_S = \left(\frac{\partial V}{\partial S} \right)_p \qquad (2.50)$$

2.10.4 The Helmholtz Free Energy

It is more convenient to control the temperature than the entropy. For that, one defines the Helmholtz free energy, $A = A(T, V)$ (again for simplicity we treat only $E(S, V)$); first we write:

$$dE = TdS - pdV = d(TS) - SdT - pdV \qquad (2.51)$$

which leads to,

$$dA = d(E - TS) = -SdT - pdV \qquad (2.52)$$

The Helmholtz free energy, $A = E - TS$ is indeed a function of T, while the other variables are unchanged, $A = A(T, V, N_i, x_i)$; A is a state function, hence, dA is an exact differential with the following relations:

$$\left(\frac{\partial A}{\partial T} \right)_V = -S \qquad \left(\frac{\partial A}{\partial V} \right)_T = -p \qquad (2.53)$$

and the Maxwell relation:

$$\left(\frac{\partial S}{\partial V} \right)_T = \left(\frac{\partial p}{\partial T} \right)_V \qquad (2.54)$$

2.10.5 The Gibbs Free Energy

The Gibbs free energy, G is a function of p and T. To obtain G, we first change the energy as:

$$dE = TdS - pdV = d(TS) - SdT - d(pV) + Vdp \qquad (2.55)$$

which leads to:

$$dG = d(E - TS + pV) = -SdT + Vdp. \tag{2.56}$$

$G = E - TS + pV$ – the Gibbs free energy is a state function, $G = G(T, p, N_i, x_i)$; thus, one obtains

$$\left(\frac{\partial G}{\partial T}\right)_p = -S \qquad\qquad \left(\frac{\partial G}{\partial p}\right)_T = V \tag{2.57}$$

and the Maxwell relation,

$$-\left(\frac{\partial S}{\partial p}\right)_T = \left(\frac{\partial V}{\partial T}\right)_p \tag{2.58}$$

2.10.6 The Free Energy, $\hat{H}(T,\mu)$

In the free energy, $\hat{H}(T,\mu)$, the variables S and N of the energy are replaced by T and the chemical potential, μ. We first write:

$$dE = TdS + \mu dN = d(TS) - SdT + d(\mu N) - Nd\mu \tag{2.59}$$

which leads to:

$$d\hat{H} = d(E - TS - \mu N) = -SdT - Nd\mu \tag{2.60}$$

where $\hat{H} = E - TS - \mu N$ and $\hat{H}(T,V,\mu,x_i)$ is a state function satisfying:

$$\left(\frac{\partial \hat{H}}{\partial T}\right)_\mu = -S \qquad\qquad \left(\frac{\partial \hat{H}}{\partial \mu}\right)_T = -N \tag{2.61}$$

and the Maxwell relation:

$$\left(\frac{\partial S}{\partial \mu}\right)_T = \left(\frac{\partial N}{\partial T}\right)_\mu \tag{2.62}$$

A summary of the various free energy derivatives is presented in a Table 2.1.

TABLE 2.1

First Derivatives of the Various Free Energies

Free Energies	Derivatives			
$S(E,V,N,x)$	$\dfrac{\partial S}{\partial E} = \dfrac{1}{T}$	$\dfrac{\partial S}{\partial V} = \dfrac{p}{T}$	$\dfrac{\partial S}{\partial N} = -\dfrac{\mu}{T}$	$\dfrac{\partial S}{\partial x} = \dfrac{X}{T}$
$E(S,V,N,x)$	$\dfrac{\partial E}{\partial S} = T$	$\dfrac{\partial E}{\partial V} = -p$	$\dfrac{\partial E}{\partial N} = \mu$	$\dfrac{\partial E}{\partial x} = -X$
$H(S,p,N,x) = E + pV$	$\dfrac{\partial H}{\partial S} = T$	$\dfrac{\partial H}{\partial p} = V$	$\dfrac{\partial H}{\partial N} = \mu$	$\dfrac{\partial H}{\partial x} = -X$
$A(T,V,N,x) = E - TS$	$\dfrac{\partial A}{\partial T} = -S$	$\dfrac{\partial A}{\partial V} = -p$	$\dfrac{\partial A}{\partial N} = \mu$	$\dfrac{\partial A}{\partial x} = -X$
$G(T,p,N,x) = E - TS + pV$	$\dfrac{\partial G}{\partial T} = -S$	$\dfrac{\partial G}{\partial p} = V$	$\dfrac{\partial G}{\partial N} = \mu$	$\dfrac{\partial G}{\partial x} = -X$
$\hat{H}(T,V,\mu,x) = E - TS - \mu N$	$\dfrac{\partial \hat{H}}{\partial T} = -S$	$\dfrac{\partial \hat{H}}{\partial V} = -p$	$\dfrac{\partial \hat{H}}{\partial \mu} = -N$	$\dfrac{\partial \hat{H}}{\partial x} = -X$

2.11 Maximal Work in Isothermal and Isobaric Transformations

Assume a system that undergoes an isothermal transformation from state a to state b. According to the first law of thermodynamics [Equation (2.5)], the mechanical work, W, done on the surroundings is:

$$W = -\Delta E + Q. \tag{2.63}$$

where

$$\int_a^b \frac{d'Q}{T} \leq S(b) - S(a). \tag{2.64}$$

Since the temperature, T, is constant, it can be taken out of the integral and:

$$Q = \int_a^b d'Q \leq T\{S(b) - S(a)\}. \tag{2.65}$$

Thus,

$$W \leq E(a) - E(b) + T[S(b) - S(a)] = E(a) - TS(a) - [E(b) - TS(b)] \tag{2.66}$$

or

$$W \leq A(a) - A(b) = -\Delta A \tag{2.67}$$

Therefore, in an isothermal transformation, the work done on the surroundings, W, can never exceed $-\Delta A$; if the transformation is reversible, $W (\equiv W_r) = -\Delta A$, while $W < W_r$ in an irreversible process. Thus, if $A(a) > A(b)$, the corresponding $W > 0$ is the maximal work done by the system on the surroundings [see Equation (2.4)]. Indeed, in the reversible expansion of an ideal gas (Section 2.8.3), we have seen that $W > 0$, since $-TS(V_1) = A(V_1) > A(V_2) = -TS(V_2)$.

Now, if $A(a) < A(b)$, $W(\equiv W_r) = A(a) - A(b) < 0$, where W_r is the *minimum* amount of *invested* (reversible) work required to bring the *system* from a to b; in this case, in an irreversible process, $W - Wr > 0$ is defined as the dissipative work. In other words:

$$W_{\text{system}} \geq A(b) - A(a) \tag{2.68}$$

where W_{system} is the amount of work done *on the system* (not on the surroundings); thus, the equality (minimum $W_{\text{system}} = W_r$) holds for equilibrium. If $A(b) > A(a)$ positive work is done on the system to shift it from a to b; in the other case (negative W_{system}, $b \to a$), work is done on the surroundings.

For an ideal gas, $W_{\text{system}} \geq TS(a) - TS(b)$, where $S(a) > S(b)$; thus, inducing a *pure* entropic change requires mechanical work. While this work can be understood in terms of the increase in the density, hence, in the pressure of the gas, it also reflects the increase in our knowledge about the distribution of the molecules, which occupy a smaller volume, $V(b)$; this picture is compatible with the interpretation of Sections 2.8.2 and 2.8.3.

Now, let us consider a system at constant T that cannot exchange energy in the form of work with the surroundings ($\Delta V = 0$ – an isochore transformation). In this case:

$$0 = W \leq A(a) - A(b) \qquad \text{or} \qquad A(b) \leq A(a) \tag{2.69}$$

Under such conditions, the free energy cannot increase during a transformation. Or if the free energy is a minimum, the system is in a state of stable equilibrium because any transformation would produce an increase in A, and this would contradict the above inequality. Thus,

$$A = \text{minimum}(A[k^*]) \tag{2.70}$$

where [k*] is the optimal set of system parameters for given T, V, and N. This should be compared to an isolated system, where at equilibrium the entropy is maximal [Equation (2.31)].

In the same way, for an *NpT* system at constant T and p, that cannot exchange energy in the form of work with the surroundings, one can show that for a transformation, $G(b) \leq G(a)$ and the state at which G is minimum is a stable equilibrium state. Also, since $\Delta G = \Delta A + p\Delta V$, in a reversible process:

$$-\Delta G = \Delta W_{res} - p\Delta V. \tag{2.71}$$

$-\Delta G$ is the maximum (net) work at constant T and p *other* than that due to a *volume change* (e.g., electrical or magnetic work).

2.12 Euler's Theorem and Additional Relations for the Free Energies

A homogeneous function of order n satisfies:

$$f(\lambda x_1, \lambda x_2, \ldots, \lambda x_t) = \lambda^n f(x_1, x_2, \ldots, x_t). \tag{2.72}$$

Taking derivatives on both sides with respect to λ leads to:

$$n\lambda^{n-1} f(x_1, x_2, \ldots, x_t) = \left(\frac{\partial f}{\partial(\lambda x_1)}\right)\left(\frac{\partial(\lambda x_1)}{\partial \lambda}\right) + \cdots \left(\frac{\partial f}{\partial(\lambda x_t)}\right)\left(\frac{\partial(\lambda x_t)}{\partial \lambda}\right)$$

$$= x_1\left(\frac{\partial f}{\partial(\lambda x_1)}\right) + \cdots + x_t\left(\frac{\partial f}{\partial(\lambda x_t)}\right) \tag{2.73}$$

Substituting $\lambda = 1$ in Equation (2.73) above leads to *Euler's theorem* for a homogeneous function:

$$nf(x_1, x_2, \ldots, x_t) = x_1\left(\frac{\partial f}{\partial x_1}\right) + x_2\left(\frac{\partial f}{\partial x_2}\right) + \cdots + x_t\left(\frac{\partial f}{\partial x_t}\right) \tag{2.74}$$

The free energies defined previously are homogenous ($n = 1$) in their *extensive* variables. For the Gibbs free energy, one obtains (Table 2.1),

$$G(T, p, \lambda N) = \lambda^1 G(T, p, N) \rightarrow G = N\left(\frac{\partial G}{\partial N}\right)_{T,p} = N\mu(T, p, N)$$

$$\rightarrow G = \sum_i \mu_i N_i \tag{2.75}$$

In a similar way, a new relation is obtained for $\hat{H}(T, V, \mu)$:

$$\hat{H}(T, \lambda V, \mu) = \lambda^1 \hat{H}(T, V, \mu) \rightarrow \hat{H}(T, V, \mu) = -p(T, V, \mu)V$$

$$\rightarrow \hat{H} = -pV \tag{2.76}$$

Finally, one obtains for the Helmholtz free energy,

$$A(T, \lambda V, \lambda N) = \lambda^1 A(T, V, N) \rightarrow A(T, V, N) = -p(T, V, N)V + \mu(T, V, N)N$$

$$\rightarrow A = -pV + \sum_i \mu_i N_i \tag{2.77}$$

2.12.1 Gibbs-Duhem Equation

According to Equation (2.56),

$$dG = -SdT + Vdp + \sum_i \mu_i dN_i \qquad (2.78)$$

and according to Equation (2.75),

$$dG = d(\sum_i \mu_i N_i) = \sum_i \mu_i dN_i + \sum_i N_i d\mu_i \qquad (2.79)$$

$$\rightarrow \quad SdT - Vdp + \sum_i N_i d\mu_i = 0 \qquad (2.80)$$

2.13 Summary

An essential thermodynamic property is entropy, which from its formal definition looks obscure. However, its meaning becomes clearer as related to an ideal gas. Thus: (1) we have demonstrated that even on the level of classical thermodynamics, the entropy for an ideal gas depends on probabilistic elements, more specifically, $S \sim \ln P$; this is a basic cornerstone in our derivation of statistical mechanics in the next chapter. Furthermore, (2) to condense an ideal gas, one has to invest mechanical work, W, against the increase of pressure. However, one can adopt an additional interpretation, with which W is needed to decrease entropy, that is, to increase our knowledge about the distribution of the molecules as the volume is decreased.

HOMEWORK FOR STUDENTS

We recommend Kubo's book [2]—a source of scientific wisdom and hundreds of *solved* problems in classical thermodynamics.

REFERENCES

1. E. Fermi. *Thermodynamics.* (Dover Publications, New York, 1956).
2. R. Kubo. *Thermodynamics, An Advanced Course With Problems and Solutions.* (North Holland, Amsterdam 1968).
3. S. Glasstone and D. Lewis. *Elements of Physical Chemistry.* (Macmillan & Co., New York, 1960).

FURTHER READING

4. F. Reif. *Fundamentals of Statistical and Thermal Physics.* (McGraw Hill, New York, 1965).
5. D. Chandler. *Introduction to Modern Statistical Mechanics.* (Oxford, New York, 1987).
6. H. B. Callen. *Thermodynamics and An Introduction to Thermostatistics.* (John Wiley & Sons, New York, 1985).

3

From Thermodynamics to Statistical Mechanics

We have already pointed out the inherent difficulty in treating thermal systems on the molecular (microscopic) level using deterministic methods that are generally applied to macroscopic mechanical systems. This difficulty is reflected by the relatively crude modeling provided by classical thermodynamics, where systems are described only by a small number of *macroscopic* parameters. Naturally, one would seek to widen the scope of thermodynamics by taking microscopic degrees of freedom into account, even approximately. This is achieved by statistical mechanics.

Thus, statistical mechanics can be viewed as an enrichment of classical thermodynamics, where the microscopic degrees of freedom are treated by statistical means. We shall derive this theory by demanding compatibility with classical thermodynamics. Our main interest is in classical systems, while quantum mechanical considerations will be limited.

3.1 Phase Space as a Probability Space

We consider a system, C, of volume, V, containing, N, particles (e.g., liquid argon) at temperature, T_C. The system is in thermal contact with a large reservoir (R) (heat bath) with a well-defined temperature, T_R. At *equilibrium*, while a very small amount of energy is constantly exchanged between R and C, $T_C = T_R$ (Figure 3.1).

From a microscopic point of view, one can envisage N particles moving constantly according to Newton's equations of motion, and the mechanical state of the system at time, t, can, in principle, be defined by the instantaneous vector $(\mathbf{x}^N(t), \mathbf{p}^N(t))$. $\mathbf{x}$ is the Cartesian coordinates vector of a particle, and $\mathbf{p} = m\mathbf{v}$ is the corresponding momentum vector, where m and $\mathbf{v}$ are the mass and velocity vector of the particle, respectively. $\mathbf{x}^N$ defines a *configuration* of the N particles, and the ensemble of all possible $\mathbf{x}^N$ points is called the *configurational space*; the ensemble of all $(\mathbf{x}^N, \mathbf{p}^N)$ points, which defines all the possible mechanical states of the system, is called *phase space*; $(\mathbf{x}^N, \mathbf{p}^N)$ is one point in phase space,

$$(\mathbf{x}^N, \mathbf{p}^N) = (x_1, y_1, z_1, \ldots, x_N, y_N, z_N, p_{x_1}, p_{y_1}, p_{z_1}, \ldots, p_{x_N}, p_{y_N}, p_{z_N}) \tag{3.1}$$

with the differential,

$$d\mathbf{x}^N d\mathbf{p}^N = dx_1 dy_1 dz_1 \cdots dx_N dy_N dz_N dp_{x_1} dp_{y_1} dp_{z_1} \cdots dp_{x_N} dp_{y_N} dp_{z_N} \tag{3.2}$$

(Clearly, on the microscopic level, the system should be described by quantum mechanics; however, for simplicity, we treat it here classically, where, quantum effects will be introduced later as the theory is developed further.)

In contrast to the case of a *macroscopic mechanical* system, it is inherently impossible to calculate the trajectory $(\mathbf{x}^N(t), \mathbf{p}^N(t))$ as a function of time for the $\sim 10^{23}$ particles because their initial velocities and coordinates are unknown. Also, the changes imposed on a trajectory by the disordered energy exchanged between the heat bath and the system are unpredictable. Furthermore, the phase space $[\mathbf{x}^N(t), \mathbf{p}^N(t)]$ provides more information that can be obtained experimentally (see discussions in Sections 10.6 and 10.9.2). The reason is that a measurement of a macroscopic thermodynamic parameter can be viewed as an *averaging* process based on contributions from a huge number of microscopic states. For example, to determine the temperature of a body, the thermometer feels the (time) *average* of the kinetic energy of a very long trajectory, consisting of a tremendous number of microscopic states $(\mathbf{x}^N, \mathbf{p}^N)$ of the tested system.

Entropy and Free Energy in Structural Biology

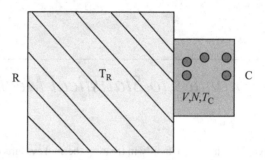

FIGURE 3.1   A system, C (defined by N, V, and T_C) is in thermal contact with a reservoir with temperature, T_R; at equilibrium $T_C = T_R$.

In other words, for a very long trajectory ($t \to \infty$), it is plausible to assume that all the microscopic states ($\mathbf{x}^N$, $\mathbf{p}^N$) have been realized, where some appear with higher frequency than others. Thus, in principle, one can try defining a probability density $P(\mathbf{x}^N, \mathbf{p}^N)$, where $P(\mathbf{x}^N, \mathbf{p}^N)d\mathbf{x}^N d\mathbf{p}^N$ is the probability that the system, C, is located within the interval $\mathbf{x}^N + d\mathbf{x}^N$ and $\mathbf{p}^N + d\mathbf{p}^N$. While $P(\mathbf{x}^N, \mathbf{p}^N)$ will contain less information than the trajectory itself, it will enable one to calculate the averages of macroscopic parameters based on microscopic degrees of freedom. More specifically, an extensive thermodynamic parameter M_{thermo} will become a statistical average (denoted by $<M>$) in statistical mechanics; for example, the thermodynamic energy, E_{thermo}, will be represented by $<E>$,

$$E_{thermo} \to < E >= \int P(\mathbf{x}^N, \mathbf{p}^N)E(\mathbf{x}^N , \mathbf{p}^N)d\mathbf{x}^N dp^N \qquad (3.3)$$

provided that the energy of interaction, $E(\mathbf{x}^N, \mathbf{p}^N)$, for the system studied is also known. For a system defined by N, V, and T, $P(\mathbf{x}^N, \mathbf{p}^N)$ and the random variables ($\mathbf{x}^N$, $\mathbf{p}^N$) and $M(\mathbf{x}^N, \mathbf{p}^N)$ define a probability space, which is called the *canonical* [or the *(NVT)*] *ensemble*. The main problem, of course, is to determine $P(\mathbf{x}^N, \mathbf{p}^N)$ simultaneously with the statistical mechanics entropy function. $P(\mathbf{x}^N, \mathbf{p}^N)$ can be obtained under certain conditions by the Liouville equation, which, however, will not be discussed in this book. Alternatively, we shall derive $P(\mathbf{x}^N, \mathbf{p}^N)$ based on three factors: (1) the second law of thermodynamics, (2) properties of the *thermodynamic* entropy, and (3) some known experimental results. $P(\mathbf{x}^N, \mathbf{p}^N)$ defines our *maximum* knowledge about the microscopic degrees of freedom—the phase space. At high T, the particles are expected to move at a very high speed, leading to a trajectory that visits the *configurational space*, $\{\mathbf{x}^N\}$ with approximately equal frequency, which means high randomness and large uncertainty, therefore high entropy. At low T, where the effect of the molecular forces is significant, some specific regions in configurational space will be more probable than others, and the entropy will decrease.

3.2 Derivation of the Boltzmann Probability

We have shown that if $P(\mathbf{x}^N, \mathbf{p}^N)$ is known, the calculation of the statistical average of extensive parameters (e.g., the energy) is straightforward. This is not the case, however, with the entropy that its local function, $\hat{S}(\mathbf{x}^N, \mathbf{p}^N)$ [equivalent to $E(\mathbf{x}^N, \mathbf{p}^N)$], is a priori unknown and should be defined. The statistical averages of the entropy and the Helmholtz free energy can formally be written as,

$$S =< S >= \int P(\mathbf{x}^N,\mathbf{p}^N)\hat{S}(\mathbf{x}^N ,\mathbf{p}^N)d\mathbf{x}^N dp^N \qquad (3.4)$$

$$A =< A >= \int P(\mathbf{x}^N,\mathbf{p}^N)[E(\mathbf{x}^N,\mathbf{p}^N)-T\hat{S}(\mathbf{x}^N,\mathbf{p}^N)]d\mathbf{x}^N dp^N \qquad (3.5)$$

where, for simplicity, from now on we shall use S and A instead of $<S>$ and $<A>$, respectively.

As pointed out above, the problem is to determine simultaneously P and $\hat{S}$, which should lead to the usual thermodynamic properties. First, we know from thermodynamics that S is extensive ($\sim N$) and positive, and for an ideal gas, S is a function of the probability density, $P=(1/V)^N$, which appears in S logarithmically (see Section 2.7). Therefore, it is plausible to *assume* that $\hat{S}[P(\mathbf{x}^N, \mathbf{p}^N)] \sim -\ln P(\mathbf{x}^N, \mathbf{p}^N)$, that is, $\hat{S}$ is a function of the probability of the $3N$ coordinates and momenta. One would expect this "extensivity" to hold also for N *interacting* particles because P becomes a product of N "partial" probability densities. For example, for three particles at positions $(\mathbf{x}_1, \mathbf{x}_2, \mathbf{x}_3)$ (ignoring for simplicity the velocities), $P(\mathbf{x}_1, \mathbf{x}_2, \mathbf{x}_3)$ can be expressed as a product of three partial probabilities,

$$P(\mathbf{x}_1, \mathbf{x}_2, \mathbf{x}_3) = P_1(\mathbf{x}_1) P_2(\mathbf{x}_2 | \mathbf{x}_1) P_3(\mathbf{x}_3 | \mathbf{x}_2, \mathbf{x}_1) \tag{3.6}$$

where

$$P_1(\mathbf{x}_1) = \int P(\mathbf{x}_1, \mathbf{x}, \mathbf{y}) d\mathbf{x} d\mathbf{y} \quad P_2(\mathbf{x}_2 | \mathbf{x}_1) = [\int P(\mathbf{x}_1, \mathbf{x}_2, \mathbf{y}) d\mathbf{y}] / P_1(\mathbf{x}_1) \tag{3.7}$$

$$P_3(\mathbf{x}_3 | \mathbf{x}_2, \mathbf{x}_1) = P(\mathbf{x}_1, \mathbf{x}_2, \mathbf{x}_3) / P_2(\mathbf{x}_2 | \mathbf{x}_1). \tag{3.8}$$

A more general entropy function, $\hat{S}$ is:

$$\hat{S} \sim -C \ln \left(P a^N b^N \right) \tag{3.9}$$

where $C>0$ and a and b are constants discussed later; being exponents of N, a and b, will keep $\hat{S}$ a function of N. Equation (3.9) probably defines the simplest function for $\hat{S}$ satisfying the above restrictions; however, we cannot prove that this function is unique. Thus, in principle, a constant d^N and more such constants could be added under the *ln* in Equation (3.9); however, we shall argue later, that for compatibility with quantum mechanics and the experiment, $\hat{S}$ includes only two parameters. Also, notice that P, as a probability density, has the dimension of $(d\mathbf{x}^N d\mathbf{p}^N)^{-1}$, which should be eliminated by a^N, or b^N: otherwise the logarithm will depend on the units of P. We shall see later that b^N is needed to make S extensive.

To determine P, we recall that for a given set of N, V, and T, the Helmholtz free energy, $A(T, V, N) = E - TS$ should be minimum with respect to other system parameters (see Section 2.11). Thus, we shall determine P by minimizing the statistical free energy, A with respect to P, ignoring $(ab)^N$ (for a moment), which does not affect P, but affects S. Also, for simplicity, we define $\mathbf{X} = (\mathbf{x}^N, \mathbf{p}^N)$ and $d\mathbf{X} = d\mathbf{x}^N d\mathbf{p}^N$, which will be used in the following derivation. Thus, we treat the free energy function $A(P)$,

$$A(P) = \int P(\mathbf{X}) [E(\mathbf{X}) + TC \ln P(\mathbf{X})] d\mathbf{X}. \tag{3.10}$$

Based on the calculus of variations, we take the derivative of $A(P)$ with respect to P and equate the result to zero,

$$\frac{\partial A}{\partial P} = \int \{ E(\mathbf{X}) + TC \ln P(\mathbf{X}) + TCP(\mathbf{X}) / P(\mathbf{X}) \} d\mathbf{X}$$

$$= \int \{ E(\mathbf{X}) + TC \ln P(\mathbf{X}) + TC \} d\mathbf{X} = 0 \tag{3.11}$$

As expected, A is minimal, as the second derivative is positive $[\partial^2 A / \partial P^2 = \int TC/P(X) > 0$ since T, C, and P are positive]. A vanishing integral requires the integrand to vanish for each $\mathbf{X}$, thus:

$$E(\mathbf{X}) + TC \ln P(\mathbf{X}) + TC = 0 \tag{3.12}$$

and the probability density is:

$$P(\mathbf{X}) = \exp[-E(\mathbf{X})/TC - 1] \qquad (3.13)$$

The constant, C, can be determined by comparing our theory to the experiment, which for an ideal gas, leads to $C = k_B$ [see discussion following Equation (4.10)]. After normalization, one obtains the Boltzmann probability density, P^B,

$$P^B(\mathbf{X}) = \frac{\exp[-E(\mathbf{X})/k_B T]}{Q} \qquad (3.14)$$

where the normalization factor, Q, is called the *canonical partition function*.

$$Q = \int \exp[-\frac{E(\mathbf{X})}{k_B T}]d\mathbf{X} \qquad (3.15)$$

3.3 Statistical Mechanics Averages

After deriving the Boltzmann probability density, one can calculate the statistical averages of extensive parameters, such as the energy, magnetization, etc., in an alternative way to the traditional thermodynamic derivatives.

3.3.1 The Average Energy

The average energy, $<E>$ becomes,

$$<E> = \int P^B(\mathbf{x}^N, \mathbf{p}^N) E(\mathbf{x}^N, \mathbf{p}^N) d\mathbf{x}^N d\mathbf{p}^N \qquad (3.16)$$

3.3.2 The Average Entropy

Calculation of the entropy is more elaborate, as $\hat{S} \sim -C\ln(P^B a^N b^N)$ [Equation (3.9)] depends on a set of undetermined parameters. First, it should be pointed out again that the probability density, P^B, has a dimension, $P^B = P^{B\prime}/d\mathbf{x}^N d\mathbf{p}^N$, where $P^{B\prime}$ is the probability to find the system between $\mathbf{x}^N$ and $\mathbf{x}^N + d\mathbf{x}^N$ and $\mathbf{p}^N$ and $\mathbf{p}^N + d\mathbf{p}^N$; this means that $\ln(P^B)$ will depend on the units of $\mathbf{x}^N \mathbf{p}^N$. Clearly, elimination of this undesired dependency on the dimensionality can be achieved by defining $a^N \sim d\mathbf{x}^N d\mathbf{p}^N$. However, for atomic particles, Δx and Δp_x are not independent, but obey the uncertainty principle of quantum mechanics, $\Delta x \Delta p_x \sim h$, where h is Planck's constant. Thus, the *infinitesimals*, $d\mathbf{x}^N d\mathbf{p}^N$, are smaller than h^{3N}, and the corresponding, $P^{B\prime}$ should be increased by $h^{3N}/(d\mathbf{x}^N d\mathbf{p}^N) = P^B h^{3N}$; $a^N = h^{3N}$ solves the dimensionality problem, as S will actually become a function of the probability $P^{B\prime}$ rather than of the probability density, P^B (for the more traditional introduction of h to the entropy, see [1], Chapter 4).

The fact that the particles in our model are indistinguishable, which can occupy all available $\mathbf{x}$ and $\mathbf{p}$ values, leads to another quantum mechanical correction. For simplicity, assume a system of only two particles, a and b with coordinates, $a(\mathbf{x}_1, \mathbf{p}_1)$ and $b(\mathbf{x}_2, \mathbf{p}_2)$. In quantum mechanics, this state is identical to the state $b(\mathbf{x}_1, \mathbf{p}_1)$ and $a(\mathbf{x}_2, \mathbf{p}_2)$, therefore, these two states should be counted only once, that is, P^B should be multiplied by 2!, and for a system of N particles by N!, which according to Stirling's approximation, $N! \sim (N/e)^N$ (for a more detailed discussion, see Section 4.1). Thus, $N!$ can be identified with the parameter b^N appearing in $\hat{S} \sim -C\ln(P^B a^N b^N)$. The need for $N!$ for keeping the entropy extensive has been recognized by Gibbs before the advent of quantum mechanics; thus, $N!$ solves the Gibbs paradox [see discussion following Equation (2.25) about the $N \ln N$ term added to the entropy of an ideal gas in thermodynamics]. We shall also show that by a comparison with experimental results for an ideal gas, $C = k_B$. The entropy for a gas or a liquid becomes:

$$S = <S> = -k_B \int P^B(\mathbf{x}^N, \mathbf{p}^N) \ln\left[P^B(\mathbf{x}^N, \mathbf{p}^N) h^{3N} N! \right] d\mathbf{x}^N d\mathbf{p}^N. \tag{3.17}$$

S is the *classical entropy*, which also relies on quantum mechanical parameters. It should be pointed out that an additional parameter (d^N) in $\hat{S}$ [Equation (3.9)] is not necessary since one can show that S defined above is compatible with the experiment and satisfies the correspondence principle of quantum mechanics, which states: the quantum entropy defined at very low temperatures should lead to the classical entropy at high temperature [see discussion following Equation (4.55)].

We note that for models of different particles or models, where the particles are anchored to positions in space [such as the harmonic oscillator, Equation (4.35)], the factor $N!$ is not needed and is thus eliminated. For a discrete system, such as the Ising model, (Section 7.3), both $N!$ and h are discarded because the probability density P^B is replaced by a set of discrete probabilities, P_i^B,

$$S = <S> = -k_B \sum_i P_i^B \ln P_i^B. \tag{3.18}$$

However, for a system of multiple polymer chains on a lattice, $N!$ is needed, while h is not.

3.3.3 The Helmholtz Free Energy

Based on the above definition of the entropy, the Helmholtz free energy in statistical mechanics becomes:

$$A = -k_B \int P^B(\mathbf{x}^N, \mathbf{p}^N)\{E(\mathbf{x}^N, \mathbf{p}^N) + k_B T \ln[h^{3N} N! P^B(\mathbf{x}^N, \mathbf{p}^N)]\} d\mathbf{x}^N d\mathbf{p}^N \tag{3.19}$$

We substitute $P^B(\mathbf{x}^N, \mathbf{p}^N)$ under the *ln* by its explicit expression [Equation (3.14)], and the expression in the curly brackets becomes:

$$E(\mathbf{x}^N, \mathbf{p}^N) + k_B T \ln\{h^{3N} N! \frac{\exp[-E(\mathbf{x}^N, \mathbf{p}^N)/k_B T]}{Q}\} = -k_B T \ln\frac{Q}{h^{3N} N!}. \tag{3.20}$$

The term on the right hand side of Equation (3.20) is constant for *any* $(\mathbf{x}^N, \mathbf{p}^N)$, and when it is taken out of the integral [Equation (3.19)]; one obtains,

$$A = <E> -TS = -k_B T \ln\frac{Q}{h^{3N} N!} = -k_B T \ln Q^\# \tag{3.21}$$

where $Q^\#$ *includes* the constant, $h^{3N} N!$, which does not appear in Q—the normalization factor of the Boltzmann probability density [Equation (3.14)]. Thus, statistical mechanics provides a prescription for a direct calculation of the Helmholtz free energy from the partition function $Q^\#$.

In summary. We have represented the *thermodynamic* Helmholtz free energy, as a statistical average, $<A(P)>$ over the microscopic (atomic) configurations with an (initially) unspecified probability density P. P has been determined by minimizing $<A(P)>$ with respect to P, which has led to the Boltzmann probability density P^B [(Equation (3.14)]. An essential ingredient in this derivation is the construction of an extensive entropy function, $\hat{S} \sim -C \ln(P^B a^N b^N)$, where its dependency on probability is based on classical thermodynamic considerations; the parameters a and b and C have been determined based on insights from quantum mechanics and the experiment. It is of interest to learn about other ways to derive the theory of statistical mechanics, see for example [2–6].

3.4 Various Approaches for Calculating Thermodynamic Parameters

The Boltzmann probability density, P^B, and the partition function, $Q^\#$, provide three alternatives for treating macroscopic systems based on their *specific* microscopic properties.

3.4.1 Thermodynamic Approach

In classical thermodynamics, all quantities are obtained as derivatives of the Helmholtz free energy, A (see Table 2.1). Hence, in statistical mechanics all these quantities can be obtained as derivatives of $-k_B T \ln Q^\#$, where $Q^\#$ is based on the *microscopic* details of the specific system studied; such direct calculations are beyond the realm of classical thermodynamics.

3.4.2 Probabilistic Approach

In statistical mechanics, a system is modeled by a probability space (ensemble); thus, a macroscopic parameter, M, can also be obtained as an expectation value,

$$< M >= \int P^B(\mathbf{x}^N, \mathbf{p}^N) M(\mathbf{x}^N, \mathbf{p}^N) d\mathbf{x}^N d\mathbf{p}^N \qquad (3.22)$$

and its variance can be calculated as well (see Chapter 5); variances are not defined in classical thermodynamics at all. However, with both thermodynamic and probabilistic approaches, an analytical derivation of averages is feasible only for very simplified models, such as the *ideal gas* and the *harmonic oscillator*, which will be treated in Chapter 4. The advantage of the probabilistic approach is the fact that it can be used with computer simulation. Thus, using Monte Carlo methods or molecular dynamics, system's configurations can, in principle, be sampled correctly according to the Boltzmann probability, and $<M>$ can be estimated by the arithmetic average, $\bar{M}$, from a sample of size n,

$$\bar{M} = \frac{1}{n} \sum_{t=1}^{n} M_{i(t)}. \qquad (3.23)$$

$M_{i(t)}$ is the value of M in configuration i obtained at time t of the simulation run. Moreover, using simulation, one can obtain averages of geometrical properties that are beyond the reach of thermodynamics, such as the end-to-end distance of a polymer, the radius of a gyration of a protein, and many other local properties; the power of Monte Carlo and molecular dynamics is that statistical averages can be obtained without calculating Q, or equivalently, without the need to know the *values* of P_i^B!! (see Section 1.16). However, the values of the probability are required for calculating the absolute entropy and free energy, which are not trivial tasks (see Chapter 10).

Finally, we mention that in addition to the two approaches discussed above, a third approach exists which is based on the "*most probable term*" (discussed in Sections 5.2 and 5.3). At this stage, the basic theory of equilibrium statistical mechanics has already been derived. In the coming chapters, we discuss additional theoretical aspects and describe various important applications.

3.5 The Helmholtz Free Energy of a Simple Fluid

We treat a large system consisting of N equal particles, with momenta $\mathbf{p}^N$ and coordinates $\mathbf{x}^N$ enclosed in a container of volume V; the system is at equilibrium with a reservoir of temperature T (see Figure 3.1). Our goal is to calculate the partition function, $Q^\#$:

$$Q^\# = \frac{1}{h^{3N}N!}\int \exp[\frac{-E(\mathbf{x}^N, \mathbf{p}^N)}{k_B T}]d\mathbf{x}^N d\mathbf{p}^N \qquad (3.24)$$

where the Hamiltonian of the system is $E(\mathbf{x}^N, \mathbf{p}^N) = E_k(\mathbf{p}^N) + E_p(\mathbf{x}^N)$; that is, E_k is the kinetic energy, which depends only on the momenta (velocities), and E_p is the potential energy due to all the particle-particle interactions. $E_p(\mathbf{x}^N)$ depends solely on the coordinates. This type of modeling where the forces *do not depend on the velocities* is very common and will be assumed throughout this book. Thus, the integrations over E_p and E_k can be separated. Moreover, the integrations over the momenta of different particles can be carried out independently, and even the components p_x, p_y, and p_z can be treated separately. Thus, we first integrate over p_x (denoted p) of a single particle with mass m, where:

$$E_k = \frac{mv^2}{2} = \frac{p^2}{2m} \qquad (3.25)$$

Using the integral [Appendix Equation (A1.1)]:

$$\int_0^\infty \exp(-ax^2)dx = \frac{1}{2}\sqrt{\frac{\pi}{a}}. \qquad (3.26)$$

we obtain:

$$\int_{-\infty}^\infty \exp(-\frac{p^2}{2mk_B T})dp = \sqrt{2\pi mk_B T}. \qquad (3.27)$$

Integrating over the $3N$ components of the momenta leads to $(\sqrt{2\pi mk_B T})^{3N}$ and $Q^\#$ becomes:

$$Q^\# = \frac{(\sqrt{2\pi mk_B T})^{3N}}{N!h^{3N}}\int \exp[-\frac{E_p(\mathbf{x}^N)}{k_B T}]d\mathbf{x}^N. \qquad (3.28)$$

Using Stirling's approximate formula, $\ln N! \approx N\ln(N/e)$, the Helmholtz free energy $A(N,V,T) = -k_B T \ln Q^\#$ becomes,

$$A = -k_B TN \ln\left\{\left(\frac{2\pi mk_B T}{h^2}\right)^{3/2}\frac{e}{N}\right\} - k_B T\ln\int\exp\left[-\frac{E_p(\mathbf{x}^N)}{k_B T}\right]d\mathbf{x}^N. \qquad (3.29)$$

A contains the parameters m and h, which are not defined in classical thermodynamics. The velocities (momenta) part of A are completely solved, and the problem of statistical mechanics is thus to solve the configurational integral for models of interest. Notice that the *configurational* partition function and the corresponding free energy are commonly denoted by Z and F, rather than by Q and A, respectively (see Section 4.9).

REFERENCE

1. T. L. Hill. *An Introduction to Statistical Thermodynamics.* (Dover, New York, 1986).

FURTHER READING

2. D. Chandler. *Introduction to Modern Statistical Mechanics.* (Oxford, New York, 1987).
3. F. Reif. *Fundamentals of Statistical and Thermal Physics.* (McGraw Hill, New York, 1965).
4. R. C. Tolman. *The Principles of Statistical Mechanics.* (Dover, New York, 1979).
5. L. D. Landau and E. M. Lifshitz. *Statistical Physics.* (Pergamon, New York, 1963).
6. K. A. Dill and S. Bromberg. *Molecular Driving Forces, Statistical Thermodynamics in Biology, Chemistry, Physics, and Nanoscience.* (Garland Science, New York, 2010).

4

Ideal Gas and the Harmonic Oscillator

4.1 From a Free Particle in a Box to an Ideal Gas

The calculation of Equations (3.28) and (3.29) for an ideal gas is trivial because the potential energy, $E(\mathbf{x}^N) = 0$, and the configurational integral lead to $\int = V^N$. Thus, the partition function is,

$$Q^{\#} = \frac{\left(\sqrt{2\pi m k_{\mathrm{B}}T}\right)^{3N}}{N!h^{3N}} V^N = \frac{1}{N!}\left(\frac{V}{\Lambda^3}\right)^N \qquad (4.1)$$

where

$$\Lambda = \left(\frac{h^2}{2\pi m k_{\mathrm{B}}T}\right)^{1/2}. \qquad (4.2)$$

Λ is a length of the order of magnitude of the de Broglie wavelength (h/mv, v = velocity) of a particle with energy $\sim k_{\mathrm{B}}T$. The Helmholtz free energy $-k_{\mathrm{B}}T\ln Q^*$ becomes,

$$A(N,V,T) = -k_{\mathrm{B}}TN \ln\left\{\left(\frac{2\pi m k_{\mathrm{B}}T}{h^2}\right)^{3/2}\frac{eV}{N}\right\}. \qquad (4.3)$$

Before deriving the various thermodynamic parameters for the (classical) ideal gas, we discuss its relation to the underlying quantum mechanical model, which is "a free particle in a box." We shall show that at a high enough temperature, the quantum model leads to Q^* [Equation (4.1)] of the classical ideal gas, as is required by the correspondence principle.

Thus, the solution of the Schrödinger equation for a *single* particle in a box of volume, V, leads to the energy levels, $\varepsilon_{l_x,l_y,l_z}$ (see [1], Chapter 11):

$$\varepsilon_{l_x,l_y,l_z} = \frac{h^2(l_x^2 + l_y^2 + l_z^2)}{8mL^2}, \qquad (4.4)$$

where l_x, l_y, and l_z are natural numbers, m is the mass, and $L^3 = V$. The partition function of a single particle is:

$$q = \sum_{l_x,l_y,l_z} \exp-[\varepsilon_{l_x,l_y,l_z}/k_{\mathrm{B}}T] \qquad (4.5)$$

and the partition function Q for N *non-interacting* particles is the product of the N single partition functions:

$$Q = q^N \qquad (4.6)$$

However, we have already argued that Q should be divided by $N!$ (Section 3.3.2). To investigate the $N!$ issue more in depth, we consider for simplicity only two particles, a and b, and two *different* energy levels, ε_i and ε_j. Thus, a typical term in Q is, $\exp-[\varepsilon_{ia}+\varepsilon_{jb}]/k_{\mathrm{B}}T$, where particle a is in level ε_i and particle b

in level ε_j; however, since the particles are *indistinguishable*, the level $\varepsilon_{ja} + \varepsilon_{ib}$, where a and b exchange the energies, i and j also appear, and it is equal to $\varepsilon_{ia} + \varepsilon_{jb}$. In quantum mechanics, this level should be considered only once, hence, a division by $N!$ (here 2!) is required. However, energy levels such as $\varepsilon_{ia} + \varepsilon_{ib}$, where a and b occupy the *same* level (i) also exist and such "bad" terms might destroy the above argument.

Still, if the number of states (from the ground state up to say, $5k_BT$) is much larger than the number of particles, N, the number of "bad" terms will be exceedingly smaller than the number of states, where the particles occupy different energy levels. Therefore, under this "classical limit condition," the effect of the bad terms can be ignored and $N!$ should be applied, that is, $Q = q^N/N!$. To check when this condition is satisfied, one integrates the energy in Equation (4.5) from the ground state up to the energy k_BT (for details see [2], Chapter 4) and finds that the condition is satisfied if:

$$\left(\frac{2\pi m k_B T}{h^2} \right)^{3/2} V \gg N \tag{4.7}$$

or

$$R = \frac{\Lambda^3 N}{V} \ll 1 \tag{4.8}$$

where Λ is the "thermal" de Broglie wavelength,

$$\Lambda = \left(\frac{h^2}{2\pi m k_B T} \right)^{1/2}. \tag{4.9}$$

V/N is the average volume occupied by a particle, and $(V/N)^{1/3}$ is the average distance between two nearest neighbor particles. Equation (4.8) asserts that the classical limit condition is satisfied if the "thermal" de Broglie wavelength, Λ, is much smaller than the smallest average distance between particles. Equation (4.7) shows that Λ increases with decreasing m and T. Indeed, for helium and hydrogen (H_2) at their boiling points (4.2 and 20.4 K), the ratio, R, is relatively high, 1.5 and 0.44, respectively, meaning that these systems should be treated by quantum mechanics. The heavier liquid argon in its (higher) boiling point ($T = 87.4$ K) can be treated classically, as the ratio is already small, 0.00054. Therefore, for an ideal gas and realistic systems at a high enough temperature the ratio is small, quantum effects can be ignored, and the classical treatment and use of $N!$ are justified.

Integration of the quantum partition function q [Equation (4.5)] using the energies defined in Equation (4.4), up to a large enough k_BT, leads to the partition function of a single particle, $q = V/\Lambda^3$, and for N ideal gas particles, to the classical partition function we have obtained in Equation (4.1). This manifestation of the correspondence principle supports our derivation of the classical partition function [Equation (4.1)]. The correspondence principle will be also demonstrated for the harmonic oscillator in Section 4.7.

PROBLEM 4.1

Derive Equation (4.7) (see [2], Chapter 4).

4.2 Properties of an Ideal Gas by the Thermodynamic Approach

With the thermodynamic approach (Section 3.4.1), the various thermodynamic properties can be obtained as derivatives of the Helmholtz free energy, A [Equation (4.3)]; the pressure is (see Table 2.1),

$$p = -\left(\frac{\partial A}{\partial V} \right)_{T,N} = Nk_B T/V \Rightarrow pV = Nk_B T, \tag{4.10}$$

meaning that we have recovered the *experimental* equation of state for an ideal gas; it should be noted that this result is compatible with our choice of the entropy parameter $C = k_B$ [see discussion following Equation (3.13)] and justifies the introduction of the $N!$ term in the partition function. The internal energy, E, is,

$$E = -T^2 \left(\frac{\partial \left(\frac{A}{T} \right)}{\partial T} \right)_{V,N} = \frac{3}{2} N k_B T. \tag{4.11}$$

As expected, E is extensive ($\sim N$); it is also in accord with the result obtained from thermodynamics, $E = C_V T + \text{constant}$ [Equation (2.14)]; the constant, which cannot be determined from thermodynamic considerations is thus determined by statistical mechanics to be zero. Also, E ($\sim T$) is the *average* kinetic energy, where each degree of freedom contributes $(1/2)k_B T$ to E. Thus, for a model where the forces do not depend on the velocities, T is determined by the kinetic energy alone; as we shall see, this is used in molecular dynamics to fix the temperature. Finally, as in Joule's experiment, E is independent of V.

According to the experiment, the heat capacity, C_V, is independent of T and V, which is also materialized, since,

$$C_V = \left(\frac{\partial E}{\partial T} \right)_{V,N} = \frac{3}{2} N k_B. \tag{4.12}$$

The entropy is,

$$S = -\left(\frac{\partial A}{\partial T} \right)_{V,N} = \frac{E - A}{T} = N k_B \ln \left[\left(\frac{2\pi m k_B T}{h^2} \right)^{3/2} \frac{V e^{5/2}}{N} \right] \tag{4.13}$$

or

$$S = \left(\frac{3}{2} N k_B \right) \ln T + N k_B \ln V - N k_B \ln N + N k_B \left[\frac{3}{2} \ln \left(\frac{2\pi m k_B}{h^2} \right) + \frac{5}{2} \right] \tag{4.14}$$

As expected, $S \sim N$ is an extensive variable. In thermodynamics, we have found [Equation (2.25.b)], $S_{IG} = C_V \ln T + N k_B \ln V - N k_B \ln N + N k_B \text{const}$, where *const* could not be determined. We see here that *const*, is determined by statistical mechanics, where the last term in Equation (4.14) contains microscopic parameters (m and h), which are unknown in thermodynamics. While $S \sim \ln T$ and $E \sim T$, that is, both parameters decrease with T, we obtain, $S(T = 0) = \infty$ meaning that S is not defined at $T = 0$, since $S(T = 0)$ should be zero according to the third law of thermodynamics. Hence, an important conclusion to remember is that the ideal gas picture holds only at high T, which is in accord with the discussion following Equation (4.9). Finally, we comment again that $N!$ (or equivalently, $\ln N$) is related to the Gibbs paradox; it has been introduced to keep the density $\rho = V/N$ constant or equivalently to keep S extensive, as discussed in the paragraph following Equation (2.25c).

The above equations for the canonical ensemble describe only the translation motion of the particles; when internal degrees of freedom exist, such as vibrations and rotations, they are treated in a similar way; for details see [2], Chapters 4, 8, and 9, and a discussion prior to Equation (4.43). One can derive easily statistical mechanics expressions for other free energies, such as the Gibbs free energy; for example, see [2] Chapter 4 and [3] Appendix 2. Still, we define below the chemical potential, which plays an important role in chemistry and structural biology and will be discussed extensively in later chapters of this book.

4.3 The chemical potential of an Ideal Gas

In thermodynamics, the chemical potential, μ, is defined as a derivative of the Helmholtz free energy, A, with respect to the number of particles, N. Thus, μ is the amount of work required for adding one molecule to a system of N molecules,

$$\mu = \left(\frac{\partial A}{\partial N} \right)_{V,T} = \frac{A(N+1) - A(N)}{1} \tag{4.15}$$

Using the Helmholtz free energy, A (Equation 4.3), one obtains for an ideal gas,

$$\mu = -k_B T \ln \left\{ \left(\frac{2\pi m k_B T}{h^2} \right)^{3/2} \frac{V}{N} \right\} = -k_B T \ln \left(\frac{2\pi m k_B T}{h^2} \right)^{3/2} + k_B T \ln \rho \tag{4.16}$$

where $\rho = N/V$ is the density. Defining $L^3 = V/N$, Equation (4.16) reads,

$$\mu(V,T) = -k_B T \ln \left(\frac{L}{\Lambda} \right)^3 = -k_B T \ln \frac{1}{\Lambda^3} + k_B T \ln \frac{1}{L^3} = \mu^0(T) + k_B T \ln \rho. \tag{4.17}$$

Since for an ideal gas, $L \gg \Lambda$ [Equation (4.8)], μ is negative, meaning that when an atom is added to the system, A is reduced. In thermodynamics, the chemical potential is defined as $\mu = \mu^0 + RT \ln \rho$, where μ^0 is called the standard chemical potential. Thus, in statistical mechanics, μ^0 (for an ideal gas) is defined explicitly in terms of microscopic parameters, $\mu^0 = -k_B T \ln \Lambda^{-3}$, which is the chemical potential for $L^3 = 1$. μ^0 is based on the density of one particle per the volume, Λ^3.

The expression L/Λ on the left hand side of Equation (4.17) is dimensionless, hence, the ln is well defined. On the other hand, in the other two equations, the expression under the ln has a dimension; thus, one should define these expressions with the same units!

Dividing each expression under the ln in the middle of Equation (4.17) by $c^0 = 1$ molar $= 1$ mol/L does not change μ, but makes each expression dimensionless. Thus, if μ is defined per mole, and the concentration c (in molar) replaces ρ, one obtains,

$$\mu(V,T) = -RT \ln \frac{1}{\Lambda^3 c^0} + RT \ln \frac{c}{c^0}. \tag{4.18}$$

Defining $c^0 = 1/1660 \text{Å}^3 = 1/V^0$ leads to:

$$\mu(V,T) = -RT \ln \frac{V^0}{\Lambda^3} + RT \ln \frac{c}{c^0} = \mu^0(T) + RT \ln \frac{c}{c^0}. \tag{4.19}$$

Substituting in Equation (4.17), $p = k_B T N/V$ leads to the Gibbs free energy per particle [Equation (2.75)],

$$\frac{G(p,T)}{N} = \mu(p,T) = -k_B T \ln \left(\frac{1}{\Lambda^3} k_B T \right) + k_B T \ln p = \mu_p^0(T) + k_B T \ln p \tag{4.20}$$

$k_B T \Lambda^{-3}$ is the pressure of one particle at T and Λ^{-3}. Notice that for an ideal gas treated in statistical mechanics, $\mu^0(T)$ and $\mu_p^0(T)$ are known explicitly, unlike μ defined in thermodynamics, where by integrating $dG = -SdT + Vdp$ at constant T leads to,

$$G - G^0 = \int_{p_0}^{p} V dp = \int_{p_0}^{p} \frac{N k_B T}{p} dp = N k_B T \ln \frac{p}{p_0}. \tag{4.21}$$

G is calculated with respect to a standard state, $G = G^0$, which is defined by the standard pressure, p^0. $\mu(T, p) = G/N$ [Equation (2.75)], therefore G per mole (or per N) leads to $\mu(T, p)$,

$$\mu(T, p) = RT \ln \frac{p}{p_0} \tag{4.22}$$

Thus, the standard pressure, p_0, replaces the $k_B T \Lambda^{-3}$ obtained from statistical mechanics in Equation (4.20).

PROBLEM 4.2

For a system of N_0 adsorption points and N adsorbing molecules ($N \ll N_0$), calculate the chemical potential. Hint: calculate # of arrangements of N molecules on N_0 points using Stirling's formula.

Answer

$\mu = k_B T \ln[N/(N_0 - N)]$.

4.4 Treating an Ideal Gas by the Probability Approach

Thus far, the thermodynamic properties of an ideal gas were obtained from derivatives of the Helmholtz free energy. However, extensive variables, such as the energy, can also be calculated based on the probability approach (see Section 3.4.2), that is, as statistical averages over phase space (the $6N$ vector $\mathbf{x}^N \mathbf{p}^N$) with the Boltzmann probability density, P^B [Equation (3.14)]. To calculate P^B for an ideal gas, we first write Q_{IG} [Equations (3.27) and (3.28)],

$$Q_{IG} = (\sqrt{2\pi m k_B T})^{3N} V^N. \tag{4.23}$$

The Boltzmann probability density becomes:

$$P^B d\mathbf{x}^N d\mathbf{p}^N = \frac{\exp-\left[\dfrac{\mathbf{p}^2}{2m k_B T}\right]}{Q_{IG}} d\mathbf{x}^N d\mathbf{p}^N = \frac{\exp-\left[\dfrac{\mathbf{p}^2}{2m k_B T}\right]}{\left(\sqrt{2\pi m k_B T}\right)^{3N} V^N} d\mathbf{x}^N d\mathbf{p}^N \tag{4.24}$$

and using the integral of Equation (A1.2) in the Appendix leads to:

$$<E> = \frac{\displaystyle\int_{-\infty}^{\infty} \frac{\mathbf{p}^2}{2m} \exp-\left[\dfrac{\mathbf{p}^2}{2m k_B T}\right] d\mathbf{p} \int_V d\mathbf{x}}{\left(\sqrt{2\pi m k_B T}\right)^{3N} V^N} = \frac{3}{2} N k_B T. \tag{4.25}$$

where for simplicity we define $\mathbf{p}^2 = p_1^2 + p_2^2 + p_3^2 + \cdots + p_{3N}^2$, $d\mathbf{p} = dp_1 dp_2 dp_3 \cdots dp_{3N}$, and $d\mathbf{x} = dx_1 dx_2 dx_3 \cdots dx_{3N}$. Indeed, we have recovered our previous result based on a thermodynamic derivative [Equation (4.11)]. To make this integration clearer, we present below the integral for two degrees of freedom. Thus, for one degree of freedom (based on p_1), Equation (4.24) defines the Boltzmann probability density, P_1^B, while according to Equation (A1.2), $E(p_1) = 1/2 k_B T$; using these, one obtains,

$$\frac{\displaystyle\int_{-\infty}^{\infty} \frac{(p_1^2 + p_2^2)}{2m} \exp-\left[\dfrac{p_1^2 + p_2^2}{2m k_B T}\right] dp_1 dp_2 \int dx_1 dx_2}{\left(\sqrt{2\pi m k_B T}\right)^2 V^{2/3}} = \frac{1}{2m} \int_{-\infty}^{\infty} \left(p_1^2 + p_2^2\right) P_1^B P_2^B dp_1 dp_2 \tag{4.26}$$

$$= \frac{1}{2m} \int_{-\infty}^{\infty} \left[p_1^2 P_1^B dp_1 + p_2^2\right] P_2^B dp_2 = \frac{2}{2m} \int_{-\infty}^{\infty} p_1^2 P_1^B dp_1 = 2 \frac{1}{2} k_B T = k_B T$$

One can show that the variance of one component of the energy is $1/8(k_{B}T)^2$. Similarly, the entropy of an ideal gas [Equation (4.14)] can be obtained based on the integral [Equation (3.17)] and the probability density P^B [Equation (4.24)]. We leave these derivations to the reader.

Finally, one can make the following changes in the probability density P^B [Equation (4.24)]: (1) dropping its coordinates' part and (2) replacing the momentum, $\mathbf{p}^2$ by the $3N$ velocity vector, $\mathbf{v}^2$, where $\mathbf{v}^2 = v_1^2 + v_2^2 + v_3^2 + \cdots + v_{3N}^2$. One thus obtains the Maxwell–Boltzmann distribution for the velocities, which is a Gaussian,

$$P_{\text{Max-Bol}} = \left[\frac{m}{2\pi k_{B}T} \right]^{3N/2} \exp-\left[\frac{m\mathbf{v}^2}{2k_{B}T} \right] \tag{4.27}$$

4.5 The Macroscopic Harmonic Oscillator

The harmonic oscillator is a basic model in all the exact sciences, with the important advantage that its free energy and other thermodynamic potentials can be solved analytically quite easily.

We shall discuss this model on three theoretical levels, starting: (1) with the macroscopic oscillator (as studied in high school), then (2) treating the *classical* microscopic oscillator by statistical mechanics, and finally (3) discussing the quantum-mechanical oscillator. Emphasizing the differences between these three models will deepen our understanding of the theoretical foundation of statistical mechanics and thermodynamics.

Suppose there is a one-dimensional macroscopic (mechanical) oscillator of mass, m, with a force constant, f, which is stretched by a maximal distance (amplitude), d; the corresponding (maximal) potential energy $E_p = E_{tot}$, where E_{tot} is the total energy of the oscillator,

$$E_p = E_{tot} = \int_0^d fx\,dx = fd^2/2 \tag{4.28}$$

To describe the dynamics of the oscillator as a function of time, t, one has to solve Newton's equation of motion; thus, the force, F, exerted on the mass as it is stretched by a distance x, and the corresponding acceleration, a are,

$$F = -fx = ma = m\left(\frac{dx^2}{dt^2} \right)$$

or

$$m\left(\frac{dx^2}{dt^2} \right) + fx = 0. \tag{4.29}$$

The solutions for x and the velocity, v, of this equation are based on the frequencies, v or ω,

$$v = \frac{(f/m)^{1/2}}{2\pi} \qquad \omega = 2\pi v = \left(\frac{f}{m} \right)^{1/2}. \tag{4.30}$$

leading to

$$x(t) = d\sin(\omega t + \theta_0) \qquad v(t) = \omega d\cos(\omega t + \theta_0) \tag{4.31}$$

where θ_0 and the amplitude, d, are constants of integration. Thus, the behavior of a macroscopic oscillator is completely deterministic, as the position, $x(t)$, and the velocity, $v(t)$, are known exactly as functions of

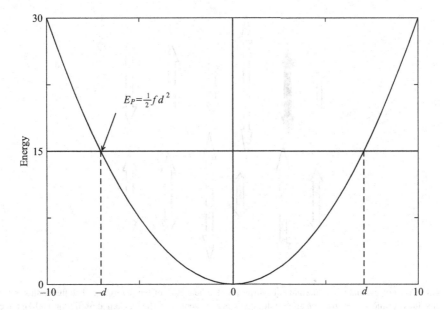

FIGURE 4.1 Potential energy of a macroscopic oscillator with $d = 10$.

the time, t. The energy, E_{tot}, is determined by the amplitude, d, where at $x = \pm d$, the energy is completely potential, $E_p = E_{tot}$, at $x = 0$, it is completely kinetic, $E_k = E_{tot}$, and at an intermediate, x, $E_k + E_p = E_{tot}$. The potential energy of a macroscopic oscillator is described in Figure 4.1.

4.6 The Microscopic Oscillator

We now consider a set of N *atoms* arranged on a lattice, as described in Figure 4.2. Each atom is affected by the force field of its neighbor atoms, and thus moves locally around its lattice position (experimental world description). These fluctuations can be modeled approximately by a microscopic oscillator with mass, m, and force constant, f, which does not interact with its neighbors, that is, an independent oscillator; each oscillator is at equilibrium with a heat bath of temperature, T, and the entire system of N oscillators is thus treated in the canonical (NVT) ensemble. This is a good model for a crystal at a high temperature, as will be discussed in more detail later.

Figure 4.3 describes a snapshot of a set of oscillators at a given time, t, where each oscillator is seen in a specific position, x. Since the oscillators are independent, one could, in principle, obtain the properties of the system by averaging over the entire (large) set of snapshots, or, alternatively, by averaging over results obtained from a specific single oscillator as a function of time. It should be pointed out that in contrast to a macroscopic oscillator, $x(t)$ and $E(t)$ are not defined as a function of time, as one only knows

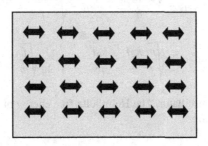

FIGURE 4.2 A set of atomic oscillators on a lattice.

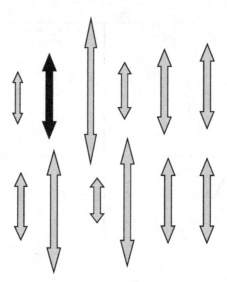

FIGURE 4.3 A set of oscillators in thermal equilibrium with a heat bath of temperature T. The figure shows a snapshot of their $x(t)$ values at time, t. The *average* amplitude, $<x^2>$, is represented by the blackened oscillator. Unlike the case of a macroscopic oscillator, $x(t)$ can be larger than $<x^2>$.

$P^B(x)$ and $P^B(E)$, which lead to the statistical averages $<E>$ and $<x^2>$, respectively [$<x> = 0$]. Also, notice that unlike a macroscopic oscillator, there is a chance that a microscopic oscillator will exceed its "amplitude", $<x^2>$, since, $P^B(x > <x>) \neq 0$.

4.6.1 Partition Function and Thermodynamic Properties

Due to independency, the partition function of the system of the N oscillators, Q_N, is equal to q^N, where q is the partition function of a single oscillator. Moreover, the components (x, y, z) are independent as well; therefore, one can calculate q_x, which leads to $Q_N = q_x^{3N}$. The factor $N!$ that was needed for an ideal gas is omitted here because the positions of the atoms are restricted on the lattice. The Hamiltonian, H_x, for one component is based on the corresponding kinetic and potential energies:

$$H_x = \frac{p_x^2}{2m} + \frac{fx^2}{2}$$

(4.32)

and the partition function is the product, $q^* = q_k q_p$, due to the fact that the kinetic and potential energy parts are independent. q_k has already been calculated for an ideal gas [Equation (3.28)], where q_p is calculated by the same integral [Equation (A1.1)], thus:

$$q_k = \frac{\sqrt{2\pi m k_B T}}{h} \qquad q_p = \sqrt{\frac{2\pi k_B T}{f}}$$

(4.33)

$$q^* = 2\pi \sqrt{\frac{m}{f}} \frac{k_B T}{h} = \frac{k_B T}{h\nu} \qquad \nu = \frac{1}{2\pi} \sqrt{\frac{f}{m}}.$$

(4.34)

where ν is the frequency of the oscillator. The Helmholtz free energy of a single oscillator, A, and of a system of $3N$ components, A_{sys}, are:

$$A = -k_B T \ln q^* \qquad \text{and} \qquad A_{sys} = -k_B T \ln\left[(q^*)^{3N}\right]$$

(4.35)

Using the thermodynamic approach, the energy is:

$$E = <E> = -T^2\left(\frac{\partial\left(\frac{A}{T}\right)}{\partial T}\right) = k_B T \tag{4.36}$$

One can show that $<E_k> = <E_p> = 1/2k_B T$. Thus, the average energy of one component of an oscillator is twice the energy of one component of an ideal gas; this is the effect of interaction, which does not exist in an ideal gas. The heat capacity is thus, $C_V = k_B$. It should be pointed out that $E = k_B T$ for a microscopic oscillator does not depend on the force constant, f, unlike the macroscopic oscillator, where $E = fd^2/2$; E depends only on T.

As expected, the energy and the heat capacity are extensive, that is, for N oscillators, $E = 3Nk_B T$ and $C_V = 3Nk_B$; this is in accord with the experimental Dulong-Petit law, which states that the molar C_V of many crystals at high T is constant and equal to $3R$. The entropy is:

$$S = \frac{E}{T} - \frac{A_{sys}}{T} = 3Nk_B[1 + \ln(k_B T/h\nu)] = 3Nk_B\left[1 + \ln\left(\frac{2\pi k_B T\sqrt{m}}{h\sqrt{f}}\right)\right] \tag{4.37}$$

Thus, as required, S is extensive, depending on N, which grows with the volume, V, because the oscillators are anchored to the lattice. This is in contrast to the case of an ideal gas, where the unrestricted particles within the volume required introducing the $\ln N!$ factor to keep the entropy extensive [Equation (4.14)]; S depends on f and m through the frequency; both E and S increase with T, and like for an ideal gas, S is not defined at $T = 0$. At low T, one has to use the quantum mechanical oscillator, which is described in Section 4.7.

To get a deeper insight into the properties of the classical microscopic oscillator, we write the Boltzmann probability density $P^B(x)$ for one component, x (the momentum part is omitted); this is the Gaussian,

$$P^B(x) = \frac{\exp[-fx^2/2k_B T]}{q_p} \tag{4.38}$$

which is maximal at $x = 0$. First, we use $P^B(x)$ for calculating the average potential energy of a single component oscillator by the probabilistic approach [using the integral Equation (A1.2)]; the expected result is,

$$<E_p> = \left(\frac{1}{2}\right)\int P^B(x) fx^2 dx = \left(\frac{1}{2}\right)k_B T. \tag{4.39}$$

Using the integral, Equation (A1.3), we also obtain $<E_p^2> = (3/4)(k_B T)^2$, which leads to the variance of the energy [Equations (1.42) and (5.3)]:

$$\sigma_{E_p}^2 = <E_p^2> - <E_p>^2 = \frac{3}{4}(k_B T)^2 - \frac{1}{4}(k_B T)^2 = \frac{1}{2}(k_B T)^2, \tag{4.40}$$

that is, the fluctuation of the potential energy is $\sigma_{E_p} = (1/2)^{1/2} k_B T$. For the potential energy of the entire system, one obtains $E_p = (3/2)Nk_B T \pm [(3/2)N]^{1/2}k_B T$. The corresponding configurational entropy, S_p, and for completeness, also the entropy due to the kinetic energy, S_k, are,

$$S_p = \frac{k_B}{2}\left(1 + \ln\frac{2\pi k_B T}{f}\right) \qquad S_k = \frac{k_B}{2}\left(1 + \ln\frac{2\pi k_B Tm}{h^2}\right). \tag{4.41}$$

The classical microscopic oscillator changes its position due to the force and the random energy delivered by the heat bath. However, unlike the macroscopic oscillator, where $x(t)$ is known [$x(t) = d\sin(\omega t + \theta_0)$], the position of a microscopic oscillator is known only with the Boltzmann probability, P^B, above. Thus, while for a macroscopic oscillator, $x \leq d$ always, for a microscopic oscillator, $P^B(x) \neq 0$ for any x, and, in principle, x can be very large. However, for a given T, there is an *average* amplitude, $<x^2(f, T)>$, and in practice, x^2 does not fluctuate much above this value,

$$< x^2(f,T) >= \int P^B(x)x^2 dx = \frac{k_B T}{f} \tag{4.42}$$

Since $<E>$ does not depend on f, two oscillators with different f at the same T will absorb the same $<E>$, but each amplitude will be proportional to the corresponding $1/f$. When T increases the average potential energy, $(1/2)k_B T$ and its variance $(1/2)k_B^2 T^2$, both increase, meaning that the average amplitude increases as well ($<x> = 0$, but $<x^2> = k_B T/f$). The position of the mass is less defined due to larger fluctuations $<x^2> = k_B T/f \pm 2^{1/2}k_B T/f$. The entropy, as expected, also increases as T is increased.

The harmonic model is combined with the ideal gas when the molecule studied has, besides the translation degrees of freedom, also internal degrees of freedom, such as vibrations; this may happen in an ideal diatomic gas, for example. Then, the molecular partition function q will include, in addition to the translational component, q_t ($=V/\Lambda^3$), also a vibrational part, $q_{vib} = q^*$ [Equation (4.34)], where $q = q_t q_{vib}^3$, and the partition function of the total system reads

$$Q^* = \frac{(q_t q_{vib}^3)^N}{N!} \tag{4.43}$$

For an extensive discussion of this issue, see [2], Chapters 8 and 9.

PROBLEM 4.3

(1) Does the dependence of the variance on T $<x^2> = k_B T/f$ explain why the entropy also increases with T? (2) Calculate in detail the energy E of N oscillators by a free energy derivative, [Equation (4.36)].

In summary: The above discussion demonstrates the usefulness of the probability approach, which enables one calculating variances. We should emphasize again that a classical microscopic oscillator is only a valid model for a system at high T (but not too high for a crystal!). At low T, one has to use the quantum-mechanical oscillator.

4.7 The Quantum Mechanical Oscillator

While most of the models studied in this book (in particular models for polymers and proteins) are treated by classical statistical mechanics, it is also important to discuss the quantum-mechanical oscillator for several reasons. First, quantum effects should be considered in treating enzymatic reactions [4] and when a harmonic analysis is carried out at low T or equivalently for large spring constants, f (as f and $1/T$ have the same effect on $Z = \int exp-[fx^2/2k_B T]$); such relatively large constants are typically used to model bond stretching (harmonic) potentials, where f can be as large as 300 kcal/(mol·Å²) or more. Also, because the quantum oscillator can be solved analytically, it would be of interest to demonstrate that the correspondence principle is satisfied, that is, the quantum description leads to the classical one at high T.

One should first solve the Schrödinger equation for a harmonic oscillator in dimension $d = 1$, $\hat{H}\psi = E\psi$; ψ is the wave function, E stands for energy values, and $\hat{H}$ is the Hamiltonian operator,

$$\hat{H} = -\frac{\hbar^2}{2m}\frac{d^2}{dx^2} + V(x) \tag{4.44}$$

The first term in Equation (4.44) is the kinetic energy operator with the Planck constant, $\hbar = h/2\pi$ and the mass, m; $V(x)$ is the potential energy operator, $V(x) = fx^2/2$, and the Schrödinger equation is,

$$\hat{H} = -\frac{\hbar^2}{2m}\frac{d^2\psi^2}{dx^2} + \frac{1}{2}fx^2\psi = E\psi \tag{4.45}$$

The solution of this equation leads to a set of energy levels [1],

$$E_n = \left(n + \frac{1}{2}\right)h\nu \qquad n = 0, 1, \ldots, \infty \tag{4.46}$$

where the frequency, ν, is defined by Equation (4.34). The partition function, q, of a single quantum oscillator is obtained by summing up the exponentials of the energy levels,

$$q = \sum_n \exp\left[-\frac{E_n}{k_B T}\right] \tag{4.47}$$

and the Boltzmann probability is:

$$P_n^B = \frac{\exp[-E_n/k_B T]}{q} \tag{4.48}$$

Specifically, using the summation formula of a geometric series one obtains,

$$q = \sum_{n=0}^{\infty} \exp-\frac{(n+1/2)h\nu}{k_B T} = \exp\left[-\frac{(1/2)h\nu}{k_B T}\right]\frac{1}{1-\exp-\dfrac{h\nu}{k_B T}} \tag{4.49}$$

and the free energy of a single oscillator, $A_{\sin}$, is:

$$A_{\sin} = -k_B T \ln q = (1/2)h\nu + k_B T \ln\left[1 - \exp\left(-\frac{h\nu}{k_B T}\right)\right]. \tag{4.50}$$

The energy $E_{\sin}$ of a single oscillator is,

$$E_{\sin} = -T^2 \frac{\partial\left(\dfrac{\partial A_{\sin}}{T}\right)}{\partial T} = -T^2\left[-\frac{(1/2)h\nu}{T^2} - k_B h\nu \frac{\exp\left(-\dfrac{h\nu}{k_B T}\right)}{k_B T^2\left[1-\exp\left(-\dfrac{h\nu}{k_B T}\right)\right]}\right] \tag{4.51}$$

$$= (1/2)h\nu + \frac{h\nu}{\exp\left(\dfrac{h\nu}{k_B T}\right)-1}$$

At $T = 0$, only the ground state $E_0 = 1/2h\nu$ will be populated. Unlike the classical microscopic oscillator, the energy of the quantum oscillator depends on f (through ν). The energy of N independent quantum oscillators is:

$$E = 3N\left[(1/2)h\nu + \frac{h\nu}{\exp\left(\dfrac{h\nu}{k_B T}\right)-1}\right]. \tag{4.52}$$

At high T, $\dfrac{hv}{k_BT} \ll 1$, and:

$$E = 3N[(1/2)hv + k_BT] \approx 3Nk_BT. \tag{4.53}$$

That is, the classical limit is obtained, where E does not depend on f [Equation (4.36)]. The entropy of $3N$ oscillators is:

$$S = -\frac{\partial A}{\partial T} = 3Nk_B \ln\left[1 - \exp\left(-\frac{hv}{k_BT}\right)\right] + 3Nk_BThv \frac{\exp\left(-\dfrac{hv}{k_BT}\right)}{k_BT^2\left[1 - \exp\left(-\dfrac{hv}{k_BT}\right)\right]} \tag{4.54}$$

$$S = -3Nk_B \ln\left[1 - \exp\left(-\frac{hv}{k_BT}\right)\right] + \frac{3Nhv}{T\left[\exp\left(\dfrac{hv}{k_BT}\right) - 1\right]} \tag{4.55}$$

Again, at high T, $\frac{hv}{k_BT} \ll 1$, and we get $S = 3Nk_B\left[1 + \ln\left(\frac{hv}{k_BT}\right)\right]$, which is the classical limit for the entropy, obtained in Equation (4.37).

Unlike a classical oscillator, the entropy of a quantum oscillator at $T = 0$ is zero, in accord with the third law of thermodynamics. Application of *strong* harmonic restraints in simulations based on classical statistical mechanics (applied to proteins, as mentioned above) should be handled with caution, as a quantum treatment might be required.

We have stated that the classical oscillator is a good model for a crystal at high T (Dulong-Petit law, $C_V = 3Nk_B$), which, however, fails at low T, where the experimental C_V and S decrease to zero. As suggested by Einstein, a better model for a crystal is the quantum oscillator. Indeed, we have already shown that $S \to 0$ as $T \to 0$, but we still have to examine the behavior of C_V,

$$C_V = \left(\frac{\partial E}{\partial T}\right)_{N,V} = 3Nk_B\left(\frac{hv}{k_BT}\right)^2 \frac{\exp\left[\dfrac{hv}{k_BT}\right]}{\left(\exp\left[\dfrac{hv}{k_BT}\right] - 1\right)^2} = 3Nk_B\left(\frac{\theta}{T}\right)^2 \frac{\exp\left[\dfrac{\theta}{T}\right]}{\left(\exp\left[\dfrac{\theta}{T}\right] - 1\right)^2} \quad \theta = \frac{hv}{k_B} \tag{4.56}$$

This C_V behaves in agreement with the experiment, since it tends to 0 as $T \to 0$,

$$C_V(T \to 0) = 3Nk_B\left(\frac{\theta}{T}\right)^2 \exp-\left[\frac{\theta}{T}\right] \tag{4.57}$$

According to this model, C_V/Nk_B is a universal function of θ/T. $\theta(v)$ might differ from one monoatomic crystal to another; however, for each crystal, T can be adjusted such that the graphs of C_V versus T/Θ for different crystals will overlap. This example of a law of "correspondence of states" is satisfied to a large extent, but not exactly. A better model for a crystal at low T is due to Debye (see [2], Chapter 5).

Summary: We have developed the theory of classical statistical mechanics for the canonical (NVT) ensemble, including the necessary quantum-mechanical elements (h and $N!$). This theory is supported by two simple, but fundamental models, the ideal gas and the microscopic oscillator, which both have analytical solutions. For the ideal gas, the theoretical results agree with the Joule's experiment and the experimental equation of state. The harmonic oscillator is in accord with the Dulong-Petit law. Both models obey the correspondence principle, where the low T quantum description becomes the classical one at high T.

4.8 Entropy and Information in Statistical Mechanics

In thermodynamics, the entropy S has been introduced as a (somewhat obscure) function, which expresses the inherent thermal disorder of systems involving heat. The significance of entropy becomes clearer in statistical mechanics in terms of the Boltzmann probability of the molecular states, which for a simple discrete system of N states, i reads [Equation (3.17)],

$$S = -k_B \sum_i P_i^B \ln P_i^B. \tag{4.58}$$

Thus, if only a single state, i, is populated, that is, $P_i = 1$, $S = 0$, and there is a complete knowledge about the state of the system (no uncertainty), which can also be defined as an absolute order; this situation never occurs in a classical system at a finite temperature, but can happen in a quantum system at $T = 0$ K (e.g., a quantum oscillator) and in somewhat artificial (lattice) models, such as the Ising model (Section 7.3); to $S \to 0$ corresponds the lowest possible energy. In the other extreme, where all states are equally probable (a random system), $P_i = 1/N$, the uncertainty concerning the population of the states is maximal (maximal disorder), and the entropy is thus maximal,

$$S = k_B \ln N. \tag{4.59}$$

This situation occurs in realistic systems at very high temperatures, where wide regions of phase space become almost equally accessible, and the corresponding energy is high. At finite temperatures, $S(T)$ and $E(T)$ receive intermediate values, which minimize the free energy, $A(T, k) = E(T, k) - TS(T, k)$ with respect to a set of parameters, k, for given NVT [see Equation (2.70) and Chapter 5].

The studies carried out thus far suggest that statistical mechanics can enrich dramatically the scope of classical thermodynamics: Thus, for an ideal gas, integration constants that cannot be evaluated in thermodynamics are determined completely by statistical mechanics. For the microscopic oscillator (which cannot be treated within the framework of thermodynamics), statistical mechanics has led to averages of S and E together with their variances, where variances are not defined at all in thermodynamics. As we shall see later, geometrical parameters of polymers, such as the end-to-end distance or the radius of gyration, can be calculated analytically, and using computer simulation, one can obtain averages of any structural motif of interest of a protein (e.g., the average distance of a specific hydrogen bond). Furthermore, one can calculate partial entropies such as the decrease in a ligand's entropy, ΔS_{ligand}, as it is transferred from the solvent environment to the active site of the protein; ΔS_{ligand} measures the decrease in the ligand's flexibility and constitutes an ingredient of the total free energy of binding (see Chapter 23 and Section 10.9.2). Again, the microscopic world is beyond the reach of classical thermodynamics. Notice that the variety of molecular level studies are all based on the *probabilistic approach* (Section 3.4.2).

Thus, while thermal systems "suffer" from the inherent "configurational uncertainty" (measured by S), statistical mechanics (and computer simulation in particular) "compensate" for this limitation by enabling the calculation of *averages* of unlimited structural, energetic, entropic, and as we shall see, also dynamic information of the entire system, as well as of parts of it. These simulations are in accord with modern molecular level experimental techniques, such as X-ray crystallography and solution and solid state NMR, which lead, for example, to the 3D structures of proteins—a microscopic information, which is far beyond the macroscopic parameters, E, p, μ, or S provided by classical thermodynamics.

4.9 The Configurational Partition Function

In a continuum system, where the forces do not depend on the velocities (momenta), the momenta part of the partition function is completely solved, and the problem is to treat the potential energy which depends on the coordinates, $\mathbf{x}^N$. The ideal gas case is trivial, as $E(\mathbf{x}^N) = 0$ and the Boltzmann probability density over the volume is uniform, $1/V$ for a single particle and $1/V^N$ for N particles. When $E(\mathbf{x}^N)$ ($\neq 0$, one is interested mainly in the *configurational* partition function denoted by Z, where the (solved) momentum part is ignored,

$$Z = \int \exp[-E(\mathbf{x}^N)/k_B T]d\mathbf{x}^N = \int \nu(E)\exp[-E/k_B T]dE. \tag{4.60}$$

It should be noted that while Q is dimensionless, Z has the dimension of $\mathbf{x}^N$; Z can also be expressed as an integration over the different energies, rather than over the coordinates, where $\nu(E)$ is the density of states for energy E; $\nu(E)dE$ is the volume in configurational space with energy between E and $E+dE$. For a discrete system, where the states are denoted by i, and $n(E_t)$ is the degeneracy of energy, E_t, one obtains:

$$Z = \sum_i \exp[-E_i/k_B T] = \sum_{E_t} n(E_t)[\exp[-E_t/k_B T]]. \tag{4.61}$$

Using Z, one defines the Boltzmann probability density:

$$P^B(\mathbf{x}^N) = \frac{\exp[-E(\mathbf{x}^N)/k_B T]}{Z} \tag{4.62}$$

where $P^B(\mathbf{x}^N)d\mathbf{x}^N$ is the probability to find the system between $\mathbf{x}^N$ and $\mathbf{x}^N + d\mathbf{x}^N$. Formally, we have created a probability space, with a $3N$ dimensional sample space, Ω where to each "point" $\mathbf{x}^N$ (which is a random variable) corresponds the energy $E(\mathbf{x}^N)$ (which is another random variable) and the Boltzmann probability density, $P^B(\mathbf{x}^N)$. The corresponding configurational Helmholtz free energy is denoted by F,

$$F = <E>_x - T<S>_x = -k_B T \ln Z \tag{4.63}$$

where $<>_x$ means averaging over the configurational space. Thus, the various thermodynamic functions can be obtained by the thermodynamic approach, as derivatives of F, or by the probabilistic approach, as statistical averages based on $P^B(\mathbf{x}^N)$. From now on, unless said otherwise, we shall be interested only in the configurational partition function and the related thermodynamic functions.

HOMEWORK FOR STUDENTS

We recommend Kubo's book *Statistical Mechanics*, [5], which like his book, *Thermodynamics* ([2], Chapter 2) contains hundreds of *solved* problems in statistical mechanics. Some problems appearing in this book are solved here, but in more ways suggested in this book (e.g., Section 5.3).

REFERENCES

1. E. Merzbacher. *Quantum Mechanics*. (John Wiley & Sons, New York, 1961).
2. T. L. Hill. *An Introduction to Statistical Thermodynamics*. (Dover, New York, 1986).
3. T. L. Hill. *Statistical Mechanics, Principles and Selected Applications*. (Dover, New York, 1987).
4. A. Warshel. *Computer Modeling of Chemical Reactions in Enzymes and Solutions*. (Wiley, New York, 1991).

FURTHER READING

5. R. Kubo. *Statistical Mechanics*. (North Holland, Amsterdam, The Netherlands, 1974).
6. D. A. McQuarrie. *Statistical Mechanics*. (HarperCollins, New York, 1976).

5

Fluctuations and the Most Probable Energy

5.1 The Variances of the Energy and the Free Energy

The *probabilistic approach* enables one to calculate fluctuations (variances) [Equation (1.42)] in thermodynamic properties. As pointed out earlier, this is an essential advantage, which widens the scope of statistical mechanics beyond the limits of classical thermodynamics, where only averages are defined (in the language of statistical mechanics). Of interest is first discussing the variance of the energy, $\sigma^2(E)$,

$$\sigma^2(E) = \int P^B(\mathbf{x}^N)[E(\mathbf{x}^N) - <E>]^2 d\mathbf{x}^N = <E^2> - <E>^2 \tag{5.1}$$

where an expectation value is denoted by $<>$. A more specific expression for $\sigma^2(E)$ can be obtained from the heat capacity, C_V, at constant volume; for simplicity, we calculate C_V for a discrete system,

$$C_V = \left(\frac{d<E>}{dT} \right)_V = \frac{1}{dT} \left(\frac{\sum_i E_i \exp(-E_i/k_B T)}{Z} \right)$$

$$= \frac{1}{k_B T^2} \left\{ \sum_i \frac{\sum_i E_i^2 \exp(-E_i/k_B T)}{Z} - \left[\frac{\sum_i E_i \exp(-E_i/k_B T)}{Z} \right]^2 \right] \right\} \tag{5.2}$$

$$C_V = \frac{1}{k_B T^2}[<E^2> - <E>^2] = \frac{1}{k_B T^2} \sigma^2(E) \qquad \text{or} $$

$$\sigma^2(E) = k_B T^2 C_V \tag{5.3}$$

In regular conditions, C_V (like E) is an extensive variable ($C_V = 3/2 N k_B$ for an ideal gas and $3N k_B$ for the microscopic oscillator), therefore, $\sigma(E) \sim N^{1/2}$, and the relative fluctuation of E decreases with increasing N,

$$\frac{\sigma(E)}{<E>} \sim \frac{\sqrt{N}}{N} = \frac{1}{\sqrt{N}}. \tag{5.4}$$

Thus, in macroscopic systems ($N \sim 10^{23}$), the fluctuation (the standard deviation) of E can be ignored because it is $\sim 10^{11}$ times smaller than the energy itself, meaning that this fluctuation will not be observed in macroscopic objects (however, we shall see in Chapter 7 that it becomes significant at a phase transition). Another consequence is that the exchange of energy (heat) with the bath is extremely small.

In contrast to the energy, the fluctuation of the Helmholtz free energy, F, is always zero, even at phase transition. Thus, we replace the Boltzmann probability under the ln in Equation (5.5) below by its explicit expression, $P_i^B = \exp[-E_i/k_B T]/Z$; F becomes,

$$F = E - TS = \sum_i P_i^{\mathrm{B}}\left[E_i + k_{\mathrm{B}} \ln P_i^{\mathrm{B}} \right] = \sum_i P_i^{\mathrm{B}}\left[E_i + k_{\mathrm{B}}T(-E_i/k_{\mathrm{B}}T - \ln Z) \right] \tag{5.5}$$

or,

$$F = \sum_i P_i^{\mathrm{B}}\left[-k_{\mathrm{B}}T \ln Z \right] \tag{5.6}$$

Thus, F is expressed formally as a statistical average with a *constant* random variable, $-k_{\mathrm{B}}T \ln Z$ for all i, which means that the fluctuation of F is zero. $\sigma^2(F) = 0$ is related to the formulation of Equation (5.5) but not to other formulations, see Equations (18.2) and (19.24). This property of F has practical advantages that will be discussed later (e.g., Section 19.3.5). The fact that $\sigma^2(F) = 0$ means that the fluctuations of E and TS are the same, and they cancel each other in $\sigma^2(F)$,

$$T^2\sigma^2(S) = \sigma^2(E) \tag{5.7}$$

One can calculate fluctuations of other thermodynamic parameters in the canonical ensemble as well as in other ensembles; more discussions about fluctuations appear in Chapter 6. However, a complete study of this important topic (in particular, in phase transitions) is beyond the scope of this book. Extensive overviews about fluctuations can be found in Chapter 4 of Hill's book [1] and in Chapter 12 of Landau & Lifshitz's book [2]. These books present different approaches for the derivation of fluctuations (see Section 11.7).

5.2 The Most Contributing Energy E^*

The fact that $\sigma(E)/\langle E \rangle$ is small has interesting consequences. Thus, like Z [Equation (4.60)], $\langle E \rangle$ can be expressed as,

$$\langle E \rangle = \int E(\mathbf{x}^N)P^{\mathrm{B}}(\mathbf{x}^N)d\mathbf{x}^N = \int E\nu(E)P^{\mathrm{B}}(E)dE, \tag{5.8}$$

where $\nu(E)$ is the density of states and $P^{\mathrm{B}}(E)$ is the Boltzmann probability of a configuration with energy E. Because $\sigma(E)/\langle E \rangle$ is extremely small, the main contribution to $\langle E \rangle$ in the above integral comes from a very narrow range of energies around a typical energy, $E^*(T) \approx \langle E \rangle$ (Figure 5.1).

Therefore, with a very good approximation, the partition function and the free energy can be expressed only by the contribution related to $E^*(T)$.

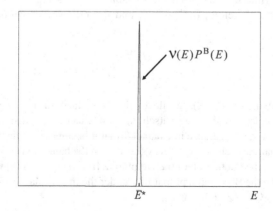

FIGURE 5.1   A sharp maximum of $\nu(E)P^{\mathrm{B}}(E)$ as a function of E at the most contributing energy, $E^*(T)$.

$$Z \equiv f_T(E) = \int \nu(E) \exp\left[-\frac{E}{k_B T}\right] dE \cong \nu(E^*) \exp\left[-\frac{E^*}{k_B T}\right] \tag{5.9}$$

$$F = -k_B T \ln Z \cong E^*(T) - k_B T \ln \nu(E^*) = E^* - TS^* \tag{5.10}$$

The entropy, $S^* = k_B \ln \nu(E^*)$, is the logarithm of the density of states of E^*, that is, each state with E^* has the same probability, $1/\nu(E^*)$. For a discrete system:

$$Z = \sum_{E_i} n(E_i) \exp\left[-\frac{E_i}{k_B T}\right] \cong n(E^*) \exp\left[-\frac{E^*}{k_B T}\right] \tag{5.11}$$

where E_i are the set of energies of the system and $n(E_i)$ their degeneracies. For a macroscopic discrete system, the number of different energies is large (~N), while only the maximal term corresponding to E^* contributes.

The product on the right-hand side of Equation (5.11) consists of two exponentials. At a very low T, the product is maximal for the ground state (GS) energy, where most of the contribution comes from the term $\exp(-E_{GS}^*/k_B T)$, while $n(E_{GS}^*) \sim 1$ ($S = 0$). At very high T the product is maximal for a high energy, where $n(E^*)$ is maximal (maximum degeneracy, which means maximum entropy), but the exponential of the energy is small. For an intermediate T, the products for $E \neq E^*$ are very close to zero, while a high "spike" occurs at $E = E^*$.

A schematic partition of the configurational space into regions of structures with equal energy appears in Figure 5.2. For six decreasing temperatures, the figure exhibits the corresponding (decreasing) energies, E^* (denoted E_6, E_5, ..., E_1), and the related decreasing $n(E_i)$ values, depicted by decreasing circle sizes. Thus, at each temperature, the system "selects" the region which maximizes the product, $n(E) \exp[-E/k_B T]$.

The fact that the contribution to the integrals comes from an extremely narrow region of energies seems at first glance advantageous, since not all space should be (numerically) integrated. However, the desired *extremely tiny* region (within the huge $3N$ dimensional configurational space) that contributes to the integrals is *unknown a priori*. Therefore, it is very difficult to estimate $<E>$, S, and other quantities by a numerical integration. In other words, dividing the space into small regions (grid) would be impractical, since the small important region would be missed and the corresponding integration would contribute actually zero (Figure 5.3). Thus, this numerical integration strategy would be a waste of computer time. For the same reason, trying to estimate the integral by a random search would be futile.

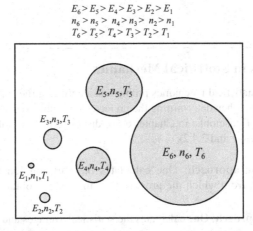

$E_6 > E_5 > E_4 > E_3 > E_2 > E_1$
$n_6 > n_5 > n_4 > n_3 > n_2 > n_1$
$T_6 > T_5 > T_4 > T_3 > T_2 > T_1$

E_5, n_5, T_5

E_3, n_3, T_3

E_6, n_6, T_6

E_4, n_4, T_4

E_1, n_1, T_1

E_2, n_2, T_2

FIGURE 5.2 An illustration of the structure of configurational space. To the six decreasing temperatures, correspond six decreasing energies, E^* (denoted E_6, E_5, ..., E_1), and six decreasing multiplicities (densities of states), $n(E_i)$, depicted by decreasing circle sizes around E^*, where the product $n(E_i) \exp(-E/k_B T)$ contributes significantly to the partition function.

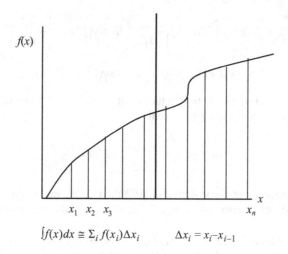

$$\int f(x)dx \cong \Sigma_i\, f(x_i)\Delta x_i \qquad \Delta x_i = x_i - x_{i-1}$$

FIGURE 5.3 Using standard numerical integration for calculating, for example, the energy of a large thermodynamic system is extremely inefficient since the contributing region to the partition function is exceedingly small and thus will be missed by any practical division of configurational space. The bold line represents the contributing region.

The success of the Monte Carlo methods lies in their ability to navigate the system very efficiently toward the contributing region, leading to a reliable estimation of various properties of interest (see discussion in Section 1.16).

Finally, we ask: What is the probability to find a system at a certain energy (not a certain coordinate, $\mathbf{x}^N$)? The answer is based on the Boltzmann probability for the *energy*,

$$P^{B}(E) \cong n(E)\exp\left[-\frac{E}{k_B T}\right]\Big/Z \tag{5.12}$$

which depends not only on the energy, but also on its degeneracy, $n(E)$. The relative population of two energies is therefore:

$$\frac{P^{B}(E_1)}{P^{B}(E_2)} = \frac{n(E_1)\exp(-E_1/k_B T)}{n(E_2)\exp(-E_2/k_B T)} = \frac{n(E_1)}{n(E_2)}\exp\left[-\frac{\Delta E}{k_B T}\right] \tag{5.13}$$

where $\Delta E = E_1 - E_2$.

5.3 Solving Problems in Statistical Mechanics

The first step in treating a statistical mechanics problem is identifying the states (configurations) of the system and their energies, and then determining the preferred ensemble to use (thus far, we have considered only the canonical NVT ensemble; in Chapter 6, we discuss other ensembles). Then, three options are available (see Sections 2.4.1 and 2.4.2).

1. **The thermodynamic approach:** One calculates the partition function, Z, and the free energy, $F = -k_B T \ln Z$, from which the properties of interest are obtained as suitable derivatives of F.

2. **The probabilistic approach:** One calculates the statistical averages and variances of the properties of interest ($<E>$, S, etc.).

3. **The most probable term approach:** It is based on calculating the leading term Z^* of Z, from which the most dominant contributions of the other properties, such as E^* and S^*, are obtained.

Example

To demonstrate these approaches, we apply them to a simple well-studied model (e.g., [3]) of N independent spins interacting with a magnetic field H. The interaction energy (potential energy) of a spin is μH or $-\mu H$, depending on whether μ, the magnetic moment, is positive or negative. Positive μ leads to the energy $-\mu H$. Calculate the various thermodynamic functions (E, F, S, etc.) at a given temperature T.

General Considerations:

The number of system configurations is 2^N because each spin has two states, $+1(-\mu)$ or $-1(\mu)$. The potential energy of system configuration i depends only on N_+ (or N_-) and N,

$$E_i = -N_+\mu H + N_-\mu H = -(N - N_-)\mu H + N_-\mu H = 2N_-\mu H - N\mu H \tag{5.14}$$

where N_+ and N_- are the numbers of $+1$ and -1 spins, respectively. The magnetization of i is,

$$M = N_+\mu - N_-\mu \tag{5.15}$$

The spins are fixed on the lattice, therefore, no kinetic energy is defined, and the factors $N!$ and h are therefore omitted. This system of independent spins is treated within the canonical (NVT) ensemble (notice, N and V are the same).

5.3.1 The Thermodynamic Approach

We have to calculate $Z = \sum \exp[-E_i/k_BT]$, where i runs over all the 2^N different configurations of the system. This summation can be calculated by a trick. Because the spins are independent (that is, they do not interact with each other), changing a spin does not affect the other spins. Therefore, the summation over the configurations of N spins can be expressed as the product, $Z = (z_1)^N$, where z_1 is the partition function of a single spin (as in the case of a microscopic oscillator). Defining,

$$\cosh(x) = [\exp(x) + \exp(-x)]/2 \tag{5.16}$$

leads to,

$$z_1 = \exp\left(-\frac{\mu H}{k_BT}\right) + \exp\left(+\frac{\mu H}{k_BT}\right) = 2\cosh\left(\frac{\mu H}{k_BT}\right) \tag{5.17}$$

where Z is:

$$Z = \left[2\cosh\left(\frac{\mu H}{k_BT}\right)\right]^N \tag{5.18}$$

and F is,

$$F = -k_BT \ln Z = -k_BTN \ln\left[2\cosh\left(\frac{\mu H}{k_BT}\right)\right]. \tag{5.19}$$

Using the definition:

$$\tanh(x) = \frac{e^x - e^{-x}}{e^x + e^{-x}} = \frac{\sinh(x)}{\cosh(x)} \tag{5.20}$$

the entropy is:

$$S = -\left(\frac{\partial F}{\partial T}\right)_H = Nk_B\left[\ln\left\{2\cosh\left(\frac{\mu H}{k_B T}\right)\right\} - \frac{\mu H}{k_B T}\tanh\left(\frac{\mu H}{k_B T}\right)\right].$$

(5.21)

Thus, at $T = \infty$, $S/N = \ln 2$, and at $T = 0$, $S = 0$. The energy is:

$$E = F + TS = -T^2\frac{\partial\left(\frac{F}{T}\right)_H}{\partial T} = -N\mu H\tanh\left(\frac{\mu H}{k_B T}\right).$$

(5.22)

At constant T, the magnetization, M [Equation (5.15)] is:

$$M = -\left(\frac{\partial F}{\partial H}\right)_T = N\mu\tanh\left(\frac{\mu H}{k_B T}\right)$$

(5.23)

Finally, the heat capacity is:

$$C = \left(\frac{\partial E}{\partial T}\right)_H = \frac{Nk_B\left(\frac{\mu H}{k_B T}\right)^2}{\cosh^2\left(\frac{\mu H}{k_B T}\right)}$$

(5.24)

5.3.2 The Probabilistic Approach

Again, we can treat first a single spin and calculate its Boltzmann probability and *average* energy. The Boltzmann probability for a $\pm$ spin is:

$$P_\pm^B = \exp\left(\mp\frac{\mu H}{k_B T}\right)/z_1$$

(5.25)

The average energy of one spin is:

$$<E_1> = \left[-\mu H\exp\left(+\frac{\mu H}{k_B T}\right) + \mu H\exp\left(-\frac{\mu H}{k_B T}\right)\right]/z_1$$

$$= -\mu H\frac{2\sinh(\mu H/k_B T)}{2\cosh(\mu H/k_B T)} = -\mu H\tanh(\mu H/k_B T)$$

(5.26)

which, for N spins, $<E>$ becomes equal to the result of Equation (5.22),

$$<E> = -N\mu H\tanh(\mu H/k_B T).$$

(5.27)

Denoting, for simplicity, $a = \frac{\mu H}{k_B T}$, and using $P_\pm^B$ [Equation (5.25)], the entropy of a single spin, s_1 is [Equation (3.18)]:

$$s_1 = -k_B[P_+^B\ln P_+^B + P_-^B\ln P_-^B] = -k_B\{\exp(+a)[a - \ln z_1] + \exp(-a)[-a - \ln z_1]\}/z_1$$

$$= k_B\{a[\exp(+a) - \exp(-a)]/z_1 - \ln z_1[\exp(+a) + \exp(-a)]/z_1\} = k_B\{a\tanh a - \ln z_1\}$$

(5.28)

Multiplying s_1 by N and replacing a and z_1 by their expressions lead to the same system's entropy as obtained in Equation (5.21).

5.3.3 Calculating the Most Probable Energy Term

First, we define the energy, E:

$$E = -N_+\mu H + N_-\mu H \quad E' = E/\mu H = -N_+ + N_-. \tag{5.29}$$

Using the equation, $N = N_+ + N_-$, one obtains,

$$N_- = (N + E')/2 \quad N_+ = (N - E')/2, \tag{5.30}$$

and the magnetization and the energy become:

$$M = (N_+ - N_-) \quad E = -MH. \tag{5.31}$$

Expressing the partition function by Equation (5.11), the number of spin configurations with energy E is $v(E) = N!/(N_+!N_-!)$. Therefore, each term in the summation of Z, denoted by $f_T(E')$, is,

$$f_T(E') = \frac{N!}{\left(\dfrac{N + E'}{2}\right)!\left(\dfrac{N - E'}{2}\right)!}\exp\left[\frac{-E'\mu H}{k_B T}\right] \tag{5.32}$$

For given N, T, and H, we seek to find the maximal term. First, we write $\ln f_T(E')$, using Stirling's formula for very large N, $\ln N! \approx N\ln N$,

$$\ln f_T(E') = N\ln N - \left(\frac{N + E'}{2}\right)\ln\left(\frac{N + E'}{2}\right) - \left(\frac{N - E'}{2}\right)\ln\left(\frac{N - E'}{2}\right) - \frac{\mu H E'}{k_B T}. \tag{5.33}$$

Then, we derive $\ln f_T(E')$ with respect to E', and equate the result to 0, which leads to,

$$\frac{\partial \ln f_T(E')}{\partial E'} = -\frac{1}{2}\ln\frac{N + E'}{2} + \frac{1}{2}\ln\frac{N - E'}{2} - \frac{\mu H}{k_B T} = 0. \tag{5.34}$$

One can show that the second derivative is negative, thus $f_T(E')$ is a maximum. One obtains,

$$\frac{1}{2}\ln\frac{N - E'}{N + E'} = \frac{\mu H}{k_B T} \Rightarrow \frac{N - E'}{N + E'} = \exp\frac{2\mu H}{k_B T} \tag{5.35}$$

$$E' = N\frac{1 - \exp\left(\dfrac{2\mu H}{k_B T}\right)}{1 + \exp\left(\dfrac{2\mu H}{k_B T}\right)} = N\frac{\exp\left(\dfrac{\mu H}{k_B T}\right)\left[\exp\left(-\dfrac{\mu H}{k_B T}\right) - \exp\left(\dfrac{\mu H}{k_B T}\right)\right]}{\exp\left(\dfrac{\mu H}{k_B T}\right)\left[\exp\left(-\dfrac{\mu H}{k_B T}\right) + \exp\left(\dfrac{\mu H}{k_B T}\right)\right]} \tag{5.36}$$

$$E' = -N\tanh\left[\frac{\mu H}{k_B T}\right] \tag{5.37}$$

The most probable energy E^* for given T, H, and N is thus,

$$E^* = -N\mu H\tanh\left[\frac{\mu H}{k_B T}\right],$$

and the magnetization is:

$$M = -N\mu\tanh\left[\frac{\mu H}{k_B T}\right].$$

The same expressions for E and M have been obtained by the other methods. Notice that this treatment, which involves the entire system, required to assume very large N, while no such conditions are needed in the two alternative solutions. For this problem, the easiest method is the thermodynamics approach. In general, the preferred approach will depend on the problem.

5.3.4 The Change of Energy and Entropy with Temperature

Table 5.1 illustrates the increasing energies (E^*), their populations (degeneracies), the typical spin configurations, and the corresponding increased temperatures. It is shown that only a single configuration (where all spins are up) leads to the lowest (negative) energy (the ground state); the corresponding temperature is $T = 0$, and as expected, the entropy is also zero. When the energy is increased, the degeneracy and T both increase, where for the maximal energy, $E = 0$, the degeneracy and the entropy are the maximal; this corresponds to $T \to \infty$. In principle, one could still define higher energies where the majority of spins are negative ($N_- > N_+$). However, this situation is not physical because it leads to negative temperatures, as an increase in dE and a decrease in dS leads to $dE/dS = -T$. In a "normal" physical system, E and S increase (decrease) together and T is positive.

Comments. The entropy can be defined in two ways: (1) as a statistical average, $S = -k_B \Sigma P_i^B \ln P_i^B$ and (2) as a function of $n(E^*)$—the degeneracy of the most probable energy (for a given T), $S \cong k_B \ln n(E^*)$. For a large system the two definitions are identical.

As a *mechanical* system, the spins "would like" to stay in the lowest energy (ground state—all spins are up), where no uncertainty exists (maximum order), that is, the entropy is zero. However, the spins interact with a heat bath at a *finite* T, where random energy flows in and out of the spin system. Thus, spins parallel to H (spin up) might absorb energy and "jump" to their higher energy level (spin down), where some of the latter spins will release their energy back to the bath by returning to their lower energy state ($\uparrow$) and vice versa. For a given T, the *average* number of excited spins ($\downarrow$) is constant, and this number increases (that is, the *average* energy increases) as T is increased. As E^* increases its degeneracy, $n(E^*)$ increases as well, meaning that the system can populate more configurations with the same E^* and probability; hence, the corresponding uncertainty increases, which is expressed in terms of a higher entropy.

Thus, the stability of a thermodynamic system can be viewed as a "competition" between two opposing preferences, the "desire" to be at the lowest potential energy and the "will" to experience maximal disorder. At $T = 0$, the potential energy wins, while at $T = \infty$, the disorder wins. At finite T, the stability becomes a compromise between these contrasting tendencies, where it is determined by the optimal combination of $E^*(T)$ and $n(E^*)$ (see Figure 5.2); this is decided by the maximal term of the partition function, $n(E^*)\exp - [E^*/k_B T]$, or equivalently by the minimal term of the free energy, $E^* + k_B T \ln n(E^*) = E^* - TS^*$. Notice, that while the (macroscopic) energy is known very accurately due to its small fluctuations, a specific configuration (state) is undefined. We only know that the system can be located with equal probability in *any* of the $n(E^*)$ states.

TABLE 5.1

Increasing Energies Above the Ground State, Their Degeneracies, the Typical Spin Configurations, and the Corresponding Temperatures

Energy*	Degeneracy	Typical T	Configuration
$-N\mu H$ (min.)	$1\ S = 0$ (min.)	$T_0 = 0$	$\uparrow\uparrow\uparrow\uparrow...\uparrow\uparrow\ \ \uparrow\ (H)$
$-(N-1)\mu H + \mu H = -N\mu H + 2\mu H$	N	Very low $T_1 > T_0$	$\downarrow\uparrow\uparrow\uparrow...\uparrow\uparrow$
$-N\mu H + 4\mu H$	$N!/[(N-2)!2!] = N(N-1)/2$	$T_2 > T_1$	$\downarrow\downarrow\uparrow\uparrow...\uparrow\uparrow$
$\vdots$	$\vdots$	$\vdots$	$\vdots$
$-N\mu H + 2k\mu H$	$N!/[(N-k)!k!]$	$T_k > T_{k-1}$	$\downarrow\downarrow\downarrow\downarrow...\uparrow\uparrow$ (k down)
$\vdots$	$\vdots$	$\vdots$	$\vdots$
$\mu H (N/2 - N/2) = 0$	$N!/[(N/2)!(N/2)!]\ S = k_B \ln N$	High T	$\downarrow\downarrow\downarrow\downarrow...\uparrow\uparrow\uparrow\uparrow$ ($N/2$ down)

REFERENCES

1. T. L. Hill. *Statistical Mechanics, Principles and Selected Applications*. (Dover, New York, 1987).
2. L. D. Landau and E. M. Lifshitz. *Statistical Physics*. (Pergamon, New York, 1963).
3. R. Kubo. *Statistical Mechanics*. (North Holland, Amsterdam, The Netherlands, 1974).

6

Various Ensembles

Thus far, we have developed the theory of statistical mechanics for the canonical (NVT) ensemble, where the various thermodynamic parameters are derivatives of the Helmholtz free energy. However, in our treatment of classical thermodynamics (Section 2.10), we have defined other free energies for different sets of thermodynamic parameters, such as the Gibbs free energy based on N, p, and T. To each of these free energies, an ensemble (probability space) is defined within the framework of statistical mechanics. Thus, the corresponding partition function and Boltzmann probability density can be derived with a procedure similar to that used for the canonical ensemble in Chapter 3.

In what follows, we describe these ensembles and the relations among them. As before, for brevity, we introduce the following notations, $(\mathbf{x}^N, \mathbf{p}^N) \equiv \mathbf{X}$, where $\mathbf{X}$ is a point in phase space.

6.1 The Microcanonical (petit) Ensemble

Assume an isolated system of N particles in a volume, V; due to isolation, the total energy of the system, which lies between E and $E + \delta E$, as well as N remain constant (however, the kinetic and potential energies, E_k and E_p ($E = E_k + E_p$) fluctuate; from now on, δE will be omitted). For this NVE system, one defines the microcanonical ensemble (also called the petit ensemble). Following the canonical ensemble treatment (but omitting many details), we seek to build a Boltzmann probability density for this ensemble. Thus, we start by defining an entropy functional based on an initially *undefined* probability density, $P(\mathbf{X})$; the average entropy $S(P)$ is,

$$S(P) = -k_B \int P(\mathbf{X}) \ln P(\mathbf{X}) d\mathbf{X} + \text{constant} \tag{6.1}$$

where the integration is over the phase space compatible with (N, V, E). For given N,V, and E, the entropy, $S(N,V,E)$ should be maximal with respect to other systems' parameters [Equation (2.31)], which in this case is the probability density, P. Thus, we maximize $S(P)$ by calculating the derivative of S with respect to P and equating the result to zero,

$$\left(\frac{\partial S}{\partial P} \right)_{N,V,E} = -k_B \int \left[\ln(P(\mathbf{X}) + 1 \right] d\mathbf{X} = 0. \tag{6.2}$$

$S(P)$ is a maximum since the second derivative, $(-1/P)$, is negative. We demand the integrand to be zero for any $d\mathbf{X}$, therefore,

$$\ln P(\mathbf{X}) + 1 = 0 \quad \text{or} \quad P(\mathbf{X}) = e^{-1} = \text{constant}, \tag{6.3}$$

and due to normalization:

$$P^B(\mathbf{X}) = \frac{1}{\Omega(N,V,E)} \qquad \Omega(N,V,E) = \int d\mathbf{X} \tag{6.4}$$

where $\Omega(N,V,E)$ is the microcanonical partition function. The entropy is [see Equation (3.17)]

$$S_{\text{micro}} = k_B \ln \Omega + \text{costant} = k_B \ln[(\Omega/(h^{3N}N!)] = k_B \ln \Omega^{\#}(N,V,E) \tag{6.5}$$

In the case of a model where the particles do not move freely, such as a lattice gas model, the momenta part of the partition function does not exist, and the partition function is defined by $Z(N,V,E)$:

$$Z(N,V,E) = \frac{1}{N!} \sum_{\mathbf{x}^N, E(\mathbf{x}^N)=E} \mathbf{x}^N \tag{6.6}$$

6.2 The Canonical (*NVT*) Ensemble

We consider a closed system defined by the number of particles, N, the volume, V, and the temperature, T. We have already defined for this system the canonical ensemble; therefore, for completeness, we only summarize the main results. The partition function, Q is,

$$Q(N,V,T) = \int \exp[-E(\mathbf{X})/k_B T] d\mathbf{X} \tag{6.7}$$

and the Helmholtz free energy is [Equation (3.21)]:

$$A = <E> -TS = -k_B T \ln \frac{Q(N,V,T)}{h^{3N}N!} = -k_B T \ln Q^{\#}(N,V,T) \tag{6.8}$$

It is of interest to discuss the relation between the canonical and the microcanonical ensembles. Thus, similar to Equation (4.60), $Q(N,V,T)$ can be expressed as an integration of $\Omega(N,V,E)\exp[-E/k_B T]$ over the energy, where $\Omega(N,V,E)$ is the partition function of the microcanonical ensemble for a given E,

$$Q(N,V,T) = \int \exp[-E(\mathbf{X})/k_B T] d\mathbf{X} = \int \Omega(N,V,E)\exp[-E/k_B T] dE \tag{6.9}$$

Because the main contribution to the first integral comes from the most probable energy, $E^* = E^*(N,V,T) \approx <E(N,V,T)>$ [Section 5.2], $Q(N,V,T)$ becomes,

$$Q(N,V,T) \approx \Omega(N,V,E^*)\exp-[E^*/k_B T] \tag{6.10}$$

where $\Omega(N,V,E^*) = v(E^*)$ defined in Equation (5.10) and

$$A = -k_B T \ln Q^{\#}(N,V,T) = <E(N,V,T)> -TS(N,V,T) \approx$$

$$\approx E^* - k_B T \ln \Omega^{\#}(N,V,E^*) = E^* - TS(N,V,E^*) \tag{6.11}$$

Thus, for a large system, a microcanonical ensemble defined by NVE^* ($E^* \approx <E(N,V,T)>$) has actually the same entropy as that of the (conjugate) canonical ensemble. In other words, as far as *statistical averages* are concerned, the two ensembles lead to the same results (however, this is not true for the fluctuations).

We have also pointed out that if the forces do not depend on the velocities, the momenta part of $Q^{\#}(N,V,T)$ can be solved analytically, and the practical task is to calculate $Z(N,V,T)$ – the contribution of the potential energy to $Q^{\#}$, which depends only on the coordinate, $\mathbf{x}^N$.

$$Q^{\#}(N,V,T) = Q_{\text{momenta}} \int \exp[-E(\mathbf{x}^N)/k_{\text{B}}T]d\mathbf{x}^N = Q_{\text{momenta}}Z(N,V,T), \tag{6.12}$$

where the integration is over the *configurational* space. Z leads to the corresponding Boltzmann probability density [Equation (4.62)]:

$$P^{\text{B}}(\mathbf{x}^N) = \frac{\exp- E(\mathbf{x}^N)/k_{\text{B}}T}{Z(N,V,T)} \tag{6.13}$$

and the configurational free energy [see Equation (4.63)],

$$F(N,V,T) = -k_{\text{B}}T \ln Z(N,V,T) = \; <E(\mathbf{x}^N)> +k_{\text{B}}T <\ln P^{\text{B}}(\mathbf{x}^N)> \tag{6.14}$$

Z and F are also used for models where the velocity part does not exist, such as the Ising and lattice gas models.

6.3 The Gibbs (*NpT*) Ensemble

We consider a thermodynamic system controlled by the temperature, T, the pressure, p, and the number of particles, N (thus, the energy, E, and the volume, V, fluctuate). The suitable free energy is the Gibbs free energy (Section 2.10.5):

$$G(N,p,T) = E - TS + pV \tag{6.15}$$

This system is described in statistical mechanics by the *NpT* ensemble (also called the Gibbs ensemble). To define a suitable Boltzmann probability density, we apply the same procedure used for the canonical ensemble (Section 3.2), consisting of minimum G, where G is expressed as a statistical average with an initially unknown probability density, $P(\mathbf{X},V)$; we also assume the existence of an entropy function $-k_{\text{B}}\ln P(\mathbf{X},V)$ (see Section 3.2),

$$G_{N,p,T}(P) = \iint P(\mathbf{X},V)[E(\mathbf{X}) + k_{\text{B}}T \ln P(\mathbf{X},V) + pV]d\mathbf{X}dV \tag{6.16}$$

$P(\mathbf{X},V)$ is determined by minimizing $G(P)$ with respect to P (for constant N, p, and T), that is, by calculating $\partial G/\partial P$ and equating it to zero; after normalization, one obtains the Boltzmann probability density:

$$P^{\text{B}}(\mathbf{X},V)_{N,p,T} = \frac{\exp\{-[E(\mathbf{X},V) + pV]/k_{\text{B}}T\}}{\Delta(N,p,T)}, \tag{6.17}$$

where the normalization factor, $\Delta(N,p,T)$ is the Gibbs partition function,

$$\Delta(N,p,T) = \iint \{\exp- [E(\mathbf{X},V) + pV]/k_{\text{B}}T\}d\mathbf{X}dV \tag{6.18}$$

and:

$$\Delta^{\#}(N,p,T) = \frac{\Delta(N,p,T)}{h^{3N}N!}. \tag{6.19}$$

Using $P^B(\mathbf{X},V)$ one can calculate various statistical averages of extensive parameters, such as the energy, $<E>$, the entropy, S, and the number of particles, $<N>$; for example:

$$S = -k_B \iint P^B(\mathbf{X},V)\ln[P^B(\mathbf{X},V)h^{3N}N!]d\mathbf{X}dV \qquad (6.20)$$

$$<V> = \iint V P^B(\mathbf{X},V)d\mathbf{X}dV = \frac{1}{\Delta}\iint V\{\exp-[E(\mathbf{X},V)+pV]/k_BT\}d\mathbf{X}dV. \qquad (6.21)$$

where $<E>$ is defined in a similar way. Thus, G becomes a statistical average expressed in terms of these averages,

$$G(N,p,T) = <E> -TS + p<V> \qquad (6.22)$$

Like the Helmholtz free energy, A, Gibbs free energy, G, is a statistical average with a *zero* fluctuation $[\sigma^2(G) = 0]$. Thus, if one replaces $P(\mathbf{X},V)$ in the brackets of Equation (6.16) by the Boltzmann probability density, P^B [Equation (6.17)], the term in the bracket becomes constant, $-k_BT\ln\Delta$, for any value of $\mathbf{X}$ and V. This means that the fluctuation in $E + pV$ is the same as the fluctuation in TS, and these fluctuations cancel each other in G. Moreover, an alternative formula for calculating G is [see discussion following Equation (3.19)],

$$G(N,p,T) = -k_BT\ln\frac{\Delta(N,p,T)}{h^{3N}N!} = -k_BT\ln\Delta^{\#}(N,p,T). \qquad (6.23)$$

In the discussion of the canonical ensemble, we have emphasized the fact that due to extremely small energy fluctuation, one can estimate a statistical average by considering a single term which is related to the most probable energy. A similar situation exists in the NpT ensemble due to the small fluctuation of the volume, V [1],

$$\frac{\sigma^2(V)}{<V>^2} = \frac{<V^2> - <V>^2}{<V>^2} = -\frac{k_BT}{V^2}\left(\frac{\partial V}{\partial p}\right)_{N,T} = \frac{k_BT\kappa}{V} \qquad (6.24)$$

where $\kappa = -V^{-1}(\partial V/\partial p)_{N,T}$ is the isothermal compressibility. For an ideal gas ($pV = Nk_BT$), the right hand side of Equation (6.24) is proportional to $1/N$ (or $\sigma(V)/<V> \sim 1/N^{1/2}$); this proportionality holds also in more complex situations as long as the system is not too close to a phase transition, for example. Thus, the relative fluctuation of V [Equation (6.24)] for a macroscopic system is extremely small, and in practice one can replace the average volume, $<V(N,p,T)>$, by its most probable value, $V^*(N,p,T)$, which, can lead to simplifications. Thus, $\Delta(N,p,T)$ can be expressed in terms of the canonical and microcanonical partition functions,

$$\Delta(N,p,T) = \int Q(N,V,T)\exp[-pV/k_BT]dV$$

$$= \iint \Omega(N,V,E)\exp\{-[E+pV]/k_BT\}dEdV \qquad (6.25)$$

where, in practice, the integration with respect to the volume can be replaced by the integrands calculated at $V^*(N,p,T)$,

$$\Delta(N,p,T) \approx Q(N,V^*,T)\{\exp-[pV^*]/k_BT\} \approx \Omega(N,V^*,E^*)\exp-[E^*+pV^*]/k_BT. \qquad (6.26)$$

$E^*(N,V^*,T)$ is the most probable energy in the canonical ensemble for N,V^*, and T. Now, using Equation (6.25), one can show that the average energy in the Gibbs ensemble, $<E(N,p,T>$, is actually equal to $<E(N,V^*,T)>$ — the average energy in the canonical ensemble, where $V^* = V^*(N,p,T)$,

$$< E(N,p,T) >_{\text{Gibbs}} = \frac{1}{\Delta} \iint E(V,\mathbf{X})\{\exp-[E(V,\mathbf{X}) + pV]/k_{\text{B}}T\}d\mathbf{X}dV \approx$$

$$\approx \frac{\exp[-pV^*]/k_{\text{B}}T}{Q(N,V^*,T)\exp[-pV^*]/k_{\text{B}}T} \int E(V^*,\mathbf{X})\exp[-E(V^*,\mathbf{X})/k_{\text{B}}T]d\mathbf{X} = \quad (6.27)$$

$$= < E(N,V^*,T) >_{\text{canonical}}$$

In a similar way, one can show that the entropy and other extensive variables are the same in these two ensembles. Because we have already shown the equivalence between the canonical and the microcanonical ensembles, this equivalence also holds between the Gibbs and the microcanonical ensembles.

As has been shown for the canonical ensemble, one can separate the momenta and coordinates (interactions) parts of the partition function,

$$\Delta^{\#}(N,p,T) = \frac{1}{h^{3N}N!} \iiint \{\exp-[E(\mathbf{x}^N,\mathbf{p}^N,V) + pV]/k_{\text{B}}T\}d\mathbf{x}^N d\mathbf{p}^N dV$$
$$= Q_{\text{momenta}}(T,N)Z(N,p,T) \tag{6.28}$$

where the velocity-dependent Q_{momenta} has already been calculated analytically [Equations (3.27–3.29)]. Hence, to obtain $\Delta^{\#}$, only the configurational partition function, $Z(N,p,T)$ (that depends on the integration over $\mathbf{x}^N$, and V), remains to be calculated,

$$Z(N,p,T) = \iint \{\exp-[E(\mathbf{x}^N,V) + pV]/k_{\text{B}}T\}d\mathbf{x}^N dV, \tag{6.29}$$

where the corresponding Boltzmann probability density, $P^{\text{B}}(\mathbf{x}^N,V)$, is:

$$P^{\text{B}}(\mathbf{x}^N,V)_{N,p,T} = \frac{\exp-[E(\mathbf{x}^N,V) + pV]/k_{\text{B}}T}{Z(N,p,T)}. \tag{6.30}$$

Thus, in practice, it remains to calculate statistical averages of thermodynamic properties, which depend on the potential energy alone. $Z(N,p,T)$ can be expressed in terms of the canonical partition function, $Z(N,V,T)$,

$$Z(N,p,T) = \int Z(N,V,T)\{\exp-[pV]/k_{\text{B}}T\}dV \tag{6.31}$$

and again, by replacing V with its most probable volume, V^*, $Z(N,p,T)$ becomes a function of the corresponding *integrand*,

$$Z(N,p,T) \approx Z(N,V^*,T)\{\exp-[pV^*]/k_{\text{B}}T\}. \tag{6.32}$$

Using the same arguments that have led to Equation (6.28), one can show that (configurational) averages of extensive properties are practically equal in the Gibbs and the canonical ensembles; for example,

$$< E(\mathbf{x}^N) >_{NpT} \approx < E(\mathbf{x}^N) >_{NV^*T} \tag{6.33}$$

6.4　The Grand Canonical (μVT) Ensemble

When a system is defined by the chemical potential, μ, the volume, V, and the temperature, T (N and E fluctuate), the suitable free energy, $\hat{H}$, is [Equation (2.60)],

$$\hat{H}(\mu, V, T) = E - TS - \mu N, \tag{6.34}$$

In statistical mechanics, such a system is described by the grand canonical ensemble. Derivation of the Boltzmann probability density, P^B, for this ensemble is similar to that described for the canonical and Gibbs ensembles. First, notice that following the same arguments used for A and G (Section 2.11), one can show that for a system in equilibrium at given T and μ, $\hat{H}$ is minimum with respect to other system parameters. Now, $\hat{H}$ [Equation (6.34)] is expressed as a statistical average with (initially unknown) probability density, $P(\mathbf{X}, N)$, and an entropy functional $\sim -k_B \ln P(\mathbf{X}, N)$,

$$\hat{H}_{\mu, V, T}(P) = \sum_{N=0}^{\infty} \int P(\mathbf{X}, N)[E(\mathbf{X}) + k_B T \ln P(\mathbf{X}, N) - \mu N] d\mathbf{X} \tag{6.35}$$

then, $\hat{H}(P)$ is minimized by equating $\partial \hat{H} / \partial P$ to zero, which leads to:

$$P^B(\mathbf{X}, N)_{\mu, V, T} = \frac{\exp[-E(\mathbf{X}, N) + \mu N] / k_B T}{\Xi(\mu, V, T)}, \tag{6.36}$$

where the normalization factor, $\Xi(\mu, V, T)$, is the grand canonical partition function,

$$\Xi(\mu, V, T) = \sum_{N=0}^{\infty} \int \{\exp[-E(\mathbf{X}, N) + \mu N] / k_B T\} d\mathbf{X}, \tag{6.37}$$

and based on the discussion for the canonical ensemble [Equations (3.20) and (3.21)],

$$\Xi^\#(\mu, V, T) = \sum_{N=0}^{\infty} \frac{1}{h^{3N} N!} \int \{\exp[-E(\mathbf{X}, N) + \mu N] / k_B T\} d\mathbf{X} \tag{6.38}$$

The entropy and other statistical averages, such as $<E>$ and $<N>$ are defined with $P^B(\mathbf{X}, N)$, where:

$$S = -k_B \sum_{N=0}^{\infty} \int P^B(\mathbf{X}, N) \ln(P^B(\mathbf{X}, N) h^{3N} N!) d\mathbf{X} \tag{6.39}$$

$$<N> = \sum_{N=0}^{\infty} \int N P^B(\mathbf{X}, N) d\mathbf{X} = \frac{1}{\Xi} \sum_{N=0}^{\infty} \int N \exp([-E(\mathbf{X}, N) + \mu N] / k_B T) d\mathbf{X} \tag{6.40}$$

where $<E>$ is defined in a similar way. Thus, $\hat{H}$ becomes a statistical average expressed in terms of these averages,

$$\hat{H}(\mu, V, T) = <E> - TS - \mu <N>. \tag{6.41}$$

Like G, $\hat{H}$ is a statistical average with a *zero* fluctuation [$\sigma^2(\hat{H}) = 0$], which can be shown by replacing $P(\mathbf{X}, N)$ by P^B [Equation (6.36)] in the brackets of Equation (6.35) (see discussion for G). This means that the fluctuation in E is equal to that of $TS + \mu N$, and these fluctuations cancel each other in $\hat{H}$. Also, one obtains the additional formula for $\hat{H}$ [see Equation (3.21)],

$$\hat{H}(\mu,V,T) = -k_B T \ln \Xi^{\#}(\mu,V,T).\tag{6.42}$$

The fluctuations of extensive thermodynamic parameters in the grand canonical ensemble are extremely small (as for the other ensembles studied earlier). This, in particular, holds for the number of particles, N, where its fluctuation is the same as that obtained for the volume in the NpT ensemble (see [1]):

$$\frac{\sigma^2(N)}{<N>^2} = \frac{<N^2>-<N>^2}{<N>^2} = -\frac{k_B T \kappa}{V} \propto \frac{1}{N}.\tag{6.43}$$

Thus, in practice, one can replace the average, $<N(\mu,V,T)>$ by its most probable value, $N^*(\mu,V,T)$, and $\Xi(\mu,V,T)$ can be expressed in terms of the canonical and microcanonical partition functions,

$$\begin{aligned}\Xi(\mu,V,T) &= \sum_{N=0}^{\infty} Q(N,V,T)\{\exp[\mu N]/k_B T\}\\&= \sum_{N=0}^{\infty}\int \Omega(N,V,E)\{\exp[-E+\mu N]/k_B T\}dE\end{aligned}\tag{6.44}$$

where a single term calculated for $N^*(\mu,V,T)$ can replace the summation with respect to N:

$$\Xi(\mu,V,T) \approx Q(N^*,V,T)\{\exp[\mu N^*]/k_B T\} \approx \Omega(N^*,V,E^*)\exp[-E^*+\mu N^*]/k_B T.\tag{6.45}$$

$E^* = E^*(N^*,V,T)$ is the most probable energy in the canonical ensemble for N^*,V, and T. Now, with the same treatment used for the Gibbs ensemble, one can show (based on Equation (6.45)), that $<E(\mu,V,T> \approx <E(N^*,V,T)>$, where $N^* = N^*(\mu,V,T)$; this equivalence between the μVT and NVT ensembles also includes the NVE ensemble, and it applies to the entropy and to other extensive variables as well.

One can calculate analytically the momenta part of the partition function, $Q_{\text{momenta}}(N,T)$ leading to:

$$\begin{aligned}\Xi^{\#}(\mu,V,T) &= \sum_{N=0}^{\infty}\frac{1}{h^{3N}N!}\iint\{\exp[-E(\mathbf{x}^N,\mathbf{p}^N)+\mu N]/k_B T\}d\mathbf{p}^N d\mathbf{x}^N\\&= \sum_{N=0}^{\infty} Q_{\text{momenta}}(N,T)Z(\mu,N,V,T)\end{aligned}\tag{6.46}$$

where:

$$Z(\mu,N,V,T) = \int\{\exp-[E(\mathbf{x}^N,N)-\mu N]/k_B T\}d\mathbf{x}^N\tag{6.47}$$

However, unlike for the NpT ensemble, the momenta and coordinates parts are connected, as they both depend on N, and thus cannot be separated unless $\Xi^{\#}(\mu,V,T)$ is calculated for N^*. However, this is not practical in simulations where N fluctuates and both parts should be considered simultaneously. On the other hand, for models where the particles do not move, the momenta part is eliminated, and the corresponding partition function is $Z(\mu,V,T)$,

$$Z(\mu,V,T) = \sum_{N=0}^{\infty}\int\{\exp[-E(\mathbf{x}^N,N)+\mu N]/k_B T\}d\mathbf{x}^N.\tag{6.48}$$

Furthermore, for a lattice model, the integration in Equation (6.48) is replaced by a summation over the lattice configurations.

6.5 Averages and Variances in Different Ensembles

We have defined four ensembles which are commonly used in statistical mechanics and simulations. Each ensemble is based on three independent thermodynamic parameters (NVE, NVT, NpT, and μVT). One can introduce other pairs, (x_i, X_i), of extensive/intensive pairs, such as magnetization and magnetic field, (M, H), already treated in Section 5.3.

The theory developed in this chapter applies to macroscopic systems ($N_{Avogdro} \sim 10^{23}$); thus, models that are treated analytically are solved in the *thermodynamic limit*, where extensive variables tend to infinity. The theory also applies to finite systems, as those studied by computer simulation; however, one should take into account finite size effects, such as non-negligible *relative* fluctuations (see Chapter 7).

As discussed above, for large systems, the different ensembles are equivalent in the sense that they lead to the same thermodynamic averages (but not to the same fluctuations). Therefore, to solve a problem, one would prefer to select the most convenient ensemble. This is demonstrated in the following problem.

Example

An adsorbent surface has N sites, and each can adsorb one gas molecule. Suppose that the surface is in contact with an ideal gas with chemical potential, μ (determined by p and T). An adsorbed molecule has energy $-\varepsilon$ ($\varepsilon > 0$) compared to a free molecule. Determine the covering ratio, $\theta = n/N$, where n is the number of adsorbed molecules. What is the relation between θ and p [2]?

Notice that the container of ideal gas is assumed to be large, that is, it is a reservoir with constant chemical potential μ_{gas}, which is not changed even if some molecules leave the container to the surface; therefore, the ideal gas part should not be optimized. Also, in the case of two phases (1 and 2) at equilibrium, it is required that the corresponding intensive parameters are equal, $T_1 = T_2$, $\mu_1 = \mu_2$, etc.

6.5.1 A Canonical Ensemble Solution (Maximal Term Method)

First, we write a single n-dependent term of the surface partition function and the corresponding free energy, $F_{surface}(n)$,

$$Z_{surface}(n) = \frac{N!}{(N-n)!n!} \exp[+n\varepsilon/k_B T] \quad \left(\text{a term of } Z\right) \tag{6.49}$$

$$F_{surface}(n) = -k_B T Z_{surface}$$

$$= -k_B T (N \ln N - n \ln n - (N-n)\ln(N-n) + n\varepsilon/k_B T]$$

The most probable n will be determined by the condition,

$$\mu_{surface} = \mu_{gas} = -k_B T \ln[k_B T / p\Lambda^3], \tag{6.50}$$

where μ_{gas}, developed in Equation (4.20), is treated as a constant and thus is not specified below,

$$\mu_{surface} = \frac{\partial F}{\partial n} = -k_B T(-\ln n + \ln(N-n) + \varepsilon / k_B T) = \mu_{gas}$$

$$-k_B T \left[\ln\left(\frac{N-n}{n}\right) + \varepsilon/k_B T \right] = \mu_{gas} \quad \rightarrow \quad -\ln\left(\frac{N-n}{n}\right) = (\mu_{gas} + \varepsilon)/k_B T$$

$$\left(\frac{N-n}{n}\right)=\left(\frac{1}{\theta}-1\right)=\exp-(\mu_{gas}+\varepsilon)/k_BT$$

$$\theta=\frac{1}{1+\exp-(\mu_{gas}+\varepsilon)/k_BT} \tag{6.51}$$

$\theta(p)$ is obtained by replacing μ_{gas} in Equation (6.51) by its expression defined in Equation (6.50).

6.5.2 A Grand-Canonical Ensemble Solution

A shorter solution is obtained by writing the grand partition function for the surface, where the summation can be simplified by the binomial expansion, $(1+x)^N=\binom{N}{0}+...\binom{N}{n}x^n+...,$

$$Z_{surface}(\mu,V,T)=\sum_{n=0}\frac{N!}{(N-n)!n!}\exp[n(\mu+\varepsilon)/k_BT]=[1+\exp[(\mu+\varepsilon)/k_BT]^N$$

$$<n>=-\frac{\partial[-k_BT\ln Z_{surface}(\mu,V,T)]}{\partial\mu}=k_BTN\frac{\ln[1+\exp[(\mu+\varepsilon)/k_BT]}{\partial\mu}=N\frac{1}{1+\exp[-(\mu+\varepsilon)/k_BT]}$$

Thus,

$$\theta=\frac{<n>}{N}=\frac{1}{1+\exp[-(\mu_{gas}+\varepsilon)/k_BT]} \tag{6.52}$$

where μ (=$\mu_{surface}$) has been replaced by μ_{gas}. As expected, Equation (6.52) is equal to Equation (6.51).

6.5.3 Fluctuations in Different Ensembles

As mentioned above, unlike statistical averages, the fluctuation (variance) of a given parameter (e.g., the energy) differs from ensemble to ensemble. Thus, we have already obtained in the canonical ensemble [Equation (5.3)],

$$<E^2>-<E>^2=k_BT^2\left(\frac{\partial<E>}{\partial T}\right)_{N,V}=k_BT^2C_V$$

while in the (N,p,T) ensemble, $\sigma^2(E)$ is,

$$<E^2>-<E>^2=k_BT^2\left[C_V-\frac{\{p(\partial V/\partial p)_{N,T}+T(\partial V/\partial T)_{N,p}\}^2}{T(\partial V/\partial p)_{N,T}}\right] \tag{6.53}$$

and in the grand-canonical ensemble,

$$<E^2>-<E>^2=k_BT^2\left[C_V-\frac{\{[p+E/V](\partial V/\partial p)_{N,T}+T(\partial V/\partial T)_{N,p}\}^2}{T(\partial V/\partial p)_{N,T}}\right]. \tag{6.54}$$

These results were taken from Chapter 4 of Hill's book [1], which provides an extensive study of fluctuations. A different derivation of fluctuations can be found in the book of Landau & Lifshitz [3], Chapter 12. The fluctuations are important in phase transitions, as discussed in the next chapter.

PROBLEM 6.1

A solid and a vapor (ideal gas of N_g molecules) are in equilibrium in a closed container of volume V at T K. The partition function of the solid of N_s atoms is $Z_s(T,N_s) = z_s(T)^{N_s}$ and that of a gas molecule is $z_g(T,V)$. Show that in equilibrium $N_g = z_g(T,V)/z_s(T)$.

PROBLEM 6.2

Show that the variance of the volume, V, in the Gibbs ensemble is:

$$\frac{<V^2> - <V>^2}{<V>^2} = -k_BT \frac{1}{<V>^2}\left(\frac{\partial <V>}{\partial p}\right)_{N,T} \qquad (6.55)$$

and that the behavior of the right hand side of Equation (6.55) for an ideal gas is $1/N$.

Summary: The theory developed thus far constitutes the basis of *equilibrium* statistical mechanics in the classical limit. This theory has been extended in the course of the years for studying complex systems in physics, chemistry, biology, and engineering, where special mathematical tools and additional concepts have been developed. We shall apply statistical mechanics to various systems, among them polymers and proteins, and will use this theory as a basis for developing computer simulation methodologies.

REFERENCES

1. T. L. Hill. *Statistical Mechanics, Principles and Selected Applications.* (Dover, New York, 1987).
2. R. Kubo. *Statistical Mechanics.* (North Holland, Amsterdam, the Netherlands, 1974).
3. L. D. Landau & E. M. Lifshitz. *Statistical Physics.* (Pergamon, New York, 1963).

FURTHER READING

4. R. C. Tolman. *The Principles of Statistical Mechanics.* (Dover, New York, 1979).
5. F. Reif. *Fundamentals of Statistical and Thermal Physics.* (McGraw Hill, Boston, MA, 1965).
6. L. E. Reichl. *A Modern Course in Statistical Physics.* (Edward Arnold, London, UK, 1980).
7. A. Munster. *Statistical Thermodynamics. Volume I.* (Springer-Verlag, Berlin, Germany, 1969).
8. D. Chandler. *Introduction to Modern Statistical Mechanics.* (Oxford, New York, 1967).

7

Phase Transitions

A phase transition occurs, for example, when a solid is heated up to become liquid and the latter is heated further and gets transferred into gas. A phase transition also occurs when a magnet loses its magnetization or a liquid crystal becomes disordered.

While phase transitions are very common, they involve a discontinuity (or divergence) in thermodynamic properties, and it was not clear until 1944 whether a phase transition can be described within the framework of equilibrium statistical mechanics. In this year, Onsager solved analytically the two-dimensional ($2d$) magnetic Ising model, where the properties of the phase transition appeared in the solution. This field has been advanced considerably during the last 80 years. It has also been established that polymers undergo a phase transition, which corresponds to a transition that occurs in a specific magnetic system. Our treatment of this topic is a very limited descriptive in essence.

7.1 Finite Systems versus the Thermodynamic Limit

Systems where particle-particle interactions or correlations are short-range behave "normally," that is, the growth of (extensive) parameters, such as the energy is proportional to the increase of the number of particles, N. Extreme examples are the harmonic oscillator and the spin system studied in Section 5.3. The oscillators (and spins) do not interact with each other, and the partition function is q^N, where q is the partition function of a single oscillator (or a single spin); correspondingly, the energy and entropy of N particles are exactly extensive. For N oscillators, the entropy is [Equation (4.37)],

$$S = 3Nk_B\left(1 + \ln\frac{k_B T}{h\nu}\right),$$

where the entropy of spins in a magnetic field reads [Equation (5.21)],

$$S = Nk_B\left[\ln\left\{2\cos h\left(\frac{\mu H}{k_B T}\right)\right\} - \frac{\mu H}{k_B T}\tan h\left(\frac{\mu H}{k_B T}\right)\right].$$

Unlike these examples, for the non-interacting ideal gas particles, we had to add the factor $N!$, which makes the entropy extensive, that is, depending on $\rho^{-1} = V/N$ under the $\ln$ rather than on V alone; thus, S becomes "purely" extensive [Equation (4.13)],

$$S = -Nk_B\ln\left[\left(\frac{2\pi mk_B T}{h^2}\right)^{3/2}\frac{Ve^{5/2}}{N}\right]$$

A different situation exists, for example, in a second-order phase transition. This is a cooperative phenomenon caused by the increased range of particle-particle interactions (e.g., polymers) or particle-particle correlations (the Ising model) as the transition point is approached. Thus, certain thermodynamic parameters can increase stronger than ~N, that is, by ~N^{1+x}, where $x > 0$.

7.2 First-Order Phase Transitions

A first-order phase transition is characterized by a discontinuity in a first derivative of the free energy, F. Examples of first derivatives are the energy, $<E> = T^2 [\partial(F/T)/\partial T]_{N, V}$, the entropy, $S = -(\partial F/\partial T)_V$, and the magnetization, M, of spins in a magnetic field, $<M> = (\partial F/\partial H)_T$.

A known example for a system undergoing first-order transition is a nematic liquid crystal. The molecules are elliptic or elongated. At low T, they are ordered in some direction in space, the interaction is strong and the average energy, $<E>$, is low (Figure 7.1a). A random arrangement occurs at high T, leading to high $<E>$ (Figure 7.1b).

At the *critical* temperature, T_c, the system co-exists in two states (phases), ordered and disordered, meaning that these states are equally stable, that is, having equal free energies, $F_{ordered} = F_{disordered}$; we denote these phases by 1 and 2, respectively, thus, $\Delta F_{1,2} = \Delta E_{1,2} - T_c \Delta S_{1,2} = 0$ or $\Delta E_{1,2} = T_c \Delta S_{1,2}$ (Figure 7.2). Therefore, this transition is characterized by a *finite* energy difference, $\Delta E_{1,2}$, called the *latent heat* and the corresponding "jump" in the entropy, $\Delta S_{1,2}$.

For a regular system, the function $f_T(E) = n(E)\exp[-E/k_BT]$ as a function of E has a *single* sharp peak at $E^*(T)$ [Equation (5.9)]. In a first-order phase transition, *two* peaks exist at T_c, around $E_{ordered}$ and $E_{disordered}$. Notice, however, that in a *small* system, the peaks will be smeared, while in a *macroscopic* system, they become very sharp. For a peptide (or a protein) *several* energies might contribute strongly to Z, and the molecule will populate all of them significantly at equilibrium (this is called *intermediate flexibility*; see Section 16.10). Because the system is relatively small, the peaks will be smeared (Figure 7.3).

Other "abnormal" systems are spin glasses, which are magnetic spin systems with competing interactions; thus, under certain conditions, the ground state is not unique (say, ferromagnetic), but is "frustrated," that is, many (random) low energy states with comparable probabilities exist, and, at low temperature, the system can visit many of them, leading to a finite entropy. A similar situation occurs for

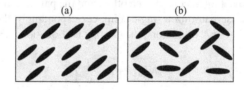

FIGURE 7.1 (a) Ordered elliptic molecules of a liquid crystal below the critical temperature, T_c; the energy and entropy are low. (b) Randomly oriented molecules at $T > T_c$; the energy and entropy are high.

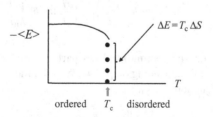

FIGURE 7.2 First-order phase transition. At T_c, $F_{ordered} = F_{disordered}$, thus $\Delta E = T_c\Delta S$; ΔE is the latent heat.

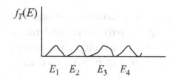

FIGURE 7.3 For a peptide, the function $f_T(E)$ might have several peaks at different energies, E_i.

atomic van der Waals clusters, which have a potential energy profile with many minima, some of which are comparable, and at low temperature might be populated significantly; thus, the system can interconvert among structurally different states.

7.3 Second-Order Phase Transitions

In a second-order phase transition, the first derivatives of the free energy (such as E and S) are continuous, while there is a discontinuity (or divergence to infinity) in a second derivative of F. Examples are the specific heat (heat capacity per particle), $C_V = 1/N(\partial E/\partial T)_V$ and the magnetic susceptibility, $\chi_T = -1/N$ $(\partial M/\partial H)_T$, where E and $M = -(\partial F/\partial H)_T$ are first derivatives of F. A second-order phase transition occurs at the critical point of a magnetic system or a liquid, where the susceptibility and the compressibility, respectively, diverge; in some respect, a polymer also behaves as a system undergoing such a transition (see Chapters 8 and 9).

To understand this phenomenon, we define a simple spin model—the Ising model on an $L \times L = N$ square lattice. Thus, on lattice site i, a spin, σ_i, is defined with two possible states, "up" and "down", $\sigma_i = +1$ or -1, respectively; therefore, the total number of spin configurations on the lattice is 2^N.

Unlike the magnetic model studied in Section 5.3, no external magnetic field is defined here, but spin, i, interacts with a *nearest neighbor* spin, j, with energy, $\varepsilon_{ij} = -J\sigma_i\sigma_j$, where the strength of the (ferromagnetic) interaction is $J > 0$,

$$\varepsilon_{ij}(++) = \varepsilon_{ij}(--) = -J$$
$$\varepsilon_{ij}(+-) = \varepsilon_{ij}(-+) = +J \tag{7.1}$$

The Ising model on a *square lattice* has been solved analytically by Onsager in 1944 [1] (see also [2] and Chapter 17 of [3]), while only approximate results based on numerical techniques (mainly exact enumeration and simulations) exist for Ising models on the cubic and other $3d$ lattices [4]. We shall only bring the main results.

In the ground state, all spins are up ($\sigma_i = +1$), their number, N_+, satisfies, $N_+ = N$, meaning that the magnetization, $M = (N_+ - N_-)/N = 1$, and the energy, $-NdJ$, is the lowest ($d = 2$); however, unlike the model studied in Section 5.3, this ground state is degenerate, that is, the same lowest energy is obtained for the configuration, where $\sigma_i = -1$ for all i, which however, has a negative magnetization, $M = -1$; the entropy is thus, $S = 0$. The (typical) temperature for this completely ordered ground state is $T = 0$, and for a large system, no transition can occur between these two phases even for $0 \le T < T_c$, due to a large configurational barrier (or equivalently, a magnetization barrier); therefore, we consider the $M = 1$ phase. As T is increased, the typical energy and entropy both increase, while the magnetization decreases monotonically from its maximal value, $M = 1$ to zero at the critical temperature, T_c; for $T > T_c$, the magnetization remains zero. The energy and entropy cross T_c *continuously*, and they keep increasing gradually to $E = 0$ and $S = Nk_B\ln2$ at $T = \infty$, where the spins are distributed randomly (Figure 7.4).

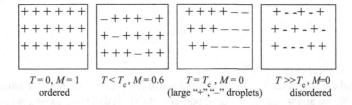

| $T = 0, M = 1$ | $T < T_c, M = 0.6$ | $T = T_c, M = 0$ | $T \gg T_c, M = 0$ |
| ordered | | (large "+","−" droplets) | disordered |

FIGURE 7.4 The three temperature regimes of a second-ordered phase transition, illustrated for the magnetization, M, of the $2d$ Ising model: ordered ($T < T_c$), T_c (droplets), and disordered, $T > T_c$.

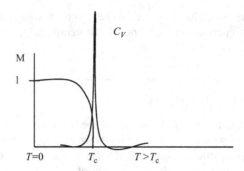

FIGURE 7.5 Schematic illustration of the decrease of the magnetization, M, from 1 ($T = 0$) to 0 at T_c and the divergence of the specific heat, C_V at T_c.

On the other hand, as the temperature approaches T_c (from both the cold and hot sides), the specific heat and the magnetic susceptibility diverge to infinity, with typical critical exponents denoted α and γ, respectively,

$$\frac{C_V}{N} \propto \frac{T_c}{(T-T_c)^\alpha} \qquad \qquad \frac{\chi_T}{N} \propto \frac{T_c}{(T-T_c)^\gamma} \tag{7.2}$$

The decrease of the magnetization (to zero) as T_c is approached from the cold side is also governed by a critical exponent, denoted β (Figure 7.5),

$$<M> \propto \frac{(T-T_c)^\beta}{T_c} \tag{7.3}$$

The divergence of C_V/N and the susceptibility χ_T/N stem from the fact that $C_V \sim \sigma^2(E)$ (Equation 5.3) and $\chi \sim \sigma^2(M)N$, where σ^2 is the variance. At non-critical temperatures, $\sigma^2(E)$ and $\sigma^2(M)N$ are extensive (proportional to N), hence, C_V/N and $\chi_{T/N}$ are finite. However, at T_c, the increase of $\sigma^2(M)N$ and $\sigma^2(E)$ with N is stronger, proportional to N^{1+x} and N^{1+y}, respectively, where $x, y > 0$ are critical exponents; therefore, C_V/N and χ_T/N diverge.

This increase in the fluctuations (σ^2) at T_c is a manifestation of the arrangement of spins on the lattice. Thus, close to T_c, the typical lattice configuration consists of large (comparable) "droplets" of same spin (+ or −), which lead to $M \approx 0$. As T_c is approached, the size of these droplets increase without limit leading to the corresponding increase in the fluctuations. Another measure of the droplet effect is the correlations between spins as a function of distance,

$$\text{corr}(i,k) = \frac{<\sigma_i\sigma_k> - <\sigma>^2}{<\sigma^2> - <\sigma>^2}. \tag{7.4}$$

In this equation, $<\sigma>$ is M. Thus, two spins on the same droplet will have the same sign, and therefore, they will be correlated. Because as $T \to T_c$, the droplet size increases, the correlation length, ξ, also increases:

$$\xi \propto \frac{T_c}{(T-T_c)^\nu} \tag{7.5}$$

where ν is another critical exponent (Figure 7.6).

The set of critical exponents α, β, γ, and ν defined above (and others) are not independent, but are connected by exact relations and inequalities. An important property is that these exponents are universal, that is, (in most cases) they only depend on the dimension. For example, Ising models defined on different 2D lattices (e.g., square, honeycomb, and triangular) have the same set of exponents; the effect of the lattices

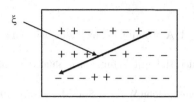

FIGURE 7.6   Correlation length, ξ, expresses the average distance between two spins on the same droplet.

TABLE 7.1

Critical Exponents for Second Order Phase Transition for Ising Models in
Dimensions, $d = 2$, 3, and $d \geq 4$[4][a]

Property	Exponent	$d = 2$	$d = 3$	$d \geq 4$ (mean field)
Specific heat, C_V	α	0	0.110(1)	0
Susceptibility, χ	γ	7/4	1.2372(5)	1
Magnetization, M	β	1/8	0.3265(3)	1/2
Correlation length, ξ	ν	1	0.6301(4)	1/2

[a] $\alpha = 0$ means logarithmic divergence. The errors are defined in parentheses, e.g., $0.110(1) =$
    0.110 ± 0.001.

appears in the different prefactors (that are not specified in Equations (7.2) through (7.5)). In Table 7.1, we
provide the critical exponents for Ising models in dimensions 2, 3, and 4; $\alpha = 0$ in $d = 2$ means that the
divergence of the specific heat in logarithmic. Notice that *approximate* values of critical exponents can be
obtained by the "mean field" approximation, where only short-range correlations are considered. These values, however, become exact for $d \geq 4$. As pointed out above, a polymer chain (like a magnetic or a fluid system at T_c) is also a critical system with characteristic critical exponents, α, β, γ, and ν (see Chapters 8 and 9).

PROBLEM 7.1

(a) The energy of a system is $E = Nk_BT^{-3}$. Can this system undergo second-order phase transition at a
finite temperature? (b) Can a one-dimensional oscillator undergo a second-order phase transition? (c)
A first-order phase transition occurs at critical temperature $T_c = 310K$, and the latent heat is 2 kcal/mol.
What is the jump in entropy? (d) What is the change in the specific heat when an ideal gas is heated from
300 K to 350 K?

PROBLEM 7.2

The average energy of a $1- d$ Ising model is $E = -NJ \tanh (J/kT)$, where $J > 0$ is a constant energy
parameter. Show that this model has a second-order phase transition only at $T = 0$ (hint; calculate C_V).

PROBLEM 7.3

A crystal consists of N atoms. If n atoms ($1 << n << N$) are displaced from lattice sites inside the crystal
to the sites on the surface, the crystal becomes imperfect. If $w >> k_BT$ is the energy necessary to displace
a single atom, show that in equilibrium at temperature T, $n \sim N \exp-[w/k_BT]$.

PROBLEM 7.4

Assume a system consisting of N independent particles—each can have two energy levels $-\varepsilon$ and ε. Find
the average energy ($E < 0$), specific heat, and entropy at temperature T. Solve this problem by three methods; compare with the example solved in Section 5.3.

REFERENCES

1. L. Onsager. Crystal statistics. I. A two-dimensional model with an order-disorder transition. *Phys. Rev.* **65**, 117–149 (1944).
2. B. Kaufmann. Crystal statistics. II. Partition function evaluated by spinor analysis. *Phys. Rev.* **76**, 1232–1243 (1949).
3. K. Huang. *Statistical Mechanics.* (John Wiley & Sons, New York, 1963).
4. H. E. Stanley. *Introduction to Phase Transitions and Critical Phenomena.* (Oxford, New York, 1971).

8

Ideal Polymer Chains

8.1 Models of Macromolecules

Proteins, nucleic acids, or polysaccharides (e.g., starch) are long molecules, consisting of specific sequences of relatively small molecular units (e.g., the 20 amino acids residues of proteins). For many proteins in vivo, the sequence determines a well-defined 3-dimensional ($3d$) structure (protein folding), which is mandatory for the protein's biological function. On the other hand, many synthetic polymers (e.g., polyethylene) and polysaccharides consist of the same repeating unit (monomer). Such a simple sequence does not lead to a definite $3d$ structure, but to a (spaghetti-like) highly flexible molecule— called a random coil. However, under unfolding conditions, proteins can also lose their structure to become flexible. Therefore, to understand the structural behavior of macromolecules (in particular, the folding-unfolding mechanism of proteins and nucleic acids), one should first study the properties of flexible polymers, which is done in this chapter and the next one.

The general model for a polymeric system consists of a set of n_p polymers and n_s solvent molecules (e.g., water) in a container of volume, V. The potential energy of the system is based on intra-polymer, polymer-polymer, polymer-solvent, and solvent-solvent interactions. However, we shall be interested mainly in the very dilute regime, where a polymer chain "meets" another one very rarely; therefore, it is appropriate to treat a *single* polymer surrounded by solvent molecules. (This level of modeling is used for treating proteins that do not aggregate, see Chapter 16).

The accuracy of the potential energy depends on the questions asked. If one is interested in the specific $3d$ structure of a protein, the potential energy function (also called force field—see Chapter 16) will be complicated, based typically on electrostatic, van der Waals, hydrogen-bond, and other interactions. An *analytical* solution of the partition function and other thermodynamic and geometrical parameters is impossible, and the only available approach is numerical, e.g., applying Monte Carlo or molecular dynamics techniques (Chapters 10, 14, and 15).

On the other hand, if one is interested in *global* properties, such as the shape and size of a *polymer*, relatively simple lattice models are sufficient, where the solvent is treated *implicitly* and only the coordinates of the polymer are considered. We shall treat polymer models of increasing complexity on a lattice starting with the simplest model called an *ideal chain* because monomer-monomer interactions do not exist and the only restriction is the chain connectivity. Notice that in such single chain models, the volume is not defined and the factor $N!$ does not appear, therefore, the ensemble is defined by all the configurations of the chain and their Boltzmann probabilities.

8.2 Statistical Mechanics of an Ideal Chain

We study an ideal chain of N bonds ($N + 1$ monomers) on a d-dimensional lattice, where the chain starts from the origin (Figure 8.1). The only condition satisfied by this model is the connectivity of its monomers, while kinetic energy and non-bonded interactions among the monomers do not exist. In other words, the *excluded volume* interaction is neglected, meaning that a lattice site can occupy any number of monomers.

We emphasize again, that no solvent molecules appear in this model. An ideal chain in continuum is called *a freely jointed chain*; it consists of constant bond lengths with no restriction on the bond-bond angles. Finally, while an ideal chain seems to be a very unrealistic model, it captures many features of real polymers, as discussed in what follows.

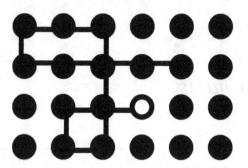

FIGURE 8.1   An ideal chain of 14 bonds (steps), that is, 15 monomers starting from the origin on a square lattice. The chain can intersect itself and go on itself.

8.2.1 Partition Function and Thermodynamic Averages

For the N-bond chain defined above, the ensemble of configurations can be considered as a sample space, where the potential energy of any configuration, i, is, $E_i = 0$; since $\exp[-E_i/k_BT] = 1$, the configurations are equally probable. This ensemble can be built step-by-step. The first bond (of size a) can be chosen in $2d$ possible directions, where this number of directions is also available for each of the remaining $N-1$ steps. The partition function, Z, is the total number of configurations, $(2d)^N$, and the Boltzmann probability and the entropy (which are also independent of the temperature, T) can be obtained trivially,

$$Z = \sum_{i=1}^{(2d)^N} 1 = (2d)^N \quad P_i^B = \frac{1}{Z} = \frac{1}{(2d)^N} \tag{8.1}$$

$$S = -k_B \sum_i P_i^B \ln P_i^B = Nk_B \ln(2d) \tag{8.2}$$

The fact that P_i^B is constant means that the fluctuation of S (which is also the free energy) is zero, as expected [see Equation (5.6)].

An essential interest in polymer physics is to calculate global geometrical properties, such as the root mean square end-to-end distance (ETED), R or the radius of gyration, R_g (Section 8.4). We denote the specific set of N directions for configuration i by the N vectors $\mathbf{a}_t$, $t = 1, 2,...,N$, of equal length, $a = |\mathbf{a}_t|$. As is evident from Figure 8.2, the ETED vector $\mathbf{R}$ is the sum of all $\mathbf{a}_t$, $\mathbf{R} = \Sigma_t \mathbf{a}_t$, where one can show (see below) that for $t \neq l$, $<\mathbf{a}_t . \mathbf{a}_l> = 0$ and $<\mathbf{a}_t . \mathbf{a}_l> = 1$ for $t = l$ [Because (in $2d$) the four scalar products of any horizontal or perpendicular unit vector, $\mathbf{i}$, with the four different unit vectors $\mathbf{1},...,\mathbf{4}$ (Figure 8.3) lead to 1, 0, and -1]. A similar proof applies to any d and to a continuum chain.

Therefore, while $<\mathbf{R}> = 0$ its fluctuation, $<R^2>$ is

$$<R^2> = <\mathbf{R} \cdot \mathbf{R}> = \sum_{t,l} <\mathbf{a}_t \cdot \mathbf{a}_l> = \sum_t <\mathbf{a}_t \cdot \mathbf{a}_t> = Na^2. \tag{8.3}$$

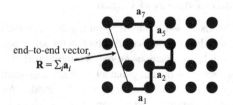

FIGURE 8.2   An illustration of an ideal chain of nine monomers or eight steps (bonds) represented by the vectors $\mathbf{a}_1, \mathbf{a}_2, ..., \mathbf{a}_8$ on a lattice. The vector connecting monomer 1 to monomer 9 is the end-to-end vector.

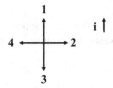

FIGURE 8.3   The four basic vectors $\mathbf{a}_i$ for a chain in two dimensions.

Thus, for any chain length, N and *any dimension d*,

$$R = \sqrt{<R^2>} = N^{1/2}a, \tag{8.4}$$

which is an important result. Thus, even though the chain can intersect itself and go on itself many times, its global shape increases with N as $N^{1/2}$; in other words, the relative number of the *open* configurations increases with N, thus dominating the statistical average, which leads to increasing R. The exponent ½ in Equation (8.4) is a critical exponent denoted by ν, related to ν defined in Equation (7.5) for the Ising model; thus, for an ideal chain $\nu = 1/2$.

8.3 Entropic Forces in an One-Dimensional Ideal Chain

Thus far, we have calculated a single thermodynamic parameter for an ideal chain—its entropy. We have also calculated the root mean square end-to-end distance as a measure for the global shape of the chain as N grows. From the problem solved below, we shall see that additional information can be drawn from the relation between *local entropies* and the ETED. We provide three solutions for this problem, which again demonstrate the flexibility in handling problems in statistical mechanics.

Example

A one-dimensional ideal chain of N bonds, each of length a, starts from the origin; the distance between the chain ends (ETED $\equiv R$) is denoted by x. Find the entropy as a function of x and the relation between the temperature and the force required to hold the chain ends at x [1,2] (Figure 8.4).
   Comments: Because the chains in the ensemble are equally probable, our focus is on the number of chain configurations $w(x)$ for a given ETED value, x. We shall see that the maximal $w(x)$ is obtained for $x = 0$, while $w(x) = 1$ for a completely stretched configuration ($x = N$); thus, $w(x)$ decreases as x is increased. $w(x)$ is a *local* partition function (related to x), which due to the lack of energy, leads to a local entropy, $S(x,N) = k_B\ln[w(x)]$ and to the corresponding Helmholtz free energy, $F(x,T,N) = -TS(x,N)$; $S(x,N)$ and $F(x,T,N)$ decrease and increase, respectively, with increasing x. Now, if a fluctuation $x > 0$ occurs, a (negative) *statistical force*, $-K$ (which increases with increasing x) is created to return x back toward $x = 0$; thus to hold the ETED at x, one needs to exert an exactly opposite force, $+K$, where $K = (\partial F/\partial x)_{T,N}$[compared with $p = (\partial F/\partial V)_T$].

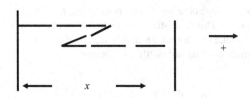

FIGURE 8.4   An $1d$ ideal chain plotted along the positive x-axis; however, the chain can go to the negative part as well.

Solution 1

We treat this problem using the most probable term method. Thus, x is defined by $x = (N_+ - N_-)a$, where N_+ and N_- are the number of bonds in the positive (+) and negative (–) directions, respectively, and $N = N_+ + N_-$; therefore,

$$N_+ = (Na + x)/2a = (N + x/a)/2 \qquad N_- = (Na - x)/2a = (N - x/a)/2. \qquad (8.5)$$

The number of chain combinations for given N_+ and N_- (that is, at a given *positive* x) is:

$$w(x) = \frac{N!}{N_+! N_-!},$$

which using the leading term of Stirling's formula ($\ln N! \sim N \ln N$), leads the entropy, $S(x)$

$$S(x) = k_B \ln w(x) = k_B (N \ln N - N_+ \ln N_+ - N_- \ln N_-). \qquad (8.6)$$

Substituting N_+ and N_- [Equation (8.5)] leads to:

$$S(x) = N k_B \left\{ \ln 2 - \frac{1}{2}\left(1 + \frac{x}{Na}\right)\ln\left(1 + \frac{x}{Na}\right) - \frac{1}{2}\left(1 - \frac{x}{Na}\right)\ln\left(1 - \frac{x}{Na}\right)\right\}. \qquad (8.7)$$

From
$$K = \frac{\partial F}{\partial x} = -T\frac{\partial S}{\partial x} \qquad (8.8)$$

one obtains

$$K = \frac{k_B T}{2a}\ln\left(\frac{1 + \dfrac{x}{Na}}{1 - \dfrac{x}{Na}}\right) = \frac{k_B T}{2a}\ln\left\{\left(1 + \frac{x}{Na}\right)\left(1 + \frac{x}{Na} + \frac{x^2}{N^2 a^2} + \dots\right)\right\}, \qquad (8.9)$$

where the denominator has been expanded according to $1/(1 - q) = 1 + q + q^2 + \dots$ for $q = x/Na < 1$, that is, for a much smaller x than the stretched chain, Na. For $x/2a \ll 1$, it is justified to take only the leading terms in the expansion (1 and $2x/Na$), thus,

$$K = \frac{k_B T}{2a}\ln\left(1 + \frac{2x}{Na} + \dots\right) = \frac{k_B T}{2a}\frac{2x}{Na} + \dots = \frac{k_B T}{Na^2}x + \dots, \qquad (8.10)$$

where we have also used the expansion, $\ln(1 + y) \cong y$ for $y \ll 1$.

Therefore, for $x \ll Na$, we have obtained the corresponding K, which obeys Hooke's law where $k_B T/Na^2$ is the force constant. This derivation applies to rubber elasticity, which is mainly caused by entropic effects. The value of x for which $S(x)$ is maximal can be found from the condition, $\partial S/\partial x = 0$; based on Equations (8.8) and (8.10) one obtains,

$$\frac{K}{T} = \frac{\partial S}{\partial x} = \frac{k_B}{2a}\ln\left(1 + \frac{2x}{Na} + \dots\right) = 0 \qquad (8.11)$$

which for $x \ll Na$, leads to $x = 0$ (the second derivative of S with respect to x is negative at $x = 0$). For $x = 0$, $N_+ = N_- = N/2$, which means that the number of chain configurations for $x = 0$ is,

$$w(x = 0) = \frac{N!}{(N/2)!(N/2)!} \tag{8.12}$$

and using Stirling's formula, leads *wrongly* (due to its approximation) to the conclusion that the number of configurations for $x = 0$ is equal to the total number of configurations (2^N),

$$\ln w(x = 0) = N \ln N - N \ln(N/2) = N \ln 2 = S/k_B. \tag{8.13}$$

Anyhow, $x = 0$ is the most probable ETED with the maximal entropy; thus, to "hold" the chain ends at a distance $x > 0$, force should be applied against the "will" of the chain to remain in its most probable state ($x = 0$). In other words, the force is required for decreasing the chain entropy. Also, Equation (8.10) shows that the force increases linearly with T, because the free energy ($-TS$) at $x = 0$ decreases, that is, the system becomes more stable at a higher T, and it is difficult to stretch it to a larger x. Finally, the force is proportional to $1/(Na^2)$, meaning that the longer is the chain, the easier it is to stretch (this is in agreement with the fact that the Gaussian shape becomes flatter with increasing N (see Section 8.6)).

It should be pointed out that Equation (8.10) can also be derived in two other ways within the framework of the "Gibbs" (K, T, N) ensemble, rather than the (N, x, T) ensemble used above (these ensembles correspond to the NpT and NVT ensembles, respectively). These solutions demonstrate again the equivalence of different ensembles for calculating averages of *large systems*, as discussed in Section 6.5.

Solution 2

The starting point is the partition function,

$$Z(K,T,N) = \sum_{n_+} \frac{N!}{(N - N_+)! N_+!} \exp \frac{Ka}{k_B T} [2N_+ - N] \tag{8.14}$$

where $(2N_+ - N)a = x$; notice that in the usual Gibbs ensemble, the summand is $\exp(-pV/k_B T)$, while here it is $\exp(+Kx/k_B T)$. Taking the logarithm of one term and using Stirling's approximation leads to:

$$\ln[Z(K,T,N)] = \frac{KaN}{k_B T} + N \ln N - (N - N_+) \ln(N - N_+) - N_+ \ln N_+ + \frac{K2a}{k_B T} N_+. \tag{8.15}$$

Derivation with respect to (N_+), and equating to 0 leads to the maximal term (for N_+, that is, for x) for given K and T; one obtains,

$$\ln \frac{(N - N_+)}{N_+} = -\frac{K2a}{k_B T}. \tag{8.16}$$

Replacing N_+ by x [Equation (8.5)] leads to:

$$\ln\left(1 - \frac{x}{Na}\right)\left(1 - \frac{x}{Na}\right) = \ln\left[1 - \frac{2x}{Na}\right], \tag{8.17}$$

which for $x/Na \ll 1$ leads to Equation (8.10),

$$K = \frac{2x k_B T}{2aNa} = \frac{k_B T}{Na^2} <x> \tag{8.18}$$

Solution 3

This solution starts again from the partition function defined in Equation (8.14), which is based on the binomial expansion,

$$(1+x)^N = \binom{N}{0} + \binom{N}{1}x + \ldots + \binom{N}{r}x^r + \ldots + \binom{N}{N}x^N,$$

becomes,

$$Z(K,T,N) = \exp-\frac{KaN}{k_BT} \sum_{N_+} \frac{N!}{(N-N_+)!N_+!}\exp\frac{K2a}{k_BT}N_+ = \exp-\frac{KaN}{k_BT}\left(1+\exp\frac{K2a}{k_BT}\right)^N. \qquad (8.19)$$

The free energy is,

$$G(K,T,N) = -k_BT \ln Z(K,T,N) = KaN - k_BTN \ln\left(1+\exp\frac{K2a}{k_BT}\right). \qquad (8.20)$$

The derivative of G with respect to K leads to $<x> = <(2N_+ - N)a>$:

$$\left(\frac{\partial G}{\partial K}\right)_{T,n} = aN - k_BTN \frac{\frac{2a}{k_BT}\exp\frac{K2a}{k_BT}}{\left(1+\exp\frac{K2a}{k_BT}\right)} = aN - k_BTN \frac{\frac{2a}{k_BT}}{\left(1+\exp-\frac{K2a}{k_BT}\right)} = aN\left[1-\frac{2}{\left(1+\exp-\frac{K2a}{k_BT}\right)}\right]. \qquad (8.21)$$

For a small h, one can approximate, $\exp(-h) \approx 1 - h + \ldots$ and $1/(1 - h) \approx 1 + h + \ldots$; therefore, for small, $\frac{K2a}{k_BT}$, one obtains:

$$<x> = aN \frac{Ka}{k_BT} \quad \text{or} \quad K = \frac{k_BT}{Na^2}<x> \qquad (8.22)$$

which is the same result obtained in Equation (8.10).

8.4 The Radius of Gyration

Besides the ETED, the global size of the entire chain can also be expressed by the radius of gyration, R_g,

$$R_g^2 = \frac{1}{N+1}\sum_{i=0}^{N} s_i^2 \qquad (8.23)$$

where s_i is the distance of monomer i (with mass m_i) from the center of mass, c_M (Figure 8.5),

$$c_M = \frac{\sum m_i \mathbf{x}_i}{\sum m_i}. \qquad (8.24)$$

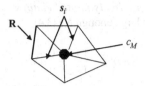

FIGURE 8.5 The center of mass, c_M, and the distances, s_i, lead to the radius of gyration, R_g; the end-to-end vector **R** is bold-faced.

It can be shown ([3], Appendix A) that for a system of equal masses:

$$R_g^2 = \frac{1}{(N+1)^2} \sum_{0 \le i \le j \le N} r_{ij}^2 \qquad (8.25)$$

where r_{ij} are distances between monomers i and j.

It should be pointed out that R_g^2 in Equation (8.25) is based on all r_{ij}, whereas the ETED is a single distance; therefore, in simulations, $<R_g^2>$ converges faster than ETED. For a sphere of radius R, the radius of gyration is:

$$\langle R_g^2 \rangle = \frac{1}{\text{Volume}} \iiint_0^R r^4 \sin\theta d\theta d\phi dr = \frac{3 \cdot 4\pi}{4\pi R^3} \int_0^R r^4 dr = \frac{3}{R^3} \frac{R^5}{5} = \frac{3}{5} R^2 \Rightarrow \qquad (8.26)$$

$$\sqrt{\langle R_g^2 \rangle} = \sqrt{\frac{3}{5}} R \approx 0.77 R$$

Thus, for a protein with approximately a spherical shape of radius, r, $\sqrt{\langle R_g^2 \rangle}$ is ~0.8r.

8.5 The Critical Exponent ν

It was shown in Equations (8.3) and (8.4) that for an ideal chain on a lattice and in continuum, $<R> = 0$, but

$$<R^2> = a^2 N \qquad <R^2>^{1/2} = aN^{1/2} \qquad (8.27)$$

where $\nu = 1/2$ is a critical exponent that characterizes the global growth of the chain with increasing N. Since the radius of gyration is also a parameter of the chain's global shape, one would expect the corresponding ν to be 1/2 as well; indeed, one can show ([3], Chapter 1) that:

$$<R_g^2> = \frac{<R^2>}{6} \Rightarrow <R_g^2>^{1/2} = \frac{aN^{1/2}}{\sqrt{6}} \qquad (8.28)$$

where the growth of $<R_g^2>$ differs from that of $<R^2>$ only by the prefactor, 1/6. The $N^{1/2}$ scaling means that the *open* chain configurations dominate both the growth of $<R^2>$ and $<R_g^2>$ for *any* dimension, $d = 1, 2, 3 \ldots$(one should recall that for the Ising model [Equation (7.5)], ν describes the increase of the spin-spin correlation length due to the increase of the spin droplets as $T \to T_c$).

Adding various *short-range* geometrical restrictions to an ideal chain might change the prefactor, but the exponent $\nu = 1/2$ will not be affected, as long as long-range interactions are not existent (e.g., the excluded volume interaction). Thus, for a *freely jointed chain* with constant bond angles, θ (called *the freely rotating chain*), one obtains (for large enough N) [3,4],

$$\langle R^2 \rangle^{1/2} = \frac{1 + \cos\theta}{1 - \cos\theta} aN^{1/2}. \tag{8.29}$$

The first factor on the right hand side of Equation (8.29) determines the stiffness of the chain. For example, for θ = 60°, $<R^2>^{1/2} = 3N^{1/2}a$, while for θ = 90°, $<R^2>^{1/2} = N^{1/2}a$ — like for an ideal chain on a lattice or a freely jointed chain without restrictions. In the θ = 60° case, the prefactor grows from 1 to 3, meaning that the size of the shortest possible loop increases as well [see Equation (14.10)]. Notice, that when the prefactor is equal to 1, the relation, $<R^2>^{1/2} = aN^{1/2}$, is satisfied already for $N = 1$, whereas for a larger prefactor, Equation (8.29) will be satisfied for a larger N.

The freely rotating chain can be rescaled into a freely jointed chain by defining a *statistical segment*, b, consisting of several a segments, where neighbor b segments along the chain are *uncorrelated*; clearly N should be decreased appropriately to M units of b segments, where $<R^2>$ is *unchanged*, thus:

$$< R^2 >^{1/2} = bM^{1/2} \qquad M < N; \qquad b > a \tag{8.30}$$

An additional condition is that the fully stretched original and rescaled chains have the same ETED length. b is called *Kuhn's statistical segment length*. For θ = 60°, one obtains $b = 28a$ and $M = N/81$ (exercise). The relation, $<R^2>^{1/2} \sim N^{1/2}a$, holds for *any global* chain parameter (e.g., the radius of gyration); it holds not only for the chain ends, but for any pair of internal monomers that are not too close. In other words, the chain's "tails" do not change the $N^{1/2}$ scaling, while the prefactor might be affected.

8.6 Distribution of the End-to-End Distance

For an ideal chain, the probability density for an ETED value, x, can be obtained *exactly* ([3], Chapter 8), and an excellent approximation for large N is the normal distribution. It is easy to obtain this Gaussian for the $d = 1$ lattice chain of N bonds studied in Section 8.3 (Figure 8.4). For simplicity, we define the bond length a to be 1. One has to calculate, $P(x)$, the ratio between the number of chains with ETED = x and the total number of chains, 2^N. We write this expression exactly, taking its logarithm and applying Stirling's formula; the result is then exponentiated back to give a Gaussian distribution. As before [Equation (8.5)], we define, N_+ and N_-,

$$N_+ = (N + x)/2 \qquad N_- = (N - x)/2 \tag{8.31}$$

The ratio for $x > 0$ is,

$$P(x) = \frac{N!}{2^N \left(\dfrac{N+x}{2}\right)! \left(\dfrac{N-x}{2}\right)!}. \tag{8.32}$$

We take the logarithm of both sides and apply Stirling's formula for a large N,

$$\ln P(x) = N \ln N - \frac{N+x}{2} \ln \frac{N+x}{2} - \frac{N-x}{2} \ln \frac{N-x}{2} - N \ln 2 \tag{8.33}$$

$$= N \ln N - \frac{N}{2} \ln N^2 \left(1 - \frac{x^2}{N^2}\right) - \frac{x}{2} \ln \left(\frac{1 + \dfrac{x}{N}}{1 - \dfrac{x}{N}}\right)$$

For $x \ll N$, $\ln(1 + x/N) \sim x/N$ and $1/(1 - x/N) = 1 + x/N + (x/N)^2 +...$, thus:

$$-\frac{N}{2}\ln N^2(1 - \frac{x^2}{N^2}) = -N\ln N + \frac{N}{2}\frac{x^2}{N^2} \tag{8.34}$$

and
$$-\frac{x}{2}\ln\left(\frac{1 + \frac{x}{N}}{1 - \frac{x}{N}}\right) = -\frac{x}{2}\ln\left(1 + \frac{x}{N}\right)\left(1 + \frac{x}{N}\right) = -\frac{x}{2}\ln\left(1 + \frac{2x}{N} +..\right) = -\frac{x}{2}\frac{2x}{N}. \tag{8.35}$$

One obtains,

$$\ln P(x) \cong \frac{x^2}{2N} - \frac{x^2}{N} = -\frac{x^2}{2N} \quad \Rightarrow$$

$$\tag{8.36}$$

$$P(x) \propto \exp\left(-\frac{x^2}{2N}\right) \cong \exp\left(-\frac{x^2}{2\langle x^2\rangle}\right).$$

Due to the approximation introduced, this Gaussian should still be normalized; thus:

$$P(x) = w(x) = \frac{1}{(2\pi\langle x^2\rangle)^{1/2}} \exp\left(-\frac{x^2}{2\langle x^2\rangle}\right). \tag{8.37}$$

The above derivation applies also to $-x$, which is reflected by the fact that x fluctuates symmetrically around $x = 0$. As expected, the maximal probability is for $x = 0$, where the entropy is maximal. It should be emphasized again that $P(x)$ is defined for a large N, and $|x| << N$. The Gaussian becomes flatter with increasing N, since the variance is $<x^2> = N$. This Gaussian is approximate since it is non-zero also for $x > Na$ — the fully stretched chain! However, it is a very good approximation for large N and $|x| << N$.

8.6.1 Entropic Forces Derived from the Gaussian Distribution

The entropy, $S(x)$, can be obtained from the Gaussian, $w(x)$, obtained in Equation (8.37), $S(x) = -k_B\ln w(x)$, which leads to the free energy, $F(x) = -TS(x)$. Since $K = \partial F(x)/\partial x$, one can easily calculate the force K from $F(x)$,

$$F(x) = -T\ln S(x) = k_B T\left[-\ln(2\pi <x^2>^{1/2}) - \frac{x^2}{2<x^2>}\right] \tag{8.38}$$

and the force required to *hold* the ETED at x is:

$$K = -\frac{\partial F}{\partial x} = k_B T\frac{x}{<x^2>} = \frac{k_B T}{Na^2}x \tag{8.39}$$

which is the same result obtained in Equation (8.10). Because the Gaussian becomes flatter as N is increased ($<x^2> = N$), the corresponding difference between the entropy at $x = 0$ and $|x| > 0$ decreases, and the force K decreases as well.

8.7 The Distribution of the End-to-End Distance Obtained from the Central Limit Theorem

For a freely jointed chain in $d = 3$, the following equation has been obtained [3]:

$$P'(r)dr = \left(\frac{3}{2\pi\langle r^2\rangle}\right)^{3/2} \exp\left[-\frac{3r^2}{2\langle r^2\rangle}\right]4\pi r^2 dr \tag{8.40}$$

$$\Big|----------- \Big| W(r)$$

$P'(r)$ is the probability for a chain starting from the origin to find its end in a spherical shell defined by the radii r and $r + dr$; it is *zero* at $r = 0$ (we replace the letter R for the ETED by r and the letter N by n). The first part $[W(r)]$ is the probability density per volume (as obtained in the $d = 1$ case), which is *maximal* at $r = 0$. The 4π comes from integrating over the angles φ and θ. The Gaussian distribution exists also for *internal* monomers i, j, which are distant enough along the chain. Next, we derive Equation (8.40) by the central limit theorem. For convenience, we start by stating the central limit theorem, which has been discussed already in Section 1.15:

If X_1, X_2,...,X_n are random variables with the mean values, μ_1, μ_1, ..., μ_n and variances $\sigma_1^2, \sigma_2^2, \cdots, \sigma_n^2$ and we define,

$$Y_n = \frac{X_1 + X_2 + \cdots X_n - \mu_1 - \mu_2 \cdots - \mu_n}{\sqrt{\sigma_1^2 + \sigma_2^2 + \cdots + \sigma_n^2}} \tag{8.41}$$

then, under a wide range of conditions, Y_n is normally distributed: for $n \to \infty$:

$$P(Y_n \leq x) \to \frac{1}{\sqrt{2\pi}} \int_{-\infty}^{x} \exp[-\frac{t^2}{2}]dt \tag{8.42}$$

with mean = 0 and variance = 1.

For a $d = 1$ chain, we change the notation of the chain length from N to n and define a random variable $X_i = \pm 1$ with equal probability ½. The expectation value is $\mu_i = 1/2 - 1/2 = 0$ and the variance is $\sigma_i^2 = 1$, therefore,

$$Y_n = \frac{X_1 + X_2 + \cdots + X_n - \mu_1 - \mu_2 \cdots - \mu_n}{\sqrt{\sigma_1^2 + \sigma_2^2 + \cdots + \sigma_n^2}} = \frac{X_1 + X_2 + \cdots + X_n}{\sqrt{n}} = \frac{x}{\sqrt{n}} \tag{8.43}$$

and Equation (8.42) becomes,

$$P\left(\frac{x}{n^{1/2}} \leq \pm t\right) = \frac{1}{\sqrt{2\pi}} \int_{-\infty}^{t} \exp\left[-\frac{t'^2}{2}\right]dt'. \tag{8.44}$$

The probability density for $x/n^{1/2} = \pm t$ is:

$$P\left(\frac{x}{n^{1/2}} = \pm t\right) = \frac{\partial}{\partial x}\left[\frac{1}{\sqrt{2\pi}} \int_{-\infty}^{\frac{x}{n^{1/2}}} \exp\left[-\frac{t'^2}{2}\right]dt'\right] = \tag{8.45}$$

$$= \frac{1}{\sqrt{2\pi}}\exp\left[-\frac{\left(\frac{x}{n^{1/2}}\right)^2}{2}\right]\frac{\partial}{\partial x}\left[\frac{x}{n^{1/2}}\right] = \frac{1}{\sqrt{2\pi n}}\exp\left[-\frac{x^2}{2n}\right]$$

and defining, $w(x) = P(x)$, we finally obtain the result of Equation (8.37),

$$w(x) = \frac{1}{(2\pi <x^2>)^{1/2}} \exp\left[-\frac{x^2}{2<x^2>}\right].$$
(8.46)

In higher dimensions, one defines $w(r_i)$ for each coordinate r_i (for $d = 3$, r_i is one of the coordinates x, y, and z), where the different $w(r_i)$ functions are uncorrelated, and thus (for large n) each will be visited *on average $n/3$ times*, therefore $<r_i^2> = n/3$ and $w(r_i)$ is:

$$w(r_i) = \frac{1}{(2\pi n/3)^{1/2}} \exp\left[-\frac{3r_i^2}{2n}\right].$$
(8.47)

$W(\mathbf{r})$ is a product of the different $w(r_i)$, and in $d = 3$, one obtains,

$$W(\mathbf{r}) = \frac{1}{(2\pi n/3)^{3/2}} \exp\left[-\frac{3r^2}{2n}\right] = \frac{1}{(2\pi <r^2>/3)^{3/2}} \exp\left[-\frac{3r^2}{2<r^2>}\right]$$
(8.48)

where $W(\mathbf{r}) = w(x)w(y)w(z)$ and $<x^2> = <y^2> = <z^2> = <r^2>/3$.

In this derivation, there is no restriction on r_i to be much smaller than n, but the derivation is for $n \to \infty$; also, in $d = 3$, $W(\mathbf{r})$ is maximal at $\mathbf{r} = 0$, while the radial distribution of r, $P'(r)$ is zero [Equation (8.40)],

$$P'(r)dr = W(\mathbf{r})4\pi r^2 dr.$$
(8.49)

The exact probability distribution was calculated analytically (see [3], Chapter 8 & Appendix F), and the difference between the Gaussian and the exact results becomes negligible (for the r values of interest) already for short chains of $N = 30$ (see figures in [3], Chapter 8).

We have already pointed out that for ideal chains on hypercubic lattices ($d \geq 1$), and for the freely jointed chain, the relation, $<R^2> = aN$ is already exact for $N = 1$; however, the corresponding Gaussian distributions hold only for large N. This is an example for the effect of system size on thermodynamic functions. In this case, the *practical* thermodynamic limit still occurs for relatively short chains. For chains with excluded volume, this limit is attained *practically* for much larger N.

PROBLEM 8.1

N monomeric units are ordered along a straight line and create a chain. Every unit can be in states where its length is a or b with energy E_α or E_β, respectively. Find the relation between the chain length, X, and the force, K, between the chain ends. Like the problem treated in Section 5.3 (involving spin up and down), there are several ways to solve this problem. See also the various ways the chain model defined in Section 8.3 has been treated, that is within the (T, K, N) and (T, X, N) ensembles. Solve this problem at least by two methods.

Solution

$X = N(aA + bB)/(A + B)$, where $A = \exp[(Ka - E_a)/k_B T]$ and $B = \exp[(Kb - E_b)/k_B T]$.

8.8 Ideal Chains and the Random Walk

There is one-to-one correspondence between an ideal chain and a random walk. On a lattice, for example, the walker starts from the origin and "jumps" to one of the nearest neighbor sites with an equal probability and continues jumping in the same way. The pathway created by the walker defines an ideal chain configuration. The decision of the walker where to go does not depend on previous jumps, that is, no memory is involved in this process. Every pathway of the walker has the same probability, $(1/2d)^N$ as that of an ideal chain. Indeed, it is known from the theory of Brownian motion [e.g., Equation (12.66)] that the average distance squared, $<S^2>$, of a walker is

$$< S^2> = 2Dt = \frac{2k_BT}{f}t \qquad (8.50)$$

where D is the diffusion constant, f is the viscosity constant, and t is time. Therefore, N corresponds to t and $<R^2>$ to $<S^2>$; it is also known that $<S^2>$ is normally distributed.

8.9 Ideal Chain as a Model of Reality

The entropy, $S(x)$, and the corresponding force, K, defined for the ETED are examples for what is known as the *potential of mean force* (PMF), which is the contribution to the free energy of configurations that satisfy a certain geometrical restriction (see Chapter 20). In our case, the restriction is a certain value of the ETED denoted by x. Derivation of the PMF (free energy) with respect to the corresponding restriction (x), leads to the *average* force required to hold the system at the restricted geometry.

Rubber elasticity stems from such entropic effects (see Chapters 13 and 21 of [5]). Rubber consists of polymer chains connected by cross-links. When the rubber is stretched, the bond angles and lengths are not changed (or broken). The main effect is the chains' ordering, which leads to the decrease of their entropy. When the rubber returns to its normal length, the energy invested in the chains' stretching is released in the form of heat ($T\Delta S$), which can be felt. In this respect, an "entropic spring" differs from the mechanical one.

Calculations of PMF values and the corresponding forces are also carried out for proteins. An example is the huge muscle protein titin (~33,000 residues), which has been stretched by atomic force microscopy, and the forces have been compared to those obtained from PMF derivatives calculated by molecular dynamics simulations (see [6]).

As we have seen, an ideal chain only satisfies one condition—the connectivity between the monomers, while the excluded volume interaction is ignored, that is, the chain unrealistically can intersect itself and go on itself. Also, no potential (attractive) energy is defined. Still, as discussed below, the ideal chain is a useful model, which has been studied extensively, and its importance is perhaps reflected by the various additional names it has acquired in the course of the years—*the Gaussian chain, a random walk*, or *the unperturbed chain*. If the chain is off-lattice (that is, in continuum), it is called *the freely jointed chain* or *the random flight chain*. When the excluded volume interaction is considered, the chain is sometimes called *a real chain*.

In spite of its simplicity, the ideal chain has several advantages. (1) Several of its properties can be solved analytically providing thus exact results, which can help understanding real chains. (2) It provides a model for rubber elasticity. (3) An ideal chain describes to a large extent a polymer at the Flory θ-point (Section 9.5) and (4) it models correctly a polymer in a *dense* multiple-polymer system. Finally, (5) an ideal chain (with short-range interactions) is the basis of the *rotational isomeric state approximation* ([3], Chapter 3).

REFERENCES

1. R. Kubo. *Statistical Mechanics*. (North Holland, Amsterdam, 1974).
2. P. G. de Gennes. *Scaling Concepts in Polymer Physics*. (Cornell University Press, Ithaca, 1979).
3. P. J. Flory. *Statistical Mechanics of Chain Molecules*. (Hanser, New York, 1988).
4. H. Yamakawa. *Modern Theory of Polymer Solutions*. (Harper & Row, New York, 1971).
5. T. L. Hill. *An Introduction to Statistical Thermodynamics*. (Dover, New York, 1986).
6. H. Lu and K. Schulten. The key event in force-induced unfolding of titin's immunoglobulin domains. *Biophys. J.* **79**, 51–65 (2000).

9
Chains with Excluded Volume

9.1 The Shape Exponent ν for Self-avoiding Walks

The ideal chain, while useful in some cases, is not a realistic model since the excluded volume (EV) interaction is not considered; when EV is taken into account, the chain is sometimes called *a real chain*. We now define the self-avoiding walk (SAW) as a model of a real chain. Assume again a chain of N steps (bonds), that is, $N + 1$ monomers anchored to the origin on a square lattice. However, unlike an ideal chain, two monomers are not allowed to share the same lattice site. More specifically, for chain configuration, i, the interaction energy is defined as $E_i = 0$, if i is a SAW, and $E_i = \infty$, if i is a self-intersecting walk, as depicted in Figure 9.1.

The partition function is:

$$Z = \sum_{RW(i)} \exp[-E_i/k_B T] = \sum_{SAWs(i)} 1_i \qquad (9.1)$$

where Z is expressed in two ways—as a summation over *all* the ideal chain configurations i [denoted here as random walks (RWs)] which are multiplied by $\exp[-E_i/k_B T]$, or as a summation over all the SAWs. Thus, Z is the total number of SAWs, which constitute a sub-group of the random walks (ideal chains). Because each SAW contributes 1 to the partition function ($E_i = 0$), the Boltzmann probability of all SAWs is the same,

$$P_i^B = \frac{1}{Z}. \qquad (9.2)$$

An analytical calculation of Z is not available. Z can be calculated exactly with a computer, but only for relatively short SAWs on lattices ($N \sim 80$ on a square lattice) by enumerating all the different SAWs. An approximate value of Z for large N can then be obtained by suitable extrapolations of the exact results or by Monte Carlo simulations.

Because of the (infinite) EV repulsion among the monomers of a SAW, the expansion of the global shape of the chain with increasing N is stronger than for an ideal chain. Thus, the increase of the end-to-end distance (ETED) ($<R^2>^{1/2}$) and the radius of gyration, $<R_g^2>^{1/2}$, are governed by an exponent $\nu \geq \frac{1}{2}$.

$$<R^2>^{1/2} = BN^\nu a \qquad\qquad <R_g^2>^{1/2} = DN^\nu a \qquad (9.3)$$

where B and D are constant prefactors that depend on the specific lattice, the model, and the dimension, d. On the other hand, ν depends only on the dimensionality, that is, a SAW on any two-dimensional lattice (e.g., a square or a triangular lattice) will have the same ν; this is a manifestation of the notion of *universality*. As demonstrated in Table 9.1, an excellent approximation for ν of SAWs is given by Flory's simple formula (see [1,2]),

$$\nu = d + \frac{3}{d+2}. \qquad (9.4)$$

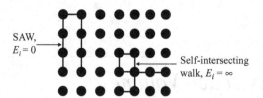

FIGURE 9.1   Two chains on a square lattice: A SAW on the left and a self-intersecting walk on the right.

TABLE 9.1

Results for the Critical Exponent, ν in Dimensions, d

d	ν (Flory)	ν (best)	Method
1	1	1	Exact (trivial)
2	0.75	0.75	Exact solution [3]
3	0.60	$\cong 0.588$	Lattice simulations [4] & theories [5,6]
$d \geq 4$	0.50	0.5	Exact

Note: The results obtained by Flory's formula [Equation (9.4)], Nienhuis exact solution [3], the best Monte Carlo simulation [4], and renormalization group calculations [5,6].

Thus, for $d = 1$, $\nu = 1$ is maximal because the chain can go left or right only; this ν also describes the increase of a rod-like molecule such as an α-helix or a DNA. As d increases, the EV interaction becomes less effective because the chain has more alternative pathways to circumvent the EV "disturbances;" therefore, ν decreases. Thus, for $d = 2$, $\nu = 0.75$ has been obtained from the *exact derivation* of Nienhuis for the honeycomb lattice, and for $d = 3$, the Monte Carlo results, $\nu \sim 0.588$ slightly deviates from Flory's results, $\nu = 0.60$. For $d \geq 4$, the spatial freedom is so large that *practically* the chain does not "feel" the EV restrictions at all, and it thus grows like an ideal chain with random walk (mean field) exponents. $d = 4$ is the upper critical dimension with logarithmic corrections to scaling,

$$< R^2(d \geq 4) >^{1/2} = CN^{1/2}a \quad \Rightarrow \quad \nu = 1/2. \tag{9.5}$$

A SAW in $d = 3$ is sometimes called a *swollen* chain (as compared to the "*thinner*" shape of an ideal chain); a SAW is a model for a polymer in a good solvent, where the solvent attracts the monomers and opens up the chain's configuration.

9.2 The Partition Function

We have seen that the partition function, Z, is the total number of SAWs. Exact enumeration studies and theoretical considerations [2] have led to the following expression for Z,

$$Z = C\mu^N N^{\gamma-1} \tag{9.6}$$

where C is a constant prefactor that depends on the model and the dimension. μ is a model-dependent *growth parameter* (an *effective* coordination number), and γ is a critical exponent that depends only on d (this μ should be distinguished from its other definitions—the chemical potential, the statistical average, etc). To understand this formula, we consider first a RW (ideal chain) on a d-dimensional lattice, where immediate return is forbidden, meaning that $2d - 1$ directions are available for each step besides the first one that has $2d$ possible directions:

$$Z_{\text{RW}} = 2d(2d-1)^{N-1} = \frac{2d}{2d-1}(2d-1)^N \tag{9.7}$$

TABLE 9.2

Results for the Exponent γ in Dimensions, d

d	γ	Method
2	$43/32 = 1.3437...$	Exact (Nienhuis) [3]
3	1.15695300 (95)	Clisby, Monte Carlo [9]
$d \geq 4$	1	Exact—random walk behavior

In this case, $C = 2d/(2d-1)$, $\mu = 2d - 1$, and $\gamma = 1$, since $N^{\gamma-1} = N^0 = 1$. It should be noted that $\mu = 2d - 1$ is the number of possible directions available to the random walker at each step, which are 3 and 5 for a square lattice and a simple cubic lattice, respectively. These results should be compared to the corresponding smaller values, $\mu \cong 2.638$ and $\mu \cong 4.68$, obtained for SAWs by an exact enumeration [7] and a Monte Carlo simulation ([8] and references cited therein). For $d = 4$, $\mu_{SAW} = 2d - 1 = 7$—as for a random walk. For large N, the μ^N term in Equation (9.6) dominates the $N^{\gamma-1}$ term. Again, for a given d, μ depends on the model (lattice), while γ is *universal*. In Table 9.2, results for $\gamma(d)$ are presented for $d = 2$, 3, and 4.

While Z (the total number of SAWs) is expressed as a simple function [Equation (9.6)], it has never been solved analytically. The difficulty lies in the existence of the *excluded volume long-range interactions* among the monomers, which allow, for example, the first monomer to interact with any other monomer including the last one. To illustrate the problem, it is useful to compare the step-by-step generation of a SAW to that of a random walk. For a RW, the transition probabilities at any step k are always the same independent of the decisions made at the past steps, 1, 2, ..., $k - 1$. On the other hand, if one seeks to build a SAW step-by-step, one has to define at step k suitable transition probabilities that take into account the already generated partial chain (the past) and all the possible continuations of the chain in the $N - k + 1$ future steps! In other words, one has to consider the whole ensemble of SAWs already for $k = 2$! (see Section 14.5.1).

In this respect, the SAW model is significantly more difficult to handle than the Ising model, which is based on short-range interactions (among nearest neighbor spins); thus, $\ln Z$ could be calculated exactly by Onsager (for $d = 2$). The difficulty arises at the critical temperature, T_c, where the *correlations* among spins become infinite, spanning the whole system; as said above, for a SAW, the *interactions* rather than the *correlations* are already *long-range*, encompassing the entire chain.

9.3 Polymer Chain as a Critical System

For a second-order phase transition in the Ising model and a fluid, we have defined several critical exponents that describe the divergence of second derivatives of the free energy as the critical temperature, T_c, is approached (Section 7.3):

$$\frac{C_V}{N} \propto \frac{T_c}{(T - T_c)^\alpha} \qquad \frac{\chi}{N} \propto \frac{T_c}{(T - T_c)^\gamma} \qquad \xi \propto \frac{T_c}{(T - T_c)^\nu} \tag{9.8}$$

In addition to the Ising model, other magnetic lattice models are known to undergo a second-order transition, where of special interest is the n-vector model in the limit of $n \to 0$. We shall not define the model here, but will only point out that there is a one-to-one correspondence between the exponents of this model and the exponents ν and γ defined above for SAWs [2]. In the magnetic model, one defines the correlation length, $\xi \sim T_c/(T - T_c)^\nu$ [Equation (7.5)], which measures the increase in the average size of a droplet of spins as T_c is approached. One can consider the radius of gyration of a chain as defining the size of a "chain droplet," $< R_g^2 >^{1/2} \sim N^\nu = 1/(1/N)^\nu$ where, $1/N$ corresponds to $(T - T_c)/T_c$. As N increases, the chain approaches its critical temperature. In the n-vector model, the contribution to ξ comes from $<R^2>$ values calculated for chains of connected spins that do not self-intersect (SAWs). Therefore, the two models have the same ν.

This correspondence is important for several reasons: First, the critical behavior (phase transition) is described by the renormalization group theory, which enables one calculating of critical exponents of magnetic systems analytically. The above correspondence means that the same exponents can also be obtained by Monte Carlo simulations of long polymers; this leads to a cross-fertilization of these seemingly different fields. Furthermore, a critical behavior occurs in many disciplines, such as high energy physics, solid-state physics, liquids, and polymers, which live under the same theoretical roof—the renormalization group theory. Finally, the above correspondence, discovered in 1972 by Pierre-Gilles de Gennes, has shown that a well-established branch of analytical approximations for polymers is incorrect. These ideas are summarized in de Gennes's book, *Scaling Concepts in Polymers* [2]; in 1983, he won the Nobel Prize in Physics. Another influential polymer scientist who is considered the "father" of the field is Paul J. Flory, mentioned earlier. He wrote two basic books on polymers [1,10] and won the Nobel Prize in Chemistry in 1974.

9.4 Distribution of the End-to-End Distance

It should first be pointed out that the growth of the ETED of two monomers, i and j, that are not end monomers is,

$$< R_{ij}^2 >^{1/2} = A |j - i|^\nu \qquad (9.9)$$

with the same exponent ν defined for the end monomers ($<R^2(1, N + 1)>^{1/2} = BN^\nu$), but with a different prefactor, that is, $A \neq B$. When i and j approach the ends, A approaches B; this might be important for analyzing fluorescence experiments involving generally two internal monomers (Figure 9.2).

The distribution function of the ETED of a chain with excluded volume is not Gaussian. The earliest attempt to define such a function is due to Domb et al. [11], based on exact enumeration of short lattice SAWs. Changes to this function have been introduced by Fisher [12], and later by des Cloizeaux [13,14] (and others), and an early review of these developments can be found in papers by Bishop et al. [15] and Valleau [16]; these authors have also tested several distribution functions against longer chains (up to $N = 400$) simulated by the Monte Carlo technique known as the "*pivot algorithm*" (see Chapter 15). Since all these studies have concluded that the distribution function defined by des Cloizeaux is the best, we only present this function below. Thus, the probability is defined for the scaled ETED, $r' = r/<r^2>^{1/2}$,

$$p(r') \sim r'^{(2+\theta)} \exp[-(Kr')^t] \qquad \theta = (\gamma - 1)/\nu \qquad t = (1 - \nu)^{-1} \qquad (9.10)$$

where $r = |\mathbf{R}|$, γ and ν are the critical exponents defined previously, and K is a parameter; clearly, the first and second terms dominate the distribution for small and large r', respectively. Equation (9.10) has been supported also by pivot algorithm simulations carried out by Pedersen et al. for freely rotating chains with EV of up to $N \sim 100,000$ bonds [17]. More recently, Caracciolo et al. applied a high level renormalization group theory for calculating the distribution of the ETED. They have found their theory (and that of des Cloizeaux) in a very good agreement with results obtained by the pivot algorithm for SAWs

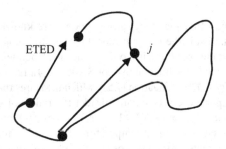

FIGURE 9.2 The end-to-end distance and the distance between two internal monomers, i and j.

of $500 \leq N \leq 32{,}000$ on the simple cubic lattice [18]. Equation (9.10) also applies to two interior monomers, i and j, where $r' = r / |i - j|^{\nu}$. As discussed later, denatured proteins belong to the same regime as SAWs. In fluorescence experiments, one measures the ETED distributions of pairs of certain residues along the chain, and the above formula can be used for analyzing the data.

9.5 The Effect of Solvent and Temperature on the Chain Size

The typical configuration of a polymer in a *good solvent* is open (swollen) due to strong attractions among the monomers and the solvent molecules. In a *bad solvent*, the monomer-monomer attractions overcome their interaction with solvent, the polymer's size decreases, and in extreme conditions, the solute molecules (polymers) will aggregate and precipitate. A simple modeling for these solvent effects (or alternatively, temperature effect) at a *high dilution* is a single SAW on a lattice, where an attraction energy $\varepsilon = -|\varepsilon|$ is defined between nearest neighbor non-bonded monomers; this is an *implicit solvation model*, where the effect of the missing solvent molecules is mimicked by ε/T (see Section 16.4). We denote this interacting SAW by ISAW (Figure 9.3).

The partition function is:

$$Z = \sum_{i(\text{SAWs})} \exp[-E_i/k_{\mathrm{B}}T] = \sum_{i(\text{SAWs})} \exp[-M_i\varepsilon/k_{\mathrm{B}}T] \tag{9.11}$$

where M_i is the number of the non-bonded nearest neighbor pairs. As the temperature is decreased, the average energy also decreases, meaning that the most probable chain configurations become more compact; thus, low temperature mimics bad solvent conditions. Notice that two (contradicting) types of long-range interactions exist here, the attractions and the EV repulsions, which act to close and open the chain, respectively. This model defines three solvent (or temperature) regimes with respect to Flory's θ temperature,

$T > \theta$: the high temperature regime (ε is constant); the energy $- E/N$ (of an ISAW) is close to zero, and the main contribution to the free energy comes from the relatively high entropy. Thus, the chain exhibits a good solvent behavior, where the open chain configurations dominate the ensemble leading to the critical exponents of a SAW [e.g., $\nu(d = 3) \sim 0.588$].

$T = \theta$: for $d = 3$, a counterbalance between the attractions and the (EV) repulsions occurs; thus, for very large N, the polymer behaves, to a large extent, as an ideal chain with the corresponding critical exponents ($\nu = 1/2$, $\gamma = 1$) (θ is called a tricritical point, as it constitutes a junction of three critical regimes.)

$T < \theta$: the solvent effect decreases further, that is, the attractions dominate the repulsions, thus leading to low energy. The polymer collapses to a compact structure with $\nu = 1/d$, while the other critical exponents are undefined. No connection exists in this phase to the n-vector model. This collapsed structure corresponds to the state of a folded protein.

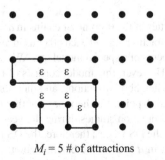

$M_i = 5$ # of attractions

FIGURE 9.3 An ISAW with an attraction, $\varepsilon = -|\varepsilon|$ between its $M_i = 5$ nearest neighbor non-bonded monomers.

TABLE 9.3

θ-point Critical Temperature and Critical Exponents for an ISAW on a Simple Cubic Lattice

	$\varepsilon/k_B\theta$	$k_B\theta/\varepsilon$	ν_θ	γ_θ	ϕ_θ	μ_θ
Scanning method	0.274(6)	3.65(8)	~0.5	1.005 (20)	0.55–058	5.058 ± 0.014
RW $d \geq 3$			0.5	1	0.5	5

Note: The results were obtained by the scanning method for maximal $N = 250$ [19]. The error is defined by parenthesis, thus, 3.65 (8) ≡ 3.65 ± 0.08.

9.5.1 θ Chains in $d = 3$

A detailed definition of the crossover of a chain from the good solvent regime (high temperature) to the collapsed region through the θ temperature is described in the next sub-section, where the critical exponent, ϕ_θ, is defined. Here, we first provide results (Table 9.3) for ISAWs on a simple cubic lattice obtained with the scanning simulation method at θ [19]. While [19] treats relatively short chains ($N = 250$), it provides a complete set of results, which are required for characterizing the θ-point phase.

Indeed, ν_θ, γ_θ, and the growth parameter, μ_θ are close to the expected random walk values, 1/2, 1, and 5, respectively. Only the results, 0.55–0.58 obtained for the crossover exponent, ϕ_θ, (see next section), deviate significantly from the expected value, 1/2. These deviations are caused by strong logarithmic corrections to scaling that exist in $d = 3$, which is the upper critical dimension in this case; thus, $< R^2 > = N[1 + A/\ln(N) + ...]$, where A is a constant (see [19]). This effect will be eliminated for $N >> 250$. Unfortunately, more recent Monte Carlo simulations of longer chains ($N = 10^4$ [20], and $N = 32,000$ [21]) and *exact enumeration* studies (using $N = 24$ [22] and $N = 27$ [23]) have been concentrated on locating θ, and, in some cases, also on identifying a lower critical temperature related to a "crystallization" transition. A thorough study by Grassberger ($N = 10^4$ [20]) using his *"pruned-enriched Rosenbluth method"* (PERM) has led to the best critical temperature thus far, $\varepsilon/k_B\theta = 0.2690 \pm 0.0003$ ($k_B\theta/\varepsilon = 3.717 \pm 0.003$), which is equal within the error bars to the temperature appearing in Table 9.3. The same central values were obtained also by Vogel et al. [21] (using a method related to PERM) based on $N = 32,000$, but with larger errors than those obtained in [20] for $N = 10^4$. The vast literature about the collapse of ISAWs on simple cubic and other $3d$ lattices can be found in references cited in [20–23].

Another aspect of the cancellation of the opposing long-range interactions in $d = 3$ (at θ) is that a *real* chain can be built step-by-step in a random walk manner, or more precisely, with a Markov chain process without memory. In most cases, three possible directions are available at step k, and the (target) bond at step $k + 1$ is chosen with the help of transition probabilities that depend only on the k and $k + 1$ bonds. This *rotational isomeric state* model developed by Flory [10] describes faithfully the experimental behavior of a chain under θ conditions. As we have already pointed out, an *exact* step-by-step build-up of a SAW (not at θ) would be extremely difficult, requiring the definition of transition probabilities that depend on the entire past $(1, ..., k)$ and all the future possible chain continuations of $k + 1, ... N$ steps!

9.5.2 θ Chains in $d = 2$

It is of interest to examine the properties of the θ-point for chains in $d = 2$, in particular, to find whether the long-range EV interactions and the attractions cancel each other as in $d = 3$. Duplantier and Saleur (DS) [24] have solved exactly the critical exponents of a special model of SAWs on a hexagonal lattice with randomly forbidden hexagons (see Table 9.4). However, this model consists of nearest neighbor attractions, as well as of a special subset of next nearest neighbor attractions, and therefore, instead of describing the usual θ point, it might describe a multicritical θ' point. Furthermore, another exactly solved model has led to still different exponents [25]. However, simulation studies during the years of ISAWs on a square lattice have led to results which are very close to the DS values; therefore, the DS model is considered now to describe a true θ point. The results show that in $d = 2$, at the balance point, no perfect cancellation between the attractions and repulsions occurs (that it, the results deviate from $\nu_\theta = 0.5$, $\gamma_\theta = 1$, and $\mu_\theta = 3$); see Table 9.4.

The literature on the θ-point is quite extensive and can be found in references cited in [24–30].

TABLE 9.4

θ-Point Critical Exponents and Temperatures for ISAWs on a Square Lattice

N	ν_θ	γ_θ	ϕ_θ	μ_θ	$\varepsilon/k_B\theta$	$k_B\theta/\varepsilon$	Method
	4/7 = 0.571...	8/7 = 1.143...	3/7 = 0.4285...				DS [24]
250	0.579 (5)	1.125 (19)	0.530 (4)	3.212 (7)	0.658 (4)	1.520 (9)	Scanning [26]
2048			0.435 (6)	3.224 (5)	0.667(1)	1.499 (2)	PERM [27,28]
3200	0.570 (2)		0.479 (6)		0.6673 (5)	1.499 (1)	Extended reptation[29]

Note: The first row contains the exact results of DS. The following rows present results obtained by simulating chains of maximal length, *N*, using various techniques. The errors are defined in the caption of Table 9.3.

9.5.3 The Crossover Behavior Around θ

As has been already pointed out, in three dimensions, at high *T*, the most probable configurations are open since the EV repulsions dominate the attractions. Thus, the chains behave as pure SAWs (with no attractions) with the corresponding critical exponents $\nu_{SAW} \cong 0.588$ and $\gamma_{SAW} \cong 1.16$. As *T* is decreased toward θ (from the hot side), the attractions increase gradually, and the chain will show a pure SAW behavior *only* if it is long enough. The closer *T* is to θ, the longer should be the chain in order to show a SAW behavior.

At a very low *T* (*T* < θ), the attractions dominate the EV repulsions (low potential energy), and the most probable configurations are compact; $\nu_{collapse} = 1/d = 1/3$. This result can be obtained by arranging the chain in a *d*-dimensional cube of side *R* and volume *V*:

$$R^d = V \Rightarrow R^d \sim N \Rightarrow R \sim N^{1/d} \qquad \text{and} \qquad \nu = 1/d \qquad (9.12)$$

What happens very close to θ? The crossover is defined by the function [2,31,32]:

$$<R^2>^{1/2} \sim N^{1/2} f(N^\phi \tau) = N^{1/2} f\left(\frac{\tau}{1/N^\phi}\right) \qquad (9.13)$$

where ϕ is a crossover exponent, $\tau = (T - \theta)/\theta$, and $1/N$ is the temperature difference in the *n*-vector model picture; thus, *f* is a function of a ratio of *two* temperature differences. Close to θ, τ is small; if *N* is not large, the product $N^\phi\tau$ will remain small, where the function *f* is approximately constant; therefore, the chain will behave as a θ chain with exponent $\nu(\theta) = 1/2$. The same applies to the cold side of θ. One concludes that short chains close enough to θ (from both sides) will show a θ-point behavior. If *N* is so large that $N^\phi\tau$ is large (that is, $1/N << \tau$), *f* becomes,

$$f = x^g \begin{cases} g = (0.588-1/2)/\phi & T > \theta \\ g = 0 & T = \theta \\ g = (1/3-1/2)/\phi & T < \theta \end{cases} \qquad (9.14)$$

and the critical behavior is,

$$\langle R^2 \rangle^{1/2} \sim N^{1/2} f(N^\phi\tau)$$
$$= \begin{cases} N^{0.588}\tau^{(0.588-1/2)/\phi} & T > \theta \\ N^{1/2} & T = \theta \\ N^{1/3}\tau^{(1/3-1/2)/\phi} & T < \theta \end{cases} \qquad (9.15)$$

Thus, for any small τ > 0, if *N* is large enough, the chain will behave as a SAW, that is, as a chain in a good solvent. On the cold side of θ, large enough *N* will lead to a collapsed chain ($\nu_{collapse} = 1/d = 1/3$).

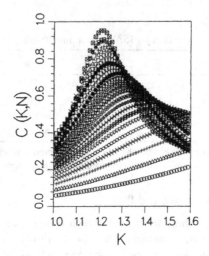

FIGURE 9.4 A log-log plot of the importance sampling results of the specific heat of trails as a function of the tempera-ture $-\varepsilon/k_B T$ and the chain length, N. (Reprinted with permission from Chang, I.S. et al., *Phys. Rev. A*, 41, 1808–1821, 1990. Copyright 1990 by the American Physical Society.)

If this condition is not satisfied (too small N), the chain will behave as a θ chain ($v = 1/2$), as pointed out above. Thus, in simulations where the practical length of a chain is limited, one should be aware of the above τ–N relation when the results are analyzed. We point out again that in the cold side ($T < \theta$), the correspondence between the n-vector model and polymers does not exist.

Approaching θ, the specific heat $C(K, N)$ is expected to diverge; however, because in practice simu-lated chains are finite the transition will not be sharp as is demonstrated in Figure 9.4, where $C(K, N)$ is calculated for different inverse temperatures, $K = \varepsilon/k_B T$. These results have been obtained for trails on a square lattice [33]. A trail is a random walk on a lattice for which two *bonds* are not allowed to overlap. However, the chain may cross itself and one may associate with each such intersection an attrac-tive energy ε; thus, trails like SAWs undergo a θ-point transition. It is evident that as N is increased, the transition becomes sharper.

9.5.4 The Blob Picture

This behavior can be accounted for by using the notion of a blob [2]. When the temperature on the hot side (SAW) decreases, neighbor monomers along the chain start to attract each other creating local clus-ters (blobs), where each grows *internally* according to $v(\theta) = 1/2$. However, the chain becomes a set of connected blobs that interact only weakly with each other. Thus, a long enough chain can be viewed as a SAW consisting of larger units—blobs (of g monomers on average), which grows with $v_{SAW} = 0.588$ and a suitable prefactor A (see Equation 9.3),

$$< R^2 >^{1/2} = A(N/g)^{v_{SAW}}. \tag{9.16}$$

In other words, a long enough chain of blobs is "hot." When θ is approached, the blob size increases; therefore, the chain length that leads to the SAW behavior should grow accordingly. If the chain is short, it will create only a single blob with a θ behavior, that is $v_\theta = 1/2$. Likewise, a long enough chain at $T < \theta$ will collapse. A schematic chain that consists of blobs is depicted in Figure 9.5.

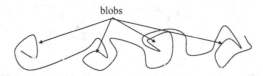

FIGURE 9.5 Blobs are formed as a chain approaching θ.

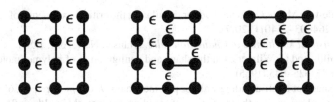

FIGURE 9.6 Three compact configurations of a polymer of length, $N = 11$, with the minimal energy, 5ε.

The ground state of a collapsed polymer is degenerate, and many structures will give the minimal potential energy; in our model, for example, the following configurations are ground states with $E = 6\varepsilon$ (Figure 9.6).

One can also obtain collapsed structures with energy which is not minimal. Globular proteins on the other hand, collapse to a unique structure that is assumed to be the most stable (that is, of minimum free energy). This makes protein folding much more difficult than "polymer folding," as one has to identify a single compact structure out of a very large set of other compact structures, with somewhat higher free energy.

9.6 Summary

In Chapters 8 and 9, we have introduced gradually the three main ingredients characterizing a *general* polymer, the chain connectivity, and the long-range interactions—the EV repulsions and the monomer-monomer attractions, which together with the temperature, mimic the effect of solvent; thus, the competition between these opposing interactions leads to the temperature regimes, $T > \theta$ (good solvent), θ, and $T < \theta$ (bad solvent). Identification of these regimes is important in protein folding experiments and simulations, in particular, to determine whether the protein has been denatured completely (that is, has become a random coil). Notice that molecules with specific short-range interactions (e.g., hydrogen bonds) along the chain do not have a collapsed ground state. Examples are the α-helical state of a polypeptide and a piece of a DNA, which have a rod-like helical ground state with $\nu = 1$. A polymer under infinite dilution (studied in Chapters 8 and 9) is not an unrealistic system, as the physical properties of complex macromolecules (e.g., proteins) are typically investigated under this condition. Finally, the scope of the polymer field is much wider than our survey, including problems related to polymers adsorbed on a surface or confined in small geometries, as well as multiple chain systems, in particular at high density; also, specific polymer systems are generally studied in continuum, mostly by computer simulation methods discussed in detail in Chapters 14 and 15, and in other parts of this book.

REFERENCES

1. P. J. Flory. *Principles of Polymer Chemistry.* (Cornell University Press, Ithaca, NY, 1953).
2. P. G. de Gennes. *Scaling Concepts in Polymer Physics.* (Cornell University Press, Ithaca, NY, 1979).
3. B. Nienhuis. Exact critical point and critical exponents of O(*n*) models in two dimensions. *Phys. Rev. Lett.* **49**, 1062–1065 (1982).
4. N. Clisby. Accurate estimate of the critical exponent ν for self-avoiding walks via a fast implementation of the pivot algorithm. *Phys. Rev. Lett.* **104**, 055702–055704 (2010).
5. J. C. Le Guillou and J. Zinn-Justin. Critical exponents from field theory. *Phys. Rev. B* **21**, 3976–3998 (1980).
6. J. C. Le Guillou and J. Zinn-Justin. Accurate critical exponents from field theory. *J. Phys. (Paris)* **50**, 1365–1370 (1989).
7. N. Clisby and I. Jensen. A new transfer-matrix algorithm for exact enumerations: Self-avoiding polygons on the square lattice. *J. Phys. A: Math. Theor.* **45**, 115202–115215 (2012).
8. N. Clisby. Calculation of the connective constant for self-avoiding walks via the pivot algorithm. *J. Phys. A* **46**, 245001–245011 (2013).

9. N. Clisby. Scale-free Monte Carlo method for calculating the critical exponent γ of self-avoiding walks. *J. Phys. A* **50**, 264003–264011 (2017).

10. P. J. Flory. *Statistical Mechanics of Chain Molecules*. (Hanser, New York, 1988).

11. C. Domb, J. Gillis and G. Wilmers. On the shape and configuration of polymer molecules. *Proc. Phys. Soc.* (London) **85**, 625–645 (1965).

12. M. E. Fisher. Shape of a self-avoiding walk or polymer chain. *J. Chem. Phys.* **44**, 616–622 (1966).

13. J. des Cloizeaux. Lagrangian theory for a self-avoiding random chain. *Phys. Rev. A* **10**, 1665–1669 (1974).

14. J. des Cloizeaux. Short range correlation between elements of a long polymer in a good solvent. *J. Phys.* **41**, 223–238 (1980).

15. M. Bishop, J. H. R. Clarke, A. Rey and J. J. Freire. Investigation of the end-to-end vector distribution function for linear polymer in different regimes. *J. Chem. Phys.* **95**, 4589–4592 (1991).

16. J. P. Valleau. Distribution of end-to-end length of an excluded-volume chain. *J. Chem. Phys.* **104**, 3071–3074 (1994).

17. J. S. Pedersen, M. Laso and P. Schurtenberger. Monte Carlo study of excluded volume effects in worm-like micelles and semiflexible polymers. *Phys. Rev. E* **54**, R5917–R5920 (1996).

18. S. Caracciolo, M. S. Causo and A. Pelissetto. End-to-end distribution function for dilute polymers. *J. Chem. Phys.* **112**, 7693–7710 (2000).

19. H. Meirovitch and H. A. Lim. Computer simulation study of the θ point in three dimensions. I. Self-avoiding walks on a simple cubic lattice. *J. Chem. Phys.* **92**, 5144–5154 (1990).

20. P. Grassberger. Pruned-enriched Rosenbluth method: Simulations of θ polymers of chain length up to 1,000,000. *Phys. Rev. E* **56**, 3682–3693 (1997).

21. T. Vogel, M. Bachmann and W. Janke. Freezing and collapse of flexible polymers on regular lattices in three dimensions. *Phys. Rev. E* **76**, 061803–061807 (2007).

22. J. H. Lee, S.-Y. Kim and J. Lee. Exact partition function zeros of a polymer on a simple cubic lattice. *Phys. Rev. E* **86**, 011802–011807 (2012).

23. C.-N. Chen, Y.-H. Hsieh and C.-K. Hu. Heat capacity decomposition by partition function zeros for interacting self-avoiding walks. *Europhys. Lett.* **104**, 20005–20006 (2013).

24. B. Duplantier and H. Saleur. Exact tricritical exponents for polymers at the theta point in 2d. *Phys. Rev. Lett.* **59**, 539–542 (1987).

25. S. O. Warnaar, M. T. Batchelor and B. Nienhuis. Critical properties of the Izergin-Korepin and solvable O(n) models and their related quantum spin chains. *J. Phys. A* **25**, 3077–3095 (1992).

26. I. Chang and H. Meirovitch. Collapse transition of self-avoiding walks on a square lattice in the bulk and near a linear wall: The universality classes of the θ and θ' points. *Phys. Rev. E* **48**, 3656–3660 (1993).

27. P. Grassberger and R. Hegger. Simulations of θ-polymers in 2 dimensions. *J. Phys.* (*Paris*) **5**, 597–606 (1995).

28. G. T. Barkema, U. Bastolla and P. Grassberger. Two-dimensional oriented self-avoiding walks with parallel contacts. *J. Stat. Phys.* **90**, 1311–1324 (1998).

29. S. Caracciolo, M. Gherardi, M. Papinutto and A. Pelissetto. Geometrical properties of two-dimensional interacting self-avoiding walks at the θ-point. *J. Phys. A* **44**, 115004–115017 (2011).

30. E. Vernier, J. L. Jacobsen and H. Saleur. A new look at the collapse of two-dimensional polymers. *J. Stat. Mech.* **2015**, P09001–P09028 (2015).

31. P. G. de Gennes. Collapse of a polymer chain in poor solvents. *J. Phys. Lett.* (*Paris*) **36**, L55–57 (1975).

32. P. G. de Gennes. Collapse of a flexible polymer chain II. *J. Phys. Lett.* (*Paris*) **39**, L299–L301 (1978).

33. I. S. Chang, H. Meirovitch and Y. Shapir. Tricritical trails on a square lattice with impenetrable linear boundary: Computer simulation and analytic bounds. *Phys. Rev. A* **41**, 1808–1821 (1990).

Section III

Topics in Non-Equilibrium Thermodynamics and Statistical Mechanics

10

Basic Simulation Techniques: Metropolis Monte Carlo and Molecular Dynamics

Non-equilibrium thermodynamics and statistical mechanics deal with time-dependent phenomena, in particular, relaxation processes toward equilibrium, which are discussed theoretically in Chapters 11–13. However, because the basic simulation methods, Metropolis Monte Carlo (MC) and molecular dynamics (MD), are of a relaxation type, it is useful to introduce them first. Applications of MC/MD provide a realistic picture of the complexity of relaxation processes (e.g., protein folding), putting in perspective the (simplified) theoretical treatments of the following chapters. Therefore, MC/MD are discussed here, rather than in Section IV, which is devoted to more advanced computer simulation methodologies.

10.1 Introduction

Equilibrium statistical mechanics provides an established theoretical framework, where the real world is modeled in terms of a probability space. However, in most cases, an exact calculation of the free energy and other statistical averages is possible only for simplified models, such as an ideal chain, the harmonic oscillator, or the two dimensional ($2d$) Ising model. For more complex models, analytical evaluations of averages (integrals) can only be carried out through *approximations* (e.g., mean field theory), which typically are not reliable in regions of interest, such as close to the critical point.

Therefore, the increase in computer power in the early 1950s has led to the advent of the new approach of computer simulation. That is, for a system defined by the number, volume, and temperature (NVT) ensemble, for example, a relatively small model with the same N/V and T, is built within the computer and then is simulated according to the rules of statistical mechanics; thus, various thermodynamic averages can be *estimated* from the generated sample and can be related to the corresponding results of the ensemble defined in the *thermodynamic limit* (N, $V \rightarrow \infty$). Formally, the model is transferred from the *probability space* (ensemble) to the *experimental world on a computer* (see Section 1.4). The simulation tool enables one to study *microscopic* properties that are beyond the reach of standard statistical mechanics, such as the average distance between two atoms of a protein.

It should be pointed out, however, that this approach is not straightforward. First, methods should be devised that enable one to generate configurations according to the Boltzmann probability. As discussed in Section 5.2, this is not a trivial task, since most of the contribution to averages comes from a tiny region in configurational space; furthermore, even if such a method has been devised, one should still establish numerical criteria to verify that in practical simulations this region has indeed been reached. Second, the simulated systems are small, while one would seek to study them in the thermodynamic limit; thus, the choice of boundary conditions might affect the results. Finite size effects are, in particular, severe in the presence of long-range interactions (e.g., Coulomb forces) or near a second-order phase transition, where long-range correlations grow with system size. Hence, procedures for analyzing such effects should be devised as well.

The two pioneering contributions in this field are the Monte Carlo (MC) method, suggested in 1953 by Metropolis et al. [1], and molecular dynamics (MD), developed in 1957 by Alder and Wainright [2,3], where both were applied originally to a system of hard spheres. These methods have been developed further in the course of years and have made a profound impact on most branches of science. MC and MD are dynamical type methods, that is, the simulation starts from an arbitrary system configuration, which undergoes consecutive changes in time until the generated sample is distributed according to the

Boltzmann probability. This relaxation process to equilibrium belongs to non-equilibrium statistical mechanics. Due to their dynamical character, these techniques enable one sampling system configurations according to the Boltzmann probability, P^B, *without providing the value of P^B*; therefore, they do not provide the absolute entropy *directly* (see discussion in Section 1.16). In this context, one should mention the early *non-dynamical* simulation methods for polymers developed by Wall and Erpenbeck, and Rosenbluth and Rosenbluth. With these methods, a polymer is generated step-by-step (starting from nothing) using transition probabilities, where their product leads to the *value* of the construction probability, hence, to the absolute entropy (see Sections 14.2 and 14.4).

Due to their popularity, MC and MD have been reviewed extensively, where general methodology issues and technical tricks to enhance efficiency have been discussed. Therefore, we limit our presentation to the basics, where advanced topics can be found, for example, in the excellent monographs [4–10]. Before describing MC/MD in detail, we discuss below general properties of sampling theory with respect to statistical mechanics.

10.2 Sampling the Energy and Entropy and New Notations

It is educational to consider first the possibility of treating a statistical mechanics system, such as liquid argon, by *simple sampling*, that is, by generating *NVT* argon configurations (see Section 1.4) at random (as in the case of the two-state coin experiment). Clearly, this would be totally useless because the chance to hit the tiny region in configurational space around the most probable energy, E^*, is nil (Section 5.2), meaning that, in practice, the results will be strongly biased. Therefore, the aim is to sample according to P^B, which will lead the system to the desired region; however, as will become evident, in some cases, one can sample also with a *non-Boltzmann* probability, P, which is sometimes called *importance sampling* as long as *P is non-random*. However, before discussing importance sampling further (see next Section 10.3), we wish to establish some notations differing somewhat from those defined in Chapter 1.

We return, for example, to N argon atoms enclosed in V at T, that is, treated in the *NVT* ensemble; the potential energy, $E(\mathbf{x}^N)$, is a random variable, that is, a function from the configuration space, $\Omega \equiv \{\mathbf{x}^N\}$, to the real line; again, ignoring the velocities, the ensemble average (denoted by $\langle\rangle$) is:

$$\langle E \rangle = \int_\Omega P^B(\mathbf{x}^N) E(\mathbf{x}^N) d\mathbf{x}^N, \tag{10.1}$$

with a standard deviation, $\sigma = (\langle E^2 \rangle - \langle E \rangle^2)^{1/2}$. For n independent (or uncorrelated) equal *random variables*, E_j, defined on the same space, Ω, one defines the arithmetic average, $\bar{E}_n$ (Equation 1.56),

$$E_n = \frac{1}{n} \sum_{j=1}^{n} E_j, \tag{10.2}$$

where $< E_n > = < E >$, but $\sigma_n = \sigma/n^{1/2}$, that is, σ_n decreases with increasing n [in Eq. (1.56) and Chapter 1 the arithmetic average is denoted with a bar, while here it is denoted for the energy by E_n; the bar will be used to denote *estimation*, see below.]. As explained in Sections 1.14–1.16, E_n is defined over the product space consisting of n-component vectors [elementary events, see Equation (1.58)] (for a discrete system of k energy values, the number of vectors is k^n). We have also shown in Sections 1.14–1.16 that the arithmetic averages of these vectors are normally distributed around $<E>$. Thus, n configurations sampled independently according to P^B, constitute only one vector (a single realization) out of the entire ensemble of vectors (k^n in the above example). However, because the arithmetic average of the generated vector is normally distributed, there is a 68% chance that it will lie within one standard deviation of $<E>$; thus, as n is increased (and $\sigma_n = \sigma/n^{1/2}$ decreases), the arithmetic average of a sampled vector is expected to provide a better and better estimation for $<E>$. In what follows, we denote the arithmetic average of a *specific* sample (a single vector) of size n with a bar,

$$\bar{E} = \frac{1}{n} \sum_{t=1}^{n} E_{i(t)} = \sum_{E} \frac{n_E}{n} E. \tag{10.3}$$

Thus, $\bar{E}$ [Equation (10.3)] is an *estimation* of the ensemble average, $<E>$ [Equation (10.1)], where E_n [Equation (10.2)] is a random variable consisting of n random variables, E_j. E_j should be distinguished from $E_{i(t)}$—the energy value of configuration, i, generated at time, t, of the sampling process. One might wonder about the (*apparent*) disappearance of the probability $P^B(\mathbf{x}^N)$ in the transfer from $<E>$ [Equation (10.1)] to $\bar{E}$ [Equation (10.3)]. However, note that $P^B(E)$—the Boltzmann probability of an *energy*, E (not of a configuration with energy, E_i), is just hidden in $\bar{E}$ as n_E/n, where n_E is the number of times energy value E appears in the summation (sample). The above efficiency of *uncorrelated* sampling (defined by $\sigma_n = \sigma/n^{1/2}$) is not shared by MC and MD due to correlations.

For the above estimation of the energy (which applies to MC and MD), one only needs to be able to sample *according to* $P^B(\mathbf{x}^N)$, while knowing the *value* of this probability density is not necessary (Section 1.16). Also, the estimation of $<E>$ is straightforward since $E_{i(t)}$ depends on configuration i alone, and can be obtained by just summing up all the interactions of i [e.g., the Lennard-Jones interactions defined in Equation (10.17)]. Clearly, this "easy" calculation applies to any thermodynamic or geometrical parameter that can be measured directly over configuration i, such as the end-to-end distance or the radius of gyration of a protein. On the other hand, estimating the *absolute* entropy is much more complex. Thus, writing S, for simplicity, in a discrete space,

$$S = -k_B \sum_i P_i^B \ln P_i^B. \tag{10.4}$$

and its estimation $\bar{S}$,

$$\bar{S} = -\frac{k_B}{n} \sum_{t=1}^{n} \ln P_{i(t)}^B. \tag{10.5}$$

demonstrates that to calculate $\bar{S}$, it is mandatory also to know the *values* of the probabilities, $P_{i(t)}^B$, which appear under the logarithm. However, $P_i^B = \exp - [E_i/k_B T]/Z$ depends not only on configuration i, but on the *entire* ensemble of configurations through the partition function, Z, which is unknown a priori and constitutes the target of our study.

In summary: There are two levels of knowledge related to probability in sampling. As pointed out in Section 1.16, dynamic type methods, such as MC/MD enable one to sample according to P_i^B, while its value is not provided, and thus, $\bar{S}$ cannot be obtained in a direct way. We have also mentioned (Section 10.1) that step-by-step construction methods based on transition probabilities do provide the value of the sampling probability, hence, the absolute entropy is obtained as a by-product of the simulation. We shall see that the step-by-step ideas can be implemented within the framework of MC/MD and the absolute entropy can thus be extracted from configurations sampled by these dynamical methods as well (see Chapter 19).

10.3 More About Importance Sampling

In the previous section, we have emphasized the difficulty to obtain the absolute entropy S, with MC/MD, since the *values* of P_i^B are not provided. Because S leads to the free energy and the partition function Z, these functions are also unknown. Below, we show that (equivalently) one can calculate Z directly by importance sampling. Thus, while Z constitutes a *summation* (rather than a statistical average), it can be converted into a statistical average by *any* probability distribution, P_i with *known values*, which is defined on the same space as Z. We first multiply and divide every exponential term in Z by P_i, transferring Z formally to a statistical average with the probability, P_i:

$$Z = \sum_i \exp[-E_i / k_B T] = \sum_i P_i \left(\frac{\exp[-E_i / k_B T]}{P_i} \right) \tag{10.6}$$

where the expression in the brackets is a random variable. Thus, $Z(P)$ can be estimated by the arithmetic average, $\bar{Z}$, from a sample of size n obtained with P_i,

$$\bar{Z} = \frac{1}{n} \sum_{t=1}^{n} \frac{\exp[-E_{i(t)} / k_B T]}{P_{i(t)}}. \tag{10.7}$$

The efficiency of this sampling depends on the variance of $Z(P)$, which is typically huge because it involves exponentials of an extensive variable (energy) and probabilities:

$$\sigma^2 = \sum_i P_i \left(\frac{\{\exp[-E_i / k_B T]\}}{P_i} - Z \right)^2 = \sum_i \frac{\exp[-2E_i / k_B T]}{P_i} - Z^2. \tag{10.8}$$

The sample size n required to "fight" such a variance is impractically large. However, for the Boltzmann probability, $P_i = P_i^B = \exp[-E_i / k_B T] / Z$, σ^2 vanishes, meaning that Z can be obtained correctly already from a single term ($n = 1$); this is not surprising because if P_i^B is known, Z is already known from the outset. This result is equivalent to the zero fluctuation of the free energy, discussed in Section 5.1. Therefore, practically, to decrease the variance, P_i should be close to P_i^B, which can be achieved only for small enough systems. For $P_i \neq P_i^B$, this procedure means that configurations are initially selected with a biased probability P_i, where the effect of this bias is corrected in the estimation stage by dividing each exponential by the same biased probability with which it has been chosen. This importance sampling procedure can be applied, in principle, to any given summation provided that adequate probability distribution can be defined. The bias/correction trick is quite common, used, for example, in Chapter 14 for polymers and in Section 20.1 in umbrella sampling.

10.4 The Metropolis Monte Carlo Method

Metropolis MC [1] is a Markov chain (Section 1.17) based on transition probabilities (TPs), which are designed to transfer a system in a non-equilibrium state at time $t = 0$ to equilibrium after t Markov steps. In the usually studied systems, where the configurational and velocity (momentum) degrees of freedom are separated, MC operates on the configurational states alone—a simplifying factor. For a complex system, the number of Markov states is huge (e.g., for an $N \times N$ Ising model, the stochastic matrix M is of size $2^N \times 2^N$). Therefore, calculating the stochastic matrix, M^t, which defines the set of probabilities at time t is feasible (numerically) only for very small systems; thus, the values of the Boltzmann probabilities (for large t) are unknown in the general case.

Instead of calculating M^t (in the probability space), one can resort to the experimental world on a computer by generating t Markov steps, which constitute a *single* term in the corresponding product space, that is, a *single* avenue (realization) out of the huge number of available avenues of length t as depicted in Figure 10.1.

If t is larger than the relaxation time, t_{relax}, configurations will reach equilibrium, that is, they will be selected with their stationary Boltzmann probabilities. As pointed out in Section 1.17.1, there are two practical options to proceed, (1) to carry out many such realizations (vectors) for a specific $t > t_{\text{relax}}$, averaging their results [e.g., for the energy, $E(t)$] obtained at t. (2) To rely on the ergodic theorem (Section 1.18),

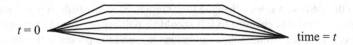

FIGURE 10.1   Illustration of the many avenues of length of time, t starting from the *same* configuration at $t = 0$, and relaxing after long t to equilibrium (see also Section 1.17.1).

carrying out a *single* very long realization (trajectory) for $t \gg t_{\mathrm{relax}}$, where averages are based on results obtained along the "production part" of the trajectory, that is, for $t > t_{\mathrm{relax}}$.

The adequate TPs, which lead to equilibrium for a given T, should satisfy the *detailed balance* condition [see Equation (1.75) and Section 11.8]:

$$\frac{T_{ij}p_{ij}}{T_{ji}p_{ji}} = \frac{p_{ij}}{p_{ji}} = \frac{P_j^{\mathrm{B}}}{P_i^{\mathrm{B}}} = \exp[-(E_j - E_i)/k_{\mathrm{B}}T]. \tag{10.9}$$

In Equation (10.9), T_{ij} is the TP for selecting a candidate configuration, j, for time $t + 1$, while the system is in configuration i at time t; p_{ij} is the TP to accept j (after i). Typically (but not always), $T_{ij} = T_{ji}$, and in this *symmetric* case, the detailed balance is applied only to p_{ij} and p_{ji}. It should be emphasized that, in practice, it is crucial to verify that $T_{ij} = T_{ji}$, otherwise the results will be incorrect. Notice, however, that if $T_{ij} \neq T_{ji}$, p_{ij} should be defined appropriately to satisfy the detailed balance condition, Equation (10.9). An example for an asymmetric T_{ij} appears in Section 14.5.6. Another example is the force bias Monte Carlo method of Pangali, Rao, and Berne, where the MC moves (for water) are biased in the direction of the forces and torques acting on the individual molecule [11,12]. Finally, we comment that detailed balance is a sufficient condition that the Boltzmann probabilities are the stationary probabilities of MC; indeed [see Equation (1.71)],

$$\sum_j P_j^{\mathrm{B}} p_{ji} = \sum_j P_j^{\mathrm{B}} \cdot p_{ij} \frac{P_i^{\mathrm{B}}}{P_j^{\mathrm{B}}} = P_i^{\mathrm{B}} \sum_j p_{ij} = P_i^{\mathrm{B}} \cdot 1 = P_i^{\mathrm{B}} \tag{10.10}$$

10.4.1 Symmetric and Asymmetric MC Procedures

First, it should be pointed out that it is important to verify that the MC procedure applied to a particular system constitutes an ergodic Markov chain, that is, one which is irreducible and aperiodic. However, for a complex system, this check is not always possible, and the *working* assumption is that the above conditions exist.

The TPs, p_{ij}, are defined up to their ratio in Equation (10.9); this freedom means that there are an infinite number of ways to determine their values. A commonly used TP is the "*asymmetric MC,*"

$$p_{ij} = \begin{cases} 1 & \text{for } \Delta E_{ji} < 0 \\ \exp[-\Delta E_{ji}/k_{\mathrm{B}}T] & \text{for } \Delta E_{ji} \geq 0 \end{cases} \quad \Delta E_{ji} = E_j - E_i. \tag{10.11}$$

Notice that if $T_{ij} \neq T_{ji}$, the second line should be changed to $p_{ij} = p_{ji}T_{ji}/T_{ij}\exp[-\Delta E_{ji}/k_{\mathrm{B}}T]$.

Operationally, these TPs are used as follows: for a system at state i at time t, one selects *at random* a candidate state j out of a set of available states, using a random number generator (see Appendix, A3). Then, the difference in energy between these states, $\Delta E_{ji} = E_j - E_i$ is calculated, and if $\Delta E_{ji} < 0$, j is accepted as the state of $t + 1$. If $\Delta E_{ji} \geq 0$, another random number, r within the segment $[0 \leq r \leq 1]$, is generated; if $r < \exp[-\Delta E_{ji}/k_{\mathrm{B}}T]$, j is accepted as the state of time $t + 1$, while in the other case, state i remains the state of $t + 1$. This procedure is repeated many times until the system has relaxed to equilibrium. Then, the process continues and statistical averages of interest are estimated from the *relaxed* sample.

A typical MC sample will consist of a series of states, such as: $i, i, i, j, j, k, l,\ldots$, meaning that due to the MC criterion [Equation (10.9)], the initial state (configuration), i, remained in the following two MC steps, and only in the third step it was changed to *j*. *j* remained in the fifth step and was transferred to *k* in the sixth step, and then was changed to *l*. It should be pointed out that the larger the groups are of same state (such as *i, i, i*), the stronger the correlations are, and the poorer is the diversity of states in the sample, which reflects the inefficiency of the MC process. Another TP, which is sometime in use is "*symmetric MC*,"

$$p_{ij} = \frac{\exp-[E_j/k_\mathrm{B}T]}{\exp-[E_i/k_\mathrm{B}T] + \exp-[E_j/k_\mathrm{B}T]} \tag{10.12}$$

where again, *j* is accepted if a random number *r*, $0 \leq r \leq 1$, satisfies $r \leq p_{ij}$. As hinted above, the TPs chosen (e.g., symmetric or asymmetric) can affect the efficiency of the simulation; thus, defining the right TPs for a given problem is the "art" of Monte Carlo. It is mandatory to verify that the detailed balance condition is satisfied.

An important advantage of Metropolis MC lies in its simplicity even for complex systems. As a dynamical method, MC is sometimes used to model relaxation and time-dependent processes, where, however, the relation between the MC time and real time should be established and justified. If the interest is in equilibrium, the first t_{relax} MC steps of the relaxation phase should be ignored. As pointed out above, at equilibrium, where the configurations are selected proportional to P_i^B, properties that are measured directly on the system, such as the energy, can be estimated from a sample of size *n*, by the arithmetic average [Equation (10.3)].

PROBLEM 10.1

Define your own MC procedure verifying that $p_{ij}/p_{ji} = \exp - (E_j - E_i)/k_\mathrm{B}T$.

10.4.2 A Grand-Canonical MC Procedure

In the case of other ensembles, Equation (10.9) holds, where $P_j^\mathrm{B}/P_i^\mathrm{B}$ is changed accordingly. For example, for the grand-canonical ensemble (Section 6.4), the TPs are changed to:

$$\frac{T_{ij}p_{ij}}{T_{ji}p_{ji}} = \frac{p_{ij}}{p_{ji}} = \frac{P_j^\mathrm{B}}{P_i^\mathrm{B}} = \exp\{-[(E_j - E_i) - \mu(N_j - N_i)]/k_\mathrm{B}T\}$$

$$= \exp\{-[\Delta E_{ji} - \mu\Delta N_{ji}]/k_\mathrm{B}T\} \tag{10.13}$$

where μ is the chemical potential. In this case, not only state *i* (at time *t*) undergoes a configurational change (becoming *j* of *t* + 1), but particles can also be added to sate *i* or removed from it. The asymmetric and symmetric MC procedures are performed with the same prescription described above [13,14]. In most cases, to ensure a high enough acceptance rate, the change in the number of particles is minimal, $\Delta N = \pm 1$. As a simple example, let us consider a lattice gas model with a number of sites (volume) *V*, temperature *T*, and chemical potential, μ (μVT ensemble; the number of *N* can change). In this model, two particles are not allowed to occupy the same lattice site (energy = ∞), where nearest neighbor particles on the lattice interact with attractive energy, $-\varepsilon$ ($\varepsilon > 0$). One can define the following procedure.

(1) With probability $T_{ij} = 1/2$, a decision is made (with a random number), (a) whether to attempt changing a particle's position or (b) to add/remove a particle to/from the lattice. (2) If (a) has been chosen, the system's energy, E_i, is calculated, a particle on the lattice is selected at random with probability $1/N$, and a site is chosen with probability $1/V$; if the site is occupied, the move is rejected, *i* becomes the state of *t* + 1, and the process returns to (1). Otherwise, a trial move of the particle to the vacant site is carried out, and the system energy E_j is calculated. Applying asymmetric MC, $\Delta E_{ji} = E_j - E_i$ is calculated, the acceptance or rejection of the trial move is determined by Equation (10.11), and the process returns to (1).

(3) If (b) is chosen, one selects a lattice site at random with probability $1/V$, where the site can be (c) occupied or (d) vacant. (4) In case (c), a *trial* removal of the particle is assumed: if $\Delta E_{ji} + \mu < 0$, the trial move is accepted, and the particle is removed from the lattice; otherwise a random number, *r*, is

generated, and if $r \leq \exp -(\Delta E_{ji} + \mu)/k_B T$, the removal is accepted, while if r is larger than the exponential, the removal is rejected; with both cases of acceptance or rejection, the process returns to (1). (5) In case (d), a *trial* addition of a particle to the chosen site is assumed: if $\Delta E_{ji} - \mu < 0$, the trial move is accepted, and the particle is added to the lattice; in the other case, the trial addition of a particle is accepted or rejected as in (3), but based on $\Delta E_{ji} - \mu$ rather than $\Delta E_{ji} + \mu$.

10.5 Efficiency of Metropolis MC

In principle, TPs can be defined between *any* two configurations of the system, that is, a trial j (after i) is selected at random from the set of *all* system configurations. In this case, the configurations i and j are expected to be very different, hence, uncorrelated. However, the simulation will become very inefficient, because the majority of configurations pertain to the high energy region, which would lead to $\Delta E_{ji} > 0$, and to the rejection of j. In other words, the acceptance rate, R_{accept}, which is the ratio between the number of MC steps accepted, n_{accept} and the number of MC steps attempted, n_{attemt}:

$$R_{accept} = \frac{n_{accept}}{n_{attemt}} \tag{10.14}$$

is expected to vanish unless the study is conducted at a very high temperature, where the most probable configurations are of high energy. Therefore, in practice, the trial configurations j should differ from i only slightly. For example, assume a dense argon system, where an argon atom is selected at random, a small cube is defined at its position (Figure 10.2), and a trial move is determined at random within the cube (the other atoms are kept intact). If the cube is large, there is a high chance that the atom in its trial position, j, will "bump" into a neighbor atom, and the move will be rejected. Therefore, the cube size should be decreased to increase the acceptance rate, where the consensus based on many studies suggests an optimal acceptance rate, R_{accept} of around 30%.

However, since configurations i and j are structurally very similar, the generated sample will be correlated, the smaller the cube, the higher the correlation. Thus, if for a sample of size n the correlation disappears after l MC steps, the standard deviation of the sample does not decrease as $\sigma_t = \sigma/n^{1/2}$, but slower, as $(n/l)^{-1/2}$,

$$\sigma_t = \frac{\sigma}{(n/l)^{1/2}} = \frac{\sigma l^{1/2}}{n^{1/2}} \tag{10.15}$$

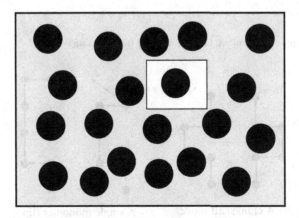

FIGURE 10.2 A two-dimensional illustration of liquid argon undergoing an MC process. Moving the chosen atom within the small enough "cube" will prevent its "bumping" into neighbor atoms.

meaning that large samples will be needed for "scrambling" the system, which is required for getting accurate estimations. This problem is severe in dense systems and near a phase transition, where the correlations increase dramatically as the critical point is approached. Therefore, one has to find the best compromise between high acceptance rate and low correlations.

For the $d = 1$ Ising model, transitions are typically allowed between configurations i and j that differ by a single spin only. Thus, the example in Figure 10.3 shows a configuration i of an Ising model of five spins; while the total number of configurations is $2^5 = 32$, only five trial configurations, j, are allowed by changing a single spin of i.

In the MC process, one of the five j configurations is selected at random (that is with the symmetric $T_{ij} = 1/5$), which is accepted or rejected according to the MC criterion in Equation (10.11).

Another interesting system is a self-avoiding walk (SAW) on a square lattice (Chapter 9). For this model, one can define two *local* configurational changes, a crankshaft move and a single monomer flip, as depicted in Figure 10.4. The crankshaft move was added after realizing that the MC process based on the single monomer flip alone is not ergodic [15–17]. However, for a long SAW, adding even larger (local) clusters will not lead to an ergodic method due to the geometrical restrictions imposed by the chain; ergodicity can be obtained by adding "transport" and "reptation" moves (see Section 15.4). Another avenue for gaining ergodicity is by designing methods based on global changes, as those imposed by the pivot algorithm reviewed in Chapter 15.

As will be discussed in detail in Section 16.9, the situation is even more severe in proteins. In addition to the connectivity effects pointed out for SAWs, the protein chain is highly dense due to attractive inter-actions (which lead to the ruggedness of the potential energy surface). Thus, moving atoms by changing their Cartesian coordinates will almost always lead to rejections, and thus to an extremely inefficient MC procedure. This stems from the need to carry out many futile expensive potential energy calculations of the entire protein (and the water around). A better efficiency can be gained by conformational moves based on internal coordinates—dihedral and bond angles (see Section 16.9). As we shall see, for proteins, MD, in general, is more efficient than MC because the atoms are not moved at random, but along the (initially) calculated gradients.

In summary: MC is a very effective simple method, which is being used extensively in all the sciences. The fact that it operates only on the coordinates part of the partition function contributes to its simplicity. Many versions of the Metropolis MC have been developed based on various cluster sizes, different defi-nitions of T_{ij} and p_{ij}, and in combination with molecular dynamics. However, even though a mathematical theorem might exist that guarantees convergence to the stationary probabilities (P_i^B), the correlations can be strong, leading to long relaxation times, and no theorem (criterion) exists which would tell us when convergence has been reached. Therefore, in practice, one looks for an apparent stability in the calcu-lated averages by carrying out several independent MC runs starting from different configurations and using different random number sequences. Finally, one should bear in mind that even if the MC process

$$i = + + - + - \ \longrightarrow \ j = - + - + - \ ; + - - + - ; + + + + - ; + + - - - ; + + - + +$$

FIGURE 10.3 Five trial spin configurations, j for MC differing from i by a single spin.

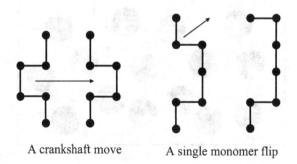

A crankshaft move   A single monomer flip

FIGURE 10.4 Left: A crankshaft MC move for a SAW. Right: A single monomer flip.

has been converged, the precision depends not only on the uncorrelated sample size, but also on the numerical precision applied (single, double, or higher precision), and the quality of the random number generator used (a simple random number generator appears in the Appendix, A3).

10.6 Molecular Dynamics in the Microcanonical Ensemble

MD is a dynamical process (like MC), where Newton's equations of motion for the entire system are solved numerically on a computer as a function of time [2,3]. Notice that unlike the MC case, with MD, the coordinates and velocities both take part in the process; after a long simulation time, the system is expected to reach equilibrium through an initial relaxation period, which depends on the initial conditions and the system's properties. Application of MD to an *NVE* system (Section 6.1) starts with an initial phase space point, that is, a set of initial coordinates and velocities which, in principle, determine the system energy for the entire duration of a simulation. In practice, however, in a long simulation, the energy will not be conserved due to round-off errors. Ignoring this fact, one has in hand a *deterministic* trajectory consisting of known initial conditions, and the complete microscopic details of the system as a function of time. Clearly, this is much more information than the experimental knowledge on which statistical mechanics has been established (Section 3.1). Moreover, one might wonder where is the entropy in such a well-defined process.

The answer to these questions lies in the basic fact that statistical mechanics operates on the level of *averages*, while specific events occurring along a single trajectory are beyond its reach. For example, statistical mechanics can provide, from a *single* MD trajectory, the average potential and kinetic energies by relying on the ergodic theorem (Section 1.18). Thus, if t_{relax} is the longest relaxation time in the system, $t >> t_{\text{relax}}$ is the production period of the trajectory from which averages can be calculated. Defining $t_{\text{product}} = t - t_{\text{relax}}$, the average potential energy, E_p, is:

$$\overline{E_p} = \lim_{t \to \infty} \frac{1}{t - t_{\text{relax}}} \int_{t_{\text{relax}}}^{t} dt' E_p(\mathbf{r}^N, t') = <E_p> \tag{10.16}$$

where the bar denotes the time average and the brackets the ensemble average. While this is the commonly used avenue, in practice, several long runs are commonly performed to check the stability of the results. A second avenue is *ensemble* averaging, where k independent trajectories with the same total energy and the same length, t_s, are generated from different sets of coordinates and velocities. If t_s is larger than t_{relax}, the potential and kinetic energies (and their fluctuations) can be averaged from the k runs at t_s. Thus, as long as an *NVE* trajectory is used for calculating averages (even of local properties, such as distances between atoms), the results will remain within the domain of statistical mechanics theory.

Now, no matter how long a trajectory is, it will never cover the total ensemble of phase points ($\mathbf{x}^N$, $\mathbf{p}^N$) with a given energy E, which define the *NVE* partition function $\Omega(N, V, E)$ [Equation (6.5)] and the entropy $S = k_B \ln \Omega(N, V, E)$ + constant. This demonstrates the difficulty to extract the entropy from a single trajectory or a sample of trajectories.

The MD method was suggested by Alder and Wainright in 1957 [2,3], as applied to hard spheres, and then has been developed further by them and others (e.g., [18]) to become an essential tool (together with MC) for studying fluids. In 1977, MD was extended by the Karplus' group [19] to a protein in vacuum described by a force field (Section 16.3), and from then, the method has been extended to various complex protein systems, becoming the main simulation tool for biological macromolecules (due to the limitations of MC pointed out in the previous section).

For simplicity, we shall explain MD as applied to an *NVE* system with constant energy, E_{tot}, composed of N argon atoms of mass, m, enclosed in an isolated container. Each pair of atoms with a distance r interact via a Lennard-Jones potential, $\phi(r)$ [Figure 10.5],

$$\phi(r) = 4\varepsilon \left[\left(\frac{\sigma}{r} \right)^{12} - \left(\frac{\sigma}{r} \right)^{6} \right] \tag{10.17}$$

Entropy and Free Energy in Structural Biology

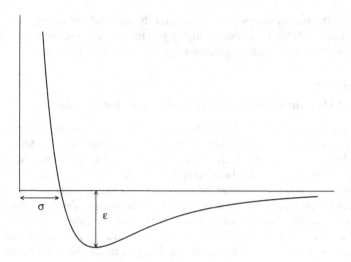

FIGURE 10.5 A Lennard-Jones potential based on the parameters ε and σ.

First, we calculate the force, F_{x_1}, in the x_1 direction between two atoms positioned at Cartesian coordinates (x_1, y_1, z_1) and (x_2, y_2, z_2); one obtains,

$$F_{x_1} = -\frac{\partial\phi}{\partial x_1} = -\frac{\partial\phi}{\partial r}\frac{\partial r}{\partial x_1} = -\frac{\partial\phi}{\partial r}\frac{\partial[(x_1 - x_2)^2 + (y_1 - y_2)^2 + (z_1 - z_2)^2]^{1/2}}{\partial x_1}$$

$$= -\frac{\partial\phi}{\partial r}\frac{x_1}{r} = 24\varepsilon\left[\frac{2\sigma^{12}}{r^{13}} - \frac{\sigma^6}{r^7}\right]\frac{x_1}{r}$$

(10.18)

where the forces in directions y_1 and z_1 are obtained in the same way.

Assume now, that at $t = 0$, atom i is positioned at coordinates $\mathbf{x}_i(0)$ and has an initial velocity, $\mathbf{v}_i(0)$ $(i = 1, N)$; one can solve numerically Newton's equations obtaining the positions $\mathbf{x}_i(t)$ and velocities $\mathbf{v}_i(t)$ at time t. This is the essence of MD. Thus, we integrate the equations of motion by the popular finite differences algorithm due to Verlet [20], where $\mathbf{r}(t + \delta t)$ is expanded in a Taylor series up to third order in δt. Omitting the index i for simplicity and denoting $\mathbf{x}_i \equiv \mathbf{r}$, one obtains,

$$\mathbf{r}(t + \delta t) = \mathbf{r}(t) + (\delta t)\mathbf{v}(t) + \frac{1}{2}(\delta t)^2\mathbf{a}(t) + \frac{1}{6}(\delta t)^3\mathbf{b}(t) + \ldots$$

(10.19)

$$\mathbf{r}(t - \delta t) = \mathbf{r}(t) - (\delta t)\mathbf{v}(t) + \frac{1}{2}(\delta t)^2\mathbf{a}(t) - \frac{1}{6}(\delta t)^3\mathbf{b}(t) + \ldots$$

and adding these equations [up to order $(\delta t)^4$] gives:

$$\mathbf{r}(t + \delta t) = 2\mathbf{r}(t) - \mathbf{r}(t - \delta t) + (\delta t)^2\mathbf{a}(t) + O[(\delta t)^4].$$

(10.20)

Thus, $\mathbf{r}(t + \delta t)$ is independent of the velocity, and the acceleration, $\mathbf{a}(t)$, is calculated from $\mathbf{F}/m$, where $\mathbf{F}$ is the resultant force exerted on atom i by all the other atoms. Calculation of $\mathbf{F}(i)$, which depends on $3(N - 1)$ equations similar to Equation (10.20) is the main time-consuming factor, as these calculations should be carried out for each of the N atoms. The velocity (which is required for calculating the kinetic energy and temperature) is,

$$\mathbf{v}(t) = [\mathbf{r}(t + \delta t) - \mathbf{r}(t - \delta t)] / 2(\delta t), \tag{10.21}$$

that is, $\mathbf{v}(t)$ is correct up to $(\delta t)^2$ only, and, therefore, its calculation (hence, the kinetic energy) is not precise enough. $\mathbf{v}(t)$ can also be estimated at the half step, $t + (1/2)\delta t$,

$$\mathbf{v}\left(t + \frac{1}{2}\delta t\right) = [\mathbf{r}(t + \delta t) - \mathbf{r}(t)] / \delta t. \tag{10.22}$$

This is an iterative process starting with initial coordinates and velocities, which both define the total energy. However, the procedure is approximate, since if δt is not small enough, numerical instabilities and drift in the total energy will occur. However, in practice, Verlet's algorithm maintains a constant energy for relatively long times (for a wider discussion about Verlet's algorithm, see, for example, [10]). While most of the computer time is spent on calculating the forces, $\mathbf{F} = \mathbf{a}m$, this time is also strongly affected by the size of δt. In particular, δt should be very small for integrating adequately the strong harmonic bond stretching potentials in proteins or other organic molecules. Therefore, in most studies of such systems, these bonds are held constant by particular procedures like SHAKE [21] and RATTLE [22]. Thus, for proteins, typically, $\delta t = 0.5 - 4$ fs (1 fs = 10^{-15} s), with the best computers to date, protein systems are limited to millisecond simulations.

Very small changes in the initial conditions will lead to different trajectories. However, as pointed out earlier, we are not interested in the trajectory, per se. We seek to generate a typical equilibrium trajectory, that is, one which would lead to the correct statistical mechanical averages of properties of interest. Many other algorithms have been developed. Some are equivalent to Verlet's method, such as the *"leap-frog"* method of Hockney [23], and *"velocity Verlet"* [24], where the calculation of $\mathbf{v}$ is more accurate, thus:

$$\mathbf{r}(t + \delta t) = \mathbf{r}(t) + \mathbf{v}(t)\delta t + \frac{1}{2}\mathbf{a}(t)(\delta t)^2$$

$$\mathbf{v}(t + \delta t) = \mathbf{v}(t) + [\mathbf{a}(t) + \mathbf{a}(t + \delta t)]\frac{\delta t}{2} \tag{10.23}$$

Here, first $\mathbf{r}(t + \delta t)$ is calculated from $\mathbf{v}(t)$ and $\mathbf{a}(t)$. $\mathbf{v}(t + \delta t)$ is calculated in two stages, where the first is at mid-step, that is, $\mathbf{v}(t + \delta t/2)$:

$$\mathbf{v}\left(t + \frac{1}{2}\delta t\right) = \mathbf{v}(t) + \frac{(\delta t)}{2}\mathbf{a}(t). \tag{10.24}$$

Finally, $\mathbf{a}(t + \delta t)$ is calculated from the corresponding force, which leads to $\mathbf{v}(t + \delta t)$:

$$\mathbf{v}(t + \delta t) = \mathbf{v}\left(t + \frac{1}{2}\delta t\right) + \frac{(\delta t)}{2}\mathbf{a}(t + \delta t) \tag{10.25}$$

A typical equilibration of the potential energy as a function of time is shown in Figure 10.6.

Several other integration methods have been developed, such as that of Beeman [25], Gear [26,27], and Nośe [28–30]. However, elaborating on these methods is beyond the scope of this book. The reader is advised to check also references [4–8,10].

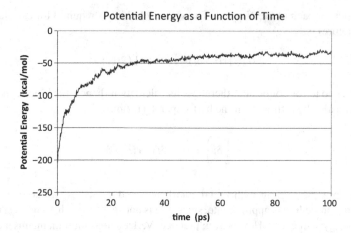

FIGURE 10.6 The figure presents an MD run of 216 argon atoms run by the velocity Verlet algorithm at constant energy (*NVE*) at $T = 200K$, meaning that the system is a dense gas (T_{critical} 150K). The time step is $\delta t = 5$ femtosecond, where each frame is taken after 20 time steps (0.1 ps), The total runtime for 2500 frames is 250 ps. To demonstrate the independence of the equilibrium behavior on the initial conditions (ergodic theorem), the simulation starts from a non-typical configuration, where the atoms are concentrated at the corner with a density of liquid argon (that is, 8 times higher than the system density after equilibration).

10.7 MD Simulations in the Canonical Ensemble

As we have seen, in the microcanonical (*NVE*) ensemble. an MD run is completely mechanistic. However, in the canonical (*NVT*) ensemble, where T replaces E, one has to provide a "thermostat," that is, a procedure for keeping T constant. Common thermostats are those suggested by Berendsen [31] and Andersen [32] (and the Nośe-Hoover thermostat [28–30]).

In the canonical ensemble, the system is in contact with a large (infinite) heat bath with constant T_{bath}, which exchanges energy with the system. At equilibrium, the average temperature of the system, T_{system}, is equal to T_{bath}, where T_{system} slightly fluctuates around its average value, the larger the system, the smaller the fluctuation. The temperature can be controlled through the kinetic energy per degree of freedom, $mv^2 = k_B T$. According to the Berendsen thermostat, at each MD time step, the velocities are rescaled by the factor:

$$\chi = \left(1 + \frac{\delta t}{t_T} \left[\frac{T_{\text{bath}}}{T_{\text{system}}} - 1 \right] \right)^{1/2} \tag{10.26}$$

where δt is the time step and t_T is a parameter. For one degree of freedom, assuming $\delta t = t_T$, one obtains:

$$mv^2 \chi^2 = k_B T_{\text{system}} \, \chi^2 = k_B T_{\text{system}} \left(1 + \left[\frac{T_{\text{bath}}}{T_{\text{sysem}}} - 1 \right] \right) = k_B T_{\text{bath}} \tag{10.27}$$

For $\delta t < t_T$, the change in the velocities is more moderate, and $t_T = 0.4$ ps has been found to be appropriate. While this procedure is easy to implement, it does not obey the canonical ensemble distribution, that is, the velocities are not distributed according to Maxwell-Boltzmann [Equation (4.27)]. This procedure can lead to inhomogeneous temperature over the system, where *local* hot and cold regions are created, while on average, T_{system} is adequate, that is, $T_{\text{system}} \sim T_{\text{bath}}$.

With the Andersen thermostat, at every predefined time interval, an atom is selected at random, and its velocity is redefined by drawing a new velocity from a Maxwell-Boltzmann distribution. Thus, during each interval, the system moves at constant energy, which is changed from interval to interval. The interval is proportional to,

$$\frac{\lambda_T}{\rho^{1/3} N^{2/3}} \qquad (10.28)$$

where λ_T is the thermal conductivity, ρ is the density, and N is the number of particles. With the Andersen thermostat, the velocities are distributed as needed—according to Maxwell-Boltzmann. However, the trajectory is not smooth, it is a collection of short microcanonical stretches. If the interval is long, the distribution is not canonical (becoming close to microcanonical). If the interval is small, the velocities are changed too frequently and the fluctuation of the kinetic energy is incorrect. In Section 12.4.2, we shall discuss the "Langevin thermostat," which is better than the Andersen and Berendsen thermostats.

MD has been applied also to other ensembles, in particular, to the NpT ensemble, where in addition to a thermostat, one has to define a barostat to control the pressure; for details see [4–8,10].

10.8 Dynamic MD Calculations

MD not only provides statistical mechanics averages, but also allows calculating dynamical properties. One can calculate from an MD trajectory time correlation functions that lead to dynamic parameters, such as diffusion coefficients. For example, the autocorrelation function of the velocity $< \mathbf{v}_i(0) \cdot \mathbf{v}_i(t) >$ can be estimated from a MD trajectory, where the velocities are measured n times in time intervals t:

$$< \mathbf{v}_i(0) \cdot \mathbf{v}_i(mt) > \;\;\rightarrow\;\; \frac{1}{(n-m-1)N} \sum_{k=m+1}^{n} \sum_{i=1}^{N} \mathbf{v}_i[kt] \cdot \mathbf{v}_i[(k-m)t] \qquad (10.29)$$

In practice, the measurements start at time $(m + 1)t$ and go back m time intervals. Contributions from all particles are considered and averaged. In the same way, one can estimate $<|\mathbf{r}_i(t) - \mathbf{r}_i(0)|^2>$ by averaging over all particles. One obtains (a Green-Kubo relationship, see Section 12.3.2):

$$\int_0^\infty < \mathbf{v}(\tau) \cdot \mathbf{v}(0) > d\tau = \lim_{t \to \infty} \frac{<|\mathbf{r}(t) - \mathbf{r}(0)|^2>}{2t} = 3D, \qquad (10.30)$$

where $\mathbf{v}$ and $\mathbf{r}$ denote particle velocity and position vectors, respectively. D is the self-diffusion coefficient. Notice that t should be large and, in practice, the length of the trajectory might be insufficient. Other correlation functions lead to the corresponding transport properties (Section 12.3.2).

10.9 Efficiency of MD

MD is a robust method, which is used extensively in a variety of systems, fluids, polymers, biological macromolecules, etc. However, as pointed out earlier, in an MD simulation, two successive snapshots at times t and $t + \delta t$ are *correlated* for typical δt values, the smaller the δt, the stronger is the correlation, and the lower is the efficiency; therefore, one seeks to maximize δt without harming the accuracy of the integration of the equations of motion. Clearly, adequate integration of the strong "spring constants," which keeps the bond lengths of a protein approximately constant requires a small time step, δt, which would lead to relatively short trajectories. To eliminate this effect, in most studies, the bond lengths are constrained by procedures, such as SHAKE [21] or RATTLE [22].

MD (like MC) is an efficient simulation method for fluids, while for proteins, MD is more efficient than MC, because the system moves according to the gradients, as pointed out earlier; still, for proteins, MD by itself is quite a limited technique. This stems from the fact that the potential energy surface of a protein is very rugged, "decorated" by a tremendous number of "valleys" and "hills," which makes it very difficult for an MD simulation to cross such energy barriers (Section 16.6). Therefore, the protein will spend a long simulation time in the first valley that it meets, and, in practice, the (desired) most stable region will never be reached. This inefficiency has led to the development of sophisticated methods, such as *"replica exchange,"* where MD is used within a theoretical framework, which drives the protein to cross energy barriers (Chapter 21).

10.9.1 Periodic Boundary Conditions and Ewald Sums

We point out that most MD (and MC) simulations are carried out with periodic boundary conditions, which can be explained simply in one dimension. Thus, assume a simulation box that starts at the origin $(0, 0, 0)$ with dimensions (x_0, y_0, z_0); if a particle leaves the box at x_0 by Δx, the program will move it from $(x_0 + \Delta x, y, z)$ back to the origin at $(\Delta x, y, z)$, keeping the velocity vector intact; thus the particle always remains inside the box. In practice, due to long-range interactions, one implements this restriction by building around the central box *identical* periodic (image) boxes, as described in the one-dimensional Figure 10.7; thus, when the empty particle is moved from the central box to the right box, a particle will move in the same direction from the left box to the central one. The same rule is applied in three dimensions. As we shall see below, to take into account the effect of long-range interaction, it is useful to construct many image boxes around the central one (Figure 10.7).

Now, calculating the long-range electrostatic interactions is time-consuming since their numbers increase with system size, as $O(N^2)$. The early attempts to decrease this factor have been by decreasing the interaction range to 8–15 Angstroms (using switching functions, which smoothly reduce the interactions to zero); however, this approximation has been found to be unsatisfactory.

A more sophisticated and efficient way to handle long-range interactions is by applying particle mesh Ewald (PME) [33–35] (or other derivatives of this procedure), where the computation efficiency is enhanced significantly becoming of the order of $O(N\log N)$. With PME, the long-range environment of the central box is defined by a large set of periodic boxes built in layers around it, where each of the N charges in the central box interacts with all the charges in the image boxes, as their number grows to infinity. The electrostatic energy is divided into two parts: a short-range contribution and a long-range one; the short-range contribution is calculated in real space, whereas the long-range contribution is calculated efficiently using a Fourier transform. PME is now implemented in all the molecular mechanics/molecular dynamics software packages for proteins (see Section 16.3).

The mathematical details of this procedure, which are quite involved, have been discussed in many papers, books, and lectures; therefore, they will not be described here, and the reader is advised to check for example, [32–35] and [4–8,10]. Another way for treating long-range electrostatic interactions is by the reaction field method, which has been less popular than PME, but has been used extensively for proteins by the van Gunsteren group at ETH, Zurich [36].

10.9.2 A Comment About MD Simulations and Entropy

Entropy is commonly stated as a measure of knowledge (through the probability, P), and one might think that increasing the knowledge about the system will reduce S (In Section 10.6, we have already discussed an example of an MD trajectory providing such extra information.). This conclusion is, of course, wrong because theoretical considerations should correspond to the experimental reality. Thus, returning

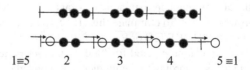

FIGURE 10.7 1-*d* periodic boundary conditions consisting of four image boxes around the central box number 3. Moving of a particle (empty circle) from the central box to the right, leads to the same movement in all boxes, and to the appearance of an (empty) particle to the left side of the central box. Boxes 1 and 5 are the same box.

to Figure 2.1a of Section 2.2, it is clear that even if the velocities and positions of all particles would be known, the maximum amount of work that could be gained would still be limited, depending on the *inherent physical disorder* of the system, which is measured by S. Thus, extra knowledge will not reduce entropy, and to remain within the domain of statistical mechanics, the information gathered from a trajectory should be appropriate, e.g., only averages should be considered. An advantage of the simulation methodology is that one can measure microscopic properties (e.g., atomic distances in a protein), which are not amenable to the experiment. Still, only *average values* are meaningful in statistical mechanics.

REFERENCES

1. N. Metropolis, A. Rosenbluth, M. Rosenbluth, A. Teller and E. Teller. Equations of state calculations by fast computing machines. *J. Chem. Phys.* **21**, 1087–1092 (1953).
2. B. J. Alder and T. E. Wainwright. Phase transition of a hard sphere system. *J. Chem. Phys.* **27**, 1208–1209 (1957).
3. B. J. Alder and T. E. Wainwright. Studies in molecular dynamics. I. General method. *J. Chem. Phys.* **31**, 459–466 (1959).
4. M. P. Allen and D. J. Tildesley. *Computer Simulation of Liquids.* (Oxford University Press, New York, 1991).
5. H. J. C. Berendsen. *Simulating the Physical World: Hierarchical Modeling from Quantum Mechanics to Fluid Dynamics.* (Cambridge University Press, Cambridge, UK, 2007).
6. D. Frenkel and B. Smit. *Understanding Molecular Simulation from Algorithms to Applications.* (Academic Press, London, UK, 2002).
7. D. C. Rapaport. *The Art of Molecular Dynamics Simulation.* (Cambridge University Press, Cambridge, UK, 2004).
8. T. Schlick. *Molecular Modeling and Simulation: An Interdisciplinary Guide.* (Springer, New York, 2002).
9. D. P. Landau and K. Binder. *A Guide to Monte Carlo Simulations in Statistical Physics.* (Cambridge University Press, New York, 2009).
10. A. R. Leach. *Molecular Modeling. Principles and Applications.* (Pearson, Prentice Hall, New York, 2001).
11. C. Pangali, M. Rao and B. J. Berne. A Novel Monte Carlo scheme for simulating water and aqueous solutions. *Chem. Phys. Lett.* **55**, 413–417 (1978).
12. M. Rao and B. J. Berne. On the force bias Monte Carlo simulation of simple liquids. *J. Chem. Phys.* **71**, 129–132 (1979).
13. G. E. Norman and V. S. Filinov. Investigations of phase transitions by a Monte-Carlo method. *High Temp.* (*USSR*) **7**, 216–224 (1969).
14. D. J. Adams. Chemical potential of hard-sphere fluids by Monte Carlo methods. *Mol. Phys.* **28**, 1241–1252 (1974).
15. P. H. Verdier and W. H. Stockmayer. Monte Carlo calculations on the dynamics of polymers in dilute solution. *J. Chem. Phys.* **36**, 227–235 (1962).
16. H. J. Hilhorst and J. M. Deutch. Analysis of Monte Carlo results on the kinetics of lattice polymer chains with excluded volume. *J. Chem. Phys.* **63**, 5153–5161 (1975).
17. H. Meirovitch. Efficient dynamical Monte Carlo method for dense polymer systems. *Macromolecules* **17**, 2038–2044 (1984).
18. A. Rahman. Correlations in the motion of atoms in liquid argon. *Phys. Rev.* **136**, A405–A411 (1964).
19. J. A. McCammon, B. R. Gelin and M. Karplus. Dynamics of folded proteins. *Nature* **267**, 585–590 (1977).
20. L. Verlet. Computer "experiments" on classical fluids. I. Thermodynamical properties of Lennard–Jones molecules of L. *Phys. Rev.* **159**, 98–103 (1967).
21. J.-P. Ryckaert, G. Ciccotti and H. J. C. Berendsen. Numerical integration of the Cartesian equations of motion of a system with constraints: Molecular dynamics of n-alkanes. *J. Comput. Phys.* **23**, 327–341 (1977).
22. H. C. Andersen. Rattle: A "velocity" version of the shake algorithm for molecular dynamics calculations. *J. Comput. Phys.* **52**, 24–34 (1983).

23. R. W. Hockney. The potential calculation and some applications. *Methods Comput. Phys.* **9**, 136–211.

24. W. C. Swope, H. C. Andersen, P. H. Berens and K. R. Wilson. A computer simulation method for the calculation of equilibrium constants for the formation of physical clusters of molecules: Application to small water clusters. *J. Chem. Phys.* **76**, 637–649 (1982).

25. D. Beeman. Some multistep methods for use in molecular dynamics calculations. *J. Comput. Phys.* **20**, 130–139 (1976).

26. C. W. Gear. *Numerical Initial Value Problems in Ordinary Differential Equations.* (Prentice Hall, Englewood Cliffs, NJ, 1971).

27. W. F. van Gunsteren and H. J. C. Berendsen. Algorithms for macromolecular dynamics and constraint dynamics. *Mol. Phys.* **34**, 1311–1327 (1977).

28. S. Nosé. A unified formulation of the constant temperature molecular-dynamics methods. *J. Chem. Phys.* **81**, 511–519 (1984).

29. S. Nosé. A molecular dynamics method for simulations in the canonical ensemble. *Mol. Phys.* **52**, 255–268 (1984).

30. W. G. Hoover. Canonical dynamics: Equilibrium phase-space distributions. *Phys. Rev. A.* **31**, 1695–1697 (1985).

31. H. J. C. Berendsen, J. P. M. Postma, W. F. van Gunsteren, A. DiNola and J. R. Haak. Molecular-dynamics with coupling to an external bath. *J. Chem. Phys.* **81**, 3684–3690 (1984).

32. H. C. Andersen. Molecular dynamics simulations at constant pressure and/or temperature. *J. Chem. Phys.* **72**, 2384–2393 (1980).

33. T. Darden, D. York and L. Pedersen. Particle mesh Ewald: An $N \cdot \log(N)$ method for Ewald sums in large systems. *J. Chem. Phys.* **98**, 10089–10094 (1993).

34. A. Y. Toukmaji and J. A. Board, Jr. Ewald summation techniques in perspective: A survey. *Comput. Phys. Commun.* **95**, 73–92 (1996).

35. P. H. Hünenberger. Ewald artifacts in computer simulations of ionic solvation and ion-ion interaction: A continuum electrostatics study. *J. Chem. Phys.* **110**, 1856–1872 (1999).

36. I. G. Tironi, R. Sperb, P. E. Smith and W. F. van Gunsteren. A generalized reaction field method for molecular dynamics simulations. *J. Chem. Phys.* **102**, 5451–5459 (1995).

11

Non-Equilibrium Thermodynamics—Onsager Theory

11.1 Introduction

The previous chapters have been devoted to systems in equilibrium, where the thermodynamic properties are *time-independent*, that is, they are invariant with respect to the sign of time, $t \rightarrow -t$; an example for time invariance is the propagation of waves,

$$\frac{1}{c^2}\frac{\partial^2 u}{\partial t^2} = \frac{\partial^2 u}{\partial x^2} + \frac{\partial^2 u}{\partial y^2} + \frac{\partial^2 u}{\partial z^2} \qquad (11.1)$$

However, most processes in Nature are irreversible, hence, irreversibility has become a wide research field developed considerably by the two Nobel laureates, L. Onsager and I. Prigogine. Therefore, we wish to widen the scope of equilibrium thermodynamics and statistical mechanics by considering non-equilibrium *time-dependent* processes. In a typical problem, a system is removed from equilibrium by external forces, and the relaxation back to equilibrium or to a stationary steady state is studied. Thus, for an irreversible process, time has a direction (see Section 2.8.1) reflected, for example, by the Fourier equation for the temperature, which, unlike Equation (11.1), is based on a first derivative of time,

$$\frac{1}{\alpha}\frac{\partial T}{\partial t} = \frac{\partial^2 T}{\partial x^2} + \frac{\partial^2 T}{\partial y^2} + \frac{\partial^2 T}{\partial z^2}. \qquad (11.2)$$

The fields of non-equilibrium thermodynamics and statistical mechanics distinguish between two regimes: *"close to equilibrium"* and *"far from equilibrium;"* we shall discuss only the close to equilibrium case. This chapter, which is devoted to Onsager's reciprocal relations [1,2], is based also on the books [3–8].

Actually, the Monte Carlo (MC) and Molecular Dynamics (MD) simulation methods discussed in Chapter 10 exhibit irreversible relaxation processes. Thus, if for example, an MC simulation of an Ising model at *high* T starts from the completely ordered configuration (all spins up—typical for $T = 0$), the typical configurations after a short simulation time will still be very different from the typical (equilibrium) random configurations at T (far from equilibrium condition); as the simulation proceeds, the system will relax to equilibrium, passing through a "close to equilibrium" state. At this state, the relaxation is slow, and local ensembles of similar configurations can be defined within short time intervals, Δt (see below).

11.2 The Local-Equilibrium Hypothesis

Definition of the close to equilibrium regime is based on the *local-equilibrium hypothesis*, which states: the system can be split into small cells, small enough that changes in thermodynamic properties within a cell (that is, gradients) can be neglected. However, the cells are sufficiently large (containing a sufficiently large number of particles) that *microscopic* fluctuations can be neglected. Moreover, all thermodynamic quantities (that is, T, S, and the Gibbs equation) are rigorously defined at each cell *as in equilibrium*. These quantities remain uniform at each cell, but *slightly* differ from cell to cell, and are changed in time; the basic unit of time, Δt, is much shorter than any *macroscopic* time evolution, but is the largest for which thermodynamic properties in a cell remain constant. Thus, the relaxation of a system to equilibrium is not very fast. Since the internal entropy increases in an irreversible process, the entropy production in time is an essential quantity of interest.

Our goal is to describe Onsager's theory, where additional assumptions related to irreversible thermodynamics (e.g., linearity) are made. We build the theory step-by-step from the basic concepts up. Thus, we start by writing again [Equation (2.6)], the first law of thermodynamics for a *closed* system:

$$dE = d'Q - pdV. \tag{11.3}$$

It should be emphasized that $d'Q$ is not an exact differential. We also write again the Gibbs equation for equilibrium [Equations (2.38) and (2.42)],

$$dE = TdS - pdV + \sum_{\gamma} \mu_{\gamma} dN_{\gamma}$$

or

$$dS = \frac{1}{T} dE + \frac{p}{T} dV - \sum_{\gamma} \frac{\mu_{\gamma}}{T} dN_{\gamma} \tag{11.4}$$

As said above, under the close to equilibrium assumption, the Gibbs equation is satisfied at each cell. Next (as in Equation 2.29), we divide the change of entropy, dS, into two components [3],

$$dS = d_e S + d_i S, \tag{11.5}$$

where $d_e S$ is the flow of entropy due to interactions with the exterior and $d_i S$ is the contribution to the entropy due to changes inside the system; $d_i S$ is called *"the entropy production."* We repeat below the second law of thermodynamic [Equations (2.29) and (2.30)],

$$d_i S = 0 \quad \text{in a reversible process} \tag{11.6a}$$

$$d_i S > 0 \quad \text{in an irreversible process} \tag{11.6b}$$

On the other hand, $d_e S$ can be positive, negative, or zero. For an *isolated system* (heat and matter cannot enter the system from outside) one obtains:

$$d_e S = 0 \quad \text{and} \quad dS = d_i S \geq 0, \tag{11.7}$$

This result is equivalent to the classical statement that entropy never decreases (Section 2.8.1).

11.3 Entropy Production Due to Heat Flow in a Closed System

Figure 11.1 presents a system consisting of two phases, I and II, with an internal contact through which heat can be transferred; the corresponding temperatures are T^I and T^{II}, where T^{II} is only *slightly* larger than T^I (a close to equilibrium condition). When II loses an amount of heat, $d_i^{II}Q$, compartment I will gain $d_i^I Q$, where $d_i^{II}Q = -d_i^I Q$ (for consistency with the literature, we write $d_i^I Q$ to emphasize that it is not an exact differential). Also, amounts of heat, $d_e^{II}Q$ and $d_e^I Q$ flow in and out of the corresponding compartments. The total change of entropy, dS, is based on the *equilibrium* formula, $S = dQ/T$, used for the various partial entropies,

$$dS = d_e S + d_i S = \frac{d_e^I Q}{T^I} + \frac{d_e^{II} Q}{T^{II}} + d_i^I Q \left(\frac{1}{T^I} - \frac{1}{T^{II}} \right) \tag{11.8}$$

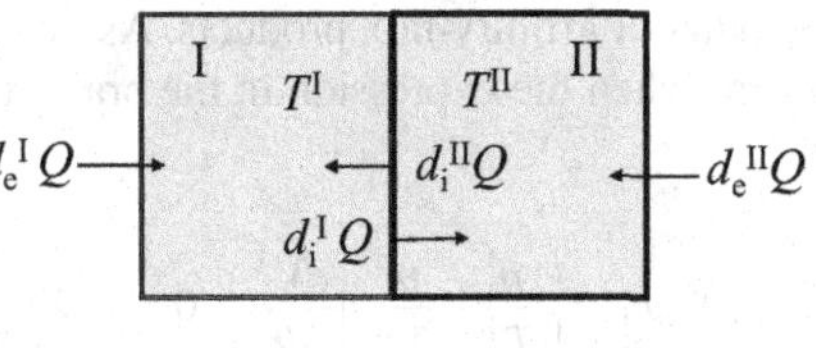

FIGURE 11.1 Two compartments I and II with temperatures T^I and T^{II} ($T^{II} > T^I$) are separated by an internal contact. Amount of heat $d_i^{II}Q$ flows from II to I $\left(d_i^{II}Q = -d_i^I Q\right)$. $d_e^{II}Q$ and $d_e^I Q$ flow in and out of II and I, respectively.

where $d_e S$ is the entropy flow through the exterior,

$$d_e S = \frac{d_e^I Q}{T^I} + \frac{d_e^{II}Q}{T^{II}} \tag{11.9}$$

and the second term is the *entropy production*, $d_i S$, which we write below per unit time,

$$\frac{d_i S}{dt} = \frac{d_i^I Q}{dt}\left(\frac{1}{T^I} - \frac{1}{T^{II}}\right) > 0 \tag{11.10}$$

Notice, that as expected, for $T^I \neq T^{II}$, $d_i S$ is *always* positive since the signs of $d_i^I Q$ and $(1/T^I - 1/T^{II})$ are always the same. In other words, the direction of the heat flow and its "rate," $d_i^I Q/dt$, are determined by the sign and size of the "force," $(1/T^I - 1/T^{II})$, which can be considered as the macroscopic cause of the irreversible process, the larger the force, the stronger is the rate. In the special case, where $d_e^{II}Q = d_e^I Q = 0$, heat will flow from II to I until $T^I = T^{II}$, where an equilibrium is reached, that is, $d_i S = 0$, and the entropy is maximal. In the general case ($d_e^{II}Q, d_e^I Q \neq 0$), the system might also reach a steady state. For a small $\Delta T = T^I - T^{II}$, one obtains:

$$\frac{\partial S_i}{\partial Q} = \frac{1}{T} - \frac{1}{T + \Delta T} \approx \frac{\Delta T}{T^2} \tag{11.11}$$

where S_i is the internal entropy of the two phases.

As mentioned above, Equation (11.10) consists of the difference $(1/T^I - 1/T^{II})$, which is considered as a generalized force, called "affinity," and denoted by X (we shall denote it here by X_T); this force is multiplied by the corresponding *induced* irreversible rate, $d_i^I Q/dt$, denoted in general by the letter J, and called "flux" (we shall denote it here by J_T), thus, Equation (11.10) becomes,

$$\frac{d_i S}{dt} = \left(\frac{1}{T^I} - \frac{1}{T^{II}}\right)\frac{d_i^I Q}{dt} = \frac{\partial S_i}{\partial Q}\frac{d_i^I Q}{dt} = X_T J_T \tag{11.12}$$

11.4 Entropy Production in an Isolated System

The above example can be extended to the more general case in which the partition between compartments I and II is movable and permeable to one component of the system. Also, the system is adiabatic as a whole, and the total energy is thus constant, $E = E^I + E^{II} = $ constant. The two compartments are characterized by different chemical potentials, μ^I and μ^{II}, with the corresponding mole numbers n^I and n^{II}, and volumes, $V = V^I + V^{II}$, with the corresponding pressures, p^I and p^{II}. Based on the Gibbs equation [Equation (11.4)], and the same considerations led to Equation (11.8), the rate of entropy production is [4],

$$\frac{d_i S}{dt} = \left(\frac{1}{T^I} - \frac{1}{T^{II}}\right)\frac{dE^I}{dt} + \left(\frac{p^I}{T^I} - \frac{p^{II}}{T^{II}}\right)\frac{dV^I}{dt} - \left(\frac{\mu^I}{T^I} - \frac{\mu^{II}}{T^{II}}\right)\frac{dn^I}{dt}. \tag{11.13}$$

This equation is defined by three pairs of affinity-flux products. As discussed for the closed system earlier, each product is larger than zero when the expression in the bracket is different from zero (positive or negative):

$$X_T J_T = \left(\frac{1}{T^I} - \frac{1}{T^{II}}\right)\frac{dE^I}{dt} > 0 \qquad X_p J_p = \left(\frac{p^I}{T^I} - \frac{p^{II}}{T^{II}}\right)\frac{dV^I}{dt} > 0 \qquad X_\mu J_\mu = \left(\frac{\mu^I}{T^I} - \frac{\mu^{II}}{T^{II}}\right)\frac{dn^I}{dt} > 0 \quad (11.14)$$

The affinities are partial derivatives of the entropy, S_i, with respect to the corresponding extensive variables; as in Equation (11.11), one can calculate derivatives for small differences $\Delta\mu$, ΔT, and Δp:

$$X_\mu = \frac{\partial S_i}{\partial n} = \frac{\mu}{T} - \frac{\mu + \Delta\mu}{T + \Delta T} \approx \frac{\mu(\Delta T)_\mu}{T^2} - \frac{(\Delta\mu)_T}{T} \qquad X_p = \frac{\partial S_i}{\partial V} = \frac{p}{T} - \frac{p + \Delta p}{T + \Delta T} \approx \frac{p(\Delta T)_p}{T^2} - \frac{(\Delta p)_T}{T} \quad (11.15)$$

Thus, the form of a general equation for the rate of entropy production, d_iS/dt, becomes a summation of all the affinity-flux products relevant for a problem; denoting the various extensive variables by x_k, one obtains,

$$\frac{d_iS}{dt} = \sum_k \frac{\partial S_i}{\partial x_k}\frac{dx_k}{dt} = \sum_k X_k J_k > 0 \tag{11.16}$$

11.5 Extra Hypothesis: A Linear Relation Between Rates and Affinities

At equilibrium, where the affinities vanish ($X_k = 0$), the rates are zero as well ($J_k = 0$). Close to equilibrium, the affinities are not too strong, and one can assume a linear relation between affinities and the related rates (fluxes), $J_k \sim X_k$; thus, for a single process one defines:

$$J_k = L_k X_k, \tag{11.17}$$

where L_k is a phenomenological coefficient *independent of time*. For two irreversible processes (mixing):

$$J_1 = L_{11}X_1 + L_{12}X_2$$
$$J_2 = L_{21}X_1 + L_{22}X_2 \tag{11.18}$$

And, in general, for the ith irreversible process out of l processes,

$$J_i = \sum_{k=1}^{l} L_{ik} X_k. \tag{11.19}$$

This equation is sometimes called the *regression equation*, where its solution depends on the initial conditions. The coefficients, L_{ik} for $i \neq k$, are called *interference coefficients*. J_k is mostly affected by X_k, and less by the other forces. J_k and L_k are functions of *all* the affinities and intensive variables (e.g., T, μ). For two irreversible processes, d_iS/dt can be expressed as a function of the forces and the coefficients,

$$\frac{dS_i}{dt} = L_{11}X_1^2 + (L_{12} + L_{21})X_1 X_2 + L_{22}X_2^2 \tag{11.20}$$

where one can show that the following relations exists [3],

$$L_{11} > 0; \qquad L_{22} > 0; \qquad (L_{12} + L_{21})^2 < 4L_{11}L_{22} \tag{11.21}$$

It should be noted that L_{11} and L_{22} are of a different dimensionality, while the dimensionalities of L_{12} and L_{21} are the same. Moreover, as discussed later, Onsager has derived the *reciprocal* relations for the interference coefficients:

$$L_{i,k} = L_{k,i} \tag{11.22}$$

Thus, for two processes,

$$\frac{dS_i}{dt} = L_{11}X_1^2 + 2L_{12}X_1X_2 + L_{22}X_2^2. \tag{11.23}$$

So far, the phenomenological theory has been presented for a discrete system, while treating realistic systems is based on a continuum theory; some of its features are described in Section 11.6.

11.5.1 Entropy of an Ideal Linear Chain Close to Equilibrium

For an isolated system, the equilibrium entropy, S_0, is maximal. If the system is (slightly) perturbed to a non-equilibrium state (close to equilibrium condition), S is decreased and can be expanded as:

$$S = S_0 - \frac{1}{2}G_{ij}y_i y_j \tag{11.24}$$

where y_i and y_j are extensive parameters that vanish at equilibrium and G_{ij} is a positive definite symmetric matrix (see Section 11.7). The derivatives, $X_i = \partial S/\partial y_i = -\Sigma_j G_{ij}y_j$, have been defined in the previous sections as forces deriving the system back to equilibrium, where the rate dy_i/dt depends linearly on the forces $\dot{y} = \Sigma_j L_{ij}X_j$ (for $dy/dt = -fy \rightarrow y(t) = c\exp{-ft}$ that is, an exponential decay).

Example

For an one-dimensional ideal chain, we have obtained that the total equilibrium entropy is $nk_B\ln2$, where n is the number of bonds and a is the bond length. However, if the chain is restricted to end-to-end distance, x, the entropy is reduced to [Equation (8.7)]:

$$S(x) = nk_B\left\{\ln 2 - \frac{1}{2}\left(1 + \frac{x}{na}\right)\ln\left(1 + \frac{x}{na}\right) - \frac{1}{2}\left(1 - \frac{x}{na}\right)\ln\left(1 - \frac{x}{na}\right)\right\} \tag{11.25}$$

while in the extreme case of a completely stretched chain ($x = na$), $S(na) = 0$, and for a small x, we have found that the force, K, required to hold the chain at x [Equation (8.10)] is:

$$K = \frac{k_B T}{Na^2}x \tag{11.26}$$

However, one can view this situation as moving the system from equilibrium to a close to equilibrium state, where Onsager' theory can provide additional information—the exponential decay (in time) of the system to equilibrium. The starting point is the non-equilibrium entropy:

$$S = S_0 - \frac{k_B x^2}{2na^2}. \tag{11.27}$$

PROBLEM 11.1

Derive Equation (11.27) from Equation (11.25) for $f = x/na \ll 1$. Now, one can calculate the force, X, based on entropic grounds,

$$X = \left(\frac{\partial S}{\partial x} \right) = -\frac{k_B}{na^2} x = -K \tag{11.28}$$

which is the same force obtained in Equation (11.26), but with an opposite direction, that is, toward $x = 0$. According to Equation (11.17),

$$\dot{x} = -L \frac{k_B}{na^2} x \Rightarrow \qquad x = C \exp \left[-L \frac{k_B}{na^2} \right] t. \tag{11.29}$$

Thus, the linearity between forces and fluxes leads to an exponential decay of x (and S) to equilibrium.

11.6 Fourier's Law—A Continuum Example of Linearity

For many everyday processes, the affinity-rate linearity is found to be valid. Known examples are the following experimental laws: (1) Fick's law, which describes the diffusion of particles in a fluid (Section 12.1), (2) Ohm's law, which states that the flow of electrical current is proportional to the gradient of the voltage, and (3) Fourier's law, discussed below, which states that the flow of energy (heat) through a conducting body is proportional to the gradient of the temperature [5],

$$\mathbf{J}_0 = -\kappa \nabla T \tag{11.30}$$

Thus, for a general *continuous* system, the energy flux, dQ/dt, is replaced by the energy current density vector, $\mathbf{J}_0$ [$\equiv (J_{0x}, J_{0y}, J_{0z})$], where J_{0i} is the energy flow through a unit area perpendicular to direction i per second; κ is the thermal conductivity (in units of mass $\cdot$ length $\cdot$ time$^{-3} \cdot T^{-1}$). To relate the continuum theory to that developed for the discrete models in the previous sections, we treat the special case depicted in Figure 11.2—the flow of heat through a metal rod with ends kept at (constant) high and low T.

Relying on the *local-equilibrium* hypothesis, we wish to express Equation (11.30) in terms of affinities and rates. Thus, the two *macroscopic* compartments defined in the discrete case (Figure 11.1) are replaced by two *microscopic* neighbor cells placed along the rod (in the z direction); still, these cells are big enough to allow an adequate definition of T and $T - \Delta T$. Now, instead of the affinity, $X = 1/T - 1/(T - \Delta T)$, defined in the discrete case, one defines (in the *general* continuous situation) the (vector) affinity, $\mathbf{X}$ [$\equiv (X_x, X_y, X_z)$] as the gradient of $1/T$, where for the rod only, the z-component, $\mathbf{X}_z = \nabla_z (1/T)$ (>0) is effective. Thus, Equation (11.30) becomes,

$$\mathbf{J}_{0z} = \kappa T^2 \nabla_z \left(\frac{1}{T} \right) = L \mathbf{X}_z \tag{11.31}$$

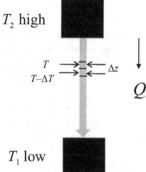

T_2 high

$T \quad \to \quad$
$T-\Delta T \quad \to \quad$ Δz

Q

T_1 low

FIGURE 11.2 The flow of heat through a metal rod with ends kept at (constant) high and low T. Two neighbor microscopic cells separated by Δz with temperatures T and $T - \Delta T$ are defined at position z.

where $\text{div}(\mathbf{J}_{0z}) = 0$, that is, $\mathbf{J}_{0z}$ is the constant independent of time (with dimension, mass $\cdot$ time^{-3}), and the system is in a steady state $[\mathbf{J}_{0z} = \kappa(T_2 - T_1)/b$, where b is the rod's length]; thus, Equation (11.31) exhibits the affinity-flux linearity with the phenomenological coefficient, L (with the dimension $T \cdot$ mass $\cdot$ length $\cdot$ time^{-3}), which changes with T along the rod

$$L = \kappa T^2. \tag{11.32}$$

According to Equation (11.23), the rate of entropy production *per volume* ($S \cdot$ time$^{-1} \cdot$ length^{-3}) is:

$$\frac{dS_i}{dt} = L\mathbf{X}_z^2 = L\left[\nabla_z\left(\frac{1}{T}\right)\right]^2 = \kappa T^2\left[\nabla_z\left(\frac{1}{T}\right)\right]^2 \tag{11.33}$$

In this example, cross terms do not appear. Clearly, if the temperature difference along the rod is increased considerably, the flow of heat will increase as well, and the *local-equilibrium hypothesis* will be violated. Thus, Equation (11.31) will become approximate as non-linear contributions become effective. However, in practice, Fick's law, Ohm's law, Fourier's law, and other transport properties are known to be valid in rather extreme conditions.

11.7 Statistical Mechanics Picture of Irreversibility

The phenomenological theory developed thus far is based on the *local-equilibrium hypothesis* and the affinity-flux linearity. The theory has been shown to gain support from several experimental phenomena. However, the complete Onsager theory (including the reciprocal relations, $L_{ik} = L_{ki}$) is based on additional assumptions, where one assumption states that the decay of a *macroscopic* disturbance close to equilibrium can be described by the decay of a *microscopic* fluctuation in equilibrium statistical mechanics. While microscopic fluctuations have already been discussed (Section 5.1), here, we present an alternative treatment due to Einstein, where their distribution is approximated by a Gaussian [6]. This theory will lead us to the reciprocal relations [7].

Assume a large isolated (microcanonical) system with N particles at equilibrium. The system's energy is limited to the shell $E + dE$, and the system's entropy is $S_{\text{syst}} = k_B\ln\Omega$, where Ω is the total phase space,

$$\Omega = \int_{E,E+dE} d\mathbf{r}^N d\mathbf{p}^N = \frac{1}{\rho_0}, \tag{11.34}$$

$\mathbf{r}^N$, $\mathbf{p}^N$ are defined in Equations (3.1) and (3.2), and $\rho_0 = 1/\Omega$ is the probability density.

Now, imagine a small volume within the larger system, which is large enough that thermodynamics is meaningful (Figure 11.3). Since the small volume is an open system interacting with the surroundings, extensive variables, such as the energy, E_{vol}, and the number of particles, N_{vol}, fluctuate around their average values in the small volume. In general, one can define a set of n extensive variables, $A_i(\mathbf{r}^N, \mathbf{p}^N)$, $i = 1$, ..., n, which are denoted by the vector $\mathbf{A}$, with averages $<A_i>$ over the small volume.

The probability density, $f(\mathbf{A})d\mathbf{A}$, that the set $\mathbf{A}(\mathbf{r}^N, \mathbf{p}^N)$ will be within the values $\mathbf{A}$ and $\mathbf{A} + d\mathbf{A}$ is:

$$f(\mathbf{A})d\mathbf{A} = \frac{1}{\Omega}\int_{\substack{\mathbf{A},\mathbf{A}+d\mathbf{A} \\ E,E+dE}} d\mathbf{r}^N d\mathbf{p}^N = \frac{\Omega(\mathbf{A})}{\Omega} \tag{11.35}$$

We shall assume that *for small fluctuations*, $(A_i - <A_i>)$, the distribution function f is a Gaussian,

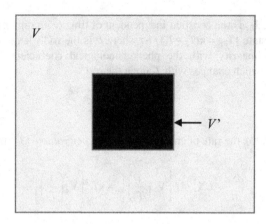

FIGURE 11.3 An *NVE* (microcanonical) ensemble, where a small volume, V', is defined in the middle of the large volume. We are interested in the fluctuations of the set of extensive variables, $\mathbf{A}(\mathbf{r}^N, \mathbf{p}^N)$, within V'.

$$f(\mathbf{A})d\mathbf{A} = c \exp\left\{-\frac{1}{2k_B}\sum_{i,j} g_{i,j}(A_j - <A_j>)(A_i - <A_i>)\right\} dA_1 dA_2 \ldots dA_n \qquad (11.36)$$

where c is a normalization constant, the g_{ij} are elements of a symmetric positive definite matrix (that is, $\Sigma g_{ij} x_i x_j$ with real x_i is positive definite), and the averages, $<A_i>$ are:

$$<A_i> = \int f(\mathbf{A}) A_i d\mathbf{A} \qquad (11.37)$$

We denote the fluctuation of A_i by α_i:

$$\alpha_i = A_i - <A_i> \qquad (i = 1, 2, \ldots, n) \qquad (11.38)$$

where the n values of α_i are represented by the vector $\boldsymbol{\alpha}$. The distribution function (11.30) becomes:

$$f(\boldsymbol{\alpha}) = c \exp\left(-\frac{1}{2k_B}\sum_{i,j} g_{i,j}\alpha_i\alpha_j\right), \qquad (11.39)$$

where the constant $c = f(\boldsymbol{\alpha} = 0)$ is determined from the normalization condition of $f(\boldsymbol{\alpha})$. The distribution function, $f(\boldsymbol{\alpha})$, is related to the entropy, $S(\boldsymbol{\alpha})$:

$$f(\boldsymbol{\alpha}) \sim \exp S(\boldsymbol{\alpha})/k_B \qquad (11.40)$$

or

$$f(\boldsymbol{\alpha}) = f(0)\exp[\Delta S(\boldsymbol{\alpha})/k_B], \qquad (11.41)$$

where

$$\Delta S(\boldsymbol{\alpha}) = S(\boldsymbol{\alpha}) - S(0) = -\frac{1}{2}\sum_{i,j} g_{i,j}\alpha_i\alpha_j \qquad (11.42)$$

In other words, for a set of *small* fluctuations, $\boldsymbol{\alpha}$, $\Delta S = S(\boldsymbol{\alpha}) - S(0)$ can be interpreted as the second-order term in a Taylor expansion of the entropy (in the α_i) around the maximal entropy, $S(0) = S_{max}$, where the first-order term vanishes and higher orders terms are neglected; the $g_{i,j}$ are,

$$g_{i,j} = \frac{\partial^2 S}{\partial \alpha_i \partial \alpha_j}\Big|_{S_{max}}. \tag{11.43}$$

Thus, $S(0)$ refers to the state of maximum probability $[f(0)]$ and can be considered as the equilibrium entropy, which is also obtained by the average $<S(\boldsymbol{\alpha})>$. However, the two definitions are in practice equivalent, as one can show that $<S(\boldsymbol{\alpha})> = S(0) - 1/2k_B n$, where n is the *small* number of extensive variables, which is negligible in comparison with the extensive, $<S(\boldsymbol{\alpha})>$ [7].

In accord with the *phenomenological* theory [Equations (11.11) and, 11.14)], we define an affinity X_i also within the present *equilibrium* theory, as the derivative of ΔS [Equation (11.42)], with respect to α_i; thus, X_i is the conjugate thermodynamic parameter to α_i. However, one should distinguish between this X_i and the macroscopic one,

$$X_i = \frac{\partial \Delta S(\boldsymbol{\alpha})}{\partial \alpha_i} = k_B \frac{\partial \ln f(\boldsymbol{\alpha})}{\partial \alpha_i} = -\sum_{k=1}^{n} g_{i,k} \alpha_k \qquad (i = 1, 2, ..., n). \tag{11.44}$$

Indeed, for $\alpha_i = \Delta E$ (using Equation (11.43) and $dS = dE/T$), we obtain the result of Equation (11.11),

$$X_i = \frac{\partial \Delta S(\boldsymbol{\alpha})}{\partial \alpha_i} = -g_{i,i} \alpha_i = -\frac{\partial}{\partial E}\left(\frac{\partial S}{\partial E}\right) \Delta E = -\frac{\partial}{\partial E}\left(\frac{1}{T}\right) \Delta E = \frac{\frac{\partial T}{\partial E}}{T^2} \Delta E \approx \frac{\Delta T}{T^2}. \tag{11.45}$$

Thus, the fluctuations of the entropy and energy in the small volume lead to a fluctuation, $\Delta T/T^2$ in the temperature.

$f(\boldsymbol{\alpha})$ and Equation (11.38) enable one to define the average $<\alpha_i X_k>$,

$$< \alpha_i X_j > = \int \alpha_i X_j f(\boldsymbol{\alpha}) d\boldsymbol{\alpha} = k_B \int \alpha_i \frac{\partial f(\boldsymbol{\alpha})}{\partial \alpha_j} d\boldsymbol{\alpha} \tag{11.46}$$

where we have used $X_j = \frac{d \ln f}{\partial \alpha_j} = \frac{1}{f} \frac{\partial f}{\partial \alpha_j}$. Integration by parts leads to:

$$< \alpha_i X_j > = [f \alpha_i]_{-\infty}^{+\infty} - k_B \int f(\boldsymbol{\alpha}) \frac{\partial \alpha_i}{\partial \alpha_j} d\boldsymbol{\alpha}, \tag{11.47}$$

which gives the important result

$$\begin{aligned} < \alpha_i X_j > &= -k_B \quad i = j \\ &= 0 \quad\quad i \neq j \end{aligned} \tag{11.48}$$

11.8 Time Reversal, Microscopic Reversibility, and the Principle of Detailed Balance

The principle of microscopic reversibility was formulated in 1924 by Richard C. Tolman [8]; it provides a dynamic description on the microscopic level (atomic or molecular) of a macroscopic system at equilibrium, which may be considered as standing still. The principle states that any molecular process and its reverse occur *on average* with equal rates. Thus, for a chemical reaction at equilibrium, the amount of reactants being converted to products per unit time is exactly matched by the amount of products being converted to reactants per unit time. The principle of microscopic reversibility, when applied to a chemical reaction that proceeds in several steps, is known as the principle of detailed balance. Basically, it states that at equilibrium, each individual reaction occurs in such a way that the forward and reverse rates are equal. The derivation is based on the fact that the equations of motion of a particle are invariant under time reversal.

We first introduce the conditional probability density in phase space, $P(\mathbf{r}'^N, \mathbf{p}'^N;\tau \mid \mathbf{r}^N, \mathbf{p}^N)$ [$\mathbf{r}^N \equiv \mathbf{x}^N$ Equation (3.1)], where $P(\mathbf{r}'^N, \mathbf{p}'^N;\tau \mid \mathbf{r}^N, \mathbf{p}^N) \, d\mathbf{r}'^N d\mathbf{p}'^N$ is the probability to find the system in the range $(\mathbf{r}'^N, \mathbf{p}'^N; \mathbf{r}'^N + d\mathbf{r}'^N, \mathbf{p}'^N + d\mathbf{p}'^N)$, at time τ, when initially (at time $\tau = 0$), it was at $(\mathbf{r}^N, \mathbf{p}^N)$. Because a classical system is completely deterministic by the equations of motion, if $(\mathbf{r}'^N, \mathbf{p}'^N)$ is a solution for the given $(\mathbf{r}^N, \mathbf{p}^N)$ and τ, it is the *only* solution and thus,

$$P(\mathbf{r}'^N, \mathbf{p}'^N;\tau \mid \mathbf{r}^N, \mathbf{p}^N) = \delta\{(\mathbf{r}'^N - \mathbf{r}^N - \Delta\mathbf{r}^N(\mathbf{r}^N, \mathbf{p}^N;\tau)\} \cdot \delta\{(\mathbf{p}'^N - \mathbf{p}^N - \Delta\mathbf{p}^N(\mathbf{r}^N, \mathbf{p}^N;\tau)\} \qquad (11.49)$$

where

$$\mathbf{r}'^N - \mathbf{r}^N = \Delta\mathbf{r}^N(\mathbf{r}^N, \mathbf{p}^N;\tau) \quad \text{and} \quad \mathbf{p}'^N - \mathbf{p}^N = \Delta\mathbf{p}^N(\mathbf{r}^N, \mathbf{p}^N;\tau) \qquad (11.50)$$

For all the points $(\mathbf{r}^N, \mathbf{p}^N)$ lying in the energy shell $(E, E + dE)$, P is normalized,

$$\int_{(E,E+dE)} P(\mathbf{r}'^N, \mathbf{p}'^N;\tau \mid \mathbf{r}^N, \mathbf{p}^N) d\mathbf{r}'^N d\mathbf{p}'^N = 1 \qquad (11.51)$$

Now, the equations of motion of the particles are invariant under "time reversal," that is,

$$\mathbf{r}^N \to \mathbf{r}^N \qquad \mathbf{p}^N \to -\mathbf{p}^N \qquad t \to -t \qquad (11.52)$$

meaning that:

$$P(\mathbf{r}'^N, \mathbf{p}'^N;\tau \mid \mathbf{r}^N, \mathbf{p}^N) = P(\mathbf{r}'^N, -\mathbf{p}'^N;-\tau \mid \mathbf{r}^N, -\mathbf{p}^N) \qquad (11.53)$$

However, instead of considering the transition from $(\mathbf{r}^N, -\mathbf{p}^N)$ to $(\mathbf{r}'^N, -\mathbf{p}'^N)$ in time $-\tau$ [in the right hand side of Equation (11.53)], one can consider the opposite transition from $(\mathbf{r}'^N, -\mathbf{p}'^N)$ to $(\mathbf{r}^N, -\mathbf{p}^N)$ in time, $+\tau$. Therefore, changing Equation (11.53) accordingly, leads to

$$P(\mathbf{r}'^N, \mathbf{p}'^N;\tau \mid \mathbf{r}^N, \mathbf{p}^N) = P(\mathbf{r}^N, -\mathbf{p}^N;\tau \mid \mathbf{r}'^N, -\mathbf{p}'^N) \qquad (11.54)$$

We return now to the extensive variables, $\mathbf{A}(\mathbf{r}^N, \mathbf{p}^N)$, defined in Section 11.7 and their fluctuations, $\boldsymbol{\alpha}$ [$\alpha_i = A_i - <A_i> (i = 1, 2, ..., n)$] [Equation (11.38)]. We are interested in the joint probability for $\boldsymbol{\alpha}$ and $\boldsymbol{\alpha}'$, where $f(\boldsymbol{\alpha})$ is defined in Equation (11.41),

$$P(\boldsymbol{\alpha}';\tau \mid \boldsymbol{\alpha}) f(\boldsymbol{\alpha}) d\boldsymbol{\alpha} d\boldsymbol{\alpha}'$$

$$= \rho_0 \int_{(\alpha,\alpha+d\alpha)(\alpha',\alpha'+d\alpha')} \int P(\mathbf{r}'^N, \mathbf{p}'^N;\tau \mid \mathbf{r}^N, \mathbf{p}^N) \, d\mathbf{r}^N d\mathbf{p}^N d\mathbf{r}'^N d\mathbf{p}'^N$$

$$= \rho_0 \int_{(\alpha',\alpha'+d\alpha')(\alpha,\alpha+d\alpha)} \int P(\mathbf{r}^N, -\mathbf{p}^N;\tau \mid \mathbf{r}'^N, -\mathbf{p}'^N) \, d\mathbf{r}'^N d\mathbf{p}'^N d\mathbf{r}^N d\mathbf{p}^N \qquad (11.55)$$

$$= \rho_0 \int_{(\alpha',\alpha'+d\alpha')(\alpha,\alpha+d\alpha)} \int P(\mathbf{r}^N, \mathbf{p}^N;\tau \mid \mathbf{r}'^N, \mathbf{p}'^N) \, d\mathbf{r}'^N d\mathbf{p}'^N d\mathbf{r}^N d\mathbf{p}^N = P(\boldsymbol{\alpha};\tau \mid \boldsymbol{\alpha}') f(\boldsymbol{\alpha}') d\boldsymbol{\alpha} d\boldsymbol{\alpha}'$$

where ρ_0 is the probability density to find the system within E and $E + dE$ [see Equation (11.34)]. All the integrals are calculated within the energy shell $(E, E + dE)$, which, for simplicity, is omitted. The second set of integrals is equal to the first set due to Equation (11.54). The equality of the third set is valid if the thermodynamic variables, $\boldsymbol{\alpha}$, are *even* functions of the momenta (e.g., energy). In this case, for any pair of momenta $(-\mathbf{p}^N, -\mathbf{p}'^N)$ that contribute to the second set of integrals, the pair $(\mathbf{p}^N, \mathbf{p}'^N)$ contribute equally, and vice versa. Therefore, one can replace each negative pair $(-\mathbf{p}^N, -\mathbf{p}'^N)$ by the positive pair $(\mathbf{p}^N, \mathbf{p}'^N)$ without changing the integration; this is done in the third integrand. Finally, the transformation to the third integral is correct in the absence of an external magnetic field (see [7]).

Thus, one obtains the property of *microscopic reversibility*, which is also known as the *principle of detailed balance*,

$$f(\boldsymbol{\alpha})P(\boldsymbol{\alpha}';\tau|\boldsymbol{\alpha}) = f(\boldsymbol{\alpha}')P(\boldsymbol{\alpha};\tau|\boldsymbol{\alpha}'). \tag{11.56}$$

Detailed balance is a fundamental principle in statistical mechanics, which actually defines a mechanism for a dynamical process at equilibrium. Furthermore, as discussed earlier, Equation (11.56) provides a prescription for the definition of transition probabilities for Markov chains that mimic the relaxation of a system from a non-equilibrium state to equilibrium. Examples are the Monte Carlo method discussed earlier in Chapter 10 and the master equation, which will be introduced in Chapter 13.

One can show (see [7]) that for thermodynamic variables, $\boldsymbol{\beta}$, that are odd functions of the momenta (e.g., electric current densities), the principle of detailed balance reads,

$$f(\boldsymbol{\alpha},\boldsymbol{\beta})P(\boldsymbol{\alpha}',\boldsymbol{\beta}';\tau|\boldsymbol{\alpha},\boldsymbol{\beta}) = f(\boldsymbol{\alpha}',\boldsymbol{\beta}')P(\boldsymbol{\alpha},-\boldsymbol{\beta};\tau|\boldsymbol{\alpha}',-\boldsymbol{\beta}') \tag{11.57}$$

It should be pointed out that both Equations (11.56) and (11.57) are derived for the case of a zero magnetic field.

11.9 Onsager's Reciprocal Relations

Equipped with the principle of detailed balance, we can derive the reciprocal relations. Thus, for two thermodynamic variables A_j and A_k (with fluctuations, $\alpha_j = A_j - A_{j0}$ and $\alpha_k = A_k - A_{k0}$, respectively), we wish first to show that:

$$< \alpha_j \alpha_k(\tau) > = < \alpha_j(\tau)\alpha_k >. \tag{11.58}$$

This can be proved by writing $< \alpha_j \alpha_k(\tau) >$ explicitly and applying the detailed balance condition,

$$< \alpha_j \alpha_k(\tau) > = \iint \alpha_j \alpha_k f(\alpha_j) P(\alpha_k;\tau|\alpha_j) d\alpha_j d\alpha_k$$
$$= \iint \alpha_j \alpha_k f(\alpha_k) P(\alpha_j;\tau|\alpha_k) d\alpha_j d\alpha_k = < \alpha_j(\tau)\alpha_k > \tag{11.59}$$

Subtracting $\alpha_j \alpha_k$ from both sides of $< \alpha_j \alpha_k(\tau) > = < \alpha_j(\tau)\alpha_k >$ and dividing by τ leads to:

$$\left\langle \alpha_j \frac{\alpha_k(\tau) - \alpha_k}{\tau} \right\rangle = \left\langle \frac{\alpha_j(\tau) - \alpha_j}{\tau} \alpha_k \right\rangle \tag{11.60}$$

where for $\tau \to 0$, one obtains:

$$< \alpha_j \dot{\alpha}_k > = < \dot{\alpha}_j \alpha_k >. \tag{11.61}$$

Notice, however, that this tendency to zero is limited by the minimal size of τ. Thus, τ should be much larger than the internal relaxation time, τ_{rel}, that for a typical gas is $\sim v_{coll}^{-1}$, where v_{coll} is the collision frequency of the gas particles ($\tau_{rel} \sim 10^{-10}$ s). On the other hand, τ should be much smaller than the characteristic relaxation time, τ_{ev}, which is the time required for a non-equilibrium system to go back to equilibrium.

Up till now, we have dealt with time correlations between the fluctuations α_j and α_k and their time derivatives *at equilibrium*. Now we make the connection with the *macroscopic* theory developed earlier. Thus, we *assume* that $\dot{\alpha}$ is governed by the same linear dynamical laws of the *macroscopic* (phenomenological) process [Equation (11.19)], that is,

$$\dot{\alpha}_k = \sum_i L_{ik} X_i \tag{11.62}$$

and inserting Equation (11.61) into Equation (11.62) leads to:

$$\sum_i L_{ik} < \alpha_j X_i > = \sum_i L_{ij} < X_i \alpha_k >. \tag{11.63}$$

We have already obtained [Equation (11.46)],

$$< \alpha_j X_i > = \begin{cases} -k_B & \text{if} \quad i = j \\ 0 & \text{if} \quad i \neq j \end{cases} \quad \text{for} \quad \mathbf{B} = 0. \tag{11.64}$$

Evaluating the terms $< \alpha_j X_i >$ in Equation (11.63) by Equation (11.64) leads to Onsager's *reciprocal relations*:

$$L_{ik} = L_{ki}. \tag{11.65}$$

This result holds for a zero magnetic field. It can be shown [7] that for $\mathbf{B} \neq 0$, the reciprocal relations read:

$$L_{ik}(\mathbf{B}) = L_{ki}(-\mathbf{B}) \tag{11.66}$$

In summary: It should be emphasized again that an essential element in the derivation of the reciprocal relations is identifying *the rate of a macroscopic* phenomenological process with the rate of its fluctuation in equilibrium statistical mechanics [Equation (11.62)].

11.10 Applications

Onsager's relations appear in many phenomena, part of them are discussed in detail in [3–8]. However, this topic is beyond the scope of this book. We shall mention only two examples — the Peltier effect and the diffusion of salt. Thus, assume a junction created by soldering two different metals. Applying voltage to drive a current can cause cooling or heating of the junction. Since $L_{volt, heat} = L_{heat, volt}$, if instead, one heats or cools the junction, an electrical current will flow. This is the basis of thermocouples, which convert temperature changes into electrical signals.

The diffusion of one salt, e.g., NaCl affects the diffusion of another salt, e.g., KCl. Table 11.1 (see [11]) shows that approximately $L_{21} = L_{12}$.

TABLE 11.1

Results for the Concentrations NaCl and KCl and the L_{ij} Values

	Concentrations			
c_{NaCl} (1)	0.25	0.5	0.25	0.5
c_{KCl} (2)	0.25	0.25	0.5	0.5
	$L_{ij}10^9RT$			
$L_{12}{}^a$	−0.75	−1.03	−0.99	−1.52
$L_{21}{}^b$	−0.729	−1.02	−0.97	−1.45
L_{12}/L_{21}	1.03	1.01	1.02	0.98

Source: Reprinted with permission from [a]Dunlop, P.J. and Gosting, L.J., *J. Phys. Chem.*, 63, 86–93, 1959; [b]Fujita, H. and Gosting, L.J., *J. Phys. Chem.*, 64, 1256–1263, 1960. Copyrights 1959, 1960, respectively, American Chemical Society.

11.11 Steady States and the Principle of Minimum Entropy Production

If a system without restraints is allowed to age, it will eventually reach equilibrium, where all the fluxes vanish, and the entropy is maximal, or the entropy production is zero. However, if constraints are applied to some of the forces, which hold them constant, the system will be prevented from reaching equilibrium, but instead might age to a non-equilibrium stationary (steady) state, where the fluxes related to the *non-restricted* forces vanish, and the rate of entropy production is minimal. This principle, due to Prigogine, is known as the principle of minimum entropy production [3]. We have already met a stationary state in the discussion of Fourier's law, where div $J_{oz} = 0$, and J_{oz} is constant independent of time.

To be more specific, assume a system of n forces, where k of them (X_i, $i = 1, 2, ..., k$) are restrained (constant), while the rest (X_i, $k < i \leq n$) are free to vary. In this case, the fluxes of the varying forces (J_i, $k < i \leq n$) will eventually vanish, where d_iS/dt is minimal, and all the parameters are independent of time. For two processes, where X_1 is constant and X_2 can vary (that is, in a steady state, its J_2 will vanish), one obtains (using $L_{12} = L_{21}$),

$$J_1 = L_{11}X_1 + L_{12}X_2$$
$$J_2 = L_{21}X_1 + L_{22}X_2 = 0$$

(11.67)

and from Equation (11.23):

$$\frac{d_iS}{dt} = L_{11}X_1^2 + 2L_{12}X_1X_2 + L_{22}X_2^2 > 0$$

(11.68)

and

$$\frac{\partial}{\partial X_2}\left(\frac{d_iS}{dt}\right) = 2\left(L_{12}X_1 + L_{22}X_2\right) = 2J_2 = 0$$

(11.69)

$d_iS/dt(X_2)$ is a minimum because its second derivative with respect to X_2 is larger than zero,

$$\frac{\partial J_2}{\partial X_2} = \frac{\partial}{\partial X_2}\left(L_{21}X_1 + L_{22}X_2\right) = L_{22} > 0$$

(11.70)

Thus, the condition, $J_2 = 0$, and *minimum* entropy production are completely equivalent, as long as the linear relations of Equation (11.67) are valid. This argument can be generalized to the case of n independent affinities.

It should be emphasized that the principle of minimum entropy production cannot be taken as universally valid. It is only valid under the following conditions: (1) the system obeys the linear phenomenological laws, (2) the Onsager reciprocal relations are valid, (3) the Onsager reciprocal coefficients are independent of the forces, and (4) entropy production is balanced by outflow and inflow of entropy from the surroundings.

An example demonstrating this principle is provided by the case discussed in Section 11.3. Thus, if the heat exchanged with the exterior is constant, the difference between T^I and T^{II} will remain constant in time that is a constant affinity, $X_T = (1/T^I - 1/T^{II})$. In Section 13.4, we present a detailed example, involving the principle of minimum entropy production.

PROBLEM 11.2

A system is in a non-equilibrium steady state. Calculating the steady-state condition has led to two solutions (for the same constraints), where the corresponding results for the entropy production are 3 and 2 kcal(mol·T·s)$^{-1}$. In which of the two states is the system expected to reside? If the constraints are removed, what would be the final state of the system, and what is the corresponding entropy production?

11.12 Summary

We have presented Onsager's theory with an emphasis on its theoretical basis and less on applications. Thus, our discussion involves mainly discrete models, while realistic systems of interest are continuous. The theory is based on the local-equilibrium hypothesis, the affinity-flux linearity, and the assumption that the decay of a *macroscopic* disturbance close to equilibrium can be described by the decay of a *microscopic* fluctuation in equilibrium statistical mechanics. Onsager's theory constitutes a part of non-equilibrium thermodynamics. In the next two chapters, the close-to-equilibrium hypothesis is used within the framework of non-equilibrium statistical mechanics as an essential factor in the derivation of equations for relaxation phenomena (e.g., Langevin and Fokker-Planck equations).

REFERENCES

1. L. Onsager. Reciprocal relations in irreversible processes. I. *Phys. Rev.* **37**, 405–426 (1931).
2. L. Onsager. Reciprocal relations in irreversible processes. II. *Phys. Rev.* **38**, 2265–2279 (1931).
3. I. Prigogine. *Introduction to Thermodynamics of Irreversible Processes.* (Interscience, New York, 1967).
4. A. Katchalsky and P. F. Curran. *Nonequilibrium Thermodynamics in Biophysics.* (Harvard University Press, Cambridge, 1965).
5. H. B. Callen. *Thermodynamics and an Introduction to Thermostatistics.* (John Wiley, New York, 1985).
6. L. D. Landau and E. M. Lifshitz. *Statistical Physics.* (Pergamon, Oxford, 1969).
7. R. de Groot and P. Mazur. *Non-equilibrium Thermodynamics* (Dover, New York, 1984).
8. R. C. Tolman. *The Principles of Statistical Mechanics.* (Dover, New York, 1979).
9. P. J. Dunlop and L. J. Gosting. Use of diffusion and thermodynamic data to test the Onsager reciprocal relation for isothermal diffusion in the system NACL-KCL-H2O at 25°. *J. Phys. Chem.* **63**, 86–93 (1959).
10. H. Fujita and L. J. Gosting. A new procedure for calculating the four diffusion coefficients of three-components systems from Gouy diffusiometer data. *J. Phys. Chem.* **64**, 1256–1263 (1960).
11. K. A. Dill and S. Bromberg. *Molecular Driving Forces, Statistical Thermodynamics in Biology, Chemistry, Physics, and Nanoscience.* (Garland Science, New York, 2010).

12

Non-equilibrium Statistical Mechanics

In this chapter and the next one, we discuss several differential equations describing the relaxation to equilibrium of systems described on the microscopic level. The diffusion, Langevin and Fokker-Planck equations (and others), discussed here have been treated extensively in the literature, and thus, our presentation follows existing derivations that have been found to be clear and succinct [1–9]. These equations belong to the "close to equilibrium" regime—a condition which is always emphasized. The present studies widen the scope of dynamics introduced initially by Monte Carlo and molecular dynamics in Chapter 10, enriching conceptually and technically, the simulation discipline.

12.1 Fick's Laws for Diffusion

In this topic, we follow the derivations of Reif [1] and Dill and Bromberg [2]. Imagine that a droplet of ink is added to a container of water. The ink will start spreading in water, and after a long enough time, t, will be distributed homogenously in the container, that is, the system will reach equilibrium. In other words, $c(\mathbf{r}, t)$, the *average* number of ink particles per unit volume at position, $\mathbf{r}$, at time, t, which initially was strongly inhomogeneous, will become independent of t and, $\mathbf{r}$ (notice that the total number of ink particles in the system is unchanged). As an average quantity, $c(\mathbf{r}, t)$ is defined within the underlying probability space $(\mathbf{r}, t)$ based on the (as yet) undefined probability density, $P(\mathbf{r}, t)$. In the experimental world, $c(\mathbf{r}, t)$, $P(\mathbf{r}, t)$, and the flux, J introduced below, can be estimated from many (same) ink experiments, or in the case of a spherical symmetry, from data gathered from different symmetrical points, $(\mathbf{r}, t)$, obtained in the same experiment. This non-equilibrium process is essentially entropy driven, as the number of configurations of ink particles in the initial droplet is exceedingly smaller than their maximal number (maximal entropy) in the entire container (see Section 2.8.1). As shown below, on the microscopic level, the particles move due to local gradients of $c(\mathbf{r}, t)$. This ink example represents many similar situations; for simplicity, we shall treat a homogeneous system of same particles, where some of them are "labeled" in some way. We shall study the movement of the labeled particles in the "background" of the unlabeled ones - a phenomenon called *self-diffusion*. In the discussion below we sometimes use ink particles as a representation of labeled molecules.

For further simplicity, we derive Fick's laws in one dimension—along the x coordinate. The flux, J_x, of the (labeled) molecules is defined as the *mean* number of particles (in the sense of the local equilibrium hypothesis) crossing a unit area perpendicular to x per unit time. Thus,

$$J_x = c(x,t)\frac{\Delta x}{\Delta t} = c(x,t)v(t) \tag{12.1}$$

where $v(t)$ is the velocity.

12.1.1 First Fick's Law

In a uniform density, $J_x = 0$; otherwise, one would expect that a small flux, J_x, will be proportional to the concentration gradient of the crossing particles. More specifically, a cell of size Δx can be divided into two sub-cells of sizes, $\Delta x'$ and $\Delta x''$ (Figure 12.1). Based on the *local-equilibrium hypothesis*, each (small) sub-cell is sufficiently large that its microscopic fluctuations (gradients) are negligible;

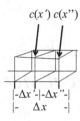

FIGURE 12.1   A one-dimensional cell of size Δx is divided into two sub-cells, $\Delta x'$ and $\Delta x''$, with slightly different average numbers of labeled particles, $c(x')$ and $c(x'')$, respectively.

hence, each of $c(x')$ and $c(x'')$ is uniform in its sub-cell, while $c(x')$ and $c(x'')$ *slightly* differ, leading to the gradient, $\Delta c / \Delta x$. Thus, one obtains the *first* Fick's law (1855),

$$J_x = -D\frac{\partial c(x)}{\partial x} \qquad (12.2)$$

$D > 0$ is the coefficient of self-diffusion with the dimension of area/time; we shall relate to D as the diffusion coefficient. Notice that without the added minus sign, the results would be incorrect. For example, if $\partial c / \partial x > 0$, the flow of particles (without the minus sign) will be incorrectly in the $+x$ direction, that is, $J_x > 0$; this is corrected by the minus sign. D is an empirical coefficient which depends on the specific parameters of the molecules studied, such as size, shape, and temperature. Thus, for a dilute gas at standard conditions, $D \sim 0.2$ cm²/sec, while for the (giant) DNA molecule in water, D is much smaller, $D \sim 1.3 \times 10^{-5}$ cm²/sec. Fick's law is an empirical kinetic law similar to Fourier's and Ohm's laws, as already mentioned in Section 11.6.

12.1.2 Calculation of the Flux from Thermodynamic Considerations

We have already shown that the flux-force linearity, $J_x = LX$, holds for the Fourier's law. A similar linearity holds for the first Fick's law. Thus, for a system of uncorrelated particles, the force can be expressed as a gradient of the chemical potential, μ. More specifically, the force $(\mu_1/T_1 - \mu_2/T_2)$ for a discreet system [Equation (11.12)] becomes the negative gradient of (μ/T) for a continuous system, where the minus sign stems from the fact that material moves from high to low μ. In our case, T is constant, and the force will be defined without the temperature, that is, as $-\text{grad}(\mu)$ (the corresponding phenomenological coefficient is denoted by L'); μ is [see Equation (4.17), where c replaces, ρ],

$$\mu(x) = \mu_0 + k_B T \text{ rate of } \ln[c(x)] \qquad (12.3)$$

and

$$\frac{d\mu}{dx} = k_B T \frac{d\ln c}{dx} = \frac{k_B T}{c}\frac{dc}{dx} \qquad (12.4)$$

Substituting dc/dx in the first Fick's law [Equation (12.2)] leads to:

$$J_x = \frac{Dc}{k_B T}\left(\frac{d\mu}{dx}\right) = \frac{Dc}{k_B T}X = L'X \qquad (12.5)$$

where $X = d\mu/dx$ is the force in continuum, and the phenomenological coefficient L' is:

$$L' = \frac{Dc}{k_B T}$$ (12.6)

The corresponding rate of entropy production is defined with the (usual) temperature-dependent, X, $X = (d\mu/dx)/T$, where $L = Dc/k_B$,

$$\frac{dS_i}{dt} = L\left(\frac{1}{T}\frac{d\mu}{dx}\right)^2$$ (12.7)

S_i is the entropy per volume (Δx). From Equation (12.6), one obtains for the diffusion coefficient as a function of temperature,

$$D = \frac{L'}{c}k_B T = uk_B T = \frac{k_B T}{\alpha}$$ (12.8)

where $u = (L'/c)$ is the mobility (in units of t/mass) and $1/u = \alpha$ is the friction coefficient (in units of mass/t).

12.1.3 The Continuity Equation

Consider a slab defined by two surfaces of area, A, which are perpendicular to the x-direction at points x and $x + dx$ (Figure 12.2). The number of particles per unit time within the slab is equal to the number of particles entering the slab at x minus those which leave at $x + dx$ (per unit time),

$$\frac{\partial}{\partial t}(cAdx) = AJ_x(x) - AJ_x(x + dx)$$ (12.9)

$$\frac{\partial c}{\partial t}dx = J_x(x) - \left[J_x(x) + \frac{\partial J_x(x)}{\partial x}dx\right]$$

and one obtains the continuity equation,

$$\frac{\partial c(x,t)}{\partial t} = -\frac{\partial J_x(x)}{\partial x}$$ (12.10)

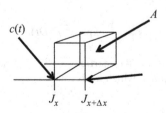

FIGURE 12.2 $c(x,\ t)Adx$ is the average number of labeled particles within the cube. Their change in time is equal to the difference in the fluxes at x and $x + dx$, $AJ_x(x) - AJ_x(x + dx)$.

12.1.4 Second Fick's Law—The Diffusion Equation

Substituting the first Fick's law [Equation (12.2)] in the continuity equation above leads to the one-dimensional diffusion equation, which is also known as the second Fick's law,

$$\frac{\partial c}{\partial t} = D \frac{\partial^2 c}{\partial x^2} \tag{12.11}$$

The limitations on Δt and Δx pointed out earlier, should be reiterated. Thus, both parameters should be very small on a macroscopic scale, but large enough on a microscopic scale to allow averaging of c. Also, notice that if D is not constant in space, one obtains the following diffusion equation,

$$\frac{\partial c}{\partial t} = \frac{\partial}{\partial x} \left[D(x) \frac{\partial c}{\partial x} \right] \tag{12.12}$$

The diffusion equation and the equation for heat conduction are basically the same. Thus, one can show that the equation for heat conduction [Equation (11.30)] can be transferred to an equation for the temperature, where c of the diffusion equation is replaced by the temperature, T, and D is replaced by a parameter, α; in three dimensions the equations are:

$$\frac{\partial c}{\partial t} = \nabla \cdot \mathbf{J} = D\nabla^2 c \qquad \frac{\partial T}{\partial t} = \alpha \nabla^2 T. \tag{12.13}$$

Analytical solutions of the diffusion equation are of different complexities depending on the problem and the boundary conditions as demonstrated by the following two examples (for more examples, see [2]).

12.1.5 Diffusion of Particles Through a Membrane

Assume a container consisting of two compartments I and II separated by a membrane of width d [2]. Particles at high concentration, c_{H}, in I move through the membrane to compartment II, which contains particles at lower concentration, c_{L}; $\Delta c = c_{\mathrm{H}} - c_{\mathrm{L}}$. In the case of large compartments (reservoirs), one can consider a steady-state flow ($\partial c/\partial t = 0$) through the membrane, where the concentration profile, $c(x)$, is of interest. Two integrations of the diffusion equation ($\partial^2 c/\partial x^2 = 0$) lead to a linear profile:

$$c(x) = A_1 x + A_2 \tag{12.14}$$

where A_1 and A_2 are integration constants, which can be obtained if the partition coefficient is known (compare with Section 11.6).

12.1.6 Self-Diffusion

We return to the droplet of ink in water discussed in the beginning of this section [1]. The amount of ink is M_0, concentrated at time, $t = 0$ (t_0), at $x = 0$; we are interested in the spread of the ink at time, $t > t_0$, in directions $\pm x$. A formal solution to the diffusion equation is:

$$c = \frac{A}{t^{1/2}} \exp(-x^2 / 4Dt) \tag{12.15}$$

where A is a constant with the dimension, (# of particles $\bullet$ $t^{1/2}$)/(volume). We assume a cylinder of infinite length and a unit cross-section (around x) which leads to:

$$M_0 = \int_{-\infty}^{\infty} c\,dx \tag{12.16}$$

Defining:

$$x^2/4Dt = \xi^2 \rightarrow dx = 2(Dt)^{1/2}d\xi \qquad (12.17)$$

leads to:

$$M_0 = 2AD^{1/2}\int \exp(-\xi^2)d\xi = 2A(\pi D)^{1/2} \qquad (12.18)$$

or

$$A = \frac{M_0}{2(\pi D)^{1/2}}. \qquad (12.19)$$

Substituting A in Equation (12.15) leads to the Gaussian:

$$c(x,t) = \frac{M_0}{(2\pi Dt)^{1/2}}\exp(-x^2/4Dt). \qquad (12.20)$$

In d-dimensions one obtains,

$$c(r,t) = \frac{M_0}{(4\pi Dt)^{d/2}}\exp(-r^2/4Dt). \qquad (12.21)$$

$c(r, t)$ defines the probability density, $P(r, t) = c(r, t)/M_0$. The boundary conditions of $c(r, t)$ should lead to M_0 at $r = 0$ and $t = 0$, while at this point $c(r, t)$ diverges. However, $P(r, t)$ becomes a Dirac Delta function, which in $d = 1$ reads:

$$\delta(x) = \frac{1}{(2\pi)^{1/2}\gamma}\exp(-x^2/2\gamma^2) \qquad \gamma \rightarrow 0 \qquad (12.22)$$

Therefore, $c(r = 0, t = 0) = M_0\delta r$ with a well-defined integral:

$$M_0\int \delta r = 1 \cdot M_0 \qquad (12.23)$$

The average distance a particle makes at time, t, is:

$$\int r^2 P(r,t)dr = <r^2> = <x^2> + <y^2> + <z^2> = 6Dt \qquad (12.24)$$

The initial concentration of molecules at $x = 0$ decreases as time, t, grows. Equations (12.20–12.22) show that at a given t, the distribution as a function of x is Gaussian, which becomes flatter and flatter as t increases (the variance ~t), meaning that the spread of the labeled molecules over the volume becomes more and more homogeneous (see Figure 12.3). However, returning to the ink example in Section 12.1, if its initial amount is too large, its initial spread will not obey the diffusion equation since the *local-equilibrium* hypothesis is violated at this early stage. The connection to an ideal polymer chain and the random walk model (Section 8.8) is clear and will be discussed further in Section 12.2.

PROBLEM 12.1

Calculate the "entropy" $-k_B\int P(r, t)\ln P(r, t)dr$ as a function of t. What kind of behaviors do you find for small and large t?

In the two Fick's laws, D is an empirical parameter. To calculate D for different systems, one needs to assume a model. For a dilute gas, one finds for D of self-diffusion (see [1], p. 485).

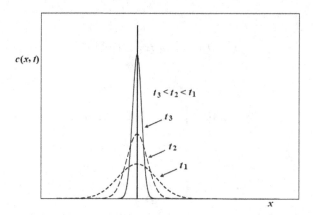

FIGURE 12.3 An one-dimensional (1-d) illustration of Equation (12.20) for labeled molecules; their Gaussian density distributions, $c(x, t)$ are drawn as a function of x for different times, t (after the ink was concentrated around $x = 0$ at $t=0$). The areas under the Gaussians are the same; they are equal to the total number of labeled molecules. The labeled molecules become more spread as time goes on.

$$D = \frac{2}{3\pi^{1/2}} \frac{1}{p\sigma_0} \sqrt{\frac{(k_{\mathrm{B}}T)^3}{m}}$$ (12.25)

where σ_0 is the mean total scattering cross-section, p is the pressure, and m is the particle mass.

In summary: In self-diffusion the particles move, not as a result of external chemical potential gradients, but due to inner entropic forces stemming from local differences in concentration. The diffusion equation is very general and can be applied to small and very large molecules (e.g., DNA), cells, etc. This equation is also used to model population problems in biology and other disciplines [3].

12.2 Brownian Motion: Einstein's Derivation of the Diffusion Equation

This section is based on Gardiner's book [4]. The notion of "Brownian motion" follows the pioneering work of the botanist Robert Brown, who in 1827 studied (under the microscope) pollen grains suspended in water and found them to execute irregular jittery fluctuations. To test whether this phenomenon was life-related, he examined further other suspensions of fine particles made of inorganic matter (glass, minerals, etc.) and discovered the same behavior, which allowed him to rule out any connection to life. Still, the origin of the above fluctuations had to be explained.

The solution to this problem was given in a 1905 paper by A. Einstein, who suggested that the movement of the pollen grains is caused by small "knocks" they receive from the moving molecules of the liquid, thus providing an indirect confirmation for the existence of atoms and molecules. The equations developed by Einstein for the Brownian motion were verified in 1908 by the experimental work of J.B. Perrin, who in1926 received the Nobel Prize for this work.

There are two parts to Einstein's theory: the first is the formulation of a diffusion equation for Brownian particles, and the second relates the diffusion constant to measurable physical quantities, which allowed Perrin to obtain Avogadro's number.

Einstein's theory is based on two assumptions: (1) as said above, the motion of a pollen grain is caused by the exceedingly frequent impacts of the surrounding liquid molecules. (2) The motion is very complicated, and the impacts can only be described in probability terms. Thus, the movement of different particles is independent, and the movements of a given particle in different time intervals, τ, are independent as well, as long as the intervals are not too short (a Markov postulate). Still, τ is very small compared to the observable time.

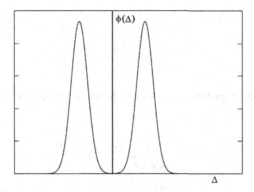

FIGURE 12.4 A typical sharp symmetric distribution function $\phi(\Delta)$.

The theory is developed in one dimension. We denote by n the number of particles in the fluid. Within time, τ, at point x, the particles' positions will be changed by Δ, where Δ can be positive or negative differing from particle to particle. More specifically, the probability density of Δ, $\phi(\Delta)$, is a sharp symmetric normalized function, $\phi(\Delta) = \phi(-\Delta)$ and $\int \phi(\Delta)d\Delta = 1$ (see Figure 12.4).

Now, $dn = n\phi(\Delta)d\Delta$, where dn is the number of particles experiencing a shift between Δ and $\Delta + d\Delta$ within τ. In one dimension, $c(x, t)$ is the number of particles per unit volume (Δx) at x at time, t. $c(x, t + \tau)dx$ is obtained by the following equation,

$$c(x,t+\tau)dx = dx \int_{-\infty}^{\infty} c(x+\Delta,t)\phi(\Delta)d\Delta. \tag{12.26}$$

This basic equation states that the number of particles per volume at x at time, $t + \tau$, is given by averaging the $c(x + \Delta, t)$ values for all Δ with respect to the probability, $\phi(\Delta)d\Delta$, at time, t; since only small Δ values are effective, only the very vicinity of x is taken into account in the integration. Because τ and Δ are very small, one can expand c with respect to both parameters,

$$c(x,t+\tau) = c(x,t) + \tau \frac{\partial c}{\partial t} \cdots$$

$$c(x+\Delta,t) = c(x,t) + \Delta \frac{\partial c(x,t)}{\partial x} + \frac{\Delta^2}{2!}\frac{\partial^2 c(x,t)}{\partial x^2} + \cdots \tag{12.27}$$

Substituting these expansions in Equation (12.26), multiplying by $\phi(\Delta)d\Delta$, and integrating leads to:

$$c + \frac{\partial c}{\partial t}\tau = c\int_{-\infty}^{\infty}\phi(\Delta)d\Delta + \frac{\partial c}{\partial x}\int_{-\infty}^{\infty}\Delta\phi(\Delta)d\Delta + \frac{\partial^2 c}{\partial x^2}\int_{-\infty}^{\infty}\frac{\Delta^2}{2}\phi(\Delta)d\Delta \tag{12.28}$$

Due to symmetry, $\phi(\Delta) = \phi(-\Delta)$, the second integral vanishes, and terms higher than the second are neglected. Defining the diffusion coefficient D,

$$\frac{1}{\tau}\int_{-\infty}^{\infty}\frac{\Delta^2}{2}\phi(\Delta)d\Delta = D \tag{12.29}$$

one obtains the diffusion equation,

$$\frac{\partial c}{\partial t} = D\frac{\partial^2 c}{\partial x^2} + \cdots \tag{12.30}$$

As before, [Fick's second law, Equation (12.11)], the solution is a Gaussian, which determines the mean displacement, $<x^2>$, at time, t:

$$c(x,t) = \frac{n}{\sqrt{4\pi D}}\frac{\exp[-x^2/4Dt]}{\sqrt{t}} \tag{12.31}$$

$$<x^2> = 2Dt$$

It should be noted that Einstein's derivation is an approximation, since it is based on a discrete time assumption that the particles' impacts occur only in time intervals, 0, τ, 2τ, 3τ, etc. This derivation is basically a random walk treatment (similar in nature to that discussed in Section 8.8 for the Gaussian polymer chain). Basically, Einstein's equation describes the decay to the equilibrium of a configurational fluctuation of pollen; due to the small difference between $c(x, t)$ and $c(x, t + \tau)$, the theory is close to equilibrium. Notice that the diffusion equation (Fick's second law), that is based on local gradients of $c(x, t)$ in the fluid seems to be more general. Still, Equation (12.26) and its solution contain many of the major concepts, which constitute the basis for more general and rigorous theories developed in the course of the years, such as the Chapman-Kolmogorov equation and the Fokker-Planck equation discussed later.

The diffusion coefficient is not a phenomenological parameter since Einstein suggested, $D = k_B T/\alpha$, where α is the friction coefficient. While we have already derived this expression for D from chemical potential considerations [Equation (12.8)], Einstein made a step further by calculating α from the macroscopic Stokes law, $\alpha = 6\pi\eta r$ (r is the particle's radius and η is the viscosity). Equivalently, $D = TR/\alpha N_A$, where R is the gas constant and N_A is Avogadro's number ($k_B = R/N_A$). Thus, by measuring D, T, and α, Avogadro's number, Å, can be obtained. As pointed out earlier, Perrin confirmed Einstein's model experimentally, which indeed has led to an estimate of Avogadro's number. Now, when Å is known, one can calculate other parameters, such as the radius of hemoglobin (assuming it is a sphere), if the D is known.

12.3 Langevin Equation

The Langevin equation constitutes an alternative way for describing the behavior of the Brownian motion of a particle in a gas or a liquid phase. The equation is more specific than Einstein's in the sense that it is based on certain forces that govern the particle behavior; thus, more (microscopic) information can be calculated (e.g., autocorrelation functions). More specifically, the Langevin equation is a Newton equation of motion for a single particle (in $d = 1$) based on two opposing effects, a frictional term $-\alpha v$ and a stochastic term $R(t)$ [1.5.9],

$$m\frac{dv}{dt} = -\alpha v + R(t) \tag{12.32}$$

Clearly, without the stochastic term, the velocity would decay to zero in contrast to the experimental evidence; therefore, we first define $R(t)$ and discuss in detail its microscopic effect. Thus, a Brownian particle (a protein, cell, etc.) is influenced by irregular, very fast (random) collisions exerted by the small solvent molecules. These forces are modeled by a stochastic process, $R(t)$,

$$<R(t)> = 0. \tag{12.33}$$

where <> means ensemble average. In the experimental world, <> means averaging the forces exerted on many particles at time, t, or averaging the forces affecting a single particle during a very short time interval, τ_1; however, τ_1 should be large enough to allow a statistically meaningful averaging. It is assumed that successive impacts are uncorrelated,

$$< R(t)R(t') > = \Gamma\delta(t - t'), \tag{12.34}$$

where Γ is a constant describing the average magnitude of the random force. In practice, Γ is defined by integration over the short time, τ_1, therefore, its dimension is (force)$^2 \cdot$time; however, the dimension of $<R(t)R(t')>$ is (force)^2 because the dimension of the δ function is t^{-1}. Since $R(t)$ is defined only by its first and second moments, it is considered to be a Gaussian distributed around zero with the width, $\Gamma^{\frac{1}{2}}$ (see Figure 12.5).

However, many of the very fast impacts (time scale, τ_1) produced by the solvent molecules are not effective (that is, they do not lead to a significant movement of the particle) due to cancellation of opposite impacts [$<R(t)> = 0$]. It takes a longer time, τ_2, for a strong fluctuation to occur, which will move the particle significantly; then the restoring (frictional) force on the particle becomes effective. Thus, the particle will move in some direction, r, with an increased velocity, $\mathbf{v}$; clearly, the particle will suffer more collisions against its movement (that is, in the $-r$ direction) than from "behind," which will act to decrease its velocity. The higher $\mathbf{v}$ is, the larger is the restoring force, which is expressed as a frictional force, $F_{\text{frictional}}$, proportional to $-\mathbf{v}$.

$$F_{\text{frictional}} = -\alpha\mathbf{v} = -\gamma m\mathbf{v} \tag{12.35}$$

In Equation (12.35), α is the friction coefficient, $\alpha = \gamma m$, where m is the particle's mass and γ is the collision frequency. γ^{-1} is proportional to the time taken for the particle to lose memory of its initial velocity (the velocity relaxation time), the larger γ is, the larger is the viscosity. Since the impacts that effectively resist the particle's movement occur on the τ_2 time scale, $\tau_2 \sim \gamma^{-1}$, meaning that γ^{-1} is much larger than τ_1—the correlation time of the random force.

The thermodynamic picture of this model is the following: the fluid provides a heat bath with T. The Brownian particles (which move independently, that is, they do not feel each other) constitute the "system," which interacts weakly with the bath, as in a canonical ensemble. Each particle gains energy from the bath (by the random forces) and releases this energy back to the bath due to dissipative (viscosity) forces. Thus, the Langevin equation models the system-bath interactions. As in an ideal gas, the long-time average of the kinetic energy is $m<v^2>/2 = (1/2)k_BT$. However, on a shorter time scale, the velocity fluctuates, and its decay to equilibrium is described by the Langevin equation. Finally, we emphasize that the acceleration/dissipation mechanism keeps Langevin's process close to equilibrium.

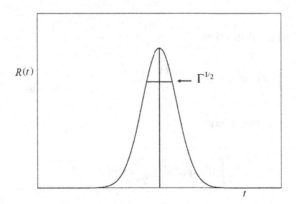

FIGURE 12.5 A stochastic function, $R(t)$, descried by a Gaussian with a standard deviation, $\Gamma^{1/2}$.

12.3.1 The Average Velocity and the Fluctuation-Dissipation Theorem

Using $\alpha = \gamma m$, the Langevin equation for a single particle in one dimension (Equation (12.32)] becomes:

$$m\frac{d^2x(t)}{dt^2} = -\gamma\frac{dx(t)}{dt}m + R(t), \tag{12.36}$$

which is simplified further by defining $V = mdx/dt$, that is,

$$\dot{V} = -\gamma V + R(t). \tag{12.37}$$

By defining $V(t = 0) = V_0$, a formal solution of the equation is:

$$V(t) = V_0\exp[-\gamma t] + \exp[-\gamma t]\int_0^t \exp[\gamma t']R(t')dt' \tag{12.38}$$

Now, V is a stochastic process, which can be averaged at time, t. (This averaging is defined within the underlying probability space (V, t) based on the (as yet) undefined probability density, $P_{V_0}(V, t)$; in the experimental world, $<V(t)>$ can be estimated by averaging $V(t)$ over many particles that all start with $V(t = 0) = V_0$ or over many runs of one particle starting from (V_0)).

Because $<R(t)> = 0$, the integral vanishes and one obtains:

$$<V(t)>_{V_0} = V_0\exp[-\gamma t] \tag{12.39}$$

meaning that V decays according to the decay constant γ^{-1}, and $<V((\infty)> = 0$. Squaring Equation (12.38) and averaging gives,

$$<[V(t)]^2> = V_0^2\exp[-2\gamma t] + \exp[-2\gamma t]\int_0^t dt'\int_0^t dt''\exp[\gamma(t' + t'')] < R(t')R(t'') > \tag{12.40}$$

and the integral becomes:

$$\int_0^t dt'\int_0^t dt''\exp[\gamma(t' + t'')] < R(t')R(t'') > \ = \int_0^t dt'\int_0^t dt''\exp[\gamma(t' + t'')]\Gamma\delta(t'' - t'). \tag{12.41}$$

Using the definition of Dirac's δ-function:

$$\int_A^B f(x)\delta(x - x_0)dx = \begin{cases} f(x_0) & \text{if} \quad A < x_0 < B \\ 0 & \text{otherwise} \end{cases} \tag{12.42}$$

enables carrying out the integration further,

$$\Gamma\int_0^t dt'\exp(2\gamma t') = \frac{\Gamma}{2\gamma}\exp[(2\gamma t) - 1] \tag{12.43}$$

and $<[V(t)]^2>$ (which decays with $(2\gamma)^{-1}$) is:

$$<[V(t)]^2 >= V_0^2 \exp(-2\gamma t) + \frac{\Gamma}{2\gamma}[1 - \exp(-2\gamma t)] \tag{12.44}$$

For $t \rightarrow \infty$, $<V(\infty)^2 > = \Gamma/2\gamma$ independent of V_0. Now, we make the connection between $<[V(t)]^2>$ and thermodynamics: using $V = mdx/dt$ and $<mv^2> = k_BT$, we obtain:

$$m^2 < \left[\frac{dx}{dt}\right]^2 > = \frac{\Gamma}{2\gamma} \quad \rightarrow \quad m < \left[\frac{dx}{dt}\right]^2 > = m < v^2 > = k_BT = \frac{\Gamma}{2\gamma m}. \tag{12.45}$$

Since γ has the dimension of t^{-1}, one can show that Γ has the correct dimension of (force)^2•time. This equation leads to the important relation between γ and Γ:

$$\Gamma = 2\gamma m k_BT, \tag{12.46}$$

which demonstrates a balance between the magnitudes of the random forces (Γ) and the viscous forces (γm), with a dependence on T as well. This is a manifestation of the fluctuation-dissipation theorem. Notice that the relation between the relaxation parameters is defined in equilibrium. Because the Brownian particles are not interacting with each other, one would expect, that like for an ideal gas, as $t \rightarrow \infty$, their velocities will be distributed according to Maxwell-Boltzmann,

$$P(v) = \left[\frac{m}{2\pi k_BT}\right]^{1/2} \exp-\left[\frac{mv^2}{2k_BT}\right] \tag{12.47}$$

Formally, the solution is for any V_0; however, physically, V_0 is relatively small, typically of the order of several times the velocity fluctuation of an ideal gas $\sim[k_BT/m]^{1/2} = [\Gamma/2\gamma m^2]^{1/2}$; therefore, the Langevin relaxation falls within the "close to equilibrium" regime.

12.3.2 Correlation Functions

It is of interest to integrate the autocorrelation function of the stochastic force. Denoting $t - t' = s$, one obtains,

$$< R(t)R(t') >_{\text{time}} = \Gamma\delta(t - t') = < R(0)R(s) > = \Gamma\delta(s) \tag{12.48}$$

$$\int_{-\infty}^{\infty} < R(0)R(s) > ds = \int_{-\infty}^{\infty} \Gamma\delta(s)ds = \Gamma = 2\gamma m k_BT = 2\alpha k_BT$$

Thus, the kinetics (relaxation to equilibrium, which is a function of γ) depend on the thermal fluctuation of R at equilibrium. This relation is an example of a general group of relations, which enable one to determine transport properties from equilibrium correlations.

In this context, it is also of interest to calculate the velocity autocorrelation function at equilibrium. Starting from the original Langevin equation for v [Equation (12.32)],

$$m\frac{dv}{dt} = -\alpha v + R(t) \tag{12.49}$$

we multiply both sides by the initial velocity $v(0) = v(t = 0)$ (to be distinguished from V_0), where $<v(0)>$ is the average velocity at equilibrium. We obtain:

$$m < v(0)\frac{dv(t)}{dt} >_{\text{at } t} = -\alpha < v(0)v(t) > + < v(0)R(t) >. \tag{12.50}$$

However,

$$< v(0)R(t) >=< v(0) >< R(t) >= 0, \tag{12.51}$$

Therefore, Equation (12.50) becomes:

$$\frac{d}{dt} < v(0)v(t) > + \frac{\alpha}{m} < v(0)v(t) >= 0 \tag{12.52}$$

which leads to an exponential solution:

$$< v(0)v(t) > = < v(0)^2 > \exp(-\alpha t / m)$$
$$= (k_{\text{B}}T / m)\exp(-\alpha t / m) = (k_{\text{B}}T / m)\exp(-\gamma t) \tag{12.53}$$

Thus, the velocity relaxation time is $m/\alpha = m/\gamma m = 1/\gamma$, as expected. Integrating the autocorrelation function leads to:

$$\int_0^\infty < v(0)v(t) > dt = (k_{\text{B}}T / m)\int_0^\infty \exp(-\alpha t / m) = \frac{k_{\text{B}}T}{\alpha} = D \tag{12.54}$$

So, like for R, the diffusion coefficient, D, which is a kinetic property is related to the (equilibrium) velocity autocorrelation, which can be obtained in a simulation from a long equilibrium trajectory. In 3-dimensions,

$$\int_0^\infty < \mathbf{v}(0)\mathbf{v}(t) > dt =3D. \tag{12.55}$$

This equation and Equation (12.48) are examples of Green-Kubo relationships between equilibrium correlation functions and kinetic quantities—transport or relaxation parameters. These relationships are based on linear response theory and are not restricted to the Langevin model; for the calculation of such correlations by molecular dynamics, see Equation (10.30).

12.3.3 The Displacement of a Langevin Particle

We write again the Langevin equation in one dimension,

$$m\frac{d\dot{x}}{dt} = -\alpha\dot{x} + R(t) \tag{12.56}$$

Since the particle moves symmetrically around the origin, its average displacement is $<x> = 0$, and its variance is thus, $<x^2>$; multiplying both sides of Equation (12.56) by x leads to,

$$mx\frac{d\dot{x}}{dt} = m\left[\frac{d}{dt}(x\dot{x}) - \dot{x}^2\right] = -\alpha x\dot{x} + xR(t) \tag{12.57}$$

Now, since x and R are uncorrelated ($<xR> = <x><R> = 0$), one may conclude (mistakenly) that averaging Equation (12.56) and $m(d\dot{x}/dt) = -\alpha x \dot{x}$ would be equivalent, as the random force would be missing in both equations! However, notice that Equation (12.57) contains the term $m\dot{x}^2$, where $<m\dot{x}^2>$ has already been calculated in Equation (12.44), where the random force has been considered. However, instead of inserting the result for $<V^2>=(m\dot{x})^2$ of Equation (12.44), we apply here an approximation, where $<m\dot{x}^2>$ is replaced by $< m\dot{x}(t = \infty)^2 >$, which is equal to k_BT in systems where the forces are independent of the velocities; Equation (12.57) becomes,

$$m\left\langle \frac{d}{dt}(x\dot{x}) \right\rangle_{time} = m\frac{d}{dt}<x\dot{x}> = k_BT - \alpha <x\dot{x}> \tag{12.58}$$

where the averaging is at time, t. This is a differential equation for $< x\dot{x} >= (1/2)(d < x^2 > /dt)$; the solution is,

$$<x\dot{x}>=C\exp(-\gamma t) + \frac{k_BT}{\alpha}. \tag{12.59}$$

Assume a set of particles each starts out at $t = 0$ at $x = 0$, where x measures the displacement from the initial position. At $t = 0$, $C + k_BT/\alpha = 0 \rightarrow C = -k_BT/\alpha$, and Equation (12.59) becomes,

$$< x\dot{x} > = \frac{1}{2}\frac{d}{dt}< x^2 > = \frac{k_BT}{\alpha}[1-\exp(-\gamma t)] \tag{12.60}$$

and after integration:

$$< x^2 > = \frac{2k_BT}{\alpha}\{t-\gamma^{-1}[1-\exp(-\gamma t)]\}. \tag{12.61}$$

Thus, (like the velocity, $< \dot{x} >$ [Equation (12.39)]), $<x^2>$ decays with a decay constant, $1/\gamma$. There are two limiting cases of interest for $<x^2>$: for very short time, t ($t < \gamma^{-1}$ or $t\gamma < 1$) and for large t; in the first case, one can expand the exponential in Equation (12.61) to second-order:

$$\exp(-\gamma t) = 1 - \gamma t + (1/2)\gamma^2 t^2 + ... \tag{12.62}$$

which leads to:

$$\{t - \gamma^{-1}[1-\exp(-\gamma t)]\} = \{t-\gamma^{-1}[\gamma t - (1/2)\gamma^2 t^2]\} = (1/2)\gamma t^2 \tag{12.63}$$

Substituting Equation (12.63) in Equation (12.61) and using $\alpha = m\gamma$, leads to:

$$<x^2> = \frac{k_BT}{\alpha}\gamma t^2 = \frac{k_BT}{m}t^2. \tag{12.64}$$

Because $x = \dot{x} t$, one can substitute x^2 in Equation (12.64) by $\dot{x}^2 t^2$, which shows that the particle moves for very short times with the thermal velocity, $\dot{x} = (k_BT/m)^{1/2}$. On the other hand, for long times ($t \gg \gamma^{-1}$), the exponent in Equation (12.61) becomes zero, leading to:

$$<x^2> = \frac{2k_BT}{\alpha}t. \tag{12.65}$$

Using for the diffusion coefficient, $D = k_B T / \alpha = k_B T / 6\pi\eta a$, one obtains:

$$<x^2> = \frac{k_B T}{3\pi\eta a} t = 2Dt \tag{12.66}$$

As mentioned above, $<v(t)>$ and $<x^2>$ decay with a relaxation constant $1/\gamma$, while $<v(t)^2>$ decays faster with $1/2\gamma$. Also, $<x^2>$ increases initially as $\sim t^2$ and after a longer time as $\sim t$. Finally, the expression for the decay of $<x^2>$ would be more complicated if we substituted Equation (12.44) in Equation (12.57), rather than the equilibrium expression $m<v^2> = k_B T$.

12.3.4 The Probability Distributions of the Velocity and the Displacement

As for an ideal chain (Section 8.7), one can consider the velocity at time, t, as a random variable, which is a sum (integral) of velocity random variables for all the previous times, $t' < t$. More specifically, let's consider the variable:

$$U(t) = V(t) - V_0 \exp[-\gamma t] \tag{12.67}$$

where, according to Equation (12.38), $V(t) = V_0 \exp[-\gamma t] + \exp[-\gamma t] \int_0^t \exp[\gamma t'] R(t') dt'$. Thus, due to Equation (12.39), $<U(t)> = 0$, and based on Equations (12.44) and (12.46),

$$<U(t)^2> = m k_B T [1 - \exp(-2\gamma t)] \tag{12.68}$$

We are interested in the statistical average, $\mu_{t,t-\Delta t}$, and the variance, $\sigma^2_{t,t-\Delta t}$, of the random variable $U_t - U_{t-\Delta t}$. Clearly,

$$\mu_{t,t-\Delta t} = <U_t - U_{t-\Delta t}> = 0 \tag{12.69}$$

and

$$\sigma^2_{t,t-\Delta t} = <(U_t - U_{t-\Delta t})^2> = <(U_t)^2> + <(U_{t-\Delta t})^2> - 2<U_t U_{t-\Delta t}> \tag{12.70}$$

To obtain $\sigma^2_{t,t-\Delta t}$, it only remains to calculate $<U_t U_{t-\Delta t}>$ (Notice that $R(t)$ is averaged over a smaller time scale (τ_1) than Δt); one obtains:

$$<U_t U_{t-\Delta t}> = <\{\exp[-\gamma(t-\Delta t)] \int_0^{t-\Delta t} \exp(\gamma t') R(t') dt'\}\{\exp[-\gamma t] \int_0^t \exp(\gamma t') R(t') dt'\}>. \tag{12.71}$$

Defining for simplicity, $f(t') = \exp(\gamma t') R(t')$, one obtains:

$$<U_t U_{t-\Delta t}> = \exp[-\gamma(2t - \Delta t)] \left\langle \int_0^{t-\Delta t} f(t') dt' \left[\int_0^{t-\Delta t} f(t') dt' + \int_{t-\Delta t}^t f(t') dt' \right] \right\rangle \tag{12.72}$$

$$= \exp[-\gamma(2t - \Delta t)] \left\langle \left\{ \int_0^{t-\Delta t} f(t') dt' \int_0^{t-\Delta t} f(t') dt' + \int_0^{t-\Delta t} f(t') dt' \int_{t-\Delta t}^t f(t') dt' \right\} \right\rangle$$

For $\Delta t \to 0$, the last integral in Equation (12.72) is ignored and:

$$<U_t U_{t-\Delta t}> = \exp[-\gamma(2t - \Delta t)] < \int_0^{t-\Delta t} f(t')dt \int_0^{t-\Delta t} f(t')dt > \sim <U_{t-\Delta t}^2 > . \tag{12.73}$$

Thus:

$$\sigma_{t,t-\Delta t}^2 = <(U_t)^2 > + <(U_{t-\Delta t})^2 > -2 <U_{t-\Delta t}^2 > \approx <(U_t)^2 > - <(U_{t-\Delta t})^2 > \tag{12.74}$$

To use the central limit theorem (CLT) (Section 1.15), we define the following random variable, which, for simplicity, is presented with discrete variables:

$$\frac{U_{t=1} + (U_{t=2} - U_{t=1}) + \dots (U_t - U_{t-1}) - \mu_1 - \mu_{2,1} - \dots - \mu_{t,t-1}}{(\sigma_1^2 + \sigma_{2,1}^2 + \dots + \sigma_{t,t-1}^2)^{1/2}} = \frac{U_t}{(\sigma_t^2)^{1/2}} \tag{12.75}$$

and more specifically:

$$\frac{U_t}{\sigma_t} = \frac{V(t) - V_0 \exp[-\gamma t]}{[mk_BT(1 - \exp[-2\gamma t])]^{1/2}} = \frac{m\{v(t) - v_0 \exp[-\gamma t]\}}{[mk_BT(1 - \exp[-2\gamma t])]^{1/2}} \tag{12.76}$$

where $v_0 = V_0/m$; applying to U_t/σ_t the same procedure used in Equation (8.47), where the derivation is carried out with respect to $v(t)$ (for constant t) leads to the Gaussian type distribution, $P(v,t|v_0)$:

$$P(v,t|v_0) = \left[\frac{m}{2\pi k_BT(1 - e^{-2\gamma t})} \right]^{1/2} \exp - \left[\frac{m(v - v_0 e^{-\gamma t})^2}{2k_BT(1 - e^{-2\gamma t})} \right]. \tag{12.77}$$

At $t = 0$, P is a δ-function. P depends on v_0 only for small t; as $t \to \infty$, P becomes the Maxwell-Boltzmann distribution [Equation (12.47)] as expected.

In $d = 3$, the general solution for the probability of a particle to be at distance $r = |\mathbf{r}|$ at time t (started at position $\mathbf{r}_0$ with velocity $\mathbf{v}_0$ at $t = 0$) is (see [6], p. 27):

$$P(r,t;\mathbf{r}_0,\mathbf{v}_0) = \left\{ \frac{m\gamma^2}{2\pi k_BT[2\gamma t - 3 + 4e^{-\gamma t} - e^{-2\gamma t}]} \right\}^{3/2} \exp - \left\{ \frac{m\gamma^2|\mathbf{r} - \mathbf{r}_0 - \mathbf{v}_0(1 - e^{-\gamma t})/\gamma|^2}{2k_BT[2\gamma t - 3 + 4e^{-\gamma t} - e^{-2\gamma t}]} \right\} \tag{12.78}$$

For $\gamma t \gg 1$, the exponents, $e^{-\gamma t}$ and $e^{-2\gamma t}$ vanish; also, because $\mathbf{r} - \mathbf{r}_0$ is of order t, much larger than v_0/γ, and 3, these terms can be ignored, and using $D = k_BT/m\gamma$, one obtains the solution for the diffusion equation:

$$P(r,t;\mathbf{r}_0) = \frac{1}{(4\pi Dt)^{3/2}} \exp(-|\mathbf{r} - \mathbf{r}_0|^2 / 4Dt) \tag{12.79}$$

where the effect of the initial velocity vanishes. $P(r,t;\mathbf{r}_0)$ is a Gaussian describing the approach to a flat (random) distribution of the particles in the bath, meaning a maximal entropy. Thus, for a large t, the phenomenological Fick's second law (based on concentration gradients), Einstein's picture (consisting of a probabilistic assumption), the Langevin equation, and an ideal chain (Sections 8.7 and 8.8) all lead to a similar picture for a long time, $<x^2> = 2Dt$, where $<x^2> =$ decays to equilibrium with a Gaussian-type probability density. However, the Langevin equation is based on a more specific physical mechanism—the interplay between the random and friction forces, and thus provides more microscopic information than the other methods. One obtains the relaxation to equilibrium of the velocity, $<v(t)>$ and $<v^2(t)>$ (governed by $\exp\text{-}\gamma t$ and $\exp\text{-}2\gamma t$, respectively), the fluctuation-dissipation theorem, Green-Kubo relationships, and a Maxwell-Boltzmann-type probability density for the velocity, $P(v,t|v_0)$ [Equation (12.77)].

It should be pointed out again that the treatment of the relaxation of the initial velocity, v_0, to equilibrium is based formally on a product space consisting of the (infinite) set of trajectories (paths), i, of intermediate velocities between $t = 0$ to t; for simplicity, they are represented by the discrete vectors $[v_i(t = 0)$, $v_i(t = 1),..., v_i(t)]$ and the corresponding probability densities $[P_i(t = 0), P_i(t = 1),...,P_i(t)]$, which lead to the averages:

$$< v(t) >= \sum_i P_i(t)v_i(t) \qquad \text{and} \qquad < v(t)^2 >= \sum_i P_i(t)[v_i(t)]^2 \qquad (12.80)$$

The significance of this picture in the experimental (or simulation) world is twofold: these averages are obtained by releasing: (1) a particle with v_0 many times or (2) many different particles with the same v_0.

12.3.5 Langevin Equation with a Charge in an Electric Field

An interesting model is the Langevin equation with the presence of an external force. As a simple example, we add to the basic Langevin equation a particle with charge, e, in a uniform electric field, E. The equation is:

$$m\frac{dv}{dt} = eE - \alpha v + R(t) \qquad (12.81)$$

Averaging both sides of this equation, using $<R(t)> = 0$, and assuming a steady state, that is, $d<v>/dt = 0$ leads to:

$$0 = m < \frac{d<v>}{dt} = eE - \alpha < v > \qquad \Rightarrow \qquad E = \frac{\alpha}{e} < v > \qquad (12.82)$$

Thus, $<v> \sim E$ and the mobility, u, defined by $u = <v>/E = e/\alpha$, where u and $D = k_B T/\alpha$ are both functions of α; one obtains the Einstein relation,

$$\frac{u}{D} = \frac{e}{k_B T} \qquad (12.83)$$

12.3.6 Langevin Equation with an External Force—The Strong Damping Velocity

We treat the situation of a Langevin equation with an external force,

$$m\frac{dv}{dt} = F(x) - \alpha v + R(t), \qquad (12.84)$$

where the acceleration term $m(dv/dt)$ can be ignored. This is justified when $\alpha v \gg m(dv/dt)$, meaning that the velocity of the particle is approximately constant [γ^{-1} is much shorter than any natural time scale associated with the motion in the potential $F(x)$]; the equation becomes:

$$\alpha v = F(x) + R(t) \qquad (12.85)$$

where, in general, $F(x) = dV/dx$, where V is the potential energy.

The basic Langevin equation, which only models the heat bath-system interactions, leads after a long time to the expected ideal gas behavior, that is, to the Maxwell-Boltzmann distribution for the velocities and to the random distribution of the particles' positions. One may ask whether the Langevin equation with an external force will lead after a long time to the Boltzmann distribution. The answer, in general, is positive, where an example is the Langevin equation with a harmonic force field solved in detail by Chandrasekhar ([6], p. 29). We shall return to this point in Sections 12.4.2 and 12.6.

Finally, the Langevin equation can be applied to models other than a Brownian particle. Thus, one can define more complex Langevin equations, e.g., with the non-linear force, ~cos2x, which is difficult to solve analytically; also, a non-Markovian equation has been treated, where the friction at time, t, depends on the history of the velocity $v(s)$ at earlier times, s, that is, the friction has a "memory." Because the Langevin is a Newton-type equation applied to a particle, it can be used in simulations of complex systems, such as proteins within the framework of molecular dynamics, as discussed below.

12.4 Stochastic Dynamics Simulations

While the Langevin equation with a complex force term cannot be solved analytically, it can be treated by computer simulation; an important example is a protein. Thus, the connectivity of the protein's chain is typically maintained by harmonic potentials between nearest neighbor atoms, and other distant-dependent interactions (e.g., Lennard-Jones interactions) are defined between any two non-bonded atoms. This set of interactions (from which the forces are derived) is generally called a "force field." The available force fields are discussed in detail in Section 16.1. The effect of the solvent on a protein of N atoms can be expressed by the friction and random Langevin forces applied to each atom. Thus, $3N$ Langevin equations are defined for all atoms, i, and their Cartesian components, x_i, y_i, and z_i; a typical equation is,

$$m_i \frac{d^2 x_i(t)}{dt^2} = \mathbf{F}_i[x_i(t)] - \gamma_i \frac{dx_i(t)}{dt} m_i + \mathbf{R}_i(t) \tag{12.86}$$

where $\mathbf{F}_i$ is the resultant force (force field) acting on atom, i, with mass, m_i. The friction coefficient, γ_i, can be defined differently for buried and exposed protein atoms. These equations are integrated numerically using molecular dynamics (MD)-type solvers. This type of simulation is called "*stochastic dynamics.*" These equations can be treated in different ways depending on the assumptions made about the relation between the frictional and random forces, where three regimes are generally defined.

1. The integration time, Δt, is much smaller than the velocity relaxation time, $\Delta t << \gamma^{-1}$, that is, the solvent does not activate (or deactivate) the particle significantly (typical, $\Delta t = 0.001$ ps; $\gamma = 50$ ps^{-1}, meaning that $\gamma^{-1} = 0.02$ ps)
2. The diffusion regime: $\Delta t >> \gamma^{-1}$ (large γ); many collisions occur within Δt and the motion is rapidly damped by the solvent (strong damping velocity)
3. The intermediate regime.

In regime (1) ($\Delta t << \gamma^{-1}$), a simple integration algorithm has been suggested by van Gunsteren et al. [7],

$$x_{i+1} = x_i + v_i \Delta t + \frac{1}{2} (\Delta t)^2 [-\gamma v_i + m^{-1}(F_i + R_i)]$$

$$v_{i+1} = v_i + (\Delta t)[-\gamma v_i + m^{-1}(F_i + R_i)] \tag{12.87}$$

where F_i, R_i, and v_i are the x components of these parameters. The average random force, R_i over Δt, is taken from a Gaussian distribution:

$$P(R) = \frac{1}{[2\pi <R^2>]^{1/2}} \exp-\left[\frac{R^2}{2 <R^2>} \right] \qquad (12.88)$$

with zero mean and variance $<R^2> = \Gamma(\Delta t)^{-1} = 2\gamma m k_B T(\Delta t)^{-1}$. Remember that $<R(0)R(\tau)> = \Gamma\delta(\tau)$, where $\delta(\tau)$ has the dimension of t^{-1} and Γ has the dimension of (force)$^2 \cdot t$ [see discussion following Equation (12.34)],

$$R_i = \int\limits_t^{t+\Delta t} R(t)dt. \qquad (12.89)$$

Thus, with the term $(\Delta t)^{-1}$ above, $<R^2>$ has the dimension of (force)^2 as required. Notice that the effect of the temperature is included in this process automatically through Γ.

In the strong damping velocity regime (2), $\gamma\Delta t \gg 1$, and the equation is,

$$\frac{dx_i(t)}{dt} = (\gamma_i m_i)^{-1}\{\mathbf{F}_i[x_i(t)] + \mathbf{R}_i(t)\}. \qquad (12.90)$$

The dynamics based on this equation is called "*Brownian dynamics.*" van Gunsteren et al. [7] obtained for this regime,

$$x_{i+1} = x_i + (\gamma m)^{-1}F_i\Delta t + X_i(\Delta t), \qquad (12.91)$$

assuming that the force is constant over the integration time step. X_i is a Gaussian distribution with zero mean and a variance of:

$$<X_i^2> = 2k_B T(m\gamma)^{-1}\Delta t = 2D\Delta t, \qquad (12.92)$$

where D is the diffusion constant and the velocities are not effective. Regime (3) has been treated by van Gunsteren and Berendsen [8], however, we shall not describe their derivation here.

12.4.1 Generating Numbers from a Gaussian Distribution by CLT

To carry out the various integrations of the equations of motion, one has to evaluate the random force by generating numbers taken from a Gaussian distribution. Many methods are available; the simplest is based on the CLT (Section 1.15), which states:

If X_1, X_2,..., X_n are uncorrelated random variables with mean values, $\mu_1, \mu_2, \cdots, \mu_n$, and variances, $\sigma_1, \sigma_2, \cdots, \sigma_n$, then, under a wide range of conditions, Y_n is normally distributed with mean = 0 and variance = 1, where:

$$Y_n = \frac{X_1 + X_2 + \cdots X_n - \mu_1 - \mu_2 \cdots - \mu_n}{\sqrt{\sigma_1^2 + \sigma_2^2 + \cdots + \sigma_n^2}} \qquad (12.93)$$

Now, if X_i, for $i = 1,2..,n$, is defined over the range [0,1] with a uniform distribution; the mean value is $\mu = 0.5$ [Equation (1.37)], and the variance is $1/12$ (Equation (1.45)]. Thus, for $n = 12$, $Y_n = [X_1 + X_2 + \ldots + X_{12} - 6]/[12 \times (1/12)]^{1/2}$, that is:

$$Y_n = X_1 + X_2 + \cdots + X_{12} - 6. \qquad (12.94)$$

Therefore, at each integration step, 12 random numbers, $\xi_1 \ldots \xi_{12}$, are generated within the range $[0,1)$ by a random number generator (see A3). According to CLT, the variable $\xi_1 + \ldots + \xi_{12-6}$ is normally distributed with a zero mean and a variance, 1. Notice that Y_n does not exceed 6, and it is not smaller than -6; however, in practice, this procedure is quite reliable. Because our variance [Equation (12.88)] is not 1, we multiply Y_n by $\Gamma^{1/2} = [2\gamma m k_B T (\Delta t)^{-1}]^{1/2}$, which becomes the value of $R_i(t)$.

12.4.2 Stochastic Dynamics versus Molecular Dynamics

To perform a stochastic dynamics (SD) simulation, one needs to determine γ. For simple molecules, it can be obtained from the viscosity, η, and the particle's radius, a, known experimentally. According to Stokes's law:

$$\mathbf{F}_{\text{frictional}} = -\gamma m \mathbf{v} = -6\pi a \eta \mathbf{v} \qquad \text{and} \qquad \gamma = 6\pi a \eta \mathbf{v} / m \qquad (12.95)$$

In other cases, one can optimize γ by comparing results obtained from SD simulations (based on different γ) to those obtained from a standard MD run in water. For a large compact molecule, such as a protein γ, values are given only to surface atoms and not to buried ones.

SD simulations can be orders of magnitude faster than MD runs with explicit solvent, not only due to the reduction in the number of atoms (SD simulations are performed in vacuum—without solvent molecules), but also because the time step Δt can be larger than that required for MD. However, for a long SD simulation, the system will relax to the equilibrium defined by the force field, **F**—the same equilibrium that would be reached by an MD simulation (in vacuum) based on **F**. The effect of SD is on the *dynamics*, which is different significantly from that obtained by MD (in vacuum), and it is closer to the dynamics obtained by MD applied to a protein in explicit solvent. SD is very important for studying the dynamics of polymers, where interesting phenomena occur over long periods of time.

As pointed out earlier, the temperature is already included in an SD simulation. So, the system is simulated in the canonical ensemble, and unlike MD, no thermostat is needed. This "Langevin thermostat" (defined through $\Gamma = 2\gamma m k_B T$) is expected to be better than the Berendsen or Andersen thermostats [see Equations (10.26) – (10.28)].

Because Langevin dynamics take the system to the equilibrium defined by the "real force," one can simulate by SD a system consisting of a protein immersed in a box of water molecules by defining γ values to water and the protein. This simulation will change the MD dynamics, but will relax the system to the same equilibrium defined by the real force field (with water), however, with a better temperature control than would be provided by Berendsen and Andersen thermostats.

An extensive body of literature about the SD approach is available, where many alternative procedures to those presented here have been developed. Our relatively limited review of this topic stems from the fact that, to date, it is customary to study proteins (our main interest) in explicit water—the ultimate model of solvation (see Chapter 16).

12.5 The Fokker-Planck Equation

In this section, we introduce the Fokker-Planck equation, which is a general second-order differential equation for the probability distribution; thus, by a proper definition of its parameters, it will describe the probabilities related to the Langevin Equation [1,9].

The Fokker-Planck equation was derived by Fokker as an approximation to the master equation (Chapter 13). We shall obtain this equation in a slightly different way following Reif [1]. Thus, having in mind the Brownian motion problem, we seek to derive an equation for the probability density of the velocity $P(v, t)$ in addition to $\langle v^2 \rangle$ provided from the Langevin equation. Notice, however, that the derivation is general, and it is applicable also to the coordinates. We start by expressing the conditional probability density, $P(v, t | v_0, t_0)$, by an integration over two intermediate transition probability densities,

$$P(v,t|v_0,t_0)dv = \int_{-\infty}^{\infty} P(v,t|v_1,t_1)dv \cdot P(v_1,t_1|v_0,t_0)dv_1 \qquad (12.96)$$

where the different times satisfy: $t > t_1 > t_0$, and the integration is carried out over all velocities v_1; this is a Chapman-Kolmogorov-type equation. Introducing the variables, $s = t_1 - t_0$ and $\tau = t - t_1$, means that $t - t_0 = s + \tau$, which leads to:

$$P(v, s + \tau | v_0) dv = \int_{-\infty}^{\infty} P(v, \tau | v_1) dv P(v_1, s | v_0) dv_1 \tag{12.97}$$

This is an integral equation, which can be converted into a differential equation by assuming that τ is small and the change $v_1 \to v$ is small as well. Denoting $v - v_1 = \lambda$ and expanding $P(v, s + \tau | v_0) dv$ to first order leads to:

$$P(v, s | v_0) + \frac{\partial P}{\partial s} \tau = \int_{-\infty}^{\infty} P(v, \tau | v - \lambda) P(v - \lambda, s | v_0) d\lambda \tag{12.98}$$

This is a general master equation (Chapter 13). Because the particle is *macroscopic*, its velocity can be changed only by a small amount during τ. Then $P(v, \tau | v - \lambda)$ is appreciable only for small λ, and the integrand can be expanded with respect to v around $P(v + \lambda, \tau | v) P(v, s | v_0)$ [that is, $P(v, \tau | v - \lambda) = P(v - \lambda + \lambda, \tau | v - \lambda)$],

$$P(v - \lambda + \lambda, \tau | v - \lambda) P(|v - \lambda, s | v_0) = \sum_{n=0}^{\infty} \frac{(-\lambda)^n}{n!} \frac{\partial^n}{\partial v^n} \left[P(v + \lambda, \tau | v) P(v, s | v_0) \right] \tag{12.99}$$

and thus

$$\frac{\partial P}{\partial s} \tau = -P(v, s | v_0) + \sum_{n=0}^{\infty} \frac{(-1)^n}{n!} \frac{\partial^n}{\partial v^n} \left[P(v, s | v_0) \int_{-\infty}^{\infty} \lambda^n d\lambda P(v + \lambda, \tau | v) \right] \tag{12.100}$$

The $n = 0$ term in the expansion is $P(v, s | v_0)$ (because by definition, the integral is equal to 1), which is cancelled. Defining M_n,

$$M_n = \frac{1}{\tau} \int_{-\infty}^{\infty} \lambda^n d\lambda P(v + \lambda, \tau | v) = \frac{<[\Delta v(\tau)]^n >}{\tau} \tag{12.101}$$

leads to:

$$\frac{\partial P(v, s | v_0)}{\partial s} = \sum_{n=1}^{\infty} \frac{(-1)^n}{n!} \frac{\partial^n}{\partial v^n} \left[M_n P(v, s | v_0) \right]. \tag{12.102}$$

For $n > 2$, $<(\Delta v)^n> \to 0$ faster than $\tau \to 0$, which justifies neglecting the higher terms ($n > 2$) in the equation. Thus, we obtain the Fokker-Planck equation,

$$\frac{\partial P}{\partial s} = -\frac{\partial}{\partial v} (M_1 P) + \frac{1}{2} \frac{\partial^2}{\partial v^2} (M_2 P) \tag{12.103}$$

This is a general equation (i.e., v should be considered as a general variable, not necessarily velocity), which should be solved for each case with respect to its corresponding boundary conditions. For Brownian motion, one obtains M_1 from the average velocity of the Langevin equation [Equation (12.39)],

$$<v> = v_0 \exp(-\gamma t). \tag{12.104}$$

By expanding the exponent to first order in τ,

$$M_1 = \frac{1}{\tau} < \Delta v(\tau) > = \frac{1}{\tau} v_0 [\exp(-\gamma(t+\tau)) - \exp(-\gamma t)] = \frac{1}{\tau} v_0 \exp(-\gamma t)[-\gamma \tau] = -\gamma v \qquad (12.105)$$

In a similar way, M_2 is obtained from the Langevin result for $<v^2>$ [Equation (12.44)]; thus:

$$M_1 = \frac{1}{\tau} < \Delta v(\tau) > = -\gamma v$$

$$M_2 = \frac{1}{\tau} < [\Delta v(\tau)]^2 > = \frac{2k_{\rm B}T}{m} \gamma \qquad (12.106)$$

and the Fokker-Planck equation for Brownian motion becomes:

$$\frac{\partial P}{\partial t} = \gamma \frac{\partial}{\partial v}(vP) + \gamma \frac{k_{\rm B}T}{m} \frac{\partial^2 P}{\partial v^2} \qquad (12.107)$$

or

$$\frac{\partial P}{\partial t} = \gamma P + \gamma v \frac{\partial P}{\partial v} + \gamma \frac{k_{\rm B}T}{m} \frac{\partial^2 P}{\partial v^2} \qquad (12.108)$$

where s is replaced by t. The Fokker-Planck equation can also be written in the form of an equation of continuity:

$$\frac{\partial P(v,t)}{\partial t} = -\gamma \frac{\partial}{\partial v}\left(-vP - \frac{k_{\rm B}T}{m}\frac{\partial}{\partial v}P\right) \qquad (12.109)$$

Comments:

1. The current density in the parenthesis has the dimension of flux ($1/x^2 t$); it is composed of a drift term ($-vP$) and a diffusion current.
2. If $P(v,t) \propto \exp-[mv^2/2k_{\rm B}T]$, the current density vanishes, meaning that the Maxwell-Boltzmann distribution is at least one equilibrium distribution solving this equation [if $dP/dt = 0$ $\rightarrow P =$ constant in time].
3. We shall argue that for long time $P(v, t)$ becomes the Maxwell-Boltzmann distribution; therefore, this is the only solution of the Fokker-Planck equation (and not just a steady-state solution).

$P(v,s|v_0)$ is obtained by solving the Fokker-Planck equation for the initial condition, $t \rightarrow 0$, $P(v,s|v_0) \rightarrow \delta(v - v_0)$. The solution is obtained by changing the variables of the original equation, transforming it thereby to the standard diffusion equation, where the solution is known. Expressing this solution back in terms of the original variables ([1], p. 581) leads to:

$$P(v,t|v_0) = \left[\frac{m}{2\pi k_{\rm B}T(1 - e^{-2\gamma t})}\right]^{1/2} \exp-\left[\frac{m(v - v_0 e^{-\gamma t})^2}{2k_{\rm B}T(1 - e^{-2\gamma t})}\right] \qquad (12.110)$$

Indeed, for $t \rightarrow 0$, $P(v,s|v_0) \rightarrow \delta(v - v_0)$, as one representation of a delta function is $\delta(x) = (2\pi\gamma)^{(-1/2)}\exp(-x^2/2\gamma^2)$; for $\gamma \rightarrow 0$ or for $t \rightarrow \infty$, one obtains the Maxwell-Boltzmann distribution. For arbitrary time, t, the solution is a Gaussian, where v is distributed around the average, $<v> = v_0 e^{-\gamma t}$ [Equation (12.39)].

Notice that the same result for $P(v, t|v_0)$ was obtained in Section 12.3.4 from the Langevin equation based on the central limit theorem; one concludes that Equation (12.109) is a Fokker-Planck equation for calculating the probability distribution that corresponds to the Langevin equation.

12.6 Smoluchowski Equation

Of special interest is a Langevin equation with a force field, $F(x)$, which is also used in SD simulations. In the strong damping velocity regime [γ^{-1} is much shorter than any natural time scale associated with the motion in the potential $F(x)$], we obtained [Equation (12.85)],

$$\alpha v = F(x) + R(t) \tag{12.111}$$

where $F(x) = -\partial V(x)/\partial x$ and V is a potential. One can show [6] that the corresponding equation for the probability distribution, $P(x,t)$, is the following *Smoluchowski equation:*

$$\frac{\partial P(x,t)}{\partial t} = -\frac{1}{m\gamma}\frac{\partial}{\partial x}[P(x,t)F(x)] + \frac{k_B T}{m\gamma}\frac{\partial^2}{\partial x^2}[P(x,t)] \tag{12.112}$$

One can cast this Smoluchowski equation in the form of an equation of continuity,

$$\frac{\partial P(x,t)}{\partial t} = \frac{\partial}{\partial x} J(x,t) \tag{12.113}$$

with a current density, $J(x,t)$, consisting of diffuse and drift terms:

$$J(x,t) = -\frac{1}{m\gamma}\left(-k_B T \frac{\partial}{\partial x} + F(x)\right) P(x,t). \tag{12.114}$$

Clearly,

$$P(x,t) \propto \exp[-[V(x)/k_B T]] \tag{12.115}$$

is a stationary solution of the Smoluchowski equation, since for this solution, $J(x,t) = 0$ identically, meaning that $\partial P(x,t)/\partial t = 0$ everywhere in space, and thus, $P(x,t)$ is constant in time.

Comments:

1. The probability for the velocity part is expected to relax to a Maxwell-Boltzmann instantaneously (large γ), and only the position–space probability distribution, $P(x,t)$, is calculated [see discussion following Equation (12.78)].

2. One would expect the Smoluchowski equation to lead the system from a non-equilibrium state to equilibrium. The fact that the Boltzmann probability is a stationary solution of this equation is thus important, even though this solution might not be unique.

3. Both the overdamping Langevin equation and the corresponding Smoluchowski equation can be generalized to any dimension – d (in Cartesians) for any coupled d-dimensional force field, that is,

$$\vec{F}(\vec{x}) = -\vec{\nabla}\vec{V}(\vec{x}) \tag{12.116}$$

where the vectors are d-dimensional. The distribution function, $P(\vec{x},t) \propto \exp[-V(\vec{x})k_B T]$, is a steady-state solution of the Smoluchowski equation. Again, this is important because the overdamping Langevin's equation is used in *Brownian dynamics* simulations, and it is imperative to show that the system relaxes to equilibrium.

4. The question of whether the Boltzmann probability is a unique stationary probability can be decided only by solving the equation analytically for a given Hamiltonian (force field). If it is not unique, different initial distributions might relax to different regions in phase space. Clearly, such analytical checks cannot be applied to a complex protein force field. Therefore, in general, one assumes that the Boltzmann probability is the unique stationary probability. Notice that the same assumption is made in molecular dynamics and Monte Carlo simulations.

12.7 The Fokker-Planck Equation for a Full Langevin Equation with a Force

Unlike the Smoluchowski equation case, where the probability distributions for the velocities and coordinates were separated, in the Fokker-Planck equation for the full Langevin equation, one calculates the probability distribution, $P(x,v,t)$, which consists of both x and v; thus, $P(x,v,t)$ is the probability density to find a particle within the intervals $[x, x + dx]$ and $[v, v + dv]$ at time, t. This Fokker-Planck equation is [5.6],

$$\frac{\partial P}{\partial t} = -v \frac{\partial P}{\partial x} - \frac{F(x)}{m} \frac{\partial P}{\partial v} + \gamma \left[\frac{\partial}{\partial v}(vP) + \frac{k_{\mathrm{B}}T}{m} \frac{\partial^2 P}{\partial v^2} \right] \qquad (12.117)$$

The previous discussion about the Smoluchowski equation applies also here. In particular, if:

$$P(x,v,t) \propto \exp{-[H(x,v)/k_{\mathrm{B}}T]} \qquad (12.118)$$

where $H(x, v) = p^2/2m + V(x)$ $(p = mv)$, then the r.h.s of the Fokker-Planck equation above vanishes identically, implying that $\partial P(x,v,t)/\partial t = 0$, and the system remains Boltzmann distributed in all subsequent times. Comments 3 and 4 of Section 12.6 apply also to the Fokker-Planck equation, Equation (12.117).

12.8 Summary of Pairs of Equations

1. Langevin equation: $m \dfrac{dv}{dt} = -\alpha v + R(t) \rightarrow$

Fokker-Plank equation: $\dfrac{\partial P}{\partial t} = \gamma P + \gamma v \dfrac{\partial P}{\partial v} + \gamma \dfrac{k_{\mathrm{B}}T}{m} \dfrac{\partial^2 P}{\partial v^2}$

2. Langevin with external force: $m \dfrac{dv}{dt} = -\alpha v + R(t) + F(x) \rightarrow$

Fokker-Planck equation: $\dfrac{\partial P}{\partial t} = -v \dfrac{\partial P}{\partial x} - \dfrac{F(x)}{m} \dfrac{\partial P}{\partial v} + \gamma \left[\dfrac{\partial}{\partial v}(vP) + \dfrac{k_{\mathrm{B}}T}{m} \dfrac{\partial^2 P}{\partial v^2} \right]$

3. Overdamping Langevin with force: $\alpha v = F(x) + R(t) \rightarrow$

Smoluchowski equation: $\dfrac{\partial P(x,t)}{\partial t} = -\dfrac{1}{m\gamma} \dfrac{\partial}{\partial x}[P(x,t)F(x)] + \dfrac{k_{\mathrm{B}}T}{m\gamma} \dfrac{\partial^2}{\partial x^2}[P(x,t)]$

REFERENCES

1. F. Reif. *Fundamentals of Statistical and Thermal Physics*. (McGraw-Hill, New York, 1965).
2. K. A. Dill and S. Bromberg. *Molecular Driving Forces, Statistical Thermodynamics in Biology, Chemistry, Physics, and Nanoscience*. (Garland Science, New York, 2010).
3. J. Crank. *The Mathematics of Diffusion*, 2nd ed. (Clarendon Press, Oxford, UK, 1993).
4. C. W. Gardiner. *Handbook of Stochastic Methods for Physics, Chemistry and the Natural Sciences*, 3rd ed. (Springer, Heidelberg, Germany, 2004).
5. N. G. Van Kampen. *Stochastic Processes in Physics and Chemistry*. (Elsevier Science, Amsterdam, the Netherlands, 1992).
6. S. Chandrasekhar. Stochastic problems in physics and astronomy, 2–91. In *Selected Papers on Noise and Stochastic Processes*. N. Wax (Ed.), (Dover, New York, 1954).
7. W. F. van Gunsteren, H. J. C. Berendsen and A. C. Rullmann. Stochastic dynamics for molecules with constraints: Brownian dynamics of n-alkanes. *Mol. Phys.* **44**, 69–95 (1981).
8. W. F. van Gunsteren and H. J. C. Berendsen. Algorithm for Brownian dynamics. *Mol. Phys.* **45**, 637–647 (1982).
9. R. Zwanzig. *Non Equilibrium Statistical Mechanics*. (Oxford University Press, Oxford, UK, 2001).

13

The Master Equation

Chapter 13 is devoted to the master equation. Formally, this equation can be derived from the Chapman-Kolmogorov equation [Equation (12.96)] as developed, for example, in several books [1–3]. However, our discussion is based on the "golden rule" in quantum mechanics, as done by Zwanzig [4] and Reif [5]. The master equation is applicable to many phenomena, such as chemical kinetics, birth-death processes, and random walks. As an example, we present a nuclear magnetic resonance problem, following [5], where the "close to equilibrium" basis of the equations is emphasized. This discussion constitutes a helpful precursor for the last section in this chapter, where the principle of minimum entropy production (and its restrictions) is demonstrated by applying a master equation to a similar system.

13.1 Master Equation in a Microcanonical System

In Section 4.7, we have introduced the Schrödinger equation for the quantum mechanical oscillator in $d = 1$ and presented its set of energy levels, $E_n = (n + 1/2)h\nu$, related to the wave functions, ψ_n ($n = 0,1,2...$); $|\psi_n(x)^*\psi_n(x)|$ is the probability density of an oscillator in energy level, E_n, to be at position x. To induce transitions among these energy levels, a suitable *small* interaction should be added to the main Hamiltonian [Equation (4.44)].

Assume now a microcanonical ensemble, that is, a large isolated system of N particles, i, $i = 1,2,... N$ with the corresponding Hamiltonians $\hat{H}_i$, each with a set of Schrödinger solutions $\{\psi_n\}$ and energies $\{E_n\}$. In principle, one can write the Hamiltonian, $\hat{H}$, of the entire system as, $\hat{H} = \hat{H}_1 + \hat{H}_2 + \hat{H}_3 + \hat{H}_j + ... + \hat{H}_N$; if the Hamiltonians, $\hat{H}_i$, are independent, a typical wave function that solves the Schrödinger equation of $\hat{H}$ will be a product of the individual $\psi_n(i)$, that is, $\psi_r = \psi_n(1)\psi_k(2)\psi_l(3) ... \psi_m(N)$ with an energy level, $E_r = E_n + E_k + E_l + ... + E_m$; r stands for a possible combination of the individual states, $r \equiv n, k, l...$ Clearly, the energy levels will be nearly continuously spaced, where many states r are related to each energy level. Again, to induce transitions among these states, a *small* interaction, $\hat{H}_{int}$, should be added to the main Hamiltonian, $\hat{H}$ ($\hat{H}_{int} << \hat{H}$), leading to the total Hamiltonian $\hat{H}_0$,

$$\hat{H}_0 = \hat{H} + \hat{H}_{int} \tag{13.1}$$

If $\hat{H}_{int} = 0$, the system would remain constantly at some r. For a finite $\hat{H}_{int}$, transitions from state α to state β and from β to α can occur, where based on quantum mechanics considerations (*the golden rule*), the transition probability rates (TPRs) (TP per unit time) for these transitions are equal:

$$W_{\alpha\beta} = W_{\beta\alpha} \tag{13.2}$$

where if $E_\alpha \neq E_\beta \rightarrow W_{\alpha\beta} = 0$. $W_{\alpha\beta}$ is a TPR defined in the probability space and should be viewed as the outcome of a huge number of experiments. Thus, an experimental definition of $W_{\alpha\beta}$ as a TPR requires considering time intervals, Δt that *are not too small*. If $P_\alpha(t)$ is the probability density of the system to be at α in time, t, the change of $P_\alpha(t)$ in time *is modeled* (approximated) by the master equation,

$$\frac{d}{dt}P_\alpha(t) = \sum_\beta P_\beta W_{\beta\alpha} - \sum_\beta P_\alpha W_{\alpha\beta} = \sum_\beta (P_\beta W_{\beta\alpha} - P_\alpha W_{\alpha\beta}) \tag{13.3}$$

The first summation term in the middle and the right sides of Eq. (13.3) is the gain of probability rate of state α due to transitions from the other states, β; the second summation is the loss of the probability rate of state α due to transitions to states, β. This equation describes irreversibility because it is not invariant under time reversal, $t \rightarrow -t$ (in this respect the master equation differs from the Schrödinger equation, which is invariant under time reversal).

We seek a *stationary* solution, $dP_\alpha(t)/dt = 0$, for all α [Equation (13.3)], which obviously is satisfied at equilibrium, since in the microcanonical ensemble by definition, $P_\alpha(eq) = P_\beta(eq)$, for all α and β; considering also the symmetry of the TPRs [Equation (13.2)], one obtains,

$$P_\beta(eq)W_{\beta\alpha} - P_\alpha(eq)W_{\alpha\beta} = P_\beta(eq)[W_{\beta\alpha} - W_{\alpha\beta}] = 0 \qquad (13.4)$$

So, the master equation describes correctly an equilibrium steady state. It should be noted that Equation (13.4) defines the *detailed balance* condition (see Equations (1.75), (10.9) and (11.50)) for α and β with the same energy,

$$P_\beta(eq)W_{\beta\alpha} = P_\alpha(eq)W_{\alpha\beta}. \qquad (13.5)$$

In other words, detailed balance is a *sufficient* condition for $dP_\alpha(t)/dt = 0$ and for equilibrium, but it is not a *necessary* condition: a system can be in a non-equilibrium steady state, where $dP_\alpha(t)/dt = 0$, but the detailed balance condition is not satisfied (e.g., see Section 13.4). In this context, it should be noted that if ergodicity is satisfied, that is, every state can be reached from any other state by a sequence of allowed transitions, equilibrium is the *only* stationary solution. Otherwise, different initial states can lead to different stationary states in long times.

Also, assuming that detailed balance [Equation (13.5)] holds at equilibrium, and using $P_\alpha(eq) = P_\beta(eq)$, leads to $W_{\alpha\beta} = W_{\beta\alpha}$. However, this relation and the W's themselves are based on quantum mechanical considerations. Therefore, one can conclude that the relation $W_{\alpha\beta} = W_{\beta\alpha}$ is generally valid, that is, in non-equilibrium situations as well; moreover, the equilibrium information is already embedded in the quantum mechanical based, $W_{\alpha\beta}$.

Finally, notice that one can define a master equation without a relation (as we did) to the underlying dynamical model, that is, based on transitions among a set of given (possibly abstract) states. For us, it is an equation that provides a suitable modeling of relaxation to equilibrium or to steady states.

13.2 Master Equation in the Canonical Ensemble

After treating the master equation under microcanonical conditions, we make a step further and study the more realistic situation of system, A, in contact with a huge heat bath, A'. Due to its size, A' has a well-defined temperature, T, that is, it remains in equilibrium in spite of its interaction with A (Figure 13.1). Our interest is in the non-equilibrium to equilibrium behavior of A. Note that the combined system, $A^0 = A + A'$ as a whole, is isolated and thus the conclusions of the previous section apply to it. The Hamiltonian of A^0 is,

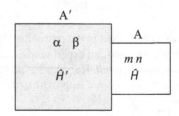

FIGURE 13.1   A reservoir A' (with Hamiltonian $\hat{H}'$ and states denoted, α and β) is in a thermal contact with system A (consisting of $\hat{H}$ and states denoted m and n). The small Hamiltonian of interaction is denoted $\hat{H}_{int}$.

$$\hat{H}^0 = \hat{H} + \hat{H}' + \hat{H}_{int} \tag{13.6}$$

where $\hat{H}^0$, $\hat{H}$, and $\hat{H}'$ are the Hamiltonians of A^0, A, and A', respectively, and H_{int} is a small interaction between A and A' ($\hat{H}_{int} \ll \hat{H}, \hat{H}'$), which leads to transitions between states of these systems. For $\hat{H}_{int} = 0$, the states, the energies, and the (time-dependent) probability densities that are related to $\hat{H}$ and $\hat{H}'$ are denoted by the letters m and α, respectively, and the states of microcanonical system, A^0 are denoted by "$m\alpha$". The master equation for A^0 is:

$$\frac{d}{dt}P_{m\alpha}(t) = \sum_{n\beta} P_{n\beta} W_{n\beta, m\alpha} - \sum_{n\beta} P_{m\alpha} W_{m\alpha, n\beta} \tag{13.7}$$

Since we assume that the heat bath, A', remains in thermal equilibrium, $P_{m\alpha}$ can be factored into a non-equilibrium probability $P_m(t)$ for A and a thermal equilibrium probability for A', ρ_α

$$P_{m\alpha} \cong P_m(t)\rho_\alpha. \tag{13.8}$$

Substituting Equation (13.8) in Equation (13.7) and summing over α leads to:

$$\frac{d}{dt}P_m(t) = \sum_n \sum_\alpha \sum_\beta \rho_\beta P_n W_{n\beta, m\alpha} - \sum_n \sum_\alpha \sum_\beta \rho_\alpha P_m W_{m\alpha, n\beta} \tag{13.9}$$

which can be written as a master equation for the system A,

$$\frac{d}{dt}P_m(t) = \sum_n P_n w_{n,m} - \sum_n P_m w_{m,n}. \tag{13.10}$$

Notice that the new TPRs (denoted by lowercase w) are no longer symmetric,

$$w_{n,m} = \sum_\alpha \sum_\beta \rho_\beta w_{n\beta, m\alpha} \qquad w_{m,n} = \sum_\alpha \sum_\beta \rho_\alpha w_{m\alpha, n\beta}. \tag{13.11}$$

Now, the equilibrium distribution ρ_β of the bath is proportional to $\exp{-[(E-E_n)/k_B T]}$, where E is the constant energy of A^0 and E_n is the energy of A in state, n. Therefore:

$$w_{n,m} = const.\exp(E_n/k_B T)\sum_\alpha \sum_\beta w_{n\beta, m\alpha} \quad w_{m,n} = const.\exp(E_m/k_B T)\sum_\alpha \sum_\beta w_{m\alpha, n\beta}, \tag{13.12}$$

Since the double summations are equal (due to symmetry in the microcanonical ensemble), one obtains the detailed balance for system A,

$$w_{n,m}\exp(-E_n/k_B T) = w_{m,n}\exp(-E_m/k_B T) \tag{13.13}$$

or

$$P_n^B w_{n,m} = P_m^B w_{m,n} \tag{13.14}$$

where P_n^B is the Boltzmann probability of state n. The TPRs $w_{n,m}$ [Equation (13.11)] are non-equilibrium entities determined solely by the *equilibrium* canonical distribution; notice, however, that only their ratio is defined, and there is a freedom to choose different suitable pairs of TPRs. Also, in the canonical case, in general, $w_{m,n} \neq w_{n,m}$, and if $E_m > E_n \rightarrow w_{n,m} < w_{m,n}$.

Considering Equation 13.10, we look again for steady-state solutions, $dP_r/dt = 0$. This condition is satisfied if the canonical distribution (hence, detailed balance) holds for system A; then, the right-hand side of Equation (13.10) vanishes. Thus, detailed balance is a sufficient (but not a necessary) condition for $dP_r/dt = 0$. In other words, while the relaxation to equilibrium is inherent in the master equation by the definition of w_{nm}, the system might end up in a non-equilibrium steady state, where $dP_n/dt = 0$, as will be demonstrated by several examples discussed below.

Finally, we mention the relation of the master equation to a Markov chain, as can be demonstrated by the simple example of the first step of a two-state Markov chain presented in Equation (1.65); thus, $dP_r(t)$ of Equation (13.10) corresponds to the discrete difference, $P_i^1 - P_i^0$:

$$P_i^1 = P_i^0 p_{ii} + P_j^0 p_{ji} \quad \rightarrow \quad P_i^1 - P_i^0 = P_j^0 p_{ji} - P_i^0 p_{ij}$$
$$P_j^1 = P_i^0 p_{ij} + P_j^0 p_{jj} \quad \rightarrow \quad P_j^1 - P_j^0 = P_i^0 p_{ij} - P_j^0 p_{ji}$$
(13.15)

Notice that the term $P_n w_{nn}$ appears in Equation (13.10) in both summations and thus gets cancelled; $P_i^0 p_{ii}$ gets cancelled here as well.

13.3 An Example from Magnetic Resonance

We now apply a master equation for treating a classical relaxation problem discussed in many nuclear magnetic resonance (NMR) textbooks. we follow the derivation of [5] with enhancements. Consider a substance containing N *non-interacting* nuclei, each having a magnetic moment, μ, where each can stay in either of two states, spin $+1/2$ (up) or $-1/2$ (down). When the substance is placed in a magnetic field, H, the spins interact with H with energies, ε_+ for spin up (parallel to the magnetic field), and ε_- for spin down, where:

$$\varepsilon_+ = -\mu H \qquad \varepsilon_- = +\mu H$$
(13.16)

The number of spins up and the number of spins down are denoted by n_+ and n_-, respectively, where $N = n_+ + n_-$; the Hamiltonian of the system is:

$$\hat{H}^0 = \hat{H}_n + \hat{H}_L + \hat{H}_{int} = \hat{H}_{main} + \hat{H}_{int}$$
(13.17)

where $\hat{H}_n$ describes the interactions between the spins and the magnetic field ($\mu-H$ interactions) defined above, leading to 2^N energy levels. $\hat{H}_L$ describes the "lattice," which includes all the other degrees of freedom of the system's atoms (besides the spin states). Since the lattice is large, it can be considered as a heat bath at equilibrium with a constant temperature, T. The spin-lattice interactions are defined by the small Hamiltonian, $\hat{H}_{int}$, which causes transitions between the spin states. In the language of the master equation, W_{+-} is the transition probability per unit time for flipping an up spin to a down spin, due to the spin-lattice interactions; from detailed balance at equilibrium, one obtains the ratio,

$$\frac{W_{-+}}{W_{+-}} = \frac{\exp-(\varepsilon_+ / k_B T)}{\exp-(\varepsilon_- / k_B T)} = \exp\left(\frac{\varepsilon_- - \varepsilon_+}{k_B T}\right) = \exp\frac{2\mu H}{k_B T}$$
(13.18)

where $\mu \sim 5 \times 10^{-24}$ erg/Gauss, $H \sim 10^4$ Gauss, and $k_B = 1.38 \times 10^{-16}$ erg/degree; therefore,

$$\frac{2\mu H}{k_B T} \approx \frac{5 \times 10^{-4}}{T} \ll 1.$$
(13.19)

$2\mu H/k_{\rm B}T \ll 1$ is generally satisfied unless T is exceedingly low. This relation is satisfied in most cases also for *electronic* moments that are 10^3 times larger than the magnetic ones. Therefore, the difference in the populations of the spin states is small. Defining $W_{+-} = W$ and expanding the exponent in Equation (13.18) to first order (a close to equilibrium condition) leads to:

$$W_{-+} = W\left(1 + \frac{2\mu H}{k_{\rm B}T}\right) \tag{13.20}$$

Another component of the experiment is the application of an *alternating* magnetic field (perpendicular to H) with angular frequency, ω, which will induce transitions ($- \to +$ and $+ \to -$) provided that $\hbar\omega \approx (\varepsilon_- -\varepsilon_+) = 2\mu H$; for $H \approx 10^4$ gauss, the required ω is $\approx 10^8$ sec^{-1}. Correspondingly, we denote by w_{+-} the TP per unit time for an up to a down transition due to the alternating field. As discussed in Section 13.1, for the microcanonical system, first-order perturbation theory in quantum mechanics leads to:

$$w_{+-} = w_{-+} \equiv w \tag{13.21}$$

where $w(\omega)$ is effective only for $\omega \approx 2\mu H/\hbar$ ($\hbar$ is Planck's constant). All the transition probabilities defined above are used in the following master equations:

$$\frac{dn_+}{dt} = n_-(W_{-+} + w) - n_+(W_{+-} + w)$$

$$\frac{dn_-}{dt} = n_+(W_{+-} + w) - n_-(W_{-+} + w) \tag{13.22}$$

Subtracting these equations leads to:

$$\frac{d}{dt}(n_+ - n_-) = -2n_+(W_{+-} + w) + 2n_-(W_{-+} + w) \tag{13.23}$$

and after defining $n \equiv n_+ - n_-$ and relying on Equation (13.20), one obtains:

$$\frac{dn}{dt} = -2(W + w)n + 4n_- W \frac{\mu H}{k_{\rm B}T} \tag{13.24}$$

This equation can still be approximated, since at relevant T, $n_- \sim n_+$, and thus, $4n_- \sim 2n_- + 2n_+ = 2N$ (a close to equilibrium condition). Finally, one obtains

$$\frac{dn}{dt} = -2(W + w)n + 2NW \frac{\mu H}{k_{\rm B}T} \tag{13.25}$$

13.3.1 Relaxation Processes Under Various Conditions

Equation (13.25) can be solved under several conditions.

Case 1: Equilibrium without a radio frequency (rf) field ($w = 0$).

At equilibrium, $n = n_{\rm eq}$, which can be obtained in two ways: first from the condition, $dn/dt = 0$, which leads to:

$$2Wn = 2NW \frac{\mu H}{k_{\rm B}T} \quad \to \quad n_{\rm eq} = N \frac{\mu H}{k_{\rm B}T} \tag{13.26}$$

and second, by expressing $n_\pm$ as canonical distribution averages, where the exponentials are expanded to first order in $x = \mu H/k_B T$, which leads to $(1 \pm x)/(1 + x + 1 - x)$ and to,

$$n_\pm = N\frac{\exp(\pm x)}{\exp(+x) + \exp(-x)} \approx N\frac{1 \pm x}{2} = \frac{1}{2}N(1 \pm x). \tag{13.27}$$

Thus:

$$n_{eq} = n_+ - n_- = N\frac{\mu H}{k_B T} \tag{13.28}$$

and Equation (13.25) becomes:

$$\frac{dn}{dt} = -2W(n - n_{eq}) - 2wn. \tag{13.29}$$

Assuming $w = 0$ and denoting $n(0) = n(t = 0)$, the solution is,

$$n(t) = n_{eq} + [n(0) - n_{eq}]\exp(-2Wt). \tag{13.30}$$

Thus, $n(t)$ decays to n_{eq} with a relaxation time, $T_1 = (2W)^{-1}$, which is called the spin-lattice relaxation time. T_1 contains structural and dynamical information of the system and thus constitutes an important parameter measured in experiments; the larger is W, the smaller is T_1, that is, the stronger is the TP per unit time from spin + to −, the shorter is the relaxation time.

It is of interest to examine the relaxation of the energy, $<E(t)>$, the entropy, $S(t)$, and the free energy, $F(t)$; we assume first an extreme case, where $n(0) = n_+ = N$ meaning that $E(0) = -\mu HN$, $S(0) = 0$, and $F(0) = -\mu HN$. For the energy, one obtains,

$$\frac{<E(t)>}{N} = -\frac{\mu^2 H^2}{k_B T} - \mu H\left(1 - \frac{\mu H}{k_B T}\right)\exp(-2Wt) = \frac{<E_{eq}>}{N} - \mu H\left(1 - \frac{\mu H}{k_B T}\right)\exp(-2Wt), \tag{13.31}$$

which demonstrates a monotonic increase with time of the energy toward equilibrium. The entropy is obtained by defining $S(t)/N = -k_B(p^+ \ln p^+ + p^- \ln p^-)$, where p^+ and p^- are the probabilities for "up" and "down" spins at time, t, respectively. Notice that only terms up to order $(n/N)^2$ have been considered; one obtains:

$$\frac{S(t)}{Nk_B} = \ln 2 - \frac{1}{2}\left(\frac{\mu H}{k_B T}\right)^2 - \left[\ln 2 - \frac{1}{2}\left(\frac{\mu H}{k_B T}\right)^2\right]\exp(-2Wt) \rightarrow \frac{S_{eq}(t=\infty)}{Nk_B} = \ln 2 - \frac{1}{2}\left[\frac{\mu H}{k_B T}\right]^2 \tag{13.32}$$

which shows the expected increase (like $<E(t)>$) of the entropy with time from zero toward its maximal value at equilibrium. The entropy is always smaller than ln2 due to the larger population of spin up ($n > 0$). However, the increase in the entropy is stronger than the corresponding increase of the energy, and the net result is the anticipated behavior of a monotonic decrease of the free energy, $F(t) = <E(t)> - TS(t)$ with t, where $F(t)$ becomes minimal at equilibrium ($t = \infty$),

$$\frac{F(t)}{N} = -\frac{\mu^2 H^2}{k_B T} - k_B T\ln 2 + \frac{k_B T}{2}\left(\frac{\mu H}{k_B T}\right)^2 + \left[-\mu H\left(1 - \frac{\mu H}{k_B T}\right) + k_B T\ln 2 - \frac{k_B T}{2}\left(\frac{\mu H}{k_B T}\right)^2\right]\exp(-4Wt). \tag{13.33}$$

Equations (13.25) and (13.30) have been derived under the condition that W_{+-} is only slightly larger than W_{-+}, hence $n_+ - n_-$ is very small as compared to N. This is a close to equilibrium condition. On the other hand, the mathematical solutions of this equation depend also on $n(t = 0)$, which (mathematically) is not restricted and can be as large as $n(t = 0) = N$ (all spins are up), as assumed in Equations (13.31–13.33);

this is a far from equilibrium situation, and thus the calculations of $E(t)$, $S(t)$, and $F(t)$ are thermodynamically meaningful only when the relaxation gets close to equilibrium. A (thermodynamically) adequate starting point would be $n = 0$, where $E(0) = 0$ and $S(0) = k_B T \ln 2$ [see also the derivation of Equation (13.36) below]. In a close to equilibrium situation, the effect of the initial state ($t = 0$) on the probabilities and averages defined at t is not strong.

PROBLEM 13.1
Derive Equation (13.32).

Case 2: The spin-lattice interactions are weak, that is, $W \approx 0$.
When $W \approx 0$, Equation (13.25), with rf becomes:

$$\frac{dn}{dt} = -2wn \tag{13.34}$$

with the solution,

$$n(t) = n(0) \exp(-2wt) \tag{13.35}$$

Thus, $n(t = \infty) = 0$ (the same population of up and down spins) is approached with the relaxation time, $\tau = (2w)^{-1}$. The amount of energy, $E_{abs} = (n_+ - n_-)\hbar\omega$ ($\hbar\omega = 2\mu H$) is absorbed by the spin system, where up spins become down spins until their numbers become equal; this situation means *saturation*, that is, no further energy is absorbed by the spins. E_{abs} is not transferred to the lattice because the spin system is isolated with no interaction with the lattice ($W \approx 0$). Therefore, the spins' temperature, $T_s = dE/dS \to \infty$ when $n_+ - n_- \to 0$.

This can be seen by defining, $n_+ = (N + n)/2$, $n_- = (N - n)/2$, and the corresponding probabilities for +/− spins, $p^+ = n_+/N = 1/2[1 + n/N]$ and $p^- = n_-/N = 1/2[1 - n/N]$. Thus, denoting $x = n/N$, the *system* entropy is:

$$-S/k_B N = p^+ \ln p^+ + p^- \ln p^- = 1/2(1+x) \ln[1/2(1+x)] + 1/2(1-x)\ln[1/2(1-x)]. \tag{13.36}$$

Since x is small, $S/k_B N$ can be expanded to second order in x, which leads to $\ln 2 - (n/N)^2$, where the system energy is $E = -n\mu H$; therefore,

$$T_s = \frac{dE}{dS} = \frac{dE/dn}{dS/dn} = \frac{\mu H N^2}{n} \to \infty \quad \text{since } n \to 0 \tag{13.37}$$

Case 3. A steady-state situation—the rf energy is transferred to the lattice.
The temperature, T, of a large lattice is almost unchanged due to its large heat capacity; in a steady-state $dn/dt = 0$, Equation (13.25) becomes:

$$W(n - n_{eq}) = -wn \tag{13.38}$$

or

$$n = \frac{n_{eq}}{1 + (w/W)} \tag{13.39}$$

meaning that $n < n_{eq}$ due to forced rf transitions. More specifically, the rf energy, which is transferred to the lattice, leads to an overpopulation of the high energy level (spin down). (We point out that a direct derivation of Equation (13.39) from the steady-state probability, p_+^s [Equation (13.42) below], is not straightforward due to the different approximations leading to these equations.)

13.3.2 Steady State and the Rate of Entropy Production

In a steady state, $dn_+/dt = 0$ or $d(n_+/N)/dt = dp_+/dt = 0$, therefore $dp_-/dt = 0$, as well. Equation (13.22) (with $w \neq 0$) becomes:

$$\frac{dn_+}{Ndt} = \frac{dp_+}{dt} = p_-(W_{-+} + w) - p_+(W_{+-} + w) \tag{13.40}$$

and using Equation (13.20),

$$\frac{dp_+}{dt} = (1 - p_+)\left(W \exp\frac{2\mu H}{k_B T} + w\right) - p_+(W + w) = 0. \tag{13.41}$$

Thus (replacing for brevity $2\mu H$ by 2ε),

$$p_+^s = \frac{W \exp(2\varepsilon/k_B T) + w}{W \exp(2\varepsilon/k_B T) + W + 2w} = \frac{\exp(2\varepsilon/k_B T) + \beta}{\exp(2\varepsilon/k_B T) + 1 + 2\beta} \tag{13.42}$$

where $\beta = w/W$ and p_+^s is the steady-state probability. To obtain the rate of entropy production, one has to calculate both the contributions of the system and the heat bath; using $p_+ = 1 - p_-$, one obtains for the system,

$$
\begin{aligned}
\frac{dS_{\text{sys}}}{k_B dt} &= -N \frac{d}{dt}[p_+ \ln p_+ + p_- \ln p_-] \\
&= -N\left[\frac{dp_+}{dt}\ln p_+ + p_+\frac{1}{p_+}\frac{dp_+}{dt} + \frac{dp_-}{dt}\ln p_- + p_-\frac{1}{p_-}\frac{dp_-}{dt}\right] \\
&= -N\left[\frac{dp_+}{dt}\ln p_+ + \frac{dp_+}{dt} - \frac{dp_+}{dt}\ln p_+ - \frac{dp_+}{dt}\right] \\
&= -N\left[\frac{dp_+}{dt}(\ln p_+ + 1 - \ln p_- - 1)\right] = -N\left[\frac{dp_+}{dt}(\ln p_+/p_-)\right]
\end{aligned}
\tag{13.43}
$$

Substituting in Equation (13.43), the expression for dp_+/dt which appears in Equation (13.40) leads to:

$$\frac{dS_{\text{sys}}}{k_B dt} = -N[\ln(p_+/p_-)][(W \exp(2\varepsilon/k_B T) + w)p_- - (W + w)p_+]. \tag{13.44}$$

Calculation of dS_{heat}/dt for the heat bath is based on the fact that the heat bath gains entropy $2\varepsilon/T$ for each downward transition of a particle $(- \rightarrow +)$ and loses the same entropy for each upward transition. Thus, at time, t, Np_+ and Np_- particles are in states $+$ and $-$, respectively. At time $t + dt$, NWp_+dt particles moved to state, $-$, and $NWp_-\exp(2\varepsilon/k_B T)dt$ [Equation (13.18)] moved to state, $+$, exchanging energy with the bath; therefore, dS_{bath}/dt is (w is cancelled),

$$\frac{dS_{\text{bath}}}{dt} = \frac{NW 2\varepsilon}{T}(\exp(2\varepsilon/k_B T)p_- - p_+). \tag{13.45}$$

It should be emphasized that the results in this subsection are based on Equation (13.22), which is less restricted (approximate) than Equation (13.25), where the assumptions of $4n_- \sim 2n_- + 2n_+ = 2N$ and the condition of Equation (13.20) have been made. In other words, the theory derived in this Section (13.3.1) expresses a larger deviation from equilibrium.

13.4 The Principle of Minimum Entropy Production—Statistical Mechanics Example

We have already discussed the principle of minimum entropy production in Section 11.11. Below, we present a statistical mechanics example of this principle due to Klein [6,7], which appears also in Kittel's book, [8], p. 165. According to this principle, the steady state of a system undergoing an irreversible process is that state for which the rate of entropy production is minimally consistent with the external constraints, which prevent the system from reaching equilibrium. Without constraints, the system will reach equilibrium, where the rate of entropy production is zero. However, this principle only gives the exact steady-state solution under the restrictive condition that the temperature be high in comparison with the relevant energy level differences.

We treat the same system presented in the previous section, on the level of approximation described by Equation (13.22)–(13.24) and discussed in Subsection 13.3.2. However, in the original paper, the basic energy levels are 0 and ε instead of our levels ε_+ and ε_-; thus, to be consistent with [7,8], the factors 2ε in the equations of Section 13.3.2 should be replaced by ε, where the probabilities p_+ and p_- are denoted by p_0 and p_ε, respectively. The steady-state equation [Equation (13.42)] becomes:

$$p_0^s = \frac{\exp(\varepsilon/k_{\rm B}T)+\beta}{\exp(\varepsilon/k_{\rm B}T)+1+2\beta} \tag{13.46}$$

where $\beta = w/W$. Now, one has to calculate the total rate of the entropy production, including both the system and the bath. Combining Equations (13.44) and (13.45) (with the new definitions) leads to,

$$\frac{dS_i}{dt} = -NW\left[(\exp(\varepsilon/k_{\rm B}T)p_\varepsilon - p_0)\ln\left(\frac{\exp(\varepsilon/k_{\rm B}T)p_\varepsilon}{p_0}\right) + \beta(p_\varepsilon - p_0)\ln\left(\frac{p_\varepsilon}{p_0}\right)\right]. \tag{13.47}$$

The minimum rate of entropy production is now obtained by minimizing dS_i/dt with respect to p_0, subject to the restriction $p_0 + p_\varepsilon = 1$. The differentiation leads to the equation,

$$[\exp(\varepsilon/k_{\rm B}T)+1]\ln[\exp(\varepsilon/k_{\rm B}T)p_\varepsilon/p_0] + 2\beta\ln(p_\varepsilon/p_0) +$$

$$[(\exp(\varepsilon/k_{\rm B}T)+\beta)p_\varepsilon - (1+\beta)p_\varepsilon]\left[\frac{1}{p_\varepsilon}+\frac{1}{p_0}\right] = 0. \tag{13.48}$$

This is an equation for p_0 (because $p_0 + p_\varepsilon = 1$), where a solution can be obtained if the following two conditions are satisfied simultaneously: (1) $p_\varepsilon/p_0 \sim 1$ and (2) $\exp(\varepsilon/k_{\rm B}T)p_\varepsilon/p_0 \sim 1$ [that is, $\exp(\varepsilon/k_{\rm B}T) \sim 1$]; more formally, ignoring terms of x^2 and higher, $p_\varepsilon/p_0 = 1 + x$ and $\exp(\varepsilon/k_{\rm B}T) = 1 + x$ for $x \ll 1$. While this guarantees that dS_i/dt is minimal, one has also to verify that the steady-state condition, $dp_0/dt = 0$, is satisfied. Indeed, these conditions are satisfied by Equation (13.46) and its equivalent, Equation (13.49),

$$\frac{p_\varepsilon^s}{p_0^s} = \frac{1+\beta}{\exp(\varepsilon/k_{\rm B}T)+\beta}. \tag{13.49}$$

Notice that $\exp(\varepsilon/k_{\rm B}T) \sim 1$ means $k_{\rm B}T \gg \varepsilon$, which is in accord with our assertion in the beginning of this section that $k_{\rm B}T \gg \varepsilon$ is the usual condition for the validity of the principle of minimum entropy production.

In practice, one has to verify that p_0 calculated by Equations (13.46) and (13.48) are the same, and that results for dS/dt based on these probabilities are the same as well. By a numerical examination, Klein has shown that the state of the minimum rate of entropy production determined by Equation (13.48) may be very close to the steady state, *even* when the above conditions do not hold. Klein's results appear in Table 13.1.

TABLE 13.1

Comparison of Results Obtained by Steady State, [Equation (13.46)] and by the Minimum Rate of Entropy Production, [Equation (13.48)], for $\exp(\varepsilon/k_B T) = 10$, and Different Values of $\beta = w/W$

	p_0			Rate of Entropy Production, Eq. (13.47)		
$\beta = w/W$	By Equation (13.46) (steady state)	By Equation (13.48) (min. S product.)	Ratio	By Equation (13.46) (steady state)	By Equation (13.48) (min. S production)	Ratio
0 (eq)	0.909	0.909	1	0	0	
1	0.846	0.861	1.019	1.594	1.570	1.015
10	0.645	0.670	1.039	6.685	6.601	1.013
100	0.521	0.526	1.009	9.821	9.802	1.002

Source: Klein, M.J., A note on the domain of validity of the principle of minimum production. In *Proceedings of the International Symposium on Transport Processes in Statistical Mechanics*, Brussels, 1956, Interscience Publishers, New York, pp. 311–318, 1958.

Note: As expected, the results for the probability, p_0, of energy level 0 ($p_0 + p_\varepsilon = 1$) decrease (toward 0.5) as β is increased, that is, as the radiation effect is enhanced (larger w); the results obtained by the two equations are very close as reflected by their ratio values, ~1, meaning that the steady state and the minimum rate of entropy production conditions are satisfied simultaneously. Inserting the two sets of values for p_0 in Equation (13.47) for the rate of entropy production leads to close results with ratios ~1. Notice that at equilibrium ($\beta = 0$), the rate of entropy production is zero, as expected. As β is increased, the steady states move away from equilibrium and the rates of entropy production increase. It should be stressed that the close results for p_0 obtained by Equations (13.46) and (13.48) have been obtained also for non-optimal values of the parameters, $\varepsilon/k_B T$ and β. These results were obtained in Table 13.1.

REFERENCES

1. C. W. Gardiner. *Handbook of Stochastic Methods for Physics, Chemistry and the Natural Sciences*, 3rd edition. (Springer, Heidelberg, 2004).
2. N. G. Van Kampen. *Stochastic Processes in Physics and Chemistry*. (Elsevier Science, Amsterdam, 1992).
3. L. E. Reichl. *A Modern Course in Statistical Physics*. (Edward Arnold Publishers, New York, 1980).
4. R. Zwanzig. *Non Equilibrium Statistical Mechanics*. (Oxford University Press, Oxford, 2001).
5. F. Reif. *Fundamentals of Statistical and Thermal Physics*. (McGraw-Hill, New York, 1965).
6. M. J. Klein and P. H. E. Meijer. Principle of minimum entropy production. *Phys. Rev.* **96**, 250–255 (1954).
7. M. J. Klein. A note on the domain of validity of the principle of minimum production. In *Proceedings of the International Symposium on Transport Processes in Statistical Mechanics*, Brussels, 1956. (Interscience Publishers, New York, pp. 311–318, 1958).
8. C. Kittel. *Elementary Statistical Physics*. (Dover, New York, 2004).

Section IV

Advanced Simulation Methods: Polymers and Biological Macromolecules

14

Growth Simulation Methods for Polymers

Metropolis Monte Carlo (MC) is a dynamical method, where an initial system configuration is changed repeatedly in the course of the simulation. Another approach has been developed for polymers, where a chain is constructed step-by-step (from nothing) with the help of transition probabilities (TPs). Thus, the *value* of the probability, P_i, of a constructed chain, i, is known: it is the product of the TPs used to generate the chain, and the *absolute* entropy, $\sim\ln P_i$, can be obtained at least approximately. This is one important advantage of chain growth techniques over MC; another advantage is the fact that the generated chains are statistically independent. For the convenience of presentation, we shall describe the growth techniques as applied to self-avoiding walks (SAWs) on a square lattice. This seemingly simple model is extremely challenging mathematically (see Chapter 9), and as we shall see, also for simulation, due to the strong *long-range* excluded-volume interactions imposed by the two-dimensional ($2d$) lattice on a SAW. We shall add attractions to the excluded-volume repulsions, where both interactions are typical to all realistic chains. Therefore, the conclusions gathered from the relatively simple SAW model apply to macromolecules in general. The SAWs model constitutes a convenient tool for testing the efficiency of various simulation techniques.

The main criteria of efficiency will be the maximum chain length that a method can handle in the bulk, and its suitability to simulate SAWs under geometrical constraints, such as an adsorbing surface. Other criteria are the applicability of a method to: (1) SAWs with attractions at low temperature, (2) multiple-chain systems, and (3) a chain in continuum. An interesting question is whether a method designed for a polymer can be extended to a bulk system, such as a fluid.

We shall not present *all* the available simulation techniques for polymers, but will concentrate on the most original and effective ideas. In this context, we mention the pioneers in this field, F.T. Wall and collaborators, and the couple, M.N. Rosenbluth and A.W. Rosenbluth, who were active during the 1950s and 1960s of the twentieth century; their contributions have become building blocks of more efficient and sophisticated techniques, as discussed in this chapter and the following one.

14.1 Simple Sampling of Ideal Chains

It is educational to discuss first the simplest polymer model—an ideal chain of N bonds ($N + 1$ monomers) on a d-dimensional hypercubic lattice (see Chapter 8). Since connectivity is the only restriction imposed on the chain (that is, it can intersect itself and go on itself), the chain can be constructed step-by-step as a random walk. Thus, the generation of the chain starts from the origin of the coordinates on the lattice; at step k ($1 \leq k \leq N$), a bond's direction, v ($1 \leq v \leq 2d$), is selected at random with TP $= 1/(2d)$ (using a random number, r, $0 < r \leq 1$); v is annexed to the partially constructed chain, and the process continues. This procedure is called *simple sampling* or *direct sampling*. The construction probability (which is the same for all chain configurations) is the Boltzmann probability:

$$P_i^{\text{B}} = \left(\frac{1}{2d}\right)^N, \tag{14.1}$$

from which the entropy, S, and the free energy, $F = -TS$, are obtained [Equation (8.2)]:

$$S = -k_B \sum_i P_i^B \ln P_i^B = N k_B \ln 2d \qquad (14.2)$$

Thus, in a free space, a chain length of millions of bonds (steps) can be generated (at any dimension) with contemporary computers, while in confined geometries, the efficiency decreases (see next section). When an attraction, ε, is defined between non-bonded nearest neighbor monomers, the construction procedure does not change, but the TP becomes proportional to $\exp[-\varepsilon'(\nu)/k_B T]$, where $\varepsilon'(\nu)$ is the interaction energy of candidate step, ν, with the partially constructed chain (that is, steps, $1 \le l \le k - 1$). An ideal chain can also be simulated by the usual MC method using, for example, the crankshaft move, the single monomer flip, and larger basic moves (Section 10.5).

14.2 Simple Sampling of SAWs

Simple sampling of an ideal chain consists of a blind selection of a direction (bond), ν (at step k), out of $2d$ possibilities, *where ν is always accepted*. Building a SAW with simple sampling would be considerably less efficient than generating an ideal chain because a blind selection of a direction, ν, could end up in an already occupied site (of the partial chain—steps, $1 \le l \le k - 1$) – a forbidden move, which violates the excluded volume condition. If the site is occupied, the partial chain is discarded, and the construction of a new chain is started (from nothing). In the other case, the randomly generated step is accepted, becoming the bond of step k, and the build-up process continues. This *"simple (direct) sampling method"* [1] is probably the earliest technique tested for SAWs, and the conclusions of this study (see below) have led to the development of more efficient techniques.

More specifically, $TP_k = 1/4$ always, and if the construction of the entire SAW i of N steps succeeds, the corresponding probability, P_i^0, is the same for any i,

$$P_i^0 = (1/4)^N = \prod_{k=1}^N TP_k \qquad (14.3)$$

However, P_i^0 is not normalized over the set of SAWs alone, but over the larger ensemble of random walks, which also includes self-intersecting walks (their probability is taken into account in the build-up process). Therefore, it is useful to express the partition function, Z, of SAWs not over the ensemble of the SAWs themselves, but over the ensemble of random walks (RWs - ideal chains). For that, we define two interaction energies for chain configuration, i, $E_i = 0$, if i is a SAW, and $E_i = \infty$, if i is a self-intersecting walk; thus:

$$Z_{SAWs} = \sum_{SAWi} 1_i = \sum_{RWi} \exp[-E_i / k_B T]. \qquad (14.4)$$

The summation over the SAWs demonstrates that they are equally probable; T is not effective as long as it is finite. Since we have in hand the *value* of P_i^0, we can apply importance sampling [Equation (10.6)], by multiplying and dividing the exponent in Equation (14.4) by P_i^0,

$$Z_{SAWs} = \sum_{RWi} P_i^0 \left[\frac{\exp[-E_i / k_B T]}{P_i^0} \right]. \qquad (14.5)$$

Z_{SAWs} [Equation (14.5)] is expressed as a statistical average of the random variable defined in the brackets. Z_{SAWs} can be estimated by calculating the arithmetic average, $\bar{Z}_{SAWs}$, of this random variable over the sample of n_{start} attempts for generating SAWs [Equation (10.7)]. Since some of these attempts have been rejected due to self-intersections (that is, $E_j = \infty$), the number of N-step (successful) SAWs generated, n_{suc}, is practically smaller than n_{start}; the estimation is,

$$\bar{Z}_{SAWs} = \frac{1}{n_{start}} \sum_{t=1}^{n_{start}} \frac{\exp[-E_{i(t)}/k_B T]}{P_{i(t)}^0} = \frac{1}{n_{start}} \sum_{t=1}^{n_{suc}} \frac{1}{P_{i(t)}^0} \tag{14.6}$$

where $i(t)$ is the chain configuration (partial or complete) obtained at attempt, t, of the process. For $d = 2$ (using $P_i^0 = 1/4^N$) Equation (14.6) becomes,

$$\bar{Z}_{SAWs} = \frac{4^N n_{suc}}{n_{start}} \tag{14.7}$$

and the estimated entropy is:

$$\bar{S} = k_B \ln \bar{Z}_{SAWs} = N k_B \ln 4 + k_B \ln \frac{n_{suc}}{n_{start}} \tag{14.8}$$

The first term on the right-hand side of Equation (14.8) is the entropy of an ideal chain, which is decreased by the second (negative) term; in general, the stronger is the excluded volume interaction, the smaller is the ratio, n_{suc}/n_{start}, and the lower is the entropy. This simple sampling method is exact because SAWs are generated with equal probability, as needed; however, the method is extremely inefficient due to the strong excluded volume interactions in $d = 2$. Thus, the probability, R_{attrit}, for generating a full-length chain decreases exponentially with increasing N:

$$R_{attrit} = \frac{n_{suc}}{n_{start}} \sim \exp{-\lambda N} \tag{14.9}$$

where λ is called the attrition constant and $\lambda \sim 0.128$ for the 3-choice square lattice (that is, immediate reversal of a step is forbidden). In practice, chains longer than $N = 100$ steps cannot be generated by this method due to the relatively large λ, which leads to a strong *sample attrition rate*, R_{attrit}. This exponential decay is obtained by the ratio of the number of SAWs, $\mu^N N^{\gamma-1}$ [Equation (9.6)], to q^N, where q is the coordination number of the lattice ($q = 4$ and $q = 3$ for a square lattice, where chain reversals are allowed or forbidden, respectively); clearly, the leading term in this ratio is, $(\mu/q)^N$. Indeed, calculation of this ratio for chains of $N = 100$ and $q = 3$ leads to $\lambda \sim 0.127$, in a very good agreement with the experimental value above. In $d = 3$, the excluded volume interaction weakens, but the sample attrition is still exponential, with λ(simple cubic) < λ(square), which enables generating SAWs of up to $N = 150$ on the simple cubic lattice. For $d \geq 4$, the method becomes much more efficient.

The strong sample attrition characterizing the simple sampling method is due to the relatively high probability for step k to create a loop (self-intersection). Thus, Wall et al. [1] have found from their simulation data that the probability, $f(0;j)$, for step k to close a loop of j steps is independent of k (besides for a very small k), and it varies with j as:

$$f(0;j) \approx \frac{m}{j^2} \tag{14.10}$$

where m is a constant, which depends on the lattice ($m = 0.36$ for the 4-choice cubic lattice). Thus, the effect of long loops is insignificant, and the failures occurring during the step-by-step construction stem from the short loops.

Estimating the average of a geometrical property, such as the end-to-end distance, R, or the radius of gyration is straightforward; for example, the statistical average $<R^2>$ is:

$$< R^2 >= \frac{\sum_{\text{RW}} R_i^2 \exp[-E_i / k_{\text{B}}T]}{Z_{\text{SAWs}}} = \frac{\sum_{\text{RW}} P_i^0 \left[\dfrac{R_i^2 \exp[-E_i / k_{\text{B}}T]}{P_i^0} \right]}{\sum_{\text{RW}} P_i^0 \left[\dfrac{\exp[-E_i / k_{\text{B}}T]}{P_i^0} \right]} \tag{14.11}$$

where $1/Z_{\text{SAWs}}$ is the Boltzmann probability of a SAW. Notice that in this case, two statistical averages over the random walks are defined in the numerator and the denominator (by multiplying and dividing by P_i^0), where each should be estimated separately. This is achieved by generating (with P_i^0) a sample of n_{suc} successful SAWs out of n_{start} SAWs attempted, and the two averages are estimated by their arithmetic averages. The estimation of $<R^2>$ is:

$$\overline{R^2} = \frac{\dfrac{1}{n_{\text{start}}} \sum_{n_{\text{start}}} \left[\dfrac{R_{i(t)}^2 \exp[-E_{i(t)} / k_{\text{B}}T]}{P_{i(t)}^0} \right]}{\dfrac{1}{n_{\text{start}}} \sum_{n_{\text{start}}} \left[\dfrac{\exp[-E_{i(t)} / k_{\text{B}}T]}{P_{i(t)}^0} \right]} = \frac{\sum_{n_{\text{suc}}} \dfrac{R_{i(t)}^2}{P_{i(t)}^0}}{\sum_{n_{\text{suc}}} \dfrac{1}{P_{i(t)}^0}} \tag{14.12}$$

For the square lattice, where $P_j^0 = 1 / 4^N$, $\overline{R^2}$ is:

$$\overline{R^2} = \frac{1}{n_{\text{suc}}} \sum_{t=1}^{n_{\text{suc}}} R_{i(t)}^2 \tag{14.13}$$

The theory developed thus far for pure SAWs can be extended to SAWs with finite interactions. Thus, if SAW i has interaction energy e_i, Equations (14.5), (14.6), (14.11), and (14.12) still hold, where the only change is that $E_i = e_i$, for a SAW (rather than zero), while $E_i = \infty$ (as before), for a self-intersecting walk. (These equations are also used in "*the Rosenbluth and Rosenbluth method*" and "*the scanning method*" discussed later in this chapter.) Another important factor in these equations is the temperature, which can determine the efficiency of the simulation. Thus, the sample created for pure SAWs [Equation (14.4)] is populated mainly by relatively open chains, and if T is low, the typical compact structures will not appear in the sample and the estimated averages for the entropy and $<R^2>$ will be biased. Thus, the lower is the temperature, the larger are the samples required for getting reliable results. Clearly, better efficiency can be gained if the interaction energy is taken into account in the definition of the TPs (see Section 14.4). While the method is applicable, in principle, to SAWs, which are subject to geometrical constraints, the stronger the constraints, the larger will be the attrition constant, and the lower will be the efficiency.

14.3 The Enrichment Method

With the simple sampling method described above, the efficiency of generating SAWs of lengths N and $N/2$ are:

$$n_{\text{suc}}(N) = n_{\text{start}} \exp - \lambda N \tag{14.14}$$

and

$$n_{\text{suc}}(N / 2) = n_{\text{start}} \exp(-\lambda N / 2) \tag{14.15}$$

In the process of generating N-step SAWs, some chains shorter than $N/2$ will be discarded (and thus will be wasted)—their number is $n_{start} - n_{suc}(N/2)$. To generate the targeted $n_{suc}(N)$ SAWs, while saving computer time, the following "*enrichment procedure*" has been suggested by Wall and Erpenbeck [2,3]. Thus, consider only half the number of the starting chains, that is, $n_{start} \rightarrow n_{start}/2$ (so the number of wasted attempts is half their number in the original procedure). Now, for each of the successful chains of length $N/2$, make two attempts to grow the remaining $N/2$ steps (for getting chains of length N). The number of successful chains of length N will be the same as before, but computer time will be reduced. The problem is that the generated chains are now *correlated*.

In a general enrichment procedure, the chain length, N, is divided into k equal segments of size $s = N/k$ and a branching number, p (that can be larger than 2), is determined. Using $n_{total} = n_{start}p^{k-1}$, the number of (correlated) SAWs of length N generated is:

$$n_{start} p^{k-1} \exp(-\lambda sk) = \frac{n_{start}}{p}[p\exp(-\lambda s)]^k \qquad (14.16)$$

Clearly, if $p\exp(-\lambda s) > 1$, an explosion in the number of chains will occur (that is, the number of chains generated will be larger than the number of chains started). Even though many chains can be produced, they are highly correlated, as pointed out above; an optimal choice is $p \sim \exp\lambda s$. As needed, the chains generated are equally probable, and thus the entropy can be calculated using Equation (14.8) (for SAWs on a square lattice). Wall and Erpenbeck studied SAWs of $N = 600$ on the square lattice and SAWs of $N = 800$ on the tetrahedral lattice, while Gans was able to increase the latter length to $N = 1700$ for SAWs on the tetrahedral lattice [4]. Clearly, with present computers, the chain length can be increased further, but not by an order of magnitude, due to the strong sample attrition and the strong correlations among the generated SAWs. The applicability of this method is similar to that discussed for the simple sampling method, while the chains are significantly longer.

As pointed out earlier, Wall and collaborators have done pioneering work in applying computer simulation to polymers. In this context, one should also mention the "*slithering snake method*" method of Wall and Mandel [5], where a monomer is removed from the tail (head) of the chain and is annexed to the head (tail); another contribution is the "*method of strides*", which will be discussed later in Section 14.5.

14.4 The Rosenbluth and Rosenbluth Method

The strong sample attrition encountered with the simple sampling method stems from the "blind" selection of directions at each step. An intuitive way to alleviate this problem to some extent is due to Rosenbluth and Rosenbluth (RR), who in 1955 [7] suggested to select only directions ν ($\nu = 1,4$ on the square lattice) that are vacant (that is, not occupied by already determined chain monomers). Thus, besides an immediate chain reversal (which is always forbidden), a maximum three vacant sites might be available on the square lattice; in this case, one site out of the available three is selected with TP = 1/3. If two directions are free, one of them is chosen with TP = 1/2, and if a single direction is free, it is always selected (with TP = 1). Obviously, if all directions are blocked, the (partial) chain is discarded and the generation of a new chain is started. (Notice, however, that a successful decision at step k might impose a long-range effect, where the constructed chain will reach a dead end at a future step, $k + m$; therefore, a perfect TP for a SAW should take into account *all* the possible continuations of the chain in the future steps $k, k + 1...N$, see Section 14.5.1). While sample attrition is not eliminated by this procedure, it is reduced in comparison with that of simple sampling. However, this increase in efficiency comes with the price that the chains are not constructed with equal probability as they should be, but with a bias. The TP at step k can be written as:

$$p(\nu_k|\nu_{k-1},...,\nu_1,f=1) \qquad (14.17)$$

where ν_k depends on all the directions $\nu_1,\nu_2,...,\nu_{k-1}$ determined in previous steps, and $f = 1$ means that the TP is defined by checking all the possible nearest neighbor vacant sites to monomer k. Thus, the probability, P_i^0, of SAW i [Equation (14.3)] is the product of all the TPs used to generate i,

$$P_i^0 = \prod_{k=1}^{N} p(\nu_k | \nu_{k-1},...,\nu_1, f = 1). \tag{14.18}$$

As said above, P_i^0 is not the same for all SAWs; it is larger for the compact SAWs than for the open ones. Thus, a fully stretched structure (a rod) has the lowest probability $(1/3)^N$, while for a compact chain, P_i^0 is larger, consisting also of many TP = 1/2 and TP = 1. Again, P_i^0 is defined not only on the group of SAWs, but also over a group of self-intersecting walks; however, this group is smaller than that defined by simple sampling (because many self-intersecting walks cannot be generated due to the restrictions imposed by the TPs of the RR method). Besides the difference between P_i^0 of simple sampling and the RR method, Equations (14.4)–(14.6), (14.9), (14.11), and (14.12) apply to both methods and to the scanning method discussed in the next section.

It is of interest to discuss the efficiency of the estimation of the partition function [Equations (14.4–14.6)] and the end-to-end distance [Equations (14.11) and (14.12)]. By definition, for an infinite sample, $\bar{Z}_{\text{SAWs}}$ is exact, $\bar{Z}_{\text{SAWs}} \rightarrow Z_{\text{SAWs}}$. For a finite sample, the quality of $\bar{Z}_{\text{SAWs}}$ depends on the variance of the random variable:

$$\frac{\exp[-E_i / k_B T]}{P_i^0} \tag{14.19}$$

defined in Equation (14.5). Clearly, this variance increases dramatically with increasing N (see Equation 10.8)], and thus huge samples are required for obtaining reliable estimations, $\bar{Z}_{\text{SAWs}}(N)$. More specifically, the majority of SAWs in the (correct) ensemble are relatively open, and thus contribute mostly to Z_{SAWs}. However, these open chains would not appear in a relatively small sample that consists typically of compact structures, which are highly favored by the biased, P_i^0. Therefore, one has to generate very large samples that will also contain enough of the unfavorable open chains. In this case, the huge $1/P_i^0$ factors of the open SAWs (with small P_i^0) will dominate the summation of Equation (14.6), leading to a reliable estimation of Z_{SAWs}. So, a compensation mechanism is active here, where the mostly biased sample of relatively compact chains generated with P_i^0 is "saved" in the calculation of $\bar{Z}_{\text{SAWs}}(N)$ by the large factors $1/P_i^0$ assigned to the minority of open chains. Notice, however, that if the open chains are not represented in the sample, $\bar{Z}_{\text{SAWs}}$ will be significantly biased. This situation does not exist with simple sampling, where all the SAWs are generated correctly with equal probability.

Estimating the average of a geometrical property, such as the end-to-end distance, R, is defined by Equations (14.11) and (14.12), where two summations, in the numerator and the denominator, should be estimated separately. Again, both estimations are exact for $n_{\text{start}} \rightarrow \infty$, while for finite n_{suc}, the precision depends on the variances of the functions in the brackets of Equation (14.11), which increase with chain length, N. Thus, if n_{suc} is large enough, the open structures will be represented in the sample leading to an adequate estimation, otherwise the estimation again will be biased.

The attrition constant λ [Equation (14.9)] of the RR method is smaller than that of simple sampling, and SAWs of up to $N \sim 150$ can be generated on a square lattice. Notice again that the generated SAWs are statistically independent, which increases the efficiency of estimating quantities of interest, as compared to Metropolis MC. The applicability of the RR method to systems with geometrical constraints is similar to that discussed for simple sampling. However, if finite interactions are defined (e.g., attractions), the RR transition probabilities should consider these interactions (unlike with simple sampling), thus:

$$\text{TP}(\nu) = \frac{\exp[-\varepsilon'(\nu) / k_B T]}{\sum_{\nu} \exp[-\varepsilon'(\nu) / k_B T]} \tag{14.20}$$

where $\varepsilon'(\nu)$ is the interaction energy of candidate step, ν (for step k), with the partially constructed chain of steps, $1,...,k-1$; the summation is over the available empty lattice sites, ν. Therefore, at low temperatures, the constructed chains will be preferentially compact as needed (unlike with simple sampling).

The RR method was used extensively until the 1970s, as applied to self-attracting SAWs, SAWs adsorbed to a surface, etc. [e.g., 8–10]. The idea of a biased sampling that is corrected later in the summations for $\bar{Z}_{SAWs}$ and $\bar{R}^2$ is an important concept that has been used in many other simulation methods developed later.

PROBLEM 14.1

Using "simple sampling" and "Rosenbluth" methods generate (on the computer) SAWs on a square lattice of N steps (bonds, i.e., $N + 1$ monomers), where $N = 20, 30, 50,$ and 100; create samples of at least 5000 chains. For both methods calculate the average end-to-end distance, $R(N)$, the attrition rate (# chains succeeded)/(# chains started), and the attrition constant, λ. Arrange these results in a table versus N. Estimate the errors from results obtained for partial samples. For both methods, estimate the best exponent, ν, from log-log plots of R vs. N. Do you see a systematic difference between the attritions, R values, and results for ν obtained by the two methods? Explain. Carry out the same analysis for the dimerization and the enrichment methods.

14.5 The Scanning Method

From the methods discussed thus far, it appears that in spite of the apparent simplicity of SAWs, their simulation is far from being trivial. In fact, to check various theories (such as the value of the exponent ν in $d = 3$), very long chains are needed, and a great deal of thought has been devoted to this problem. The RR method, which scans the immediate vicinity of step k ($f = 1$ steps ahead) constitutes an improvement over simple sampling; however, the chain can still get trapped in a dead end during construction. With the scanning method, suggested by Meirovitch [11–13], this problem is alleviated further by scanning $f > 1$ steps ahead. Thus, not only the attrition and the bias both weaken, but the reliability of the results can be determined from simulations based on an increased future scanning parameter, f.

Also, the method defines several ways for calculating the absolute entropy (and free energy), among them, an upper bound and a lower bound, which approach each other as f is increased; thus a "self-checking" mechanism is provided for the accuracy of S without knowing its correct value a priori. The scanning method enables one locating transition temperatures very efficiently, and we shall demonstrate that increasing f leads to the high applicability of the method, not only to polymer models, but also to bulk systems, such as fluids. Finally, the scanning method constitutes the basis for "the hypothetical scanning method," which is a general approach for extracting the absolute entropy and free energy from MC and MD samples (see Chapter 19). Due to its richness, the scanning method will be described with some detail.

While practically the scanning method appears as an extension of the RR method, the starting points of view of these two methods are different. With the scanning method (which was initiated as applied to the Ising model [14]), the initial question has been: How can one generate SAWs step-by-step with *zero* sample attrition, where all chains are built with the same Boltzmann probability, $1/Z_{SAW}$. We show below that this can be achieved, if at each step, the complete future (based on steps $k, k + 1,...,N$) is fully scanned, as compared to the single step scanning of the RR method.

14.5.1 The Complete Scanning Method

Assume a process where a SAW is generated step-by-step with TPs on the square lattice. Thus, at step k of the process, $k - 1$ directions (bonds), $\nu_1,...,\nu_{k-1}$ ($\nu_i = 1,...,4$ on the square lattice) will have already been constructed. To determine the direction, ν_k, we calculate the future partition functions, Z_k^ν, where Z_k^ν is the number of *all* possible chain continuations in the *remaining* $N - k + 1$ future steps that start in direction ν at step k. Z_k^ν is a *complete future partition function*, which leads to TP(ν) for direction ν,

$$TP(\nu) = p(\nu \mid \nu_{(k-1)},...,\nu_1, N - k + 1) = Z_k^\nu / \sum_{\nu=1}^{4} Z_k^\nu. \tag{14.21}$$

Due to the complete future scanning, sample attrition is avoided, the probability P_i^0 is defined over the group of SAWs alone, and it is the same for all SAWs, i; in other words, P_i^0 becomes *exact*, $P_i^0 = P_i^B$,

$$P_i^0 = \prod_{k=1}^{N} p(\nu_k \mid \nu_{(k-1)},...,\nu_1) = \frac{Z_1^{\nu_1}}{Z_{\text{SAW}}} \frac{Z_2^{\nu_2}}{Z_1^{\nu_1}} \frac{Z_3^{\nu_3}}{Z_2^{\nu_2}} \cdots \frac{Z_{N-1}^{\nu_{N-1}}}{Z_{N-2}^{\nu_{N-2}}} \frac{1}{Z_{N-1}^{\nu_{N-1}}} = \frac{1}{Z_{\text{SAW}}} = P_i^B. \qquad (14.22)$$

$P_i^0 = P_i^B$ can also be deduced from the fact that already in the construction of the first step of the chain, the whole future is scanned, that is, Z_{SAW} is calculated, hence P_i^B is known. Thus, for calculating P_i^B, generating a *single* SAW (or even only $Z_1^{\nu_1}$) is sufficient, which is a manifestation of the zero fluctuation of the free energy [Equation (5.6)]. However, a sample of SAWs is still needed for estimating the average end-to-end distance and the radius of gyration. Clearly, the complete scanning is equivalent to exact enumeration techniques, which are limited to short chains, in particular, as the lattice coordination number is increased. Therefore, one has to resort to partial rather than complete scanning.

14.5.2 The Partial Scanning Method

The partial scanning method is based on scanning f steps ahead, rather than the whole future ($N - k + 1$ steps). Because the partial scanning method will become our practical tool, we omit, for simplicity, the word *partial* and call it just, "*the scanning method.*" The heart of the method is that at step k of the process, we calculate the future partition functions, $Z_k^{\nu}(f)$, where $Z_k^{\nu}(f)$ is the number of possible chain continuations in f future steps that start in direction ν at step k. $Z_k^{\nu}(f)$ is a partial future partition function, and f is the scanning parameter. TP(ν) (on the square lattice) for direction ν is defined by Equation (14.21), where f replaces $N - k + 1$ and $Z_k^{\nu}(f)$ replaces Z_k^{ν}. TP(ν) depends on the whole past $\nu_1,...,\nu_{k-1}$ and on all of the f future steps (see Figure 14.1). After the construction of SAW i has been completed, its probability, $P_i^0(f)$, is the product of the values of TP(ν), as in Equation (14.18). Again (as with RR), P_i^0 is normalized over a space that also includes self-intersecting walks, and P_i^0 is larger for the compact structures than for the open ones. As f is increased, this bias and the number of self-intersecting chains both decrease and P_i^0 becomes flatter.

Notice that in practice, f is limited ($f \ll N$), where on a square lattice $f \leq 20$ using present-day computers. Therefore, if N is large, say, $N \sim 1000$ (or even much smaller), partial scanning cannot prevent the chain from getting trapped during construction, and thus, as with RR, if a self-intersection occurs, the partial chain is discarded, and the generation of a new one is started. The ratio $n_{\text{suc}}/n_{\text{start}}$ increases with f, that is, the attrition constant λ decreases [Equation (14.9)]. Therefore, much longer SAWs can be simulated with the scanning method than with the RR technique, where on the square and simple cubic lattices, $N \sim 1000$ and $N \sim 2000$, respectively.

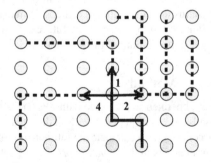

FIGURE 14.1 The 4th step in the scanning construction of a SAW of $N = 20$ steps. At this stage, three steps have already been determined (bold lines), and they will remain unchanged during the process. One has to decide which direction to select for step 4, $\nu = 1$, 2 or 4? (arrows). Using a future scanning parameter $f = 5$, one calculates, $Z_k^{\nu}(f)$—all the possible continuations of the chain in five future steps starting in direction ν (some of these future chains are depicted by dotted lines). The probability TP(ν) of direction, ν, is defined by Equation (14.21), where $Z_k^{\nu}(f)$ replaces Z_k^{ν}.

As for RR, the bias inherent in the scanning method is removed by an *importance sampling* procedure [Equation (14.6)], which leads to an unbiased estimation of S (or Z),

$$\bar{S} = k_B \ln \frac{1}{n_{\text{start}}} \sum_t^{n_{\text{start}}} \frac{\exp[-E_{i(t)}/k_B T]}{P_{i(t)}^0(f)}. \tag{14.23}$$

For SAWs in $d = 2$, $R \sim N^{0.75}$; therefore, relatively open SAWs dominate the partition function, and they should appear in a finite sample. For that, $P_i^0(f)$ should be close enough to P_i^B, otherwise the variance of $Z_{\text{SAW}}[P_i^0(f)]$ [Equation (14.19)] will be very large, and the results for a finite sample are expected to be biased. Unlike RR, with the scanning method, one can improve $P_i^0(f)$ [$P_i^0(f) \to P_i^B$] by increasing f, which, in general, is more efficient than increasing the sample size n_{start}.

Estimation of properties, such as the end-to-end distance, R, are carried out in the same way as described for the RR method [Equations (14.11) and (14.12)], where the only difference is that $P_i^0(f = 1)$ used with RR is replaced by $P_i^0(f)$.

14.5.3 Treating SAWs with Finite Interactions

Finite interactions are treated similar to the RR treatment [Equation (14.20)]. Thus, adequate TPs can be obtained by calculating the interaction energy $E_{j(v)}^k(f)$ of future SAWs j (that starts from v) with itself and with the partial chain constructed in steps $1,...,k - 1$ (if the chain is adsorbed to a surface, for example, the adsorption energy of the future chain should also be included in $E_{j(v)}^k(f)$; the corresponding Boltzmann factors contribute to the future partition function, thus:

$$Z_k^v(f,T) = \sum_{j(v)} \exp[-E_{j(v)}^k(f)/k_B T] \tag{14.24}$$

where the summation is over the future SAWs starting in direction v and $E_{j(v)}^k(f) = 0$ for a self-intersecting future walk. $TP(v_k) = p(v| v_{(k-1)} ...., v_1, f, T)$ is defined by Equation (14.21), where $Z_k^v(f,T)$ replaces Z_k^v; hence, the probability of chain i, $P_i^0(f,T) = \prod TP(v_k)$ becomes a function of T as well and the estimated free energy is:

$$\overline{F(T)} = -k_B T \ln \frac{1}{n_{\text{start}}} \sum_{t=1}^{n_{\text{start}}} \frac{\exp[-E_{i(t)}/k_B T]}{P_{i(t)}^0(f,T)} \tag{14.25}$$

where $E_{i(t)}$ is the energy of SAW i generated at time, t, of the simulation; $E_{i(t)}$ is zero for a self-intersecting walk. The discussion following Equation (14.21) also applies here.

14.5.4 A Lower Bound for the Entropy

Based on $P_i^0(f)$, one can define a probability distribution, $P_i(f)$, over the entire group of SAWs (unlike $P_i^0(f)$ that is defined over a group, which also includes self-intersecting walks) [15]. Thus,

$$P_i(f) = \frac{P_i^0(f)}{\sum_{\text{SAWs}} P_i^0(f)} = \frac{P_i^0(f)}{A} \tag{14.26}$$

where $A < 1$ defined over the SAWs is:

$$A = \sum_{j(\text{walks})} P_j^0(f) \exp[-E_j/k_B T] = \sum_{\text{SAWs } i} P_i^0(f). \tag{14.27}$$

A can be estimated by $\overline{A}$ from the sample of n_{start} attempted SAWs, where n_{suc} of them were successfully generated:

$$\overline{A} = \frac{1}{n_{\text{start}}} \sum_{t=1}^{n_{\text{suc}}} 1_{i(t)} = \frac{n_{\text{suc}}}{n_{\text{start}}}. \tag{14.28}$$

This leads to an approximate entropy, $S'(f)$ [or in the case of finite interactions, to a free energy, $F'(f)$],

$$S'(f) = -k_{\text{B}} \sum_{\text{SAWs}} P_i(f) \ln P_i(f) \tag{14.29}$$

which by definition constitutes a *lower bound* for the correct (maximal) entropy, S, defined by the Boltzmann probability,

$$S = -k_{\text{B}} \sum_{\text{SAWs}} P_i^{\text{B}} \ln P_i^{\text{B}}, \tag{14.30}$$

the larger is f, the higher is $S'(f)$, and $S'(f) \to S$ as $f \to N - k + 1$. $S'(f)$ is estimated by $\overline{S'(f)}$

$$\overline{S'(f)} = -\frac{k_{\text{B}}}{n_{\text{suc}}} \sum_{t=1}^{n_{\text{suc}}} \ln P_{i(t)}(f). \tag{14.31}$$

We have pointed out that the fluctuation of the correct free energy, F, is zero [Equation (5.6)], hence, for SAWs (where $S = -F$), the fluctuation of S is also zero. However, an approximate $S'(f)$ has a non-zero fluctuation, $\sigma(f)$, which is expected to decrease as f is increased. Therefore, if a parameter m is added to the scanning method, its optimal value can be obtained by maximizing S' or minimizing $\sigma(f)$, with respect to m.

14.5.5 A Mean-Field Parameter

In practice, because of limits on the size of the parameter, f, the scanning can be considered as a localized procedure around step k, and one would like to consider also the global effect of the partial generated chain; this can be achieved by defining a mean field parameter. Thus, Equation (14.21) becomes,

$$p(\nu \mid \nu_{(k-1)},...,\nu_1,f,m) = Z_k^{\nu}(f)m^{-\nu \bullet x} / \sum_{\nu=1}^{4} Z_k^{\nu}(f)m^{-\nu \bullet x} \tag{14.32}$$

where ν is a unit vector in direction v and x ($x(k)$) is a unit vector, which points from the center of mass of the partial chain ($\nu_{(k-1)},...,\nu_1$), toward monomer k. Clearly, for $m < 1$, a direction ν, which points toward the outer (inner) part of the chain will generally lead to $\nu \bullet x > 0$ (<0), which means that $p(\nu \mid \nu_{(k-1)},...,\nu_1,f,m)$ will be larger (smaller) than $p(\nu \mid \nu_{(k-1)},...,\nu_1,f)$. Therefore, the preference given by $P_i^0(f)$ to the compact SAWs is weakened in $P_i^0(f,m)$. In other words, the effect of $m < 1$ is to open the chain, that is, to increase the end-to-end distance and the radius of gyration. The parameter m can be optimized by running several short simulations based on different m values, where the m value that maximizes the corresponding entropy, $\overline{S'(f,m)}$ [Equation (14.31)] is the optimal value, m^*, which is used in the production runs. A somewhat less effective way to obtain m^* is by minimizing the fluctuations of S'. To save computer time, $x(k)$ can be calculated only every several steps. An adequate application of m to other models requires understanding the physical behavior of the models. The mean field parameter has been applied to several models, among them, polymers adsorbed on surfaces (see [16,17]); it has been found that m^* is most effective for $f = 1$, where its effectiveness decreases as f is increased. The results for m^* obtained for various models (see Table 14.1) range between $m^* = 0.65$ for $f = 1$ and ~ 0.9 for $f = 7$. In what follows, for simplicity, we shall omit m^* from the equations.

14.5.6 Eliminating the Bias by Schmidt's Procedure

The efficiency of the method is determined, not by measuring n_{suc}, the number of SAWs generated, but by the effective number of SAWs that define an *unbiased* sample, that is, those that are distributed according to the correct Boltzmann probability (if the sample contains only compact chains, it is biased). For that, we define an MC procedure suggested by Schmidt [18–21] that enables one to extract from a sample selected by the biased probability $P_i(f)$, an *effectively* smaller sample in which SAWs are weighted correctly according to P_i^{B}. Assume that a sample of n biased SAWs has been generated by the scanning method ($n = n_{\text{suc}}$ is used for simplicity). We process this sample with an MC procedure as follows:

1. The first SAW generated (i) is always accepted.
2. The second SAW (j) is accepted with probability, p_{ij}. If j is accepted, the sample contains i and j. If j is rejected the sample contains i twice (with probability $1 - p_{ij}$).
3. The third SAW is checked against the second one in the same way, and the process continues until the whole biased sample of size n has been examined and an unbiased sample of *accepted* SAWs has been obtained; the number of *different accepted* SAWs is denoted by n_{accept}, where, in practice, $n_{\text{accept}} < n$. The TPs are obtained by the usual MC prescription [Equation (10.9)]:

$$\frac{T_{ij} p_{ij}}{T_{ji} p_{ji}} = \frac{P_j^{\text{B}}}{P_i^{\text{B}}} = \exp[-(E_j - E_i)/k_{\text{B}}T] \tag{14.33}$$

Generally, $T_{ij} = T_{ji}$, where T_{ij} is the probability to select a *candidate* state j at time $t + 1$ after accepting state i at t. In the above procedure, however, $T_{ij} \neq T_{ji}$ since $T_{ij} = P_j^0(f)$ and $T_{ji} = P_i^0(f)$. For SAWs, the energy is zero, therefore one obtains,

$$\frac{P_j^0(f) p_{ij}}{P_i^0(f) p_{ji}} = 1 \qquad \text{or} \qquad \frac{p_{ij}}{p_{ji}} = \frac{P_i^0(f)}{P_j^0(f)} \tag{14.34}$$

Therefore, defining $A_{ij} = P_i^0(f)/P_j^0(f)$, one can write the *asymmetric* MC [Equation (10.11)],

$$\begin{aligned} p_{ij} &= A_{ij} \qquad \text{for} \qquad A_{ij} < 1 \\ p_{ij} &= 1 \qquad \text{for} \qquad A_{ij} \geq 1 \end{aligned} \tag{14.35}$$

Notice the difference of Equation (14.35) from the equation in the usual MC, for which $A_{ij} = P_j^{\text{B}}/P_i^{\text{B}}$. Also, for a model where interaction energy is defined, A_{ij} becomes:

$$A_{ij} = P_i^0(f)/P_j^0(f) \exp[-(E_j - E_i)/k_{\text{B}}T] \tag{14.36}$$

Thus, the correct measure of efficiency is not $n_{\text{suc}}/n_{\text{start}}$, but $n_{\text{accept}}/n_{\text{start}}$, and we define the acceptance rate as,

$$R_{\text{accept}} = \frac{n_{\text{accept}}}{n_{\text{start}}}. \tag{14.37}$$

For the Boltzmann probability, P_i^B, $R_{accept} = 1$, whereas for a biased probability, $P_i^0(f)$, generally, $R_{accept} < 1$. However, one would expect R_{accept} to increase as the approximation improves. It is of interest to examine the properties of the accepted sample. Thus, in the beginning of the process, the acceptance rate is relatively high because most of the chains are expected to be compact with comparable (high) probabilities. Then, an open chain j might appear, which will certainly be accepted (because $P_i^0(f) > P_j^0(f)$ or $A_{ij} > 1$). The following compact chains in the original sample will be rejected until a second open chain appears with a high probability to be accepted and so on. Thus, R_{accept} will decrease until it has reached a point of convergence; thus, the initial set of the accepted sample prior to this point should be discarded in calculations of statistical averages and the effective accepted sample size is the converged part, $n_{conv} < n$. Notice that some SAWs in the accepted sample will appear more than once (on average, an accepted SAW will be repeated n/n_{accept} times). Denoting a compact and an open chain by "c" and "o," respectively, the original and accepted samples might look as shown in Figure 14.2.

One can estimate a geometrical property, such as $<R^2>$ not only from Equations (14.11) and (14.12), but also directly from the unbiased accepted sample (where R is denoted by R_a—"a" stands for accepted),

$$\overline{R_a^2} = \frac{1}{n_{conv}} \sum_{t=1}^{n_{conv}} {}'R_{i(t)}^2.$$

(14.38)

Σ' means summation over the accepted sample, and n_{conv} (unlike n_{accept}) includes the repetitions. In contrast to R^2, the entropy can be estimated (by S_a) from the accepted sample only approximately, since it is defined by the biased $P_i(f)$ [Equation (14.26)]:

$$\overline{S_a} = -\frac{k_B}{n_{conv}} \sum_{t=1}^{n_{conv}} {}'\ln P_{i(t)}(f)$$

(14.39)

In fact (due to Jensen's inequality [22]), S_a constitutes an *overestimation* of the correct S, which can also be deduced from the fact that $P_i(f)$ of an *open* accepted chain is, in general, smaller than P_i^B. Thus, S_{ave}, the average of $\overline{S_a}$ and $S'(f)$ [Equation (14.31)],

$$S_{ave} = [\overline{S_a(f)} + \overline{S'(f)'}]/2,$$

(14.40)

which is bound from both sides, is expected to provide a better estimation for S than S_a and S' individually; S_{ave} approaches S as f is increased.

14.5.7 Correlations in the Accepted Sample

One might consider n_{accept} as the effective sample size of the accepted sample. However, the accepted SAWs are not statistically independent (in contrast to the SAWs in the original sample based on importance sampling), and this might further decrease the effective sample size. Thus, one needs to study the autocorrelation function, $R(t)$ [Equation (1.85)], of the properties of interest over a particular sample of

Original   c c c o c o c c c o

Accepted   c c c o o o o o o o

FIGURE 14.2 The majority of chains in the original scanning sample are closed (c), while the accepted sample contains predominately open chains (o).

size n_{accept} consisting of the accepted SAWs, that is, *without repetitions*. The number of uncorrelated SAWs, n_{uncor}, is expected to satisfy $n_{uncor} < n_{accept}$. Such studies have indeed been carried out for many models; for the squared radius of gyration and the squared end-to-end distance, the autocorrelation function, $R(t)$, is ~0 already for $t = 1$, meaning that $n_{accept} = n_{uncor}$. For the energy and entropy, $R(t = 1)$ can be as large as 0.5, becoming zero already for $t = 4$. These higher correlations stem from the fact that the acceptance test for SAW j depends on i through $\exp[\ln P_i^0(f) - \ln P_j^0(f)]$, where i is the last SAW accepted. Obviously, as $P_i^0(f)$ improves $P_i^0(f) \rightarrow P_i^B$, the acceptance rate $\rightarrow 1$, and the correlation vanishes (for details see [21] and references cited therein). Therefore, in practice, it is reasonable to adopt n_{accept} as an *average* effective sample size for the Schmidt procedure.

However, using R_{accept} [Equation (14.37)] as a criterion of efficiency requires analyzing the accepted sample further. First, if the original sample does not contain open SAWs, the correction mechanism is meaningless and the sample will remain biased. If the original sample contains only a single open chain, the accepted sample with high probability will consist of this open chain alone! Beyond these extreme cases, one should always verify that the relatively high R_{accept} in the beginning of the process has not been converged to a too small value, meaning that not enough *different* open chains appear in the sample. A more reliable measure of efficiency than R_{accept} would be the number of accepted SAWs generated per unit of computer time.

In summary: Both importance sampling and the Schmidt procedure remove the bias introduced by the scanning method. However, since the Schmidt procedure is a stochastic process, the averages obtained by importance sampling are expected and indeed have been found to be the more stable ones. On the other hand, the Schmidt procedure provides n_{accept}, which enables one to estimate very conveniently the quality of a biased sample.

14.5.8 Criteria for Efficiency

Reliable results with the scanning method can be obtained in two ways: by using a sufficiently large sample size, n_{start}, or a large enough future scanning parameter f. Clearly, increasing both n_{start} or f costs computer time, and one should devise a measure of efficiency that would take these effects into account. As pointed out above, a suitable such measure is the Central Processing Unit (CPU) time required for generating say, 1000 accepted SAWs; we denote this time by t_{1000}, where the lower the t_{1000} the better is the efficiency. This test is carried out on relatively small samples prior to the production run. For a given f, one optimizes the mean field parameter, m, as discussed earlier.

Table 14.1 presents an efficiency analysis for SAWs on a square lattice. For each chain length, N results are shown for different scanning parameters f, optimal mean field parameter, m^*, attrition rate, R_{attrit} [Equation (14.9)], acceptance rate, R_{accept} [Equation (14.37)], and t_{1000}. The table reveals that, as expected, for every N, m^* becomes less effective ($m^* \rightarrow 1$) as the approximation improves (as f is increased); for example, for $N = 50$ and 100, m^* increases from 0.68 to 0.88 as f increases from 1 to 4. Correspondingly, R_{attrit} and R_{accept} increase significantly with f. The results for t_{1000} also show the expected trend, where their minimal (optimal) values are obtained for larger f as N is increased. Thus, for $N = 50$ and 100, the minimal t_{1000} (4.9 and 20 CPU seconds, respectively) is obtained for $f = 2$, while for $N = 200$ and 400, the values $t_{1000} = 122$ and 1550 seconds are obtained for $f = 3$ and 4, respectively.

It is of interest to note that while the RR method is efficient for $N = 50$ and 100, it hardly can handle $N = 130$, as R_{accept} drops down to 0.003, while experience has shown that simulations based on $R_{accept} < 0.01$ lead to unstable results.

It should be pointed out that the results presented in Table 14.1 have been obtained before 1990 on outdated computers in present terms; therefore, the absolute values of t_{1000} are not relevant, and only their relative values are of interest. Analyses similar to that of Table 14.1 have been carried out for other lattice models, among them are SAWs and trails on square and simple cubic lattices, with and without attractions, where in some models the chains are also adsorbed at a surface (see [21,23] and references cited therein); other models treated are branch polymers [24], many chain systems [25], and a continuum model of decaglycine [26]. In these studies, critical exponents, transition temperatures, entropies, etc. have been calculated.

TABLE 14.1

Results for SAWs on a Square Lattice Obtained with the Scanning Method

f	m^*	n_{suc}/n_{start}	n_{accep}/n_{start}	t_{1000} (s)
		$N = 50$		
1	1.00	0.58	0.12	7.9
1	0.68	0.76	0.21	6.6
2	0.80	0.88	0.45	**4.9**
3	0.82	0.95	0.55	6.8
4	0.88	0.97	0.65	12.0
		$N = 100$		
1	1.00	0.22	0.01	106
1	0.68	0.49	0.05	46
2	0.80	0.67	0.20	**20**
3	0.82	0.81	0.33	**21**
4	0.88	0.87	0.43	35
		$N = 130$		
1	1.00	0.12	0.003	466
2	0.80	0.54	0.11	43
		$N = 150$		
1	1.00	0.08	~0.00011	12,500
2	0.80	0.48	0.08	66
		$N = 200$		
2	0.80	0.32	0.05	177
3	0.82	0.52	0.09	**122**
4	0.88	0.63	0.15	180
5	0.88	0.72	0.34	303
		$N = 400$		
3	0.82	0.19	~0.008	2170
4	0.88	0.25	0.03	**1550**
5	0.88	0.39	0.04	2867

Source: Meirovitch, H., *Int. J. Mod. Phys.* 1, 119–145, 1990.

Note: f is the scanning parameter, m^* is the optimized mean-field parameter [Equation (14.32)], n_{start} and n_{suc} are the numbers of SAWs started and succeeded, respectively [Equation (14.7)], n_{accep} is the number of SAWs accepted [Equation (14.37)], and t_{1000} (s) is the CPU time required for generating 1000 accepted SAWs.

14.5.9 Locating Transition Temperatures

Transition temperatures (e.g., the Flory θ-point) have been determined very efficiently with the scanning method. Thus, studying a model by importance sampling at a given temperature, T, will generate a set of probabilities $P_i^0(f,T)$, which can be used for obtaining results at a neighbor temperature T_1; Equation (14.25) becomes:

$$\bar{F}(T_1) = -k_B T_1 \ln \frac{1}{n_{start}} \sum_{t=1}^{n_{start}} \frac{\exp[-E_{i(t)}/k_B T_1]}{P_{i(t)}^0(f,T)} \qquad (14.41)$$

where a similar expression can be written for the estimation of the end-to-end distance, R [Equation (14.12)]:

$$R^2(T_1) = \frac{\sum_{t=1}^{n_{start}} \dfrac{R_{i(t)}^2 \exp[-E_{i(t)} / k_B T_1]}{P_{i(t)}^0(f,T)}}{\sum_{t=1}^{n_{start}} \dfrac{\exp[-E_{i(t)} / k_B T_1]}{P_{i(t)}^0(f,T)}} \quad (14.42)$$

This is a good approximation as long as T_1 is not very different from T. Typically, a single run at a given temperature has led to results at 20 neighbor temperatures, which have been used for locating θ and studying the behavior of thermodynamic properties as θ is approached (see [21,23] and references cited therein).

In this context, it should be pointed out that a different procedure, which enables one to get results for a range of temperatures from a single simulation carried out by MC at a given T was suggested by Ferrenberg and Swendsen [27]. Thus, the probability of energy, E, is:

$$P_T^B(E) = \frac{1}{Z(T)} n(E) \exp[-E / k_B T] \quad (14.43)$$

where $n(E)$ is the degeneracy of E. The Boltzmann probability of E (not i !) at temperature, T_1, can be written in terms of the probability at T, by the reweighting equation,

$$P_{T_1}^B(E) = \frac{n(E) \exp\left[-\dfrac{E}{k_B T_1}\right]}{\sum_E n(E) \exp\left[-\dfrac{E}{k_B T_1}\right]} = \frac{\dfrac{1}{Z(T)} n(E) \exp\left[-\dfrac{E}{k_B T}\right] \exp\left[-\dfrac{E}{k_B T_1} + \dfrac{E}{k_B T}\right]}{\sum_E \dfrac{1}{Z(T)} n(E) \exp\left[-\dfrac{E}{k_B T}\right] \exp\left[-\dfrac{E}{k_B T_1} + \dfrac{E}{k_B T}\right]}$$

$$= \frac{P_T^B(E) \exp\left[-\dfrac{E}{k_B T_1} + \dfrac{E}{k_B T}\right]}{\sum_E P_T^B(E) \exp\left[-\dfrac{E}{k_B T_1} + \dfrac{E}{k_B T}\right]} \quad (14.44)$$

Thus, one generates a histogram for the energy, E, based on an MC simulation at T, and from the probabilities $P_T^B(E)$ calculates the probabilities at different temperatures T_1 and the required averages, such as $\overline{R^2(T_1)}$. Both procedures work well for temperatures, T_1, that are in the vicinity of T, otherwise the energies that contribute to T will not contribute to T_1, that is, $P_{T_1}^B(E) \to 0$. However, Equation (14.44) leads to less information than Equation (14.43), which provides the absolute entropy (and free energy) at the different temperatures. We shall show later in Section 20.4 that methods based on histograms (developed again by Ferrenberg and Swendsen) enable one to calculate differences in free energies of different temperatures.

14.5.10 The Scanning Method versus Other Techniques

Monte Carlo methods consisting of *local* configurational changes are highly correlated and are not ergodic for SAWs (see Section 10.5). With the scanning method, this problem is avoided because the chains are statistically independent (within P^0), and all of them, in principle, can be generated. For chains in the bulk (free space), such as SAWs without self-attractions, there are more efficient techniques than the scanning method, which is limited to ~1000 and 2000 SAWs steps on the square lattice and the simple cubic lattice, respectively. In particular, with "*the pivot algorithm*" (Chapter 15) and "*the dimerization method*" of Alexandrowicz (see Section 14.6), SAWs of thousands of monomers have been studied effectively. However, these methods are based on global moves, which make them inefficient for treating chains that are subject to geometrical constraints (such as a surface), self-attracting chains near the θ-temperature, and in the collapse region. The scanning method, on the other hand, is a very versatile technique, which can handle such chain models efficiently due to its local construction nature. However, with the scanning method chains are generated with a bias, therefore, one has to verify that the bias is removed from the constructed samples. Criteria for checking the existence of the bias have been outlined in the sections above.

It should also be noted that the RR method has been extended to self-attracting SAWs by McCrakin, Guttman, and Mazur [8], who also employed a parameter (different from the mean field parameter described above), but did not provide a criterion for optimization. To increase future scanning, Wall, Rubin, and Isaacson suggested the method of strides [6]. Thus, at step k of the chain construction, all the future SAWs of f steps (called strides) are exactly enumerated and one of them is selected at random and is annexed to the partial chain of $k - 1$ steps. Thus, the chain grows by strides of f steps at a time, rather than by one step, as done with the scanning method. This means that the first step of a stride is selected on the same level of accuracy as with the scanning method, while the accuracy of the subsequent steps of the stride is monotonically worsened toward the end of the stride. With the scanning method, on the other hand, all steps are selected with the same level of approximation. Indeed, it has been shown that for $f = 4$, the attrition constant for SAWs (on a square lattice) is 8 times smaller with the scanning method than with the method of strides.

14.5.11 The Stochastic Double Scanning Method

Because exact enumeration of the future chains is time consuming, the scanning parameter f is limited, and for a large chain length, N, $f << N$. This means that sample attrition will still be severe and the successful chains will be generated with a considerable bias. In an attempt to alleviate this problem, the exact enumeration set of future chains (for SAWs on a square lattice) has been replaced by a smaller sample of chains generated at random [13]. The hope has been that the smaller number of random chains will save computer time, allowing, thereby, to increase f. However, this hope has not materialized, as exact enumeration on a square lattice has been found to be much more efficient than the generation of chains at random. It should be noted, however, that randomly generated chains would be important in applications of the method to continuum models of peptides, for example. Indeed, this "*double scanning method*" was applied successfully to a continuum model of decaglycine [28].

14.5.12 Future Scanning by Monte Carlo

The double scanning method can be generalized by replacing the *random* generation of future chains with a long Metropolis Monte Carlo simulation; this idea is due to Dr. R.P. White [29]. More specifically, at step k, an arbitrary configuration of an f-step future chain is initially constructed, and then changed by a long Metropolis MC simulation, where the already determined steps $1,2,...,k - 1$ are held fixed. From the generated (future) sample, which is based on l MC steps, the TPs are obtained by calculating the number of times, m_{ν} the first step of the future chain (k) has visited each of the four possible directions ν ($\nu = 1,..,4$); the corresponding *stochastic* TPs are,

$$\text{TP}_k(f,l) = p(\nu \mid \nu_{(k-1)},...,\nu_1,f,l) = \frac{m_\nu}{\sum_\nu m_\nu} \tag{14.45}$$

where m_ν [$=m_\nu(f, l)$] replaces $Z_k^\nu(f)$ [Equation (14.21)]. This procedure is called the "*partial* MC scanning method" due to the fact that $f << N$; thus, sample attrition is not avoided, and the corresponding probability $P_i^0(f,l)$ is defined over a group of chains that also contain self-intersecting walks.

If the whole future is scanned by MC (that is, $f = N - k + 1$ steps), the method is called "the *complete* MC *scanning method*", for this method, sample attrition is avoided since the entire future is scanned, and the corresponding probability, $P_i(l)$, [$P_i(l) = P_i^0(N - k + 1, l)$] is defined over the group of SAWs only.

Still, $P_i(l)$ is biased, and its normalization is compromised; however, its accuracy can be improved *at will* by increasing the sample size, l, where $P_i(l) \to P_i^B$ as $l \to \infty$.

The partial and complete MC scanning methods, while very versatile (in principle), are expected to be time consuming for generating large systems; therefore, both methods have not been tested. However, the common element of these methods—the (stochastic) MC growth procedure—constitute the basis for a general approach for extracting the absolute entropy from MC and MD samples, called "*the hypothetical scanning* MC (or MD) *method*," which is practical for continuum systems; this method will be discussed in Chapter 19.

14.5.13 The Scanning Method for the Ising Model and Bulk Systems

The scanning method is designed to handle a linear system and naturally has been introduced above, as applied to chain models. However, historically, the method was applied first to the $2d$ Ising model by viewing it as a linear entity, where spins are added to an initially *empty* lattice step-by-step, line-by-line [14]. It should be pointed out that such linearization of the Ising model was not new at that time. Kikuchi [30] was the first to suggest a similar linear picture, where *approximate* TPs were calculated analytically. Alexandrowicz [31] developed this approach further for computer simulation by defining TPs based on a set of parameters $\{\alpha_i\}$. Thus, several relatively large Ising configurations were initially generated step-by-step with different sets of parameters α_i, leading to system probabilities, $P(\alpha_i)$, entropy, $S(\alpha_i)$, and free energy, $F(\alpha_i)$, where the optimal set was the one that minimized the free energy; using the optimal set, a production run of Ising lattices was produced from which thermodynamic quantities of interest were calculated.

However, with these two approaches, it is difficult to improve the sets of TPs, while with the scanning method, the TPs can be improved systematically by increasing the number of future spins that are considered at each construction step. More specifically, as seen in the Figure 14.3, at step k of the construction process, two regions are defined on the lattice of size N, the $k - 1$ spins already determined (the past) and the empty part of the lattice (the future). To apply the scanning method, one defines a rectangle consisting of f sites over the *empty* part around step k. To calculate the probability for σ_k, one calculates two partition functions $Z_k(+, f, T)$ and $Z_k(-, f, T)$ for spin "+" and spin "−", respectively. Thus, holding spin k at $+ (-)$, all the configurations, j, of the (future) spins in the rectangle are calculated, while the previously defined

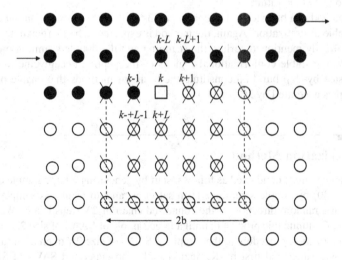

FIGURE 14.3 An illustration of the kth step of the spiral construction of the square Ising lattice of size $N = L \times L$ for $L = 10$. In the spiral boundary conditions, the last spin (L) of line n is a nearest neighbor of the first spin (1) of the next $(n + 1)$th line (demonstrated by arrows). Full circles denote lattice spins already specified with spins in the previous steps of the process, while open circles are the still empty lattice sites. The dashed lines define a rectangle of $f = 2b \times (b + 1)$ lattice sites for an approximation based on $b = 3$. The crossed full circles and the (border) slashed circles take part in the calculation of $Z_k(+, f, T)$ and $Z_k(-, f, T)$.

spins ("past spins") at sites $1,...,k-1$ are held fixed. For each configuration, j, one calculates the corresponding energy, E_j, due to interactions of the future spins among themselves and with the past spins, which leads to $\exp - [E_j/k_B T]$. These exponentials are summed up for all j leading to the future partition functions, $Z_k(+, f, T)$ and $Z_k(-, f, T)$. The TP is thus,

$$p(\sigma_k, f, T) = p(\sigma_k | \sigma_1, ..., \sigma_{k-1}, f, T) = \frac{Z_k(\sigma_k, f, T)}{Z_k(+, f, T) + Z_k(-, f, T)} \qquad (14.46)$$

f – the size of the future rectangle, determines the approximation, the larger is f, the better is the approximation, which is limited only by the computer capabilities to handle a large f. After $p(\sigma_k, f, T)$ has been calculated, the spin at step k, σ_k, is determined by a random number. The probability $P_i(f, T)$ of lattice configuration, i, is the product, $P_i(f, T) = \Pi p_k(\sigma_k, f, T)$ for $k = 1,...,N$. The corresponding entropy, obtained from a sample of size n, $S' = -k_B \Sigma_n P_i(f, T)\ln P_i(f, T)$, constitutes an *underestimation* of the correct entropy, (as in Equations (14.29) and (14.31), where n replaces n_{suc}) and no sample attrition occurs for the Ising model.

For the Ising model, this inherently approximate scanning method leads to excellent results ($S' \rightarrow S$) not too close to the critical temperature, T_c. As T_c is approached, the approximation worsens due to the increase in the range of the spin-spin correlations, but the results are still very good [14]. For example, close to the critical temperature for $K = J/k_B T = 0.43$, at the hot side, the free energy is $F/Nk_B T = 0.915232(1)$ and 0.915260 for the scanning and the exact results, respectively, while for $K = 0.455$, at the cold side, the corresponding results are 0.951042(1) and 0.951043. It should be pointed out, however, that the exact Metropolis MC method is simpler and more effective than the scanning method for calculating properties that are measured directly on the system, such as the energy, magnetization, and their fluctuations. The advantage of the scanning method lies in its ability to obtain the absolute entropy and free energy, which are not calculable directly by MC.

As for SAWs, one can replace the exact enumeration of $Z_k(+)$ and $Z_k(-)$ by stochastic calculations based on MC simulations (Section 14.5.12). If the entire future is considered (that is, $f = N - k + 1$), the corresponding "complete MC scanning method" can be applied to the Ising model. Alternatively, one can use a "partial MC scanning method," where MC is applied to a small square ($f < N - k + 1$) of future spins rather than to the whole future.

The scanning method can be extended to *continuum 3d* bulk systems or to continuum chain models, by applying a suitable discretization. Again, the method is expected to be inefficient for generating large systems; thus, as already mentioned earlier, the scanning and the double scanning methods have only been used to study the stable conformational regions of the peptide decaglycine in vacuum [26,28]. The ideas of the step-by-step build-up constitute the basis for methods that enable one to extract the entropy from samples generated by MC or MD.

14.6 The Dimerization Method

The dimerization procedure is conducted as follows: start by generating a large sample of relatively short SAWs (say, of $N = 30$) by simple sampling. Then select at random from this sample a pair of chains and connect them at a random direction; if the connected chain of $2N$ steps is a SAW, keep it on a file, otherwise discard it. Continue this procedure until a large sample of SAWs of size $2N$ has been obtained. Then apply this dimerization procedure to the sample of SAWs of size $2N$ to create a sample of $4N$, etc.

This procedure was suggested first by K. Suzuki [32], who generated SAWs of 8192 steps on the diamond lattice. The method was discovered independently by Z. Alexandrowicz, who called it "dimerization" and derived the theory behind its high efficiency [33,34]. Thus, considering first the simple sampling method, he calculated the attrition constant, λ_J, by summing up (for a given construction step k) the loop probabilities, $f(0;j) = m/j^2$ [Equation (14.10)], from the shortest possible loop of size $j = J$ to the longest one of $k + 1$ steps. One obtains from this calculation:

$$\lambda_J = \frac{m}{J} \tag{14.47}$$

Thus, as has been pointed out by Wall et al. [see discussion prior to Equation (14.10)], λ_J is affected only by the shortest loop. The attrition rate, R_{attrit} [Equation (14.9)], is obtained as a product of the probabilities (for not to close a loop), $1 - mJ^{-1}$ for all the chain steps; since for small x, $\exp(x) \sim 1 - x$, one obtains,

$$\prod_{k=J}^{N}\left(1 - \frac{m}{J}\right) \approx \exp(-mJ^{-1}N) = \exp(-\lambda_J N). \tag{14.48}$$

While in the above derivation every step contributes the same probability, if two SAWs of size $N/2$ each are connected, the size of the shortest loop changes with k as $k + 1$ and Equation (14.48) becomes,

$$\frac{n_{suc}(N)}{[n_{suc}(N/2)]^2} = \prod_{k=J}^{N/2}\left(1 - \frac{m}{k+1}\right) \cong \exp\left\{-m\int_J^{N/2}(k+1)^{-1}dk\right\} = \text{const}(N^{-m}) \tag{14.49}$$

which constitutes a considerably higher efficiency than the exponential decay of simple sampling, the RR method, and the scanning method. A different derivation of Equation (14.49) is based on the approximate equation for the number of SAWs (partition function) $\sim C\mu^N N^{\gamma-1}$ [Equation (9.6)], which leads to the same behavior as Equation (14.49):

$$R_{attrit}\left(\frac{N}{2};N\right) \approx \frac{n_{suc}(N)}{[n_{suc}(N/2)]^2} = \frac{C\mu^N N^{\gamma-1}}{(C\mu^{N/2}(N/2)^{\gamma-1})^2} = \left(\frac{2^{2\gamma-2}}{C}\right)N^{1-\gamma} \tag{14.50}$$

The exponent $\gamma-1$ is ~ 0.34 for $d = 2$ and ~ 0.16 for $d = 3$, which, however, differs from $m = 0.36$ obtained by Wall et al. [1] for the 4-choice cubic lattice. For this lattice, SAWs of up to $N = 8192$ steps were generated by dimerization [33,34], and this chain length could certainly be increased using present computers. However, based on a heuristic argument, Madras and Sokal [35] have shown that the expected CPU time, T_N, for generating an N-step SAW by dimerization is:

$$T_N \approx N^{c_1 \log_2 N + c_2} \tag{14.51}$$

where

$$c_1 = \frac{\gamma - 1}{2}, \qquad c_2 = \frac{5 - 3\gamma}{2} + \log_2 C. \tag{14.52}$$

This means that the dimerization method will become unfeasible for very large N, due to the increase of the exponent in Equation (14.51). For the square lattice, for example, $c_1 \approx 0.17$ and $c_2 \approx 0.72$, which lead to the exponent 3 for $N = 11,000$. Equation (14.51) is obtained by realizing that $1/R_{attrit}$ is the average number of trials required for obtaining an N-step SAW by concatenating pairs of $N/2$ SAWs. Iterating this, dimerization k times from N_0 to N ($k = \log_2(N/N_0)$) leads to Equation (14.51).

The dimerization method provides the absolute entropy, which can be calculated based on Equation (14.8). Since the samples consist of the same basic units, this "inbreeding" might introduce some biases, which can be reduced by increasing the samples' sizes. Indeed, calculation of the exponent ν based on data of [34] leads to $\nu \approx 0.592$ for the 4-choice cubic lattice, while the accepted value is ~ 0.5877. This might suggest that the dimerized SAWs are slightly too extended, however, the culprit might be the simplified analysis, which ignores corrections to scaling; in fact, using the pivot algorithm, Madras and Sokal [35] have arrived to the same estimation based on SAWs of up to $N = 3000$ (where corrections to scaling have been considered). The method is designed to handle global structural properties of a polymer in a free space, such as the end-to-end distance and the radius of gyration. On the other hand, because the chains are changed globally, the distributions of local structures, such as a specific angle or a local blob are not expected to be simulated properly. Finally, the method is not suitable for treating SAWs under geometrical constrains or SAWs with attractions at low *T*, because the construction criterion is optimal for pure SAWs.

REFERENCES

1. F. T. Wall, L. A. Hiller Jr. and D. J. Wheeler. Statistical computation of mean dimensions of macromolecules. I. *J. Chem. Phys.* **22**, 1036–1041 (1954).
2. F. T. Wall and J. J. Erpenbeck. New method for the statistical computation of polymer dimensions. *J. Chem. Phys.* **30**, 634–640 (1959).
3. F. T. Wall, S. Windwer and P. J. Gans. Monte Carlo methods applied to configurations of flexible polymer molecules. In *Methods of Computational Physics*, B. Alder, S. Fernbach and M. Rotenberg (Eds.), vol. I, 217–243 (Academic, New York, 1963).
4. P. J. Gans. Self-avoiding random walks. I. simple properties of intermediate-length walks. *J. Chem. Phys.* **42**, 4159–4163 (1965).
5. F. T. Wall and F. Mandel. Macromolecular dimensions obtained by an efficient Monte Carlo method without sample attrition. *J. Chem. Phys.* **63**, 1462–1465 (1975).
6. F. T. Wall, R. J. Rubin and L. M. Isaacson. Improved statistical method for computing mean dimensions of polymer molecules. *J. Chem. Phys.* **27**, 186–188 (1957).
7. M. N. Rosenbluth and A. W. Rosenbluth. Monte Carlo calculation of the average extension of molecular chains. *J. Chem. Phys.* **23**, 356–359 (1955).
8. F. L. McCrackin. Configuration of isolated polymer molecules adsorbed on solid surfaces studied by Monte-Carlo computer simulation. *J. Chem. Phys.* **47**, 1980–1986 (1967).
9. J. Mazur and F. L. McCrackin. Monte Carlo studies of configurational and thermodynamic properties of self-interacting linear polymer chains. *J. Chem. Phys.* **49**, 648–665 (1968).
10. F. L. McCrackin, J. Mazur and C. M. Guttman. Monte Carlo studies of self-interacting polymer chains with excluded volume. I. squared radii of gyration and mean-square end-to-end distances and their moments. *Macromolecules* **6**, 859–887 (1973).
11. H. Meirovitch. A new method for simulation of real chains: Scanning future steps. *J. Phys. A* **15**, L735–L740 (1982).
12. H. Meirovitch. Computer simulation of self-avoiding walks: Testing the scanning method. *J. Chem. Phys.* **79**, 502–508 (1983). Erratum: *J. Chem. Phys.* **81**, 1053–1053 (1984).
13. H. Meirovitch. Statistical properties of the scanning simulation method for polymer chains. *J. Chem. Phys.* **89**, 2514–2522 (1988).
14. H. Meirovitch. An approximate stochastic process for computer simulation of the Ising model at equilibrium. *J. Phys. A* **15**, 2063–2075 (1982).
15. H. Meirovitch. The scanning method with a mean-field parameter: Computer simulation study of the critical exponents of self-avoiding walks on a square lattice. *Macromolecules* **18**, 563–569 (1985).
16. S. Livne and H. Meirovitch. Computer simulation of long polymers adsorbed on a surface: I. corrections to scaling in an ideal chain. *J. Chem. Phys.* **88**, 4498–4506 (1988).
17. H. Meirovitch and S. Livne. Computer simulation of long polymers adsorbed on a surface: II. Critical behavior of a self-avoiding walk. *J. Chem. Phys.* **88**, 4507–4515 (1988).
18. K. E. Schmidt. Using renormalization-group ideas in Monte Carlo sampling. *Phys. Rev. Lett.* **51**, 2175–2178 (1983).
19. H. Meirovitch. Comment on using renormalization-group ideas in Monte-Carlo sampling. *Phys. Rev. Lett.* **53**, 952–952 (1984).
20. H. Meirovitch. Scanning method as an unbiased simulation technique and its application to the study of self-attracting random walks. *Phys. Rev. A* **32**, 3699–3708 (1985).
21. H. Meirovitch. The scanning simulation method for macromolecules. *Int. J. Mod. Phys.* **1**, 119–145 (1990).
22. E. Parzen. *Modern Probability Theory and Its Applications*; Chapter 10, p. 434. (John Wiley & Sons, New York, 1960).

23. I. Chang and H. Meirovitch. Collapse transition of self-avoiding walks on a square lattice in the bulk and near a linear wall: The universality class of the θ and θ' points. *Phys. Rev. E* **48**, 3656–3660 (1993).

24. H. Meirovitch. A new simulation of branched polymers. *J. Phys. A* **20**, 6059–6073 (1987).

25. H. Meirovitch. Entropy, pressure and chemical potential of multiple chain systems from computer simulation. I: Application of the scanning method. *J. Chem. Phys.* **97**, 5803–5815 (1992).

26. H. Meirovitch, M. Vásquez and H. A. Scheraga. Stability of polypeptides conformational states: II. The free energy of the statistical coil obtained by the scanning simulation method. *Biopolymers* **27**, 1189–1204 (1988).

27. A. M. Ferrenberg and R. H. Swendsen. New Monte Carlo technique for studying phase transitions. *Phys. Rev. Lett.* **61**, 2635–2638 (1988).

28. H. Meirovitch, M. Vásquez and H. A. Scheraga. Free energy and stability of macromolecules studied by the scanning method. *J. Chem. Phys.* **92**, 1248–1257 (1990).

29. R. P. White and H. Meirovitch. Absolute entropy and free energy of fluids using the hypothetical scanning method. II. Transition probabilities from canonical Monte Carlo simulations of partial systems. *J. Chem. Phys.* **119**, 12096–12105 (2003).

30. R. Kikuchi. A theory of cooperative phenomena. *Phys. Rev.* **81**, 988–1003 (1951).

31. Z. Alexandrowicz. Stochastic models for the statistical description of lattice systems. *J. Chem. Phys.* **55**, 2765–2779 (1971).

32. K. Suzuki. The excluded volume effect of very-long-chain molecules. *Bull. Chem. Soc. Japan* **41**, 538–538 (1968).

33. Z. Alexandrowicz. Monte Carlo of chains with excluded volume: A way to evade sample attrition. *J. Chem. Phys.* **51**, 561–565 (1969).

34. Z. Alexandrowicz and Y. Accad. Monte Carlo of chains with excluded volume: Distribution of intersegmental distances. *J. Chem. Phys.* **54**, 5338–5345 (1971).

35. N. Madras and A. D. Sokal. The pivot algorithm: A highly efficient Monte Carlo method for the self-avoiding walk. *J. Stat. Phys.* **50**, 109–186 (1988).

15

The Pivot Algorithm and Hybrid Techniques

15.1 The Pivot Algorithm—Historical Notes

The pivot algorithm is a dynamic Metropolis Monte Carlo (MC) procedure for chains, based on global moves; the algorithm consists of four stages described below for a Self-Avoiding Walk (SAW) on a square lattice. (1) An initial configuration of an N-step (bond) SAW that starts from the origin is defined on the lattice. (2) A step k, $2 \leq k \leq N$, is selected at random, that is, with probability $1/(N-1)$ (since due to chain symmetry, the first step can be kept constant); thus, the chain is divided into two parts, the fixed part consisting of steps $1,2,...,k-1$ and the moving part of steps $k, k+1,...,N$. (3) Four structural moves are defined for step k – the first step of the moving part: two $90°$ rotations, a $180°$ rotation, and the possibility of k to remain in its present state (identity). (4) A *trial* move (out of four) is selected at random, and step k together with the entire moving part connected to it are placed *temporarily* in their new positions. If no intersections between the two parts are found, the trial move is accepted, becoming permanent. In the other case, the trial move is rejected, that is, the moving part remains in its original configuration. In the next MC step, a new step k is chosen, etc. This procedure satisfies the detained balance condition.

This algorithm was suggested for lattice polymer models by Lal in 1969 [1]. In 1972, Stellman and Gans [2] extended Lal's method to a continuum chain model, pointing out that the method had actually been initiated by Fluendy in 1963, as applied to relatively short alkenes [3]. In 1976, the method was used by Olaj and Pelinka for studying lattice polymers [4] and by Freire and Horta to treat *n*-alkanes [5]; in 1985, the algorithm was rediscovered by MacDonald et al. [6], who used it for studying lattice polymers. In 1988, Madras and Sokal (MS) [7] became interested in the efficiency of this method, as applied to SAWs on lattices, and in their 77-page paper, they called it *"the pivot algorithm"*. A detailed description of all the ingredients of this paper is beyond the scope of this book; we shall only present the main conclusions.

15.2 Ergodicity and Efficiency

MS prove that the algorithm described above is ergodic on the square lattice. They point out that more structural moves can be defined on this lattice, such as two diagonal reflections, which together with the $180°$ rotation also lead to an ergodic algorithm. They analyze the conditions for ergodicity on the cubic and hypercubic lattices ($d \geq 3$), which will not be discussed here.

The first question one would like to answer is about the attrition rate for the pivot algorithm. Since the method is actually based on connecting two global pieces of the chain, one would expect a similar behavior to that found for the dimerization method, that is, $R_{attrit} \sim N^{-p}$ [Equations (14.9) and (14.49)], where p is a dimension-dependent exponent. A similar analysis to that carried out in Equation (14.50) would suggest that $p = \gamma - 1$ (~ 0.34 for the square lattice); however, by averaging simulation results for different MC moves, MS have obtained the smaller value, $p \approx 0.192$. On the cubic lattice, the picture is similar, $p \approx 0.107$ versus $\gamma - 1 \approx 0.162$.

Finally, MS calculate the amount of work required for storing the chain's coordinates and for checking the intersections between the fixed and the moving parts. Clearly, a successful pivot move requires a *complete* examination for self-avoidance along the chain, while a failed pivot move involves, on average, much less checks; this also stems from the fact that intersections between monomers that are close to the pivot spot are the most probable and can be checked first [see Equation (14.10)]. Based on heuristic arguments, MS predict that the amount of work required for treating a failed pivot move is $T_N \sim N^{1-p}$, and since the average number of failed moves is $\sim N^p$, one obtains the total amount of work, E, for a successful move to be $\sim N$,

$$E = O(N^{1-p})N^p + N \tag{15.1}$$

where the last N on the right hand side in Equation (15.1) is the time required for writing the coordinates of the accepted N-step SAW.

Another element affecting efficiency is the *equilibrium* autocorrelation time, τ, for a global chain property, G (e.g., the radius of gyration). Thus, if i and j are two chain configurations along the MC trajectory separated on average by n *successful* pivot moves, $G(i)$ and $G(j)$ will be uncorrelated [see Equation (1.84)]. Because the pivot moves lead to drastic configurational changes, one would expect n to be small, $n = 5$–10 and thus $\tau \sim nN^p$. Indeed, for the square lattice, the MS simulations lead to $\tau \sim N^q$, where $q = 0.205 \pm 0.015$ is only slightly larger than their estimation for p. Notice, however, that for a local parameter, such as a specific angle, the autocorrelation time is expected to be $\sim N$ times larger, that is, $\tau \sim N^{p+1}$.

This discussion brings us to the problem of the initial relaxation time, τ_R, that is, the time (in number of observables) required for a system placed initially in an arbitrary configuration to reach equilibrium; this *thermalization* period is important because the related non-equilibrated configurations should be discarded from the production run. Clearly, τ_R depends on the longest relaxation time in the system, thus $\tau_R \sim gN^{p+1}$, where MS consider g to be a small number, say $g = 10$, which leads to $\tau_R = \sim 600,000$ observables for $N = 10,000$. However, MS already find $200,000$ to be an adequate initial sample size for a global parameter; they also discuss situations where shorter initializations will be sufficient, as well as cases which need longer thermalizations (e.g., $\tau_R \sim 10N^2$).

An alternative option for initialization suggested by MS is an *equilibrium* start, which can be obtained by generating an initial chain configuration by the dimerization method. While in this way the relaxation process is avoided, this choice can be more expensive than relaxation since $T_N \approx N^{c_1 \log_2 N + c_2}$ [Equations (14.51) and (14.52)], meaning that for a very large N, the initialization time will dominate the production run.

In a 1995 pivot paper, Li, Madras, and Sokal generated the longest SAWs of $N = 80,000$ on the square and the simple cubic lattices [8]. In addition to the exponent ν, they studied universal amplitude ratios, such as $<R_g^2>/<R^2>$ and the "hyper-scaling relation." Clearly, a further increase in the efficiency of the method can only be achieved by improving the way the chain's coordinates are handled and the intersections are checked. Thus, Kennedy [9] recognized that it is not necessary to write down the SAW for each successful pivot move, and from a clever use of geometric constraints, he developed an algorithm that broke the $O(N)$ barrier; he also improved the MS result, from $T_N = O(N^{0.81})$ to $O(N^{0.38})$ for a SAW on a square lattice, and thus was able to increase the length of the simulated SAW up to $N = 10^6$. A more dramatic improvement in efficiency is due to Clisby [10], who presented arguments that for very large N, $T_N = O(1)$ for SAWs on both a square lattice and a simple cubic lattice, while for the maximal chain length studied by him, ($N = 3.3 \times 10^7$), $T_N = O(\log N)$. This chain length allowed him to obtain the best estimate thus far for the exponent ν for the simple cubic lattice, $\nu = 0.587597(7)$.

15.3 Applicability

Since the global pivot moves lead to ergodicity (for SAWs), it has been important to extend the algorithm to other systems. Thus, the method was applied to polymer rings [11] and to SAWs with attractions near the θ temperature, where the chain's shape is still relatively open and the pivot moves are thus efficient [12]. However, in the latter case, it was necessary to add local moves to treat correctly the autocorrelation of the energy. Clearly, the pivot algorithm is inefficient in the collapsed region and when geometrical constraints are applied. Also, due to its global (unrealistic) moves, the algorithm is not suitable to model dynamics.

While the basic implementation of the pivot algorithm (as described in the beginning of this chapter) is quite straightforward, exploiting its full potential requires investing in sophisticated programming, as discussed in [9,10]. However, elaborating on this subject is beyond the scope of this book.

15.4 Hybrid and Grand-Canonical Simulation Methods

The simulation methods described in Chapters 10, 14, and 15 can be considered as the basic methods in the polymer field (reviewed in detail by Kremer and Binder [13]), as some of them have also been used as building blocks for more sophisticated techniques. Thus, Rapaport [14,15] combined the ideas of the enrichment and dimerization techniques (Sections 14.3 and 14.6) and studied SAWs of up to 2400 steps on the square, triangular, simple cubic, and body-center cubic lattices, concentrating on the critical exponent ν and its corrections to scaling. Grassberger [16] has suggested the *"pruned-enriched Rosenbluth method,"* which is based on the Rosenbluth & Rosenbluth (RR) and the enrichment methods (Sections 14.3 and 14.4). The pruned-enriched Rosenbluth method has been shown to be efficient for a θ-polymer on the simple cubic lattice, where SAWs (with attractions) of up to $N = 10,000$ steps were simulated; clearly, a better efficiency can still be obtained by replacing RR with the scanning method. The scanning procedure was used within an MC study of ring polymers. Thus, a segment of n monomers is chosen at random and all the configurations (loops) connecting to its end points are scanned, where one of them (denoted k) is chosen at random; k is accepted or rejected according to the MC prescription [17].

In Section 10.5, we have emphasized the non-ergodicity of simulation methods based on the configurational changes of local structures, such as a crankshaft (kink) or larger clusters. However, if one adds "transport" and "reptation" moves, ergodic procedures can be devised for $d = 2$ or 3, depending on the composition of such moves [18–23]. An example of a "kink-transport move" is shown in Figure 15.1, where a kink is cleaved from the chain and attached at a pair of neighboring sites somewhere else along the chain.

Using a method of this type, Caracciolo et al. have studied the θ-point behavior of SAWs of $N = 3200$ on a square lattice [23].

All the methods described thus far treat chains in the canonical ensemble, that is, at constant T and N. However, grand canonical methods have also been developed, where the chemical potential is constant and monomers are added to the chain or removed from it, leading to fluctuations in N. In this category, we mention the methods of Reynold and Redner [24] and Berretti and Sokal [25]; in the latter paper an interesting technique for calculating the entropy is introduced based on the (statistical) method of maximum likelihood. Other methods in this category appear in [18,19,26].

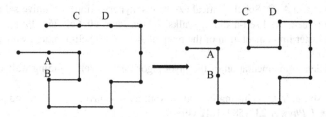

FIGURE 15.1   A kink has been cleaved from AB and attached at CD. The new kink is permitted to occupy one or both of the sites abandoned by the old kink.

15.5 Concluding Remarks

We have evaluated the performance of the various simulation methods as applied to a *single* open SAW on an infinite lattice (that is, a polymer in a good solvent at infinite dilution); our main criterion of efficiency has been the maximal chain length that a method can handle. For theoretical reasons (see Chapter 9), calculation of the critical exponents of this model (in particular ν) has attracted a great deal of interest, leading to the development of a large number of simulation techniques, among them, the highly efficient dimerization and pivot algorithms. The model of SAWs demonstrates the inherent difficulty in treating long-range interactions, which are shared by all realistic macromolecular systems, in particular, proteins. However, a protein is affected by both repulsions and attractions, where the latter drive the chain (through a θ-point) into its native collapsed state (bad solvent). Such dense structures cannot be treated efficiently by the dimerization and the pivot procedures because their global moves will produce a very high rejection rate. These methods are also useless for dense many-chain systems or a single polymer subject to a geometrical constraint, such as an adsorbing surface or a narrow tube. In these situations, methods that are based on local configurational changes are required, such as grand canonical techniques or MC procedures consisting of local structural changes (e.g., crankshafts) and more complex "transport" and "reptation" moves.

However, the above methods still might not have the required flexibility. On the other hand, RR and the scanning method are ideal for systems with constraints, because they handle the attractions in a natural way and their "future feelers" help avoiding geometric obstacles. In this category, it is also important to mention the MC-based bond-fluctuation model, which can handle effectively dense multiple-polymers and has been used extensively in the literature [27,28]. Finally, this chapter and the previous one provide only a basic picture of the rich area of simulation methodologies for polymers. The simulation difficulties encountered for polymers constitute a precursor for the problems involved in the simulation of biological macromolecules.

REFERENCES

1. M. Lal. Monte Carlo computer simulation of chain molecules. I. *Molec. Phys.* **17**, 57–64 (1969).
2. S. D. Stellman and P. J. Gans. Efficient computer simulation of polymer conformation. I. geometric properties of the hard-sphere model. *Macromolecules* **5**, 516–526 (1972).
3. M. A. D. Fluendy. Calculation of cyclization probabilities and other configuration properties of alkane-type chains by a Monte Carlo method. *Trans. Faraday Soc.* **59**, 1681–1694 (1963).
4. J. J. Freire and A. Horta. Mean reciprocal distances of short polymethylene chains: Calculation of the translational diffusion coefficient of *n*-alkanes. *J. Chem. Phys.* **65**, 4049–4054 (1976).
5. O. F. Olaj and K. H. Pelinka. Pair distribution function and pair potential of lattice model chains under theta conditions, 1. Numerical evaluation. *Makromol. Chem.* **177**, 3413–3425 (1976).
6. B. MacDonald, N. Jan, D. L. Hunter and M. O. Steinitz. Polymer conformations through 'wiggling'. *J. Phys. A* **18**, 2627–2631 (1985).
7. N. Madras and A. D. Sokal. The pivot algorithm: A highly efficient Monte Carlo method for the self-avoiding walk. *J. Stat. Phys.* **50**, 109–186 (1988).
8. B. Li, N. Madras and A. D. Sokal. Critical exponents, hyperscaling, and universal amplitude ratios for two- and three-dimensional self-avoiding walks. *J. Stat. Phys.* **80**, 661–754 (1995).
9. T. Kennedy. A faster implementation of the pivot algorithm for self-avoiding walks. *J. Stat. Phys.* **106**, 407–428 (2002).
10. N. Clisby. Efficient implementation of the pivot algorithm for self-avoiding walks. *J. Stat. Phys.* **140**, 349–392 (2010).
11. E. J. J. van Rensburg, S. G. Whittington and N. Madras. The pivot algorithm and polygons: Results on the FCC lattice. *J. Phys. A* **23**, 1589–1612 (1990).
12. M. C. Tesi, E. J. J. van Rensburg, E. Orlandini and S. G. Whittington. Interacting self-avoiding walks and polygons in three dimensions *J. Phys. A* **29**, 2451–2463 (1996).
13. K. Kremer and K. Binder. Monte Carlo simulation of lattice models for macromolecules. *Comput. Phys. Rep.* **7**, 259–310 (1988).

14. D. C. Rapaport. On three-dimensional self-avoiding walks. *J. Phys. A* **18**, 113–126 (1985).
15. D. C. Rapaport. Two-dimensional polymers: Universality and correction to scaling. *J. Phys. A* **18**, L39–L42 (1985).
16. P. Grassberger. Pruned-enriched Rosenbluth method: Simulations of θ polymers of chain length up to 1,000,000. *Phys. Rev. E* **56**, 3682–3693 (1997).
17. S. Medalion, E. Aghion, H. Meirovitch, E. Barkai and D. A. Kessler. Size distribution of ring polymers. *Sci. Rep.* **6**, 27661–27668 (2016).
18. B. Berg and D. Foerster. Random paths and random surfaces on a digital computer. *Phys. Lett. B* **106**, 323–326 (1981).
19. C. Aragão de Carvalho and S. Caracciolo. A new Monte-Carlo approach to the critical properties of self-avoiding random walks. *J. Phys. (Paris)* **44**, 323–331 (1983).
20. T. Pakula. Cooperative relaxations in condensed macromolecular systems. 1. A model for computer simulation. *Macromolecules* **20**, 679 (1987).
21. J. Reiter. Monte Carlo simulations of linear and cyclic chains on cubic and quadratic lattices. *Macromolecules* **23**, 3811–3816 (1990).
22. S. Caracciolo, M. S. Causo, G. Ferraro, M. Papinutto and A. Pelissetto. Bilocal dynamics for self-avoiding walks. *J. Stat. Phys.* **100**, 1111–1145 (2000).
23. S. Caracciolo, M. Gherardi, M. Papinutto and A. Pelissetto. Geometrical properties of two-dimensional interacting self-avoiding walks at the θ-point. *J. Phys. A* **44**, 115004–115017 (2011).
24. S. Redner and P. J. Reynolds. Position-space renormalisation group for isolated polymer chains. *J. Phys. A* **14**, 2679–2703(1981).
25. A. Berretti and A. D. Sokal. New Monte Carlo method for self-avoiding walk. *J. Stat. Phys.* **40**, 483–529 (1985).
26. C. Aragão de Carvalho, S. Caracciolo, and J. Fröhlich. Polymers and $g|\varphi|^4$ theory in four dimensions. *Nucl. Phys. B* **215**, 209–248 (1983).
27. I. Carmesin and K. Kremer. The bond fluctuation method: A new effective algorithm for the dynamics of polymers in all spatial dimensions. *Macromolecules* **21**, 2819–2823 (1988).
28. H. P. Deutsch and K. Binder. Interdiffusion and self-diffusion in polymer mixtures: A Monte Carlo study. *J. Chem. Phys.* **94**, 2294–2304 (1991).

16

Models of Proteins

16.1 Biological Macromolecules versus Polymers

In this chapter, we discuss models of increasing complexity for proteins and other biological macro-molecules. First, it is important to make the distinction between biological chain molecules and simple (synthetic) polymer chains. The main interest in polymers is in their global properties such as the end-to-end distance and the radius of gyration, where a lot of work has been devoted to understanding their asymptotic growth and spatial distributions (Chapter 9). Also, of interest is the increase in entropy with increasing the chain length. Therefore, for such problems, simplified lattice models are adequate (see Chapters 8 and 9). On the other hand, a biological macromolecule, such as a protein, ribonucleic acid (RNA), or a carbohydrate is of a *finite* size, and it is composed of a specific sequence of chemical units that determine its biological activity. Thus, a detailed modeling is required for understanding the molecule's function.

Concentrating on proteins, we remind the reader that a protein is defined by a *specific* sequence of the 20 occurring *amino acid residues*, which determine to a large extent its specific three dimensional ($3d$) collapsed (native) structure, and thus its biological activity. In this respect, a protein differs from a polymer, which can reside in many collapsed structures of comparable free energy (see Figure 9.6). While many of the protein properties can be obtained experimentally (e.g., its $3d$ structure can be determined by X-ray crystallography), it is highly desirable to be able to predict them by theoretical-computational methods. Thus, typical questions of interest are: (1) how to predict the protein's native $3d$ structure (protein folding) based solely on its sequence of residues, or (2) what is the free energy of binding of a ligand to the protein's active site, and (3) how does an enzyme catalyze a chemical reaction? These questions are much more specific than those asked for polymers, and answering them would require a detailed treatment on the quantum mechanics level, which, however, is unfeasible for such large systems. Therefore, empirical potential energy functions (also called *force fields*) have been developed, where the effect of the electrons is averaged out. Similar force fields are also available for small organic molecules and other macromolecules, such as deoxyribonucleic acid (DNA) and RNA.

16.2 Definition of a Protein Chain

Before discussing force fields, we provide the reader with some details about a protein chain and its amino acids components. First, each amino acid residue consists of three connected backbone atoms, N, C^α, and C', where a hydrogen is connected to N and C^α, and an oxygen is connected to C' by a double bond. Two residues are connected by a strong peptide bond between the $C'=O$ of the first residue and the N–H of the second residue. The "personality" of a residue is determined by the side chain connected to C^α. The 20 different side chains can be divided into three groups, non-polar (hydrophobic), polar, and charged. Their names appear in Table 16.1.

The $3d$ structure (also called conformation) of a protein is defined by the Cartesian coordinates of all the atoms. However, a more economical structure definition can be obtained by internal coordinates. Thus, three dihedral angles, φ, ψ, and ω are defined for the backbone, where φ defines a rotation around the N–C^α bond, ψ around the C^α–C' bond, and ω around the peptide bond; typically, ω fluctuates only slightly around $180°$ (in the trans position). Dihedral angles, χ, are also defined for the side chains; each dihedral angle is determined by the positions of four suitable atoms, while bond angles are defined by

TABLE 16.1

Full Names, 3-Letter Codes, and 1-Letter Codes of the 20 Occurring Amino Acids

Name	3	1	Name	3	1
Glycine	Gly	G	Proline	Pro	P
Alanine	Ala	A	Valine	Val	V
Leucine	Leu	L	Isoleucine	Ile	I
Methionine	Met	M	Cysteine	Cys	C
Phenylalanine	Phe	F	Tyrosine	Tyr	Y
Tryptophan	Trp	W	Histidine	His	H
Lysine	Lys	K	Arginine	Arg	R
Glutamine	Gln	Q	Asparagine	Asn	N
Glutamic Acid	Glu	E	Aspartic Acid	Asp	D
Serine	Ser	S	Threonine	The	T

three neighbor atoms. If one assumes that the bond lengths and bond angles are constant (which is a very good approximation), the number of parameters (angles) defining a protein conformation is significantly reduced as compared to the number of the Cartesian coordinates, which can lead to a significant gain in computer time. This is only a short descriptive discussion about the variables determining the protein structure. More details can be found, for example, in the books [1–4] and in the discussion related to Figure 19.3.

It should also be pointed out that a collapsed protein structure is typically "decorated" by *secondary local* structures, such as a rod-like "*α-helix*" stabilized by hydrogen bonds. A liner segment of the chain can change direction by 180°, creating a "*turn*," which leads to a "*β-hairpin*" stabilized by segment-to-segment hydrogen bonds; when the two segments belong to distant parts of the chain, they create a "*β-sheet*." Another local structure is a "*surface loop*", which takes part in protein-ligand recognition processes; for more details, see [1–4].

Typically (but not always), proteins are studied at high dilution, where practically, a molecule does not "feel" the other molecules around; thus, as for polymers, only a *single* protein chain is treated. Below, three types of protein models of increasing complexity are presented: (1) a protein in vacuum, (2) a protein in implicit solvent, and (3) a protein immersed in a box of explicit water.

16.3 The Force Field of a Protein

The typical potential energy function (force field), E_{FF}, for a protein is defined below, without elaborating on the methods used to derive its parameters; in vacuum, E_{FF} is,

$$E_{FF} = \sum_{bonds} K_r(r - r_{eq})^2 + \sum_{angles} K_\theta(\theta - \theta_{eq})^2 \quad \text{connectivity}$$

$$+ \sum_{dihedrals} \frac{V_n}{2}[1 + \cos(n\varphi - \gamma)] \qquad \text{torsional} \qquad (16.1)$$

$$+ \sum_{1 < j} \frac{A_{ij}}{R_{ij}^{12}} - \frac{B_{ij}}{R_{ij}^6} + \frac{q_i q}{\varepsilon R_{ij}} \qquad \text{non-bonded}$$

where K_r, K_θ, r_{eq}, θ_{eq}, γ, n, V_n, A_{ij}, B_{ij}, and q_i are parameters optimized by applying this function to a large amount of experimental data and results obtained from quantum mechanical *ab initio* calculations (for small molecules). The first line of Equation (16.1) defines the connectivity of the protein chain by harmonic potentials: r is the distance between nearest neighbor atoms along the chain and r_{eq} is their equilibrium distance. K_r and K_θ are (strong) spring constants. In the second line, a torsional potential is defined for each torsional rotation, φ, of four atoms about a central bond; V_n, n, and γ are parameters. For specific angles, φ, the potential is minimal, and the corresponding side chain conformations (defined by the angle, χ) are called rotamers. The last line of Equation (16.1) presents the non-bonded interactions. They consist of Lennard-Jones interactions (depending on the parameters A_{ij} and B_{ij}) and electrostatic interactions, where R_{ij} is the distance between atoms i and j, q_i is the partial charge of atom i, and ε is a dielectric constant. The electrostatic interactions are mainly responsible for the generation of hydrogen bonds. This force field describes a protein in vacuum ignoring solvent effects. However, the screening of the Coulomb potentials by water can partially be mimicked by increasing ε within the range, $\varepsilon = 1 - 80$.

This force field depends on Cartesian coordinates, and thus is suitable for molecular dynamics (MD) simulations (Sections 10.6–10.9). Equation (16.1) constitutes the prototype for most of the available force fields (which still might differ by the values of the parameters, K_r, K_θ, r_{eq}, etc. and might include some functional modifications). The most popular force fields are imbedded in the molecular dynamics/molecular mechanics (MD/MM) programs for proteins, DNA, etc., such as AMBER(UCSF) [5], CHARMM (Harvard) [6], OPLS (Yale) [7], and GROMOS (Berendsen-van Gunsteren) [8]. Other MD/MM programs enable one to use several of the above force fields (e.g., GROMACS (Groningen) [9], NAMD (Illinois) [10], and TINKER (J. Ponder) [11]). A force field depending on internal coordinates is ECEPP (Cornell), developed by Scheraga's group [12]. Discussing the different versions of these force fields and the related programs developed during the years is beyond the scope of this book.

Proteins interact with water and with small organic molecules; therefore, potential energy functions for water-water, water-protein, and protein-small molecules have been developed as well (see [22–25] and Section 16.5). It should be emphasized, however, that all force fields are far from being perfect, and their improvement is critical for the progress of computational structural biology (see Section 23.8.6). Advances can be achieved in several avenues, among them, optimizing further the force field parameters and devising more realistic electrostatic interactions, which also consider polarizability. Developing better procedures for handling the long-range effects of electrostatic interactions is also a challenge (Section 10.9.1).

16.4 Implicit Solvation Models

Proteins "live" in water, therefore, considering the protein-water interactions is crucial. However, an MD simulation of a protein in a box of water molecules is time consuming and thus simplified models have been developed, where solvent effects are considered *implicitly*. A typical (simple) model for the total energy, E_{total}, consists of the force field energy, E_{FF}, and an implicit solvation term, $E_{solvation}$ [13],

$$E_{total} = E_{FF}(\varepsilon = 2r) + E_{solvation} = E_{FF}(\varepsilon = 2r) + \sum_i A_i \sigma_i \qquad (16.2)$$

In this equation, the Coulomb potential between two charges with a distance r is $q_i q_j / [(2r)r]$, where $\varepsilon = 2r$ is a distance-dependent dielectric function. A_i is the solvent accessible surface area of atom i; A_i depends on the protein's $3d$ conformation and is thus calculated for all conformations in the sample or practically for a smaller group of significantly different conformations. σ_i is an atomic solvation parameter, which is positive for a hydrophobic (non-polar) atom and negative for a hydrophilic (polar) atom; hence, in low energy conformations, the hydrophobic (hydrophilic) atoms will be mostly buried (exposed to the solvent). Various sets of atomic solvation parameters have been derived, mostly based on the experimental free energy of transfer of *small* molecules from the gas phase to water [14,15]. Equation (16.2) with some modifications has been widely used, where in some studies the atomic solvation parameters have been determined by optimization with respect to crystal (or nuclear magnetic resonance) structures of cyclic peptides or loops in proteins (see [16–18] and references cited therein).

More sophisticated implicit solvation models have also been devised, such as the generalized Born surface area (GB/SA) model of Still's group [19,20], where the electrostatic interactions are modeled more reliably than with Equation (16.2); Several versions of the GB/SA model have been developed, where their parameters are commonly optimized against properties of small molecules—experimentally determined solvation energies or free energies obtained by the Poisson-Boltzmann (PB) equation; in general, the more complex is the model, the better is the agreement with PB results at the expense of an increase in computer time (see references 51, 55, 56, and 87–96 in [16]). The quality of some of these models, as applied to loops, has been checked by comparing their results to calculations obtained by finite difference PB including a hydrophobic term [21]. While a great deal of work has been carried out with implicit solvation models (including important PB studies), we shall not elaborate on this topic further because the focus in this book is on explicit solvation models, as discussed in the next section.

16.5 A Protein in an Explicit Solvent

The protein is immersed in a "box" of thousands of water molecules. Several models for water have been developed during the years. We shall concentrate here on the simple highly used rigid model, "three-site transferrable intermolecular potential" (TIP3P) of Jorgensen's group [22]. In this model, the oxygen-hydrogen (O=H) distance is 0.9572 Å, the angle H–O–H is 104.52 degrees, a charge, $-0.834e$, is placed on the oxygen, and half of it, $-0.417e$, on each of the hydrogens; thus, the molecule has neutral charge. Also, the Lennard-Jones parameters $\sigma = 3.15061$Å and $\varepsilon = 0.6364$ kJ/mol [Equation (10.17)] are defined for two oxygens of different water molecules. The energy of interaction, E_{mn}, between two water molecules m and n is thus (Figure 16.1):

$$E_{m,n} = \sum_{i(m)} \sum_{j(n)} \frac{q_i q_j e^2}{r_{ij}} + 4\varepsilon \left[\left(\frac{\sigma}{r_{oo}} \right)^{12} - \left(\frac{\sigma}{r_{oo}} \right)^6 \right] \tag{16.3}$$

where q_i runs over the three charges and r_{ij} are the charge-charge distances. It should be noted that the O–O repulsion provides protection from the occurrence of "bad" overlaps between the hydrogens of a water molecule and atoms of a neighbor molecule. More sophisticated models, TIP4P [22] and TIP5P [23] with 4 and 5 interaction sites, respectively have been developed as well, with the expense of increasing computer time. Another popular set of water models are SPC (simple point-charge) [24], and SPC/E (SPC/Extended) [25].

For all these models, Lennard-Jones parameters are also defined between the oxygen of a water molecule and the protein's atoms. To take into account long-range electrostatic effects, the system is usually studied in periodic boundary conditions with particle mesh Ewald and sometimes with reaction field methods (see Section 10.9.1). While a protein in explicit solvent is the most realistic model, most of the computer time is spent on treating the water-water interactions, which are not of much interest. This is why "cheaper" implicit solvation models have been initially developed.

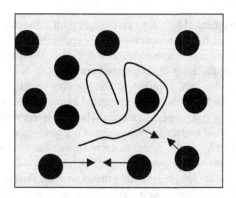

FIGURE 16.1 A protein dissolved in explicit water. The arrows denote protein-water and water-water interactions.

16.6 Potential Energy Surface of a Protein

For this discussion, it is convenient to consider a protein described by a force field in vacuum or in an implicit solvation. A folded protein at $T = 300$ K resides in a collapsed state (the native structure), that is, in a well-defined $3d$ structure (unlike the many collapsed structures available for a synthetic polymer). The potential energy surface of a protein is very rugged, "decorated" by a tremendous number of local energy minima, which constitute the "bottoms" of the corresponding local energy wells. More accurately, denoting the coordinates defining an energy minimum i by $\mathbf{x}_i \equiv \mathbf{x}_i^N$, the related *local energy well*, Ω_i is the local conformational region around $\mathbf{x}_i$, which constitutes the basin of attraction of $\mathbf{x}_i$; practically, a conformation $\mathbf{x}$ pertains to this basin if a *local* minimization of its energy will end up at $\mathbf{x}_i$.

One might identify the native structure with the global energy minimum (GEM) and its local well, which, however, are extremely difficult to locate due to their tiny size compared to the entire conformational space; furthermore, the existence of trillions of minima far and close to the GEM makes this "hunting" problem even harder. To date, this is an unsolved mathematical problem in global optimization. This situation is depicted schematically in Figure 16.2.

Actually, the situation is more complex, since the *free energy* rather than the energy is the correct criterion of stability. Thus, for local energy well, i, one can define a partition function, Z_i, the related absolute free energy, F_i, and the absolute entropy, S_i,

$$Z_i = \int_i \exp(-E / k_{\mathrm{B}}T)d\mathbf{x}$$

$$F_i = -k_{\mathrm{B}}T \ln Z_i = <E>_i -TS_i \tag{16.4}$$

$$S_i \sim -k_{\mathrm{B}} \int_i P_i^{\mathrm{B}}(\mathbf{x}) \ln P_i^{\mathrm{B}}(\mathbf{x})d\mathbf{x}$$

These integrations are carried out over the coordinates of local energy well, $i(\Omega_i)$, and the Boltzmann probability density, $P_i^{\mathrm{B}}(\mathbf{x})$, is normalized over i only. The *absolute* F_i and S_i are defined up to additive constants, which are cancelled in differences, ΔF_{ij} and ΔS_{ij}, between local energy wells i and j at the same temperature. Also, one obtains *approximately*:

$$S_i \sim k_{\mathrm{B}} \ln \Omega_i. \tag{16.5}$$

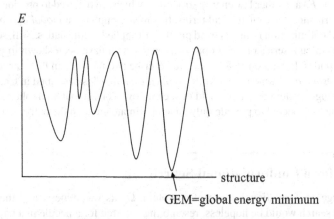

GEM=global energy minimum

FIGURE 16.2 An illustration mimicking the tremendous number of energy minima and the GEM.

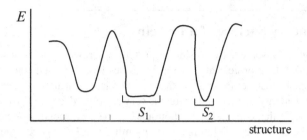

FIGURE 16.3   Energy well 2 (the GEM) is lower than energy well 1; however, well 1 is wider (higher entropy) and therefore is the most stable (lowest free energy).

Clearly, the lower F_i is, the higher is the stability, which is affected not only by lower E_i, but also by higher S_i. In Figure 16.3, E_2 is lower than E_1, hence, E_2 is the GEM; however, S_1 is larger than S_2, therefore, local energy well 1 is more stable than 2. The partition function of the entire conformational space, Z, is of less interest than the local partition functions, Z_i, which determine the relative stabilities of local energy wells.

16.7 The Problem of Protein Folding

The aim of protein folding is to predict the native structure of a protein based on its known sequence of amino acid residues and a force field (in our discussion, in vacuum or with implicit solvation). One can envisage a process where a set of highly stable local energy wells is found first, their F_i values are calculated, and the most stable local well is identified as that with the lowest F_i. However, this is an extremely difficult task, since one needs to first locate by conformational search methods a group of highly stable local energy wells out of trillions of them, and then to calculate their entropies, S_i, which lead to F_i; as discussed later in Chapters 18 and 19, calculation of S_i is a non-trivial task. An approximate way for calculating the absolute S_i is by assuming that the local energy well is a parabola, that is, $E(\Omega_i)$ is a harmonic potential around $\mathbf{x}_i$. In this case:

$$S_i \sim k_B \ln[\det(\text{Hessian})] \tag{16.6}$$

where Hessian is the matrix of second derivatives of the potential with respect to the coordinates $\mathbf{x}_i$ at the minimum (see Section 18.2). This discussion demonstrates the immense problems involved in protein folding already for a *simplified* model of a protein based on implicit solvent.

These difficulties have led to a more modest avenue of research, which concentrates on finding the GEM (rather than the lowest F_i) and other low energy structures, which are believed to provide information about the correct folded structure. In mathematical terms, this means applying a *global* optimization technique, while to date, global optimization is an unsolved problem in applied mathematics. Still, some methods have been suggested (based on a series of *local* optimizations), which have been shown to perform efficiently at least for short peptides. In Section 16.8 below, we describe two methods in this category, *"Monte Carlo minimization"* and *"simulated annealing,"* while *"replica exchange"* is described in Chapter 21. We stress again that representing a protein by a force field in vacuum (with or without an implicit solvent) is not realistic enough and thus is expected to provide only an approximate guide in the native structure search.

16.8 Methods for a Conformational Search

Because of the huge conformational space of a protein, Ω, its ruggedness, and the small $\Omega_i(\text{GEM})$, applying a random search would be hopeless, resembling a search for a needle in a haystack. Still, some "smart" methods based on the Monte Carlo idea are able to lead the search more efficiently toward the desired region in Ω; two such methods are described below. However, these methods (and others) are based on energy minimization, which constitutes a central instrument in structural biology (e.g., as the basis of harmonic analysis). Therefore, we devote a short discussion to this topic describing the steepest descents method.

16.8.1 Local Minimization—The Steepest Descents Method

Assume for simplicity a one-dimensional function, $f = f(x)$, where $f(x_0)$ is known at an arbitrary point, x_0. We ask: what is the closest x_{min} to x_0 for which $f(x_{min})$ is a local minimum? $f(x_{min})$ can be found by various methods, such as steepest descents, conjugate gradients, or Newton-Raphson. The x values belonging to the local well around x_{min} define the basin of attraction of x_{min}. To make progress, we define the *vector* gradient of a *general* function, $f(x, y, z)$,

$$\text{gradient} = \nabla f = \frac{\partial f(x,y,z)}{\partial x}\mathbf{i} + \frac{\partial f(x,y,z)}{\partial y}\mathbf{j} + \frac{\partial f(x,y,z)}{\partial z}\mathbf{k} \tag{16.7}$$

where $\mathbf{i}$, $\mathbf{j}$, and $\mathbf{k}$ are the usual perpendicular unit vectors. The gradient at a given point is tangent to the surface at this point, where at minimum and maximum points of the function, the gradient is zero; if $f(x, y, z) = \text{constant}$, the gradient is perpendicular to this surface. For example, for the circle, $x^2 + y^2 = 1$, the gradient is $(2x, 2y)$ and thus, at $x = 0$, $y = 1$, the gradient is $(0,2)$, pointing in the $+y$ direction, that is, perpendicular to the circle at $x = 0$.

We shall discuss now a simple minimization technique, the steepest descents method, as applied to the function $f(x, y) = x^2 + 5y^2$ with minimum at $(0,0)$ and gradient $(2x, 10y)$. The search starts at an arbitrary point, $\mathbf{a} = (x_0, y_0)$, and the minimum is reached by defining a set of coordinates, which approach the minimum in several iterations. Each iteration of the search is based on two steps: (1) determining a direction to move and (2) deciding how much to move along this direction.

1. The move is done in the opposite direction to the gradient $(-\nabla f)$, which in our case is $(-2x, -10y)$.

$$x' = x_0 - \alpha \frac{\partial f}{\partial x}\bigg|_{x_0, y_0} = x_0 - \alpha 2 x_0$$

$$y' = y_0 - \alpha \frac{\partial f}{\partial y}\bigg|_{x_0, y_0} = y_0 - \alpha 10 y_0 \tag{16.8}$$

2. One has to find, α, the optimal amount of move along $(-\nabla f)$, which satisfies, $f[x'(\alpha), y'(\alpha)] < f(x_0, y_0)$ and $f[x'(\alpha), y'(\alpha)]$ is minimal along this line; this process is called a "*line search*."

In practice, one first calculates a (search) line along the gradient at point $\mathbf{a}$, and selects two points $\mathbf{b}$ and $\mathbf{d}$ on this line, which are located on both sides of point $\mathbf{a}$ and both satisfy, $f(\mathbf{b}) > f(\mathbf{a})$ and $f(\mathbf{d}) > f(\mathbf{a})$. Then, in several function evaluations, the minimum point $\mathbf{c}$ between $\mathbf{b}$ and $\mathbf{d}$ on the $\mathbf{b}$–$\mathbf{d}$ line is found. Point $\mathbf{c}$ belongs to an ellipsoidal contour, $n = 8$ (see Figure 16.4); the line $\mathbf{b}$–$\mathbf{d}$ is tangent to this contour at point $\mathbf{c}$.

In the next iteration, the gradient at point $\mathbf{c}$ is calculated, and the above process is repeated several times until the minimum is reached. Notice that the gradient at point $\mathbf{c}$ is perpendicular to the line $\mathbf{b}$–$\mathbf{d}$. Also, if $f(x, y)$ is circular, that is, $f(x, y) = mx^2 + my^2$, the minimum is found in one iteration!

The line search, requiring 3–10 function evaluations, can be computationally expensive for a force field of a protein. To increase efficiency, one can stop the line search at the first time the function f becomes lower than $f(\mathbf{a})$; however, in this case, the gradients will not be perpendicular.

The steepest descent method is robust far from the minimum, but becomes inefficient close to the minimum since the gradients tend to zero; therefore, the steepest descent is sometimes replaced there by

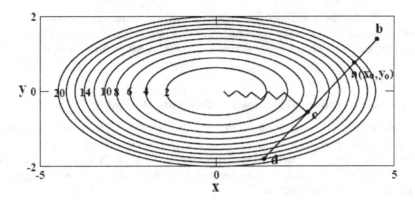

FIGURE 16.4 A contour surface of the function $x^2 + 5y^2 = n$, where n increases by two units ($n = 2,4...$). The gradient from the initial point $a(x_0, y_0)$ defines the line search direction. Note that the minimum point c occurs precisely at the point where the line is tangent to the contour, $n = 8$, implying that the subsequent gradient is orthogonal to the previous one.

more efficient procedures, such as conjugate gradients or limited memory Broyden–Fletcher–Goldfarb–Shanno (L-BFGS), which are based on second derivatives of the target function. The efficiency of these methods (and others) for peptides and proteins has been found to be system-dependent to a large extent. Therefore, we shall not discuss these methods further. An excellent review on optimization methods for proteins has been written by Schlick [26]; the performance of various minimization techniques for protein/peptide systems has also been studied by us in [27,28].

16.8.2 Monte Carlo Minimization

For simplicity, we describe the Monte Carlo minimization (MCM) method of Li and Scheraga [29], as applied to a peptide where bond angles and lengths are kept constant; thus, a conformation is defined solely by the dihedral angles. The search starts from *any* energy minimized structure, i, with energy, E_i^{min}. A small number of dihedral angles, n, are selected *at random*, and their values are redefined *at random* within the range $[-180°, 180°]$. The energy of the new structure is minimized (locally), leading to a trial structure, j, with E_j^{min}; j is accepted (or rejected) according to an asymmetric Metropolis transition probability, p_{ij} [Equation (10.11)],

$$p_{ij} = \begin{cases} 1 & \text{for } \Delta E_{ji} < 0 \\ \exp[-\Delta E_{ji} / k_B T] & \text{for } \Delta E_{ji} \geq 0 \end{cases} \qquad \Delta E_{ji} = E_j^{min} - E_i^{min}. \qquad (16.9)$$

T is a temperature parameter (not a thermodynamic temperature due to the minimized structures) that affects the efficiency strongly, and k_B is the Boltzmann constant. The process is repeated many times, where the generated minimized energies are compared. For an efficient MCM, the number of treated angles, n, at each step should be small, where in some studies $n = 1$ was used.

With MCM energy barriers are crossed efficiently; when E_j^{min} is smaller than E_i^{min}, j is always accepted and the search goes downhill, while in the other case, the higher energy avoids trapping of the system in a low potential well, which is different from the GEM. The method is very efficient. For small peptides, such as Met and Leu enkephalin (Tyr-Gly-Gly-Phe-Met and Tyr-Gly-Gly-Phe-Leu) described by the ECEPP force field, the GEM has been found *on average* after 3000 minimizations, as compared to 50,000–100,000 minimizations required with a random search. MCM has been found to be significantly more efficient than simulated annealing (see Section 16.8.3 below) and has been used not only for proteins, but also for Lennard-Jones clusters and other systems.

Notice that the structures generated by Equation (16.9) are energy minimized, hence, they are not distributed according to the Boltzmann probability like those obtained in a standard Metropolis process (that is, without minimizations), and the averaged MCM energy is not the thermodynamic average.

An attempt has been made [30] to calculate at each minimum also the harmonic entropy, S_i^m and to carry out MCM with respect to the harmonic free energy $F_i^{min} = E_i^{min} - TS_i^{min}$. However, the free energy values are much closer to each other than the related energies because lower energy typically corresponds to lower entropy; therefore, the free energy provides a much weaker guidance than the energy for the process to reach the lowest free energy minimum.

16.8.3 Simulated Annealing

This approach [31] is carried out with MD or Monte Carlo (MC). One starts simulating the system at high T (e.g., 1500 K), where the system can cross most energy barriers. During the simulation, the temperature is gradually decreased to 300 K; the hope is that during the process, the system will have become caught in the GEM region. Then, every say, 1 ps, the energy of the current structure is minimized locally, and the set of minimized energy structures generated is expected to include the GEM. With MC, one can carry out the whole simulation with minimized energies rather than energies.

Notice, that the method is stochastic, and thus there is no guarantee that the GEM will be found. The efficiency increases as the relaxation to $T = 300$ slows. Simulated annealing is not an efficient method if the only guidance toward the GEM is provided by the energy.

The method has been used for structure determination of proteins, where proton-proton distance constraints [obtained by nuclear magnetic resonance] are added to the force field, E_{FF}. E_{tot} is,

$$E_{tot} = E_{FF} + \sum_i a_i(r_i - b_i)^2 \qquad (16.10)$$

where b_i is the ith constraint and a_i is a parameter. For such optimizations (with extra guidance by constraints), simulated annealing works well. MCM was found to be more efficient than simulated annealing for finding the GEMs of Met and Leu enkephalin modeled by ECEPP [32]. Thus, while simulated annealing is not very efficient, it has been a popular method due to its simple implementation.

16.9 Monte Carlo and Molecular Dynamics Applied to Proteins

In principle, protein folding can be obtained by carrying out long Monte Carlo (MC) or MD simulations. First, it should be pointed out that defining MC trial moves based on Cartesian coordinates will lead almost always to rejections; this is because a *random* conformational change in a compact structure is expected to cause overlaps of atoms and a strong increase in the harmonic potentials, which keep the chain connectivity. Thus, for proteins, this MC procedure is extremely inefficient because of the need to carry out many expensive (and futile) potential energy calculations of the *entire* chain (see also Sections 10.5 and 10.6). A better efficiency can be gained with an MC procedure based on internal coordinates—dihedral and bond angles. Such a procedure has been implemented in Jorgensen's program BOSS [7,33], which is used regularly by his group; one can also attempt applying force-biased MC (Section 10.6). However, for proteins, MD is the method of choice, since the atoms are moved, not at random, but in an optimal direction along the calculated gradients. Thus, most protein simulations have been carried out by MD, which is implemented in all the MC/MD software packages mentioned in Section 16.3 above.

Returning to protein folding, one would expect to fold a protein by a *very long* MD simulation, hoping to reach the most stable region in conformational space, and to visit the most stable energy wells in this region; however, in practice, this is not feasible at room temperature due to high energy barriers (Figure 16.2). The system will get trapped for a long simulation time in a deep energy well, and the move toward the more stable parts of conformational space will be extremely slow. In this context, we point out that the time scales available to MD with present computers (microseconds) are much smaller than the experimental time scales of protein folding (seconds and above). Protein folding is an active area of research, which, however, is beyond the scope of this book.

16.10 Microstates and Intermediate Flexibility

Thus far, we have been interested in local energy wells and the GEM, as related to protein folding. However, MD studies have shown that a protein will stay in a local energy well for a very short time (as short as several fs), while spending a considerable amount of time in a "wider" energy well, which consists of many local ones (Figure 16.5) [34,35]. Therefore, the wider wells, called "*microstates*" (denoted by Ω_m rather than by Ω_i), are of a greater physical significance than the local wells, Ω_i. An example of a microstate is the conformational region spanned by the local fluctuations of an alpha helix of a peptide. Thus, one can define average energy, $E(\Omega_m)$, entropy, $S(\Omega_m)$, and free energy, $F(\Omega_m)$, for a microstate, Ω_m, using Equation (16.4), where the integration is carried out over the microstate rather than over the local energy well. It should be emphasized that the microstates are defined also for a protein in explicit water, where the integration defined in Equation (16.4) includes also water.

There are problems which are less challenging than protein folding, but involve what we call "*intermediate flexibility*," where a flexible protein segment (e.g., a side chain or a surface loop), populates significantly several different microstates in thermodynamic equilibrium; thus, unlike a large regular thermodynamic system, where the most probable energy, $E^*(T)$, dominates the partition function ($(Z \sim \nu(E^*) \exp - [E^*(T)/k_{\mathrm{B}}T]$, see Section 5.2) in intermediate flexibility, several different microstates, possibly with different energies have comparable most significant contribution, Z_m to Z. Intermediate flexibility can also be found in a cyclic peptide or a ligand (small molecule) bound to an enzyme. It is of interest, for example, to know whether the conformational change adopted by a loop (a side chain, ligand, etc.) upon protein-ligand binding has been induced by the bound ligand (induced fit [36,37]) or, alternatively, the free loop already interconverts among different microstates, where one of them is selected upon binding (selected fit [38]). This analysis requires calculating the relative populations, p_m/p_n, of microstates m and n, which can be obtained from the ratio of the corresponding partition functions $Z(\Omega_m)/Z(\Omega_n)$; such calculations are also needed for a correct analysis of nuclear magnetic resonance and X-ray data of flexible macromolecules [39,40].

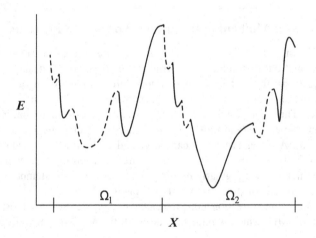

FIGURE 16.5 Schematic one-dimensional representation of part of the energy surface of a peptide or a protein, as a function of a coordinate, X. The two large potential energy wells are defined over the corresponding microstates denoted, Ω_1 and Ω_2. Each microstate consists of many localized potential wells denoted intermittently by solid and dashed lines. The partition function, Z_m, of microstate, m, is obtained by integrating $\exp[-E/k_{\mathrm{B}}T]$ over Ω_m, where $F_m = -k_{\mathrm{B}}T \ln Z_m$ is the microstate's free energy. The figure suggests that the second microstate is the more stable among the two, due to lower energy and higher entropy (Ω_2 is larger than Ω_1), hence, lower free energy. If F_2 is also the global free energy minimum of a protein, Ω_2 is expected to describe the native microstate (assuming a perfect force field), and a simulation started from Ω_2 will keep the protein in this microstate for a long time. On the other hand, a peptide can populate significantly several of the most stable microstates in thermodynamic equilibrium.

16.10.1 On the Practical Definition of a Microstate

Calculating populations, p_m, or ratios, p_m/p_n, by various techniques cannot be achieved without first establishing a *practical* definition of a microstate, which is, however, not trivial. Therefore, we elaborate below on this important issue that has been given considerable thought by us over the course of the years (see [16–18] and references cited therein). For simplicity, let us consider an N-residue peptide with rigid geometry, that is, with constant bond lengths and bond angles; thus, its backbone conformation is solely defined by the dihedral angles, φ_k and ψ_k, $k = 1,2,\ldots, N$ (ω_k, of the peptide bond is fixed at 180°.) For a helical microstate (Ω_h), these angles are expected to vary within relatively small ranges $\Delta\varphi_k$ and $\Delta\psi_k$ around $\varphi_k = -60°$ and $\psi_k = -50°$, respectively (we ignore for a moment the side chains). However, if N is not too small, the correct limits of Ω_h in the $[\varphi_k,\psi_k]$ space are unknown even for this simplified model since they constitute a complicated narrow "pipe" contained within the (larger) region defined by the product, $\Delta\varphi_1 x \Delta\psi_1 x \Delta\varphi_2 x \Delta\psi_2 \ldots \Delta\varphi_N x \Delta\psi_N$ due to the strong correlations among the dihedral angles. Obviously, these correlations are taken into account naturally by an exact simulation method, such as MD, and thus, in practice, Ω_h can be defined (or more correctly, represented) by a *local* MD sample of conformations initiated from an α-helical structure, as mentioned earlier.

However, this definition should be used with caution. Thus, a short simulation will span only a small part of Ω_h, and this part will grow constantly as the simulation continues; correspondingly, the calculated average potential energy, E_h, and the entropy, S_h (obtained by any method) will both increase and the free energy, F_h, is expected to change as well. As the simulation time is increased further, side chain dihedrals will "jump" to different rotamers, which according to our definition should also be included within Ω_h; for a long enough simulation, the peptide is expected to "leave" the α-helical region and move to a different microstate. Thus, *in practice*, the microstate size and the corresponding thermodynamic quantities can depend on the simulation time used to define the microstate. In some cases, one can better define Ω_h by discarding structures with dihedral angles beyond predefined $\Delta\varphi_k$ and $\Delta\psi_k$ values or structures that do not satisfy a certain number of hydrogen bonds; one can also apply energetic restraints on coordinates, where this bias should be removed. However, these restrictions are somewhat arbitrary. In practice, we shall be interested in differences $\Delta F_{m,n}$ and $\Delta S_{m,n}$, etc, between microstates Ω_m and Ω_n. Therefore, one should calculate these differences for an increasing simulation time until stability in the results has been obtained for a long enough time. One should bear in mind that in practice there is always some arbitrariness in the definition of a microstate, which affects the calculated averages [41]. This arbitrariness is severe with some methods and can be controlled (minimized) by others, as is discussed later in Chapters 18 and 19.

REFERENCES

1. D. Whitford. *Proteins. Structure and Function.* (Wiley, New York, 2005).
2. C. I. Branden and J. Tooze. *Introduction to Protein Structure.* (Garland, New York, 1991).
3. G. E. Schulz and R. H. Shirmer. *Principles of Protein Structure.* (Springer-Verlag, New York, 1979).
4. I. Bahar, R. L. Jernigan and K. A. Dill. *Protein Actions, Principles and Modeling.* (Garland, New York, 2017).
5. D. A. Case, T. E. Cheatham, III, T. Darden, et al. The Amber biomolecular simulation programs. *J. Comput. Chem.* **26**, 1668–1688 (2005).
6. B. R. Brooks, C. L. Brooks, III, A. D. Mackerell, Jr, et al. CHARMM: The biomolecular simulation program. *J. Comput. Chem.* **30**, 1545–1614 (2009).
7. W. L. Jorgensen and J. Tirado-Rives. Molecular modeling of organic and biomolecular systems using BOSS and MCPRO. *J. Comput. Chem.* **26**, 1689–1700 (2005).
8. M. Christen, P. H. Hünenberger, D. Bakowies, et al. The GROMOS software for biomolecular simulation: GROMOS05. *J. Comput. Chem.* **26**, 1719–1751 (2005).
9. B. Hess, C. Kutzner, D. Van Der Spoel and E. Lindahl. GROMACS 4: Algorithms for highly efficient, load-balanced, and scalable molecular simulation. *J. Chem. Theory Comput.*, **4**, 435–447 (2008).
10. J. C. Phillips, R. Braun, W. Wang, et al. Scalable molecular dynamics with NAMD. *J. Comput. Chem.* **26**, 1781–1802 (2005).
11. J. W. Ponder. TINKER – software tools for molecular design. version 6.1 (2012).

12. M. J. Sippl, G. Némethy and H. A. Scheraga. Intermolecular potentials from crystal data. 6. Determination of empirical potentials for O–H...O = C hydrogen bonds from packing configurations. *J. Phys. Chem.* **88**, 6231–6233 (1984).

13. L. Wesson and D. Eisenberg. Atomic solvation parameters applied to molecular dynamics of proteins in solution. *Protein Sci.* **1**, 227–235 (1992).

14. D. Eisenberg and A. D. McLachlan. Solvation energy in protein folding and binding. *Nature* **319**, 199–203 (1986).

15. T. Ooi, M. Oobatake, G. Némethy and H. A. Scheraga. Accessible surface areas as a measure of the thermodynamic parameters of hydration of peptides. *Proc. Natl. Acad. Sci. USA* **84**, 3086–3090 (1987).

16. A. Szarecka and H. Meirovitch. Optimization of the GB/SA solvation model for predicting the structure of surface loops in proteins. *J. Phys. Chem. B.* **110**, 2869–2880 (2006).

17. B. Das and H. Meirovitch. Optimization of solvation models for predicting the structure of surface loops in proteins. *Proteins* **43**, 303–314 (2001).

18. C. Baysal and H. Meirovitch. Free energy based populations of interconverting microstates of a cyclic peptide lead to the experimental NMR data. *Biopolymers* **50**, 329–344 (1999).

19. W. C. Still, A. Tempczyk, R. C. Hawley and T. F. Hendrickson. Semianalytical treatment of solvation for molecular mechanics and dynamics. *J. Am. Chem. Soc.* **112**, 6127–6129 (1990).

20. D. Qiu, P. S. Shenkin, F. P. Hollinger and W. C. Still. The GB/SA continuum model for solvation. A fast analytical method for the calculation of approximate Born radii. *J. Phys. Chem. A* **101**, 3005–3014 (1997).

21. K. C. Smith and B. Honig. Evaluation of the conformational free energies of loops in proteins. *Proteins* **18**, 119–132 (1994).

22. W. L. Jorgensen, J. Chandrasekhar, J. D. Madura, R. W. Impey and M. L. Klein. Comparison of simple potential functions for simulating liquid water. *J. Chem. Phys.* **79**, 926–935 (1983).

23. M. W. Mahoney and W. L. Jorgensen. A five-site model liquid water and the reproduction of the density anomaly by rigid, non-polarizable models. *J. Chem. Phys.* **112**, 8910–8922 (2000).

24. H. J. C. Berendsen, J. P. M. Postma, W. F. van Gunsteren and J. Hermans. Interaction models for water in relation to protein hydration. In *Intermolecular Forces*, B. Pullman (Ed.), 331–342 (Reidel, Dordrecht, the Netherlands, 1981).

25. H. J. C. Berendsen, J. R. Grigera and T. P. Straatsma. The missing term in effective pair potentials. *J. Phys. Chem.* **91**, 6269–6271 (1987).

26. T. Schlick. Optimization methods in computational chemistry. In *Reviews in Computational Chemistry* **3**, 1–71. K. B. Lipkowitz and D. B. Boyd, eds., VCH Publishers, New York (1992).

27. C. Baysal, H. Meirovitch and I. M. Navon. Performance of efficient minimization algorithms as applied to models of peptides and proteins. *J. Comput. Chem.* **20**, 354–364 (1999).

28. B. Das, H. Meirovitch and I. M. Navon. Performance of hybrid methods for large-scale unconstrained optimization as applied to models of proteins. *J. Comput. Chem.* **24**, 1222–1231 (2003).

29. Z. Li and H. A. Scheraga. Monte Carlo-minimization approach to the multiple-minima problem in protein folding. *Proc. Natl. Acad. Sci. USA* **84**, 6611–6615 (1987).

30. M. Vásquez, E. Meirovitch and H. Meirovitch. A free energy based Monte Carlo minimization procedure for macromolecules. *J. Phys. Chem.* **98**, 9380–9382 (1994).

31. S. Kirkpatrick, C. D. Gelatt, Jr. and M. P. Vecchi. Optimization by simulated annealing. *Science* **220**, 671–680 (1983).

32. H. Meirovitch and M. Vásquez. Efficiency of simulated annealing and the Monte Carlo minimization method for generating a set of low energy structures of peptides. *J. Molec. Struct.* (Theochem) **398–399**, 517–522 (1997).

33. J. P. Ulmschneider and W. L. Jorgensen. Monte Carlo backbone sampling for polypeptides with variable bond angles and dihedral angles using concerted rotations and a Gaussian bias. *J. Chem. Phys.* **118**, 4261–4271 (2003).

34. R. Elber and M. Karplus. Multiple conformational states of proteins – A molecular dynamics analysis of myoglobin. *Science* **235**, 318–321 (1987).

35. F. H. Stillinger and T. A. Weber. Packing structures and transitions in liquids and solids. *Science* **225**, 983–989 (1984).

36. E. D. Getzoff, H. M. Geysen, S. J. Rodda, H. Alexander, J. A. Tainer and R. A. Lerner. Mechanisms of antibody binding to a protein. *Science* **235**, 1191–1196 (1987).

37. J. M. Rini, U. Schulze-Gahmen and I. A. Wilson. Structural evidence for induced fit as a mechanism for antibody-antigen recognition. *Science* **255**, 959–965 (1992).

38. K. L. Constantine, M. S. Friedrichs, M. Wittekind, et al. Backbone and side chain dynamics of uncomplexed human adipocyte and muscle fatty acid-binding proteins. *Biochemistry* **37**, 7965–7980 (1998).

39. D. M. Korzhnev, X. Salvatella, M. Vendruscolo, A. A. Di Nardo, A. R. Davidson, C. M. Dobson and L. E. Kay. Low populated folding intermediates of Fyn SH3 characterized by relaxation dispersion NMR. *Nature* **430**, 586–590 (2004).

40. E. Z. Eisenmesser, O. Millet, W. Labeikovski, et al. Intrinsic dynamics of an enzyme underlies catalysis. *Nature* **438**, 117–121 (2005).

41. H. Meirovitch and T. F. Hendrickson. The backbone entropy of loops as a measure of their flexibility. Application to a Ras protein simulated by molecular dynamics. *Proteins* **29**, 127–140 (1997).

17

Calculation of the Entropy and the Free Energy by Thermodynamic Integration

We have already pointed out (Section 10.2) the inherent difficulty in the *direct* calculation of the *absolute* values of the entropy, *S*, and the free energy, *F*, with dynamical methods, such as the Metropolis Monte Carlo (MC) and molecular dynamics (MD) (In our discussions, we refer mostly to the Helmholtz free energy, *F*, but the main conclusions apply also to the Gibbs free energy, *G*, and other types of free energies.). While the calculation of *S* and *F* (and differences in these quantities) is fundamental in most of the exact sciences, it has a special importance in structural biology. *S* is a measure of order, where changes in the *S* of water lead to the hydrophobic interaction—an essential driving force in protein folding. *F* constitutes the criterion of stability, which defines the binding affinities of protein-protein and protein-ligand interactions; *F* also quantifies many other important processes, such as enzymatic reactions, electron transfer, ion transport through membranes, and the solvation of small molecules. Thus, the calculation of *F* is essential for studying the structure and function of peptides, proteins, nucleic acids, and other biological macromolecules.

Since *S, F* and the energy, *E*, are extensive variables, that is, increase with the number of particles, *N*, for bulk systems (e.g., the Ising model or a system of Lennard-Jones particles), one is interested, in general, in their values per particles (e.g., E/N); therefore, the corresponding fluctuations *decrease* as $\sim N^{-1/2}$, the larger is *N*, the higher is the accuracy (notice that *approximate* free energy has a non-zero fluctuation; see Section 5.1). On the other hand, in protein systems, one is typically interested in the contributions to the total *F* and *S* (not per *N*) of small sub-structures, such as a side-chain, a loop, or a ligand. Since these contributions are typically small (only of several kilocalories per mole [kcal/mol]), their calculation with adequate accuracy is, in principle, feasible. However, such estimations become more difficult as the partial systems of interest grow in size, leading to increased fluctuations ($\sim N^{1/2}$).

While it is difficult to extract the absolute entropy from a single MC run, *differences* in entropy and free energy between two states can be obtained straightforwardly by thermodynamic integration (TI) over thermodynamic quantities, such as the temperature or the energy. Furthermore, if the *absolute* value of *S* (*F*) of a reference state is known, one can also obtain by TI the *absolute S* (*F*) of the state of interest. However, the TI process might be lengthy due to the need to carry out many different simulations for evaluating the intermediate integrands and problems arise if a phase transition occurs along the integration path.

Also, in structural biology, there is an interest to determine the relative stability of microstates of a peptide, for example, by calculating the difference in their free energies. Thus, if two such microstates have significant structural variance (e.g., an α-helical and a β-hairpin states of a peptide), defining an integration path from alpha to β based on thermodynamic quantities (e.g., energy) is impractical and some geometrical constraints should be added to the process. In such cases, methods for calculating the *absolute* free energies, F_α and F_β, are desirable, as they lead directly to $\Delta F = F_\alpha - F_\beta$, and the integration process can be avoided; such methods will be discussed in the next two chapters. Still, for "normal" cases, TI is a robust and very general technique due the fact that it relies on thermodynamic parameters (e.g., *T* or *E*), while specific structural parameters (such as dihedral angles) are not involved. In this chapter, we discuss several aspects of TI with some detailed examples.

17.1 "Calorimetric" Thermodynamic Integration

In classical thermodynamics, entropy and free energy *differences* (and as we have pointed out above, in some cases their absolute values) can be obtained by TI over *thermodynamic variables*, such as the temperature. Thus, the entropy *difference*, ΔS, for a system at temperatures T_1 to T_2 is:

$$\Delta S = S(T_2) - S(T_1) = \int_{T_1}^{T_2} \frac{\partial S}{\partial T} dT. \tag{17.1}$$

Expressing $S = (F - <E>)/T$ [where $F = -k_B T \ln Z$ and $Z = \Sigma \exp(-E_i/k_B T)$] and taking the derivative of S with respect to T leads to $T^{-1}\partial <E>/\partial T = C_V/T$, where C_V is the heat capacity; one obtains,

$$\Delta S = \int_{T_1}^{T_2} \frac{C_V}{T} dT \tag{17.2}$$

Equation (17.2) enables one to calculate the *absolute* entropy at T_2 if the absolute entropy at T_1 is known. Note that for a *classical* system, $T_1 = 0$ cannot be chosen because C_V is finite at this temperature and the integral is thus undefined. In a computer simulation, Equation (17.2) leads to ΔS since the calculation of C_V is straightforward.

The difference in $-\ln Z$ between two temperatures (at constant N and volume, V) can be obtained by integrating $-\partial(\ln Z)/\partial\beta = <E>$ ($\beta = 1/k_B T$) with respect to β. Thus:

$$k_B\{-\ln Z(T_2) - [-\ln Z(T_1)]\} = -\int_{T_1}^{T_2} \frac{E}{T^2} dT \tag{17.3}$$

Another integration between two densities ρ_1 and ρ_2 ($\rho = N/V$) at constant T leads to the corresponding difference in $\ln[Z(\rho)]$,

$$\frac{k_B T}{N}\left[-\ln Z(\rho_2) + \ln Z(\rho_1)\right] = \int_{\rho_1}^{\rho_2} \frac{p}{\rho^2} d\rho \tag{17.4}$$

where p is the pressure. Other equations based on thermodynamic integration are also known; for example, per definition, $\Delta S = \int dE/T$. Notice that while the integrations in Equations (17.2) and (17.3) are carried out over extensive variables, in simulations, the integration of Equation (17.3) is the more efficient one, because the energy converges faster than the heat capacity. Notice also that Equations (17.3) and (17.4) calculate the partition function and thus are not "pure" classical thermodynamics entities, but are related to the wider field of statistical mechanics. As discussed below, statistical mechanics opens the way for a rich arena of TI procedures.

Indeed, differences in free energy can be obtained not only due to changes in thermodynamic variables, but also due to changes in parameters of the Hamiltonian, such as the masses or the Lennard-Jones parameters, σ and ε. This idea, initiated by Kirkwood in 1933, was generalized in 1954 by Zwanzig; for didactic reasons, we start with the formula of Zwanzig.

17.2 The Free Energy Perturbation Formula

Assume a system of N particles with two different potential energies (Hamiltonians), $E_1(\mathbf{x})$ and $E_2(\mathbf{x})$, where, for simplicity, we replace $\mathbf{x}^N$ by $\mathbf{x}$; we wish to calculate the difference, ΔF, between the corresponding free energies F_1 and F_2,

$$\Delta F = \Delta F_{1,2} = F_2 - F_1 = -k_B T \ln \frac{Z_2}{Z_1} = -k_B T \ln \frac{\int\limits_{\Omega} \exp[-E_2(\mathbf{x})/k_B T]d\mathbf{x}}{\int\limits_{\Omega} \exp[-E_1(\mathbf{x})/k_B T]d\mathbf{x}} \tag{17.5}$$

Multiplying the integrand in the numerator by $\exp[-E_1(\mathbf{x})/k_B T]\exp[+E_1(\mathbf{x})/k_B T]$ $(=1)$ leaves it unchanged. Denoting $\Delta E(\mathbf{x}) = E_2(\mathbf{x}) - E_1(\mathbf{x})$ leads to:

$$\Delta F = -k_B T \ln \frac{\int\limits_{\Omega} \exp[-E_1(\mathbf{x})/k_B T]\exp[-\Delta E(\mathbf{x})/k_B T]d\mathbf{x}}{\int\limits_{\Omega} \exp[-E_1(\mathbf{x})/k_B T]d\mathbf{x}} =$$

$$= -k_B T \ln \int\limits_{\Omega} P_1^B(\mathbf{x})\exp[-\Delta E(\mathbf{x})/k_B T]d\mathbf{x} \tag{17.6}$$

or

$$\Delta F = -k_B T \ln < \exp[-\Delta E(\mathbf{x})/k_B T] >_1 \tag{17.7}$$

Thus, the integral due to Zwanzig [1] is a statistical average of the random variable $\exp[-\Delta E(\mathbf{x})/k_B T]$ with a Boltzmann probability density, $P_1^B(\mathbf{x})$ [based on the potential $E_1(\mathbf{x})$]. Estimating this average by MC or MD will lead to sampled configurations with energies distributed around the typical energy, $E_1^*(T)$ (Section 5.2), where the integral can be approximated by the product, $v(E_1^*)P^B(E_1^*)\exp - [E_2(E_1^*) - E_1^*]$; $E_2(E_1^*)$ is the value of the potential E_2, in the conformational region defined by E_1^*. However, if the difference, $E_2(E_1^*) - E_1^*$ is large, its exponent will control the above product, leading it to zero. The conformational region mostly contributing to the integral, which is shifted from the E_1^* region, will never be reached in a practical simulation based on $P^B(E_1)$; thus, the estimated ΔF will be biased. Another measure of this bias is the fluctuation in $\Delta E(\mathbf{x})$, which grows as $\Delta E(\mathbf{x})$ increases ($\sim N^{1/2}$), leading to a huge fluctuation in the exponential. This problem limits the applicability of the method to relatively small systems (see Figure 17.1).

To obtain a reliable estimation of ΔF, the potential energies, $E_1(\mathbf{x})$ and $E_2(\mathbf{x})$, should be close enough, a condition that generally is not satisfied. A standard remedy to this problem is obtained by defining several intermediate potentials, $E_i(\mathbf{x})$, leading from $E_1(\mathbf{x})$ to $E_2(\mathbf{x})$; thus, one calculates $\Delta F_{i,\,i+1}$ (between neighbor energies i and $i + 1$), where $\Delta F_{1,2}$ is the sum of the intermediate free energies. A continuous transition between E_1 and E_2 is defined typically by a parameter λ, $0 \le \lambda \le 1$, which leads to a hybrid potential energy, $E(\mathbf{x}, \lambda)$,

$$E(\mathbf{x},\lambda) = (1-\lambda)E_1(\mathbf{x}) + \lambda E_2(\mathbf{x}). \tag{17.8}$$

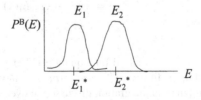

FIGURE 17.1 Probability distributions concentrated around the *distant* energies E_1^* and E_2^*. In this case, the factor $\exp - (E_2 - E_1)/k_B T$ leads the integral [Equation (17.7)] to zero because the most contributing conformational region to the integral cannot be reached in a practical simulation based on $P^B(E_1)$. Therefore, the results for ΔF [Equation (17.7)] are biased and the method is limited to small systems.

Thus, $E(\mathbf{x}, \lambda = 0) = E_1(\mathbf{x})$ and $E(\mathbf{x}, \lambda = 1) = E_2(\mathbf{x})$, as required. The range $[0,1]$ can be sub-divided into smaller ranges $(\Delta\lambda_1, \Delta\lambda_2, \ldots)$ called *"windows"*. One can carry out several simulations based on $\Delta\lambda_i$, where the variances are small and the estimation of the integrals is reasonable. The results for $\Delta F(\Delta\lambda_i)$ are then accumulated, leading to $\Delta F = \Sigma_i F(\Delta\lambda_i)$. This method, which is highly used in structural biology, is called the *"free energy perturbation"* (FEP).

In the next short paragraph, expansion of Equations (17.7) and (17.8) in terms of small $\Delta\lambda$ leads to an important result for Jarzynski's theory treated in Section 21.6; therefore, this paragraph can be skipped in first reading.

One can view the parameter, λ, as an external force acting on the energy function (Hamiltonian); thus, within this picture, it is of interest to find the relation between λ and the work, W, done *on the system* [$W \equiv W_{\text{system}}$, see Equation (2.68)]. We know from thermodynamics, [Equation (2.70)] that for a system in equilibrium at constant T, $\Delta F = \Delta W$, where ΔW is the *minimal* amount of work done on a system for imposing a change, ΔF. Thus, for a *small window*, $\Delta\lambda$, and based on Equation (17.7), ΔF can be expanded,

$$\Delta F = \Delta W = -k_B T \ln < \exp[-E(\mathbf{x}, \lambda + \Delta\lambda) - E(\mathbf{x}, \lambda)]/k_B T >_\lambda =$$

$$= -k_B T \ln < \exp[-\Delta E(\mathbf{x}, \Delta\lambda)/k_B T] >_\lambda \approx \tag{17.9}$$

$$\approx -k_B T \ln[1 - < \Delta E(\mathbf{x}, \Delta\lambda) >_\lambda / k_B T] \approx < \Delta E(\mathbf{x}, \Delta\lambda) >_\lambda$$

Thus, since $\Delta W \approx < \Delta E(\mathbf{x}, \Delta\lambda) >_\lambda$, one can interpret the *instantaneous* function $E(\mathbf{x}, \lambda + \Delta\lambda) - E(\mathbf{x}, \lambda)$ (at some $\mathbf{x}$) as a (non-equilibrated) work function satisfying:

$$E(\mathbf{x}, \lambda + \Delta\lambda) - E(\mathbf{x}, \lambda) = \Delta W(\mathbf{x}, \lambda) \geq \Delta W \tag{17.10}$$

As pointed out above, this interpretation of the work function will become useful in Jarzynski's theory.

Finally, Zwanzig's perturbation scheme can also be applied to changes in the temperature, where the Hamiltonian is kept constant. Thus, for T_1 and T_2, one obtains [see Equation (17.6)],

$$\frac{F(T_2)}{T_2} - \frac{F(T_1)}{T_1} = -k_B \ln \frac{Z_2}{Z_1} = -k_B \ln < \exp- \left[\frac{E(\mathbf{x})(1/T_2 - 1/T_1)}{k_B} \right] >_1 \tag{17.11}$$

17.3 The Thermodynamic Integration Formula of Kirkwood

As shown in Equation (17.9), for very small $\Delta\lambda_i$, one obtains $\Delta F \approx < \Delta E(\mathbf{x}, \Delta\lambda_i) >$, where the summation of this expression over windows, i, can be replaced by the integral:

$$\Delta F = \int_{\lambda=0}^{\lambda=1} \frac{\partial < E(\mathbf{x}, \lambda) >_\lambda}{\partial\lambda} d\lambda = \int_{\lambda=0}^{\lambda=1} < E_2(\mathbf{x}) - E_1(\mathbf{x}) >_\lambda d\lambda \tag{17.12}$$

The right-hand side of Equation (17.12) is based on Equation (17.8). This formula was suggested by Kirkwood in 1933 [2]; it is called in the literature *"thermodynamic integration,"* and it is denoted by TI like the other versions of thermodynamic integration discussed above; however, in this book, applying TI to protein systems means using Kirkwood's equation (17.12). TI and FEP have been tested and used extensively in various disciplines; for proteins, they constitute the basis for "thermodynamic cycles" discussed later in Section 17.6.

17.4 Applications

The TI approach is very robust and has been used extensively during the years. Below, we describe some of the early pioneering work and present several detailed examples.

In the area of simple fluids, Equation (17.4) has been used extensively for calculating entropy and free energy differences. In 1967, McDonald and Singer were the first to calculate by MC such differences for a Lennard-Jones fluid of 32 particles [3]. Hansen and Verlet were the first to use Equation (17.4) for calculating the absolute free energy of 864 Lennard-Jones particles from MC simulations [4]. They evaluated F as a function of the density for three isotherms, and their integration scheme has been used constantly for fluids since then (e.g., see [5–8]).

Binder used both Equations (17.2) and (17.3) for calculating absolute entropies in an antifromagnet Ising spin system in the presence of an external magnetic field [9]. His MC simulations started at $T = \infty$ or $T = 0$, where the entropy is known. The perturbation formula defined by Equation (17.11) [which corresponds to Equation (17.3)] was first used by McDonald and Singer [3,10] and later by Torrie and Valeau in their umbrella sampling papers [11]. Equation (17.11) constitutes the basis of the "Monte Carlo recursion method" of Li and Scheraga, which breaks the simulation into small temperature windows. These authors calculated the free energy of 32 Lennard-Jones particles [12] and 64 water molecules [13] by starting MC simulations at $T = \infty$, where the free energy is known [$\sim\ln(V^N)$]. While the results and the efficiency of these studies are comparable to those obtained by other techniques, they demonstrate the limitation of the method to handle large systems, which requires increasing the number of windows. More recent studies of liquid argon and water are described in Chapter 19.

17.4.1 Absolute Entropy of a SAW Integrated from an Ideal Chain Reference State

Since the absolute entropy of an ideal chain is known [Equation (8.2)], one can use it as a reference state for calculating the absolute S of a self-avoiding walk (SAW) on a lattice by TI [14]. For this purpose, a unitless energy function, E, is defined where:

$$E = \sum_j \varphi_j. \tag{17.13}$$

φ_j is the "overlap value" at lattice site j, and the summation is carried out over all sites. The overlap value is defined as follows. For a lattice site that is occupied by only a single monomer (or is unoccupied), $\varphi_j = 0$. For a doubly occupied site (that is, a single overlap), $\varphi_j = 1$; a triply occupied site contributes, $\varphi_j = 2$, and so on. The value of E is thus always an integer. For a SAW, E must be zero.

The above defined energy function is used to describe a *general* chain, which can exist at any arbitrary finite temperature. The partition function for the general chain ensemble is given by:

$$Z = \sum_{id} \exp[-E_i / T] \tag{17.14}$$

where the sum is carried out over all *ideal chain* configurations, i, and where we have introduced a unitless temperature, T. Note that at high (infinite) T, the Boltzmann factor is unity and the partition function approaches that of the ideal chain reference state (that is, $Z_{id} = 4 \times 3^{N-1}$, where immediate reversal is forbidden). At low T ($T = 0$), only zero energy configurations will contribute to the summation and the partition function becomes that of the SAW, Z_{SAW}.

The difference, $\ln Z_{SAW} - \ln Z_{id}$, can be evaluated by integration over T, or over $1/T$, using the derivative relations,

$$\frac{d\ln Z}{dT} = \left(\frac{1}{Z}\right) \sum_{id} \frac{E_i}{T^2} \exp[-E_i / T] = \left\langle \frac{E}{T^2} \right\rangle \tag{17.15}$$

and

$$\frac{d\ln Z}{d(1/T)} = \left(\frac{1}{Z}\right)\sum_{id} -E_i \exp[-E_i/T] = -\langle E \rangle. \tag{17.16}$$

The corresponding integrals are respectively:

$$\ln\left[\frac{Z_{SAW}}{Z_{id}}\right] = \int_{\infty}^{0} \frac{d\ln Z}{dT} dT \text{ and } \ln\left[\frac{Z_{SAW}}{Z_{id}}\right] = \int_{0}^{\infty} \frac{d\ln Z}{d(1/T)} d(1/T). \tag{17.17}$$

Therefore, this is basically a *calorimetric* thermodynamic integration procedure (within the framework of statistical mechanics), where a single potential energy is defined while T is being changed. To avoid integrating in the ∞ limit in [14], a two-stage procedure has been applied based on both integrations:

$$\ln\left[\frac{Z_{SAW}}{Z_{id}}\right] = \int_{0}^{1/T^*} \frac{d\ln Z}{d(1/T)} d(1/T) + \int_{T^*}^{0} \frac{d\ln Z}{dT} dT, \tag{17.18}$$

where T^* is an intermediate temperature. The left hand term thus quantifies the change in $\ln Z$ for going from an ideal chain to the general chain at T^*, and the right hand term is the change from this point to the SAW. In the implementation of [14], $1/T^* = 1.32$ ($T^* = 0.75757575$), and 100 evenly spaced temperatures were simulated by MC for each integration and a total of nine such integrations have been performed. The integrated quantity in each integration step is the average energy, $<E>$ and $<E>/T^2$.

SAWs of up to 599 steps on the square lattice were studied. A simple trapezium integration was adequate. The efficiency of the method could be improved by optimizing T^* or by employing less simple-minded quadrature techniques. The efficiency of the process can also be affected considerably by the MC procedure employed. On the square lattice, "crankshaft" moves are in most cases rejected due to the strong excluded volume interactions, while corner moves (single monomer flips) have a somewhat higher acceptance rate (see Section 10.5). Therefore, the MC procedure applied is based on 50% corner moves (that provide local conformational changes) and 50% "pivot" moves that induce global changes (Section 15.1). For $N = 29$, 6×10^7 MC moves were carried out at each temperature, where run lengths of 10^8, 10^8, 6×10^7, 5×10^7, 3.2×10^7, and 2.4×10^7 steps were used for $N = 49, 99, 149, 249, 399$, and 599, respectively.

In Table 17.1, TI results for the absolute entropy *per bond* of SAWs of $N = 29$–599 are presented and compared to those obtained by a series expansion formula that is based on extrapolating exact enumeration data for relatively short chains [15,16]. The table reveals that the results for S_{TI} and S_{series} are equal within the error bars for all N, with comparable errors for $N = 49, 99$, and 149. However, for $N = 29$, the error in S_{series} is significantly larger than that of S_{TI} and for $N > 149$, error(S_{TI})–error(S_{series}) increases constantly with N.

TABLE 17.1

Results for the Entropy Per Bond (S/N) for SAWs on a Square Lattice

N	29	49	99	149	249	399	599
S_{TI}/N	1.016145 (3)	1.000897 (3)	0.987727 (3)	0.982742 (3)	0.978358 (4)	0.975655 (8)	0.97404 (1)
S_{series}/N	1.01615 (1)	1.000899 (5)	0.987730 (3)	0.982740 (2)	0.978360 (1)	0.975652 (1)	0.974025 (1)

Source: Reprinted with permission from White, R.P. and Meirovitch, H., *J. Chem. Phys.*, 123, 214908, 2005. Copyright 2005 by the American Institute of Physics.

Note: S_{TI} and S_{series} are results obtained by thermodynamic integration and series expansion, respectively. The errors appear in parenthesis, for example: 0.97404 (1) means 0.97404 ±0.00001.

17.4.2 Harmonic Reference State of a Peptide

In structural biology, the potential energy, $E(\mathbf{x}_1, \mathbf{x}_2, \ldots, \mathbf{x}_N)$, of a peptide of N atoms can be minimized with respect to the coordinates, where the minimized energy is obtained at the set of coordinates, $\{\mathbf{x}_i^0\}$. One can then define a harmonic potential, E_H, based on independent (Einstein) oscillators expanded to second order around $\{\mathbf{x}_i^0\}$,

$$E_H = \sum k_i (\mathbf{x}_i - \mathbf{x}_i^0)^2 \tag{17.19}$$

where the k_i are spring constants and x_i is the instantaneous coordinates of atom i; this potential leads to the (calculable) harmonic entropy and free energy which constitute convenient (well used) reference values (see Equation (4.41) and Chapter 18). The related early work of Stoessel and Nowak [17] should be pointed out. To obtain the absolute S and F for the potential (force field), E (and T), they defined the hybrid Hamiltonian, H, which depends on E_H and E, $H(\lambda) = (1 - \lambda)E + \lambda E_H$ $(0 \le \lambda \le 1)$. Thus, for $\lambda = 0$, the peptide's atoms interact by the force field E, where E_H is not effective; in the other extreme, the atoms are held together only by the harmonic potential. For decaglycine in an α-helical microstate, the estimated error in the free energy, F (~2 kcal/mole), is relatively high. More recently [18], a similar idea has been implemented somewhat differently, where $H(\lambda) = E + (1/2)\lambda E_H$ and the free energy of the final state $(\lambda = 1)$ is calculated by a normal mode analysis. For the pentapeptide Met-enkephalin, the maximum error in the absolute F of a microstate is again relatively high, ± 1.5 kcal/mol. For comparison, using "*the hypothetical scanning MC (HSMC) method*" and its MD version (HSMD), discussed in Chapter 19, errors of ~0.2 kcal/mol were obtained for ΔS (ΔF) between microstates of decaglycine [19] and the decapeptide, $NH_2(Val)_2(Gly)_6(Val)_2CONH_2$ [20].

In summary: Calculating the absolute S or F by TI or FEP works very well as long as the target state is not structurally restricted. This indeed happens for argon, water, and the detailed TI process described for SAWs; TI has also been found to be very efficient for calculating the chemical potential (see Section 22.9). Still, as the system size grows, the increase in fluctuations can make the integration process quite expensive in terms of computer time, due to the need to increase the sample size and the number of λ values studied. Also, if the target state is defined *structurally*, such as a microstate of a peptide (say, an α-helix), the energy alone as an integration parameter does not provide an effective "guidance" toward this state, which is reflected by the mediocre TI results presented in the previous paragraph [17,18] (this subject is discussed further in Section 17.5.2 below). As pointed out above, a much better accuracy has been obtained by methods that calculate separately the absolute entropies S_m and S_n, where ΔS_{mn} is obtained directly from the difference, $\Delta S_{mn} = S_m - S_n$, and the integration from m to n is avoided (see Chapter 19).

17.5 Thermodynamic Cycles

Understanding the mechanisms of basic biological processes requires in many cases the knowledge of the binding free energy of a ligand (or ligands) to the active site of a protein. In particular, in rational drug design, one seeks to calculate the *relative* free energy of binding, that is, the difference between the free energies of binding ΔF_1 and ΔF_2 of two ligands L1 and L2 to the same protein, P. Experimentally, ΔF_1 (ΔF_2) is the free energy required to transfer L1 (L2) from the solvent (where the ligand is separated from the protein) to the active site; however, this physical transformation is difficult to carry out computationally. Instead, one can recruit either FEP or TI for carrying out *non-physical* perturbations. Thus, using FEP (TI) one calculates ΔF_p and ΔF_s—the free energies for converting L1 into L2 in the protein and in the solvent environment, respectively. Because the free energy is a state function, its accumulation along a closed thermodynamic cycle is zero; thus, ΔF_p and ΔF_s lead to $\Delta F_1 - \Delta F_2$ (see Figure 17.2).

$$\Delta F_1 - \Delta F_2 - (\Delta F_s - \Delta F_p) = 0 \quad \rightarrow \quad \Delta\Delta F = \Delta F_1 - \Delta F_2 = \Delta F_p - \Delta F_s \tag{17.20}$$

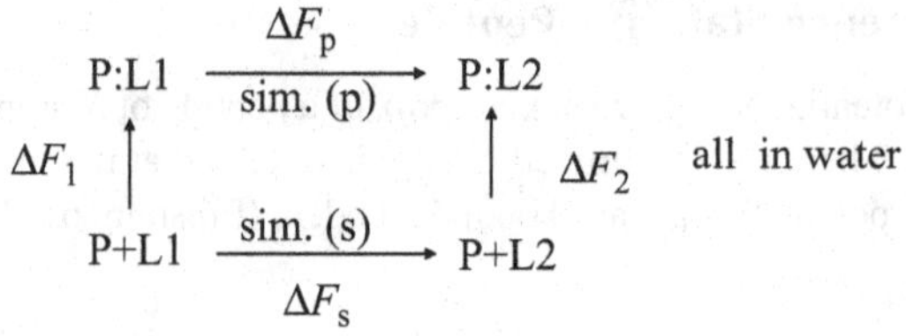

$$P:L1 \xrightarrow[\text{sim. (p)}]{\Delta F_p} P:L2$$

$$\Delta F_1 \uparrow \qquad\qquad \uparrow \Delta F_2 \qquad \text{all in water}$$

$$P+L1 \xrightarrow[\Delta F_s]{\text{sim. (s)}} P+L2$$

FIGURE 17.2 A thermodynamic cycle for the binding of two ligands L1 and L2 to a protein, P. In the experiment, the ligands are transferred from the solvent to the active site, where one measures the difference $\Delta\Delta F = \Delta F_{L1 \to P:L1} - \Delta F_{L2 \to P:L2}$. In simulations, the nonphysical transformation L1 $\to$ L2 is carried out in the protein and in solution and the corresponding free energy differences, $\Delta F_p = \Delta F_{P:L1 \to P:L2}$ and $\Delta F_s = \Delta F_{L1 \to L2}$ are calculated. Because the free energy of the entire cycle is zero, the desired $\Delta\Delta F$ is obtained in terms of the non-physical free energy differences $\Delta\Delta F = \Delta F_p - \Delta F_s$.

Before discussing more aspects of this procedure, it would be of interest to follow the historical development of the thermodynamic cycle idea in proteins. Thus, the non-physical transformations were introduced by Warshel in 1981, who studied ionization in acidic residues in proteins (pKa calculations) [21]. Although the cycle included non-physical transformations, they were not carried out by perturbation techniques. A year later [22], Warshel used the perturbation method together with umbrella sampling (see Chapter 20) to study the solvation free energy contribution to an electron transfer coordinate using two spheres for a donor and acceptor in water. The perturbation, however, was performed along a physical path. In 1984 [23], Tembe and McCammon combined all these ingredients into a method that has made a large impact on the computational chemistry community. Treating a simplified model, they suggested using thermodynamic cycles for calculating the relative binding of two ligands to the active site of an enzyme by converting one ligand to the other by non-physical perturbations, as described above.

We return now to the thermodynamic cycle [Equation (17.20)]. Typically, this process involves annihilation and creation of atoms, which can be treated by a linear hybrid potential $E(\mathbf{x},\Omega,\lambda)$ $(0 \leq \lambda \leq 1)$ [Equation (17.8)]:

$$E(\mathbf{x},\Omega,\lambda) = (1-\lambda)E_1(\mathbf{x},\Omega) + \lambda E_2(\mathbf{x},\Omega), \tag{17.21}$$

where $\mathbf{x}$ denotes the Cartesian coordinates of the *extended* ligand based on L1 and L2. Thus, if L2 has extra atoms, their coordinates are included in $\mathbf{x}$. Ω denotes the coordinates of the environment—solvent and protein. For a ligand in *solvent*, $E(\mathbf{x},\Omega,\lambda)$ includes the ligand-ligand and ligand-solvent interactions (but not the solvent-solvent ones). Similarly, for a ligand bound to a *protein*, the ligand-ligand, ligand-protein, and ligand-solvent interactions are included in $E(\mathbf{x},\Omega,\lambda)$, while protein-protein, protein-solvent, and solvent-solvent interactions are not included.

The fact that only the ligand-ligand and ligand-environmental interactions are considered is very important because typical ligands are relatively small, and during a calculation, only several atoms of a ligand are created, annihilated, or their potential is changed. Therefore, only a very small part of the system's energy is considered in the calculation of $\Delta\Delta F$, and the corresponding energy fluctuation (variance) involved is relatively small; thus, $\Delta\Delta F$ can be estimated reliably, even for a *very large* protein-ligand-solvent system. In this context, it should be emphasized that it would not be feasible to estimate $\Delta\Delta F$ [which is typically of the order of only several kcal/mol] by calculating the differences in the *total* energy and entropy of the two entire systems due to their large size, which induces large fluctuations ($N^{1/2}$) in these quantities. Notice also that in most applications, $\Delta\Delta G$ rather than $\Delta\Delta F$ is calculated. Using FEP, the integration is carried out by Equation (17.6), while TI is carried out by Equation (17.12),

$$\Delta F = \int_{\lambda=0}^{\lambda=1} < E_2(\mathbf{x},\Omega) - E_1(\mathbf{x},\Omega) >_\lambda d\lambda - \int_{\lambda=0}^{\lambda=1} < \Delta E >_\lambda d\lambda \tag{17.22}$$

where, as before, the interval $0 \leq \lambda \leq 1$ is divided into m sub-intervals of size $\Delta\lambda = 1/m$ defined by $\lambda_0 = 0$, $\lambda_1 = \Delta\lambda,\ldots,\lambda_{\bar{m}} = 1$. The meaning of $<\Delta E>_\lambda$ is that ΔE is averaged by a simulation based on $E(\mathbf{x},\Omega,\lambda)$; this, however, might lead to singularities in ΔF at the end points, $\lambda = 0$ ($\lambda = 1$), as discussed below.

Thus, one has to distinguish between the simulation and the integration process [Equation (17.22)]. Thus, assume that L2 consists of more atoms than L1, and thus E_1 does not operate on these extra atoms, which sometimes are called ghost atoms, since with respect to E_1, they can move freely over the whole system. Now, the simulation is started with a very small $\Delta\lambda$, thus, $\Delta\lambda E_2(\mathbf{x},\Omega)$ is very small, and the interaction of the extra atoms with the other system's atoms will be very weak. In other words, these atoms will be highly flexible, overlapping (at least partially) other atoms of the protein and the solvent. However, under this condition, the integration [Equation (17.22)] requires averaging $\Delta E = <E_2(\mathbf{x}, \Omega) - E_1(\mathbf{x}, \Omega)>_\lambda$, where E_2 *is not* multiplied by $\Delta\lambda$!; therefore, E_2 will be very high (due to the above bumping of atoms), leading to the explosion of the integration; a similar situation can occur at the other end ($\lambda = 1$) (Figure 17.3).

To demonstrate the problem further, assume a simple example of a Lennard-Jones potential, where we want to integrate the free energy of argon from an ideal gas ($\lambda = 0$) into a full interaction ($\lambda = 1$) (see Section 19.5). One might define the following *simple* scaling of the interaction:

$$\varphi\left(r_{ij},\lambda\right) = \lambda 4\varepsilon \left[\frac{\sigma^{12}}{r_{ij}^{12}} - \frac{\sigma^6}{r_{ij}^6} \right]. \tag{17.23}$$

Denoting the expression in the brackets by [ ], the derivative of φ with respect to λ is 4ε[ ]. For $\Delta\lambda$ (that is, treating the $\lambda = 0$ end), the energy in the Boltzmann factor is $4\varepsilon\Delta\lambda$[ ], and the simulation thus allows (for small $\Delta\lambda$) the atoms to approach each other closely; however, these configurations would lead to a very high averaged derivative, 4ε[ ], which is the energy without the $\Delta\lambda$ factor.

One way to alleviate this problem is to use a non-linear scaling based on λ^k, $k > 1$, or a more complicated scaling which includes an additional parameter δ, as suggested by McCammon's group [24],

$$\varphi\left(r_{ij},\lambda\right) = \lambda 4\varepsilon \left[\frac{\sigma^{12}}{\left(r_{ij}^2 + \delta(1-\lambda)\right)^6} - \frac{\sigma^6}{\left(r_{ij}^2 + \delta(1-\lambda)\right)^3} \right] \tag{17.24}$$

As can be seen, for $\lambda = 0$, the interaction as well as the derivative $= [\ ] + \lambda\{6\delta(\ )^{-7} - 3\delta(\ )^{-4}\}$ will not explode due to δ ([ ] and ( ) stand for the brackets and large parenthesis in Equation (17.24)). The other end ($\lambda = 1$) is not problematic. In this way, one can define other soft-core potentials. Equation (17.24) was used in our calculation of the absolute S and F of argon and water from an ideal gas reference (see Chapter 19).

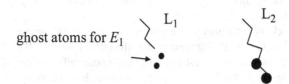

FIGURE 17.3 Ligand L2 has two more atoms (black spheres) than ligand L1, meaning that they are not seen by the potential energy, E_1 (ghost atoms). For a small $\Delta\lambda$, $\Delta\lambda E_2(\mathbf{x}, \Omega)$ will be small, enabling the black atoms to partially overlap other atoms of the system during simulation; however, this will lead to very large $E_2(\mathbf{x}, \Omega)$ and to the explosion of the integration defined in Equation (17.22).

17.5.1 Other Cycles

In the above example, the idea of a thermodynamic cycle has been used for comparing the binding free energy of two ligands to the same protein. However, this idea can be used in other ways. For example, to compare the stabilities of a native and a mutated protein, $F_{n,\text{folded}} - F_{n,\text{unfolded}}$ should be compared to $F_{m,\text{folded}} - F_{m,\text{unfolded}}$, where "$n$" stands for "native" and "m" for "mutated." Instead, one carries out two non-physical mutations in the folded and unfolded proteins. The unfolded protein is commonly approximated by a short chain of three (or more) residues consisting of the mutated residue, which is flanked by its neighbor residues in the protein. In another application of thermodynamic cycles, one seeks to compare the free energy of solvation of small molecules A and B. In this case, two non-physical conversions of A $\rightarrow$ B are carried out, in water and in vacuum,

$$\Delta\Delta F = \Delta F_{\text{solvatio,A}} - \Delta F_{\text{solvatio,B}} = \Delta F_{\text{A}\rightarrow\text{B,water}} - \Delta F_{\text{A}\rightarrow\text{B,vacuum}} \tag{17.25}$$

One can also calculate the absolute free energy of binding, a topic that will be discussed in detail in Chapter 23.

17.5.2 Problems of TI and FEP Applied to Proteins

A basic weakness of TI and FEP, which has already been pointed out, is that they are energy-dependent, and thus can handle a system with structural changes very poorly. A typical example is calculating the free energy difference, ΔF_{mn}, between microstates m and n with significant structural variance (e.g., a helix and a hairpin of a peptide); in this case, the integration $\int_m^n dF$ becomes difficult (due to the complex path), and for a large molecule unfeasible. Thus, one has to impose suitable "reaction coordinates" or other geometrical constraints, which impose a bias that should be compensated; typically, the integration will consist of a large number of windows (λ).

A relatively old study that demonstrates this problem is the work of Wallqvist and Covell [25]. They studied, by MD, the free energy profiles of the conformational change of dodecane ($CH_3(CH_2)_{10}CH_3$) from an extended to a hairpin state in vacuum and in a box of 717 water molecules. The reaction coordinate used was the end-to-end distance, R and the free energy differences were calculated by FEP. Even this relatively simple system (without side chains) required defining 500 small windows, each being 0.02 Å. Clearly, a hairpin to helix transformation would be much more complex, and, in general, for models of peptides and proteins, defining an adequate reference state with a suitable integration path to the state of interest might not be straightforward.

This difficulty is also reflected in the calculation of differences in the free energy of binding described above (see Figure 17.2). Thus, whereas the mutation of L1 to L2 in water is relatively well controlled by TI, the simulation carried out in the protein environment might not converge for very long times due to conformational changes occurring constantly in the entire protein (e.g., "jumps" of side chains among their rotamers, etc.); in other words, the microstate of P:L2 (and to some extent also of P:L1) keeps changing as the simulation time increases. Also, sometimes the TI of a mutation does not lead to the required size and shape of the active site of P:L2, or to the correct number of water molecules in it; the position of L2 in this active site (after integration) can also deviate significantly from the experimental position [26,27]. Thus, the results would keep changing as computer time goes on, and it is not clear when to stop the simulation.

A common way to check the reliability of the results is by increasing computer time. Another way is to carry out, in addition to $\lambda = 0 \rightarrow 1$, also the reverse transformation $\lambda = 1 \rightarrow 0$ (that is, L2 $\rightarrow$ L1) to verify consistency. A comparison of simulation results to the experiment suffer not only from convergence problems, but also from deficiencies in the force field used. For example, a thermodynamic cycle was defined in which oxidized Asn47 of azurin is mutated to oxidized Leu and both amino acid side chains are reduced; in this case, $\Delta\Delta G$ can be calculated by two different transformations (say, a and b). The TI study of Mark and van Gunsteren [28] has found (in kcal/mol) $\Delta\Delta G(a) = -5.7$ and $\Delta\Delta G(b) = -3.6$, where the experimental value is $\Delta\Delta G = -2.6$. Another example is the conversion of Guanosine- 2'- O- monophosphate (2'GMP) to Guanosine- 3'- O- monophosphate (3'GMP) in solution and in ribonucleas T_1. Using TI, Mackerell et al. [29]

obtained for the forward transformation, ΔG(protein) = 12.69 ± 0.19, ΔG(solvent) = 6.24 ± 0.15, thus, $\Delta\Delta G$ = 6.45 ± 0.3 kcal/mol. On the other hand, for the reverse transformation, they obtained ΔG(protein) =−7.65 ± 0.19, ΔG(solvent) = −3.25 ± 0.17, and $\Delta\Delta G$ = 4.40 ± 0.4 kcal/mol; therefore, the free energy difference between the two transformations is ~2.05 kcal/mol. While these studies are relatively old, the discrepancies between the different transformations are typical. We shall return to the accuracy issue in the calculation of the free energy of binding in Chapter 23.

In summary: Thermodynamic cycles are used extensively in computational structural biology. The small fluctuations of the relatively small interacting parts, allow calculating small free energy differences in large systems (a large protein in large container of water). In fact, much more complicated cycles than those presented here have been used and various integration procedures have been tested (e.g., [30–35]). However, the results typically depend on the simulation time (and the force field used), and it is therefore difficult to assess their accuracy, as demonstrated above. We have pointed out in the beginning of this chapter that one of the strengths of TI and FEP is the fact that they operate on thermodynamic parameters, such as temperature or energy without the need to consider structural properties of the system investigated; this advantage leads to the easy implementation of TI and FEP, and to their high efficiency and wide applicability. However, for proteins and peptides, which reside in stable microstates, it is difficult for the thermodynamic parameters alone to provide the required guidance to the integration process to "travel" from a reference microstate to a target one. As has been already pointed out, better efficiency has been gained by adding suitable reaction coordinates or geometrical restraints, which, however, complicate the calculations. In the coming chapters, we shall introduce methods designed to overcome the weaknesses of FEP and TI.

REFERENCES

1. R. W. Zwanzig. High-temperature equation of state by a perturbation method. I. Nonpolar gases. *J. Chem. Phys.* **22**, 1420–1426 (1954).
2. J. G. Kirkwood. Statistical mechanics of fluid mixtures. *J. Chem. Phys.* **3**, 300–313 (1935).
3. I. R. McDonald and K. Singer. Machine calculation of thermodynamic properties of a simple fluid at supercritical temperatures. *J. Chem. Phys.* **47**, 4766–4772 (1967).
4. J.-P. Hansen and L. Verlet. Phase transitions of the Lennard-Jones system. *Phys. Rev.* **184**, 151–161 (1969).
5. J.-P. Hansen. Phase transition of the Lennard-Jones system. II. High-temperature limit. *Phys. Rev. A* **2**, 221–230 (1970).
6. D. Levesque and L. Verlet. Perturbation theory and equation of state for fluids. *Phys. Rev.* **182**, 307–316 (1969).
7. B. Smit, K. Esselink and D. Frenkel. Solid-solid and liquid-solid phase equilibria for the restricted primitive model. *Mol. Phys.* **87**, 159–166 (1996).
8. R. K. Bowles and R. J. Speedy. The vapor pressure of glassy crystals of dimers. *Mol. Phys.* **87**, 1349–1361 (1996).
9. K. Binder. Monte Carlo study of entropy for face-centered cubic Ising antiferromagnets. *Z. Phys. B* **45**, 61–69 (1981).
10. R. McDonald and K. Singer. Calculation of thermodynamic properties of liquid argon from Lennard-Jones parameters by a Monte Carlo method. *Discuss. Faraday Soc.* **43**, 40–49 (1967).
11. G. M. Torrie and J. P. Valleau. Nonphysical sampling distributions in Monte Carlo free-energy estimation: Umbrella sampling. *J. Comp. Phys.* **23**, 187–199 (1977).
12. Z. Li and H. A. Scheraga. Monte Carlo recursion evaluation of free energy. *J. Phys. Chem.* **92**, 2633–2636 (1988).
13. Z. Li and H. A. Scheraga. Computation of the free energy of liquid water by the Monte Carlo recursion method. *Chem. Phys. Lett.* **154**, 516–520 (1989). Erratum: ibid. **157**, 579 (1989).
14. R. P. White and H. Meirovitch. Calculation of the entropy of random coil polymers with the hypothetical scanning Monte Carlo method. *J. Chem. Phys.* **123**, 214908–214911 (2005).
15. A. J. Guttmann and I. G. Enting. The size and number of rings on the square lattice. *J. Phys. A* **21**, L165–L172 1988.

16. A. R. Conway, I. G. Enting, and A. J. Guttmann. Algebraic techniques for enumerating self-avoiding walks on the square lattice. *J. Phys. A* **26**, 1519–1534 (1993).

17. J. P. Stoessel and P. Novak. Absolute free energies in biomolecular systems. *Macromolecules* **23**, 1961–1965 (1990).

18. M. D. Tyka, A. R. Clarke and R. B. Sessions. An efficient path-independent method for free energy calculations. *J. Phys. Chem. B* **110**, 17212–17220 (2006).

19. S. Cheluvaraja and H. Meirovitch. Calculation of the entropy and free energy by the hypothetical scanning Monte Carlo Method: Application to peptides. *J. Chem. Phys.* **122**, 054903–054914 (2004).

20. S. Cheluvaraja and H. Meirovitch. Calculation of the entropy and free energy of peptides by molecular dynamics simulations using the hypothetical scanning molecular dynamics method. *J. Chem. Phys.* **125**, 024905–024913 (2006).

21. A. Warshel. Calculations of enzymic reactions: Calculations of pKa, proton transfer reactions, and general acid catalysis reactions in enzymes. *Biochemistry* **20**, 3167–3177 (1981).

22. A. Warshel. Dynamics of reactions in polar solvents. Semiclassical trajectory studies of electron-transfer and proton-transfer reactions. *J. Phys. Chem.* **86**, 2218–2224 (1982).

23. B. L. Tembe and J. A. McCammon. Ligand receptor interactions. *Comput. Chem.* **8**, 281–284 (1984).

24. M. Zacharias, T. P. Straatsma and J. A. McCammon. Separation-shifted scaling, a new scaling method for Lennard-Jones interactions in thermodynamic integration *J. Chem. Phys.* **100**, 9025–9031 (1994).

25. S. Miyamoto and P. A. Kollman. Absolute and relative binding free energy calculations of the interaction of biotin and its analogs with streptavidin using molecular dynamics/free energy perturbation approaches. *Proteins* **16**, 226–245 (1993).

26. S. Miyamoto and P. A. Kollman. What determines the strength of noncovalent association of ligands to proteins in aqueous solution. *Proc. Natl. Acad. Sci. USA* **90**, 8402–8406 (1993).

27. A. Wallqvist and D. G. Covell. Free-energy cost of bending n-dodecane in aqueous solution. Influence of the hydrophobic effect and solvent exposed area. *J. Phys. Chem.* **99**, 13118–13125 (1995).

28. E. Mark and W. F. van Gunsteren. Decomposition of the free energy of a system in terms of specific interactions: Implications for theoretical and experimental studies. *J. Mol. Biol.* **240**, 167–176 (1994).

29. A. D. MacKerell, M. S. Sommer and M. Karplus. pH dependence of binding reactions from free energy simulations and macroscopic continuum electrostatic calculations: Application to 2'GMP/3'GMP binding to ribonuclease T_1 and implications for catalysis. *J. Mol. Biol.* **247**, 774–807 (1995).

30. W. F. van Gunsteren. Methods for calculation of free energies and binding constants: Successes and problems. In *Computer Simulations of Biomolecular Systems: Theoretical and Experimental Applications*, W. F. van Gunsteren and P. K. Weiner (Eds.), vol. 1, 27–59 (ESCOM, Leiden, the Netherlands, 1989).

31. D. L. Beveridge and F. M. DiCapua. Free energy via molecular simulation: Applications to chemical and biomolecular systems. *Annu. Rev. Biophys. Biophys. Chem.* **18**, 431–492 (1989).

32. C. L. Brooks, M. Karplus and B. M. Pettitt. Proteins: A theoretical perspective on structure, dynamics and thermodynamics. In *Advances in Chemical Physics*, Ilya Prigogine and S. A. Rice (Eds.), vol. 71, 1–259 (John Wiley & Sons, New York, 1990).

33. T. P. Straatsma. Free energy by molecular simulations. In *Reviews of Computational Chemistry*, K. B. Lipkowitz and D. B. Boyd (Eds.), vol. 9, 81–127 (VCH, New York, 1996).

34. H. Meirovitch. Calculation of the free energy and the entropy of macromolecular systems by computer simulation. In *Reviews of Computational Chemistry*, K. B. Lipkowitz and D. B. Boyd (Eds.), vol. 12, 1–74 (VCH, New York, 1998).

35. T. P. Straatsma and J. A. McCammon. Computational alchemy. *Annu. Rev. Phys. Chem.* **43**, 407–435 (2003).

18

Direct Calculation of the Absolute Entropy and Free Energy

An important objective in structural biology is to be able to compare the stabilities of proteins, peptides, and partial systems, such as loops and side chains, in different microstates. In the previous chapter, we have introduced thermodynamic integration and free energy perturbation as robust techniques for calculating free energy differences, ΔF. The strength of these methods stems from the fact that they operate on thermodynamic parameters (e.g., T, E, C_V), while structural factors are not considered at all. However, this simplicity becomes a weakness if one seeks to calculate $\Delta F_{m,n}$ between, say two microstates, m and n, of a peptide (e.g., a helix and a hairpin) (for the definition of microstates, see Section 16.9); in this case, the thermodynamic parameters of thermodynamic integration and free energy perturbation do not provide an efficient guidance for the transformation from m to n, and geometrical constraints should be added, which complicates the calculation. An alternative approach would be to calculate the *absolute* free energy, F_m and F_n (or the entropies, S_m and S_n) for each microstate, where simply, $\Delta F_{m, n} = F_n - F_m$, and the need to carry out the integration from m to n is avoided. In this chapter and the next one, we discuss methods for calculating the absolute S and F that stand alone and those that are combined with thermodynamic integration.

18.1 Absolute Free Energy from $<\exp[+E/k_BT]>$

Our aim is to express the partition function as a statistical average. One can write the identity:

$$\frac{1}{Z} = \frac{1}{V^N} \frac{\int \exp[-E(\mathbf{x}^N)/k_BT]\exp[+E(\mathbf{x}^N)/k_BT]d\mathbf{x}^N}{Z} \tag{18.1}$$

which is correct since the integral over the entire volume, V, is equal to V^N. Therefore, the free energy depends on the statistical average of the random variable, $\exp[+E/k_BT]$, with the Boltzmann probability density,

$$F = k_BT \ln \frac{1}{V^N} \int \frac{\exp[-E(\mathbf{x}^N)/k_BT]}{Z} \exp[+E(\mathbf{x}^N)/k_BT]d\mathbf{x}^N =$$

$$= k_BT \ln \frac{1}{V^N} <\exp[+E/k_BT]> \tag{18.2}$$

Because F depends only on the coordinates (the velocities part is ignored), the expression under the *ln* is not dimensionless and F is thus defined up to an additive constant; this method was suggested by Salsburg et al. [1]. However, estimating the average in Equation (18.2) by a Monte Carlo (MC) simulation is practical only for very small systems (ca. 10 particles) because of the large fluctuation of the random variable $\exp[+E/k_BT]$. More specifically, at a finite temperature, an MC simulation will select configurations from a narrow region around the most contributing energy, $E^*(T)$, where the random variable is relatively small, while, for example, the high energy regions, where $\exp[+E/k_BT]$ is very large, will not be visited even for a very large sample. Therefore, this method is extremely inefficient at room temperature and will work only at very high T, where the Boltzmann probability is represented more faithfully by $1/V^N$.

18.2 The Harmonic Approximation

In the harmonic approximation, the Boltzmann probability distribution is replaced by a multidimensional Gaussian. This approximation was introduced for peptides by Gō, Gibson, and Scheraga [2–4]. To obtain a harmonic potential for a peptide, the force field is expanded in a Taylor series around an energy minimized structure, i, with energy, E_i^{min}, where only the quadratic terms are considered (the first terms vanish). In other words, one assumes a multidimensional parabola around E_i^{min}. The absolute entropy is:

$$S_i^{har} = -\frac{k_B}{2}\ln[Det(Hess)] \tag{18.3}$$

where Hess stands for the Hessian matrix, that is, the matrix of second derivatives of the potential energy with respect to the coordinates (internal or Cartesian) at minimum energy E_i^{min}. Hess $= [\partial^2 E / \partial x_i \partial x_j]_{min}$, i, $j = 1,2,...,N$ (Figure 18.1).

We are interested in estimating the difference in the entropy between two microstates, m and n; for that, a microstate is represented by a suitable potential energy well (see Sections 16.6 and 16.10). In detail, to estimate $\Delta S_{m,n}$, one carries out two minimizations of the potential energy starting from structures that pertain to m and n. The harmonic entropy calculated around the minimized structure, $i(m)$, represents the entropy, S_m^{har}, of microstate, m, and $\Delta S_{m,n}$ is approximated by $\Delta S_{m,n}^{har}$,

$$\Delta S_{m,n}^{har} \equiv \Delta S_{i,j}^{har} = -\frac{k_B}{2}\ln\frac{Det[Hess(i(m)]}{Det[Hess(i(n)]} \tag{18.4}$$

Thus, $\Delta S_{m,n}^{har}$ is defined by $\Delta S_{i,j}^{har}$, which depends on the minimized structures i and j chosen; clearly, one would seek to represent m and n by their *global* minimized structures. Notice that S_i^{har} does not depend on the temperature, T, unlike the harmonic entropy, S_p, obtained for an one-dimensional (1d) oscillator [Equation (4.41)],

$$S_p = \frac{k_B}{2}\left(1 + \ln\frac{2\pi k_B T}{f}\right) \tag{18.5}$$

where the spring constant, f, represents the Hessian matrix. However, in the entropy difference between energy wells i and j, the temperature cancels out,

$$\Delta S(p)_{i,j} = \frac{1}{2}k_B \ln\frac{f_i}{f_j} \tag{18.6}$$

and only the spring constants, f_i and f_j, which define the convexities of the parabolas affect $\Delta S(p)_{i,j}$. Thus, the physical significance of Equation (18.3) is only in entropy differences [Equation (18.4)].

Notice that S_i^{har} is *negative*, that is, it decreases the positive entropy related to the velocity part of the partition function [which we have not discussed, see Equation (4.41)]. S_i^{har} should be calculated only for a true minimum, where the Hessian is positive definite; thus, it is crucial to determine the minimum

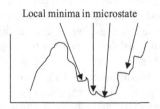

FIGURE 18.1 A microstate is "decorated" by many local energy minima.

energy, $E_i^{\min}$, with high precision, and using a minimizer with analytical derivatives is preferred. The total *harmonic energy* of microstate, m, (including the kinetic part) is:

$$E_{i(m)}^{\text{har}} = E_{i(m)}^{\min} + Nk_BT / 2 \tag{18.7}$$

and the harmonic free energy is:

$$F_{i(m)}^{\text{har}} = E_{i(m)}^{\min} + Nk_BT / 2 - TS_{i(m)}^{\text{har}} \tag{18.8}$$

N is the number of degrees of freedom, which might be different (due to approximations) for Cartesian or internal coordinates. Again, Equation (18.8) should be used for calculating the difference in the free energy of the same molecule residing in different microstates at the same temperature. For Cartesian coordinates, the vibrational entropy has been calculated frequently by normal mode analysis, that is, from the eigenvalues of the mass-weighted Hessian matrix of second derivatives (e.g., [5]). This analysis leads to sets of vibrational frequencies $\nu_k(i)$ and $\nu_k(j)$. For N atoms, one obtains:

$$\Delta S_{m,n}^{\text{har}} \equiv \Delta S_{i,j}^{\text{har}} = k_B \ln\left[\left(\prod_{k=1}^{3N-6} \nu_k(i)\right)\left(\prod_{k=1}^{3N-6} \nu_k(j)\right)^{-1}\right] \tag{18.9}$$

In vacuum, the six frequencies related to translation and rotation vanish and thus are discarded. The frequencies depend on the masses, which are cancelled in Equation (18.9) and $\Delta S_{i,j}^{\text{har}}$ [Equation (18.4)] is recovered [see also Equation (4.37)].

The harmonic approximation ignores anharmonic effects; therefore, its reliability increases at low temperatures, where most of the contribution comes from the bottom part of the potential energy and anharmonicity is low. An advantage of the harmonic approximation is that it can be applied to large molecules, such as proteins and no simulation is required!

At a very low T, one has to use the quantum mechanical (QM) oscillator, which was first used for peptides by Hagler et al. [6]. The QM oscillator is required also for calculating high frequency modes, such as those related to the bond stretching potentials. Based on Section 4.7, the entropy is a summation over the contributions of the different energy modes, k [Equation (4.55)]:

$$TS_i^{\text{QM}} = \sum_k \left\{ \frac{h\nu_k}{\exp(h\nu_k / k_BT) - 1} - k_BT \ln\left[1 - \exp\left(\frac{h\nu_k}{k_BT}\right)\right] \right\}. \tag{18.10}$$

Notice that the QM entropy, unlike the classical one, depends on T and the mass, m, is not cancelled in ΔS_{ij} [see Equation (4.34)]. As has been shown earlier [see discussion following Equation (4.55)], at high T, expanding the exponents of Equation (18.10) to first order leads to the classical S. While calculating both entropies is efficient because simulations are not required, minimizing the energy to $E_i^{\min}$ can be time consuming (minimization methods are discussed in Section 16.8.1). The method is suitable only for molecule models in vacuum.

18.3 The M2 Method

The harmonic approximation has been used extensively, not only in structural biology, but in organic chemistry, polymers, etc. In this context, one should mention *"the second generation mining minima"* (M2) method of Gilson's group [7–10]. With M2, low energy minimized structures within a microstate are initially identified and their entropy (and free energy) is calculated with a method that considers both harmonic and an-harmonic effects; the contributions of the individual wells is then accumulated, leading to an estimation for the microstate's entropy.

18.4 The Quasi-Harmonic Approximation

The need for developing the mining minima method mentioned above demonstrates the limitation of the harmonic approximation, which treats only a localized energy well. Indeed, as discussed in Section 16.10, in a molecular dynamics (MD) simulation, a peptide will stay in a local well a very short time, while spending much longer times in a microstate, which consists of many local energy wells. Therefore, the microstate, Ω_m, is of a greater physical significance than a local energy well, and it is thus of interest to develop methods for calculating the entropy of a microstate from MD or MC samples.

With the "*quasi-harmonic*" (QH) method [11], the probability distribution over a *microstate* is assumed to be a multivariate Gaussian. This probability distribution is based on the covariance matrix (in the same manner the harmonic approximation is based on the Hessian).

The calculation is carried out as follows: The microstate is simulated, say by MD, and a sample of conformations is retained for analysis. Then, typically (but not always) the structures are transformed from Cartesian to internal coordinates (denoted by $\mathbf{q}$), where the Jacobian is neglected because its effect on entropy *differences* is expected to be negligible. The next step is to assume that the original Hamiltonian, $E(\mathbf{x}^N)$ is replaced by:

$$E(\mathbf{x}^N) \rightarrow \quad \frac{1}{2}\mathbf{q}^+\mathbf{Fq} \quad (\mathbf{q} + \text{transposed}) \tag{18.11}$$

where the elements of the matrix $\mathbf{F}$ are:

$$\mathbf{F}_{m,n} = k_\mathrm{B}T[\sigma^{-1}]_{m,n} \tag{18.12}$$

and the matrix σ consists of the variances and covariances, that is, its elements $\sigma_{m,n}$ are:

$$\sigma_{m,n} = <(q_m - <q_m>)(q_n - <q_n>)> = <\Delta\sigma_m\Delta\sigma_m>. \tag{18.13}$$

q_m are the *internal* coordinates, that is, the elements of $\mathbf{q}$. The QH probability distribution is:

$$P^{\mathrm{QH}}(\mathbf{q}) \approx \exp\left[-\frac{1}{2}\mathbf{q}^+\sigma^{-1}\mathbf{q}\right] \tag{18.14}$$

and the QH entropy is obtained from the determinant of the matrix σ.

$$S^{\mathrm{QH}} = \frac{1}{2}Nk_\mathrm{B} + \frac{1}{2}k_\mathrm{B}\ln[2\pi^N\,\mathrm{Det}\sigma] \tag{18.15}$$

where N is the number of internal coordinates. S^{QH} [unlike S_i^{har} of Equation (18.3)] depends on the temperature through σ. The significance of these equations for an $1d$ oscillator [Equation (4.42)] is,

$$\sigma_{1,1} = <x^2> = \frac{k_\mathrm{B}T}{f} \quad \rightarrow \quad F_{1,1} = \frac{k_\mathrm{B}T}{<x^2>} = \frac{k_\mathrm{B}Tf}{k_\mathrm{B}T} = f \tag{18.16}$$

The partition function, q, of the oscillator due to the potential energy and the corresponding free energy, F are [Equation (4.33)]:

$$q = \left[\frac{2\pi k_\mathrm{B}T}{f}\right]^{1/2} \quad \rightarrow \quad F = -\frac{k_\mathrm{B}T}{2}\ln\left[\frac{2\pi k_\mathrm{B}T}{f}\right] \tag{18.17}$$

and the entropy is the same as that defined in Equation (18.15) for $N = 1$,

$$S = -\frac{\partial F}{\partial T} = \frac{k_B}{2} + \frac{k_B}{2} \ln(2\pi < x^2 >) \tag{18.18}$$

Based on Jensen's inequality (Appendix, A4), QH provides an upper bound for the entropy. The QH method is an important contribution to the arsenal of methods for calculating S, which has been used extensively during the years ([12–14] and references cited therein). It should be emphasized, however, that QH is limited to a peptide in vacuum or a peptide (protein) in solvent, where only the contribution of the peptide to the total entropy is calculated. Clearly, the method is applicable to a microstate that is close to a Gaussian, where anharmonic effects are not too severe. In this respect, the method might work also for a bulk system, such as argon particles since the distribution of the coordinates around the center can be considered as a flat Gaussian. Indeed, such calculations have been carried out by van Gunsteren's group quite successfully [13,14]. Note that because σ is an $N \times N$ matrix, the molecular size that can be treated is limited. Also, convergence of the matrix elements is slow, requiring large samples, as discussed below.

A systematic study of the QH performance has been carried out by Gilson's group [15]. They treated linear alkanes and a host-guest system (urea receptor with the ethylene urea ligand), comparing the QH results to those obtained by their M2 method mentioned above. The conclusions of this study are that QH can be accurate for a highly populated *single* energy well (microstate), where the calculation is based on internal coordinates; the use of Cartesians, however, leads to errors of several kcal/mol. When the simulation covers several energy wells, the errors of QH (in internal coordinates) can increase to tens of kcal/mol and are significantly larger with QH(Cartesians). Also, while errors sometimes get cancelled in entropy differences, the host-guest studies have shown that the errors in ΔS^{QH} are substantial. Finally, the convergence of the QH results is slow, and in the host-guest system, convergence has not been obtained even with 6 ns MD runs—in accord with previous studies.

These conclusions probably apply to other versions of QH, where σ is defined in Cartesian coordinates, such as the ad-hoc quantum mechanical approximation of Schlitter [13,16] and the exact derivation of quantum mechanical QH [17]; the performance of these two methods has been compared [18]. QH has been developed further by including anharmonic effects and higher order correlations [19–21]. A different version of QH has been suggested by Wang and Brüschweiler [22,23], which enables one estimating the contribution of different potential wells, for example, rotameric states.

In summary: QH constitutes an important theoretical contribution to computational structural biology, and due to the problems stated above, it should be used with caution. Indeed, the applicability of QH and its convenience to use have made it a popular method for peptides and proteins.

18.5 The Mutual Information Expansion

With the "*mutual information expansion*" (MIE), developed by Gilson's group [24], Kirkwood's superposition approximation is applied to a molecule described by internal coordinates, $(x_1,...,x_m)$, with a probability density $\rho(x_1,...,x_m)$; the first approximation for $\rho(x_1,...,x_m)$ is:

$$\rho(x_1,...,x_m) \approx \rho_1(x_1)\rho_1(x_2).....\rho_1(x_m) \tag{18.19}$$

where $\rho_1(x_i)$ are marginal probability densities. The corresponding approximation for the entropy, S_1 (up to a Jacobian), is:

$$S_1 = -k_B \sum_i \int \rho_1(x_i) \ln \rho_1(x_i) \tag{18.20}$$

TABLE 18.1

Entropy (*TS*) Results (kcal/mol) for Alkans in *T* = 1000 K obtained by M2 and Three Approximations of -*TS*$_i$ of MIE[a]

	M2	-*TS*$_1$	-*TS*$_2$	-*TS*$_3$	Time (ns)
Butane	24.09	19.13	23.62	24.51	150
Nonane	73.08	61.55	73.64	88.53	150
Cyclohexane	61.38	34.74	51.17	57.93	50

[a] This table is part of Table V of the source below.

Source: Reprinted permission from Killian, B.J. et al., *J. Chem. Phys.* 127, 024107, 2007. Copyright 2007 by the American Institute of Physics.

The next approximation for $\rho(x_1,...,x_m)$ is a product of the $\binom{m}{2}$ pairs $\rho_2(x_i, x_j)$ and the *m* values of $\rho_1(x_i)$, (for three particles, $\rho(x_1,x_2,x_3) = [\rho_2(x_1,x_2)\rho_2(x_1,x_3)\rho_2(x_2,x_3)]/[\rho_1(x_1)\rho_1(x_2)\rho_1(x_3)]$); the corresponding approximation, S_2, depends on these probabilities. The approximation S_3 also depends on the $\binom{m}{3}$ triplets $\rho_3(x_i, x_j, x_k)$. ρ_1, ρ_2, ρ_3, and perhaps ρ_4 can reliably be estimated from a long MD trajectory. Thus, unlike QH, MIE can be improved by considering correlations of higher order than second, and the molecule's structure is not limited to a quasi-harmonic *well*.

A feel for the performance of MIE can be obtained from Table 18.1, where some MIE results taken from [24] are compared to M2 results, which are considered to be exact. The two-body approximation (S_2) is converged and appears to be satisfactory besides for cyclohexane. On the other hand, the three-body results (S_3) are not converged, and significantly longer trajectories than 50 nanoseconds (ns) are required to attain convergence. The required samples are expected to increase further at 300K, where the conformational search capability of MD is reduced. As QH, MIE is not suitable for treating an explicit solvent.

18.6 The Nearest Neighbor Technique

The "*nearest neighbor*" (NN) technique is based on ideas from statistics [25]. Thus, assume an *s*-dimensional Cartesian system, where each configuration, *i*, is defined by an *s*-dimensional vector, $\mathbf{x}^i = (x_1^i,...,x_s^i)$. The system is simulated according to a probability density $P(\mathbf{x}^i)$ (e.g., by MC), and one seeks to calculate the *value* of $P(\mathbf{x}^i)$ from *n* snapshots $\mathbf{x}^i$, $1 \leq i \leq n$, taken from the trajectory. Thus, for each vector, $\mathbf{x}^i$, one calculates the Euclidian distance, $R_{i,m}$, from its nearest neighbor vector, *m*, in the sample and the related *s*-dimensional sphere of volume $V(R_{i,m})$.

$$R_{i,m} = \sqrt{\sum_{l=1}^{s}(x_l^i - x_l^{(i,m)})^2} \tag{18.21}$$

Due to the large sample (*n*), *k* vectors are expected to satisfy the minimal distance condition, Equation (18.21); The probability density is:

$$P(\mathbf{x}^i) = \frac{k/n}{V(R_{i,m})} \tag{18.22}$$

where $P(\mathbf{x}^i)$ leads to the absolute conformational entropy, which, as S^{QH}, is also an upper bound (see next chapter). The performance of NN is demonstrated by results for three NN approximations, $S(t)$ obtained in [25] for *R, S*, tartaric acid—a molecule with seven rotatable bonds used as variables. The results are based on a sample of *n* = 14.4 million vectors obtained from a 20 ns MD simulation of 72 molecules in a box at 485K (*NVT* ensemble). The results (in kcal\mol per k_B), $S(1) = 6.45$, $S(2) = 5.39$, and $S(3) = 5.04$ are compared to the QH value, 12.32 ± 0.01. While all these results are upper bounds, the QH entropy is significantly larger than the NN values that only slightly decrease as the approximation improves. It should be pointed out that for the penta-peptide Leu-enkephalin (with 24 dihedral angles as variables, Section 16.8.2), convergence of the NN results could not be achieved. NN is more general than QH, but it is limited to small systems and taking into account explicit water is not practical.

18.7 The MIE-NN Method

In the original MIE, the probability distributions (of one-, two-, and three-body terms) are calculated by a histogramming procedure, which is now replaced (for a better efficiency) by NN, leading to a hybrid, MIE-NN [26]. The MIE-NN results for the entropy of R, S, tartaric acid (based on 14.4 million samples), $S(1) = 6.449$, $S(2) = 5.141$, $S(3) = 5.007$, $S(4) = 5.176$, and $S(5) = 5.075$ show an oscillatory behavior; these results are comparable to those obtained by MIE-histograms, $S(1) = 6.454$, $S(2) = 5.127$.

18.8 Hybrid Approaches

The limitations of QH have led to attempts to devise more accurate techniques for calculating the absolute S, which can be applied to relatively large systems, such as small proteins. One step in this direction has been made recently by Hensen et al. who have developed a new methodology consisting of three building blocks [27,28]. Thus, they replace the NN method described above, which is effective only for a small number of coordinates, with adaptive anisotropic ellipsoidal kernels that capture the configurational density in sufficient detail for up to 45-dimensional spaces. They generate minimally coupled sub-spaces of internal degrees of freedom by applying a linear orthogonal transformation to Cartesian coordinates in such a way that the mutual information among the resulting coordinates is minimized. The new coordinates are subsequently clustered according to their degree of correlation, where correlations among different clusters are neglected. Each oversized cluster with dimensionality, $d > 15$ is sub-divided into smaller sub-clusters with a maximum dimensionality, $d = 15$, and its configurational entropy is computed as a sum of the estimated entropy of its components (sub-clusters), and then corrected by means of MIE. For the stiffest degrees of freedom resulting from the orthogonal transformation, Hensen et al. also propose to employ a generalized QH-Schlitter formula that accounts for their quantum mechanical nature. Entropy results obtained with the combined method are shown to be smaller than the corresponding QH values that constitute upper bounds, where proteins of up to 67 amino acid residues have been treated.

REFERENCES

1. Z. W. Salsburg, J. D. Jacobson, W. Fickett and W. W. Wood. Application of the Monte Carlo method to the lattice gas model. I. Two dimensional triangular lattice. *J. Chem. Phys.* **30**, 65–72 (1959).
2. K. D. Gibson and H. A. Scheraga. Minimization of polypeptide energy, V. Theoretical aspects. *Physiol. Chem. Phys.* **1**, 109–126 (1969).
3. N. Gō and H. A. Scheraga. Analysis of the contribution of internal vibrations to the statistical weights of equilibrium conformations of macromolecules. *J. Chem. Phys.* **51**, 4751–4767 (1969).
4. N. Gō and H. A. Scheraga. On the use of classical statistical mechanics in the treatment of polymer chain conformation. *Macromolecules* **9**, 535–542 (1976).
5. S. Wlodek, A. G. Skillman and A. Nicholls. Ligand entropy in gas phase upon solvation and protein complexation, fast estimation with quasi-Newton Hessian. *J. Chem. Theory Comput.* **6**, 2140–2152 (2010).
6. A. T. Hagler, P. S. Stern, R. Sharon, J. M. Becker and F. Naider. Computer simulation of the conformational properties of oligopeptides. Comparison of theoretical methods and analysis of experimental results. *J. Am. Chem. Soc.* **101**, 6842–6852 (1979).
7. C. E. Chang and M. K. Gilson. Tork: Conformational analysis method for molecules and complexes. *J. Comput. Chem.* **24**, 1987–1998 (2003).

8. W. Chen, C. E. Chang and M. K. Gilson. Concepts in receptor optimization: Targeting the RGD peptide. *J. Am. Chem. Soc.* **128**, 4675–4684 (2005).

9. W. Chen, M. K. Gilson and M. J. Potter. Modeling protein-ligand binding by mining minima. *J. Chem. Theory Comput.* **6**, 3540–3557 (2010).

10. W. M. Huang, W. Chen, M. J. Potter and C. A. Chang. Insights from free-energy calculations: Protein conformational equilibrium, driving forces, and ligand-binding modes. *Biophys. J.* **103**, 342–351 (2012).

11. M. Karplus and J. N. Kushick. Method for estimating the configurational entropy of macromolecules. *Macromolecules* **14**, 325–332 (1981).

12. O. L. Rojas, R. M. Levy and A. Szabo. Corrections to the quasiharmonic approximation for evaluating molecular entropies. *J. Chem. Phys.* **85**, 1037–1043 (1986).

13. H. Schäfer, A. E. Mark and W. F. van Gunsteren. Absolute entropies from molecular dynamics simulation trajectories. *J. Chem. Phys.* **113**, 7809–7817 (2000).

14. H. Schäfer, X. Daura, A. E. Mark and W. F. van Gunsteren. Entropy calculations on a reversibly folding peptide: Changes in solute free energy cannot explain folding behavior. *Proteins* **43**, 45–56 (2001).

15. C. E. Chang, W. Chen and M. K. Gilson. Evaluating the accuracy of the quasiharmonic approximation. *J. Chem. Theory. Comput.* **1**, 1017–1028 (2005).

16. J. Schlitter. Estimation of absolute and relative entropies of macromolecules using the covariance matrix. *Chem. Phys. Lett.* **215**, 617–621 (1993).

17. I. Andricioaei and M. Karplus. On the calculation of entropy from covariance matrices of the atomic fluctuations. *J. Chem. Phys.* **115**, 6289–6292 (2001).

18. J. Carlsson and J. Åqvist. Absolute and relative entropies from computer simulation with applications to ligand binding. *J. Phys. Chem. B* **109**, 6448–6456 (2005).

19. R. Baron, W. F. van Gunsteren and P. H. Hünenberger. Estimating the configurational entropy from molecular dynamics simulations: Anharmonicity and correlation corrections to the quasi-harmonic approximation. *Trends Phys. Chem.* **11**, 87–122 (2006).

20. R. Baron, P. H. Hünenberger and J. A. McCammon. Absolute single-molecule entropies from quasi-harmonic analysis of microsecond molecular dynamics: Correction terms and convergence properties. *J. Chem. Theory Comput.* **5**, 3150–3160 (2009).

21. K. W. Harpole and K. A. Sharp. Calculation of configurational entropy with a Boltzmann quasiharmonic model: The origin of high-affinity protein_ligand binding. *J. Phys. Chem. B* **115**, 9461–9472 (2011).

22. J. Wang and R. Brüschweiler. 2D entropy of discrete molecular ensembles. *J. Chem. Theory Comput.* **2**, 18–24 (2006).

23. D. W. Li, M. Khanlarzadeh, J. Wang, S. Huo and R. Brüschweiler. Evaluation of configurational entropy methods from peptide folding-unfolding simulation. *J. Phys. Chem. B* **111**, 13807–13813 (2007).

24. B. J. Killian, J. Y. Kravitz and M. K. Gilson. Extraction of configurational entropy from molecular simulations via an expansion approximation. *J. Chem. Phys.* **127**, 024107–024116 (2007).

25. V. Hnizdo, E. Daria, A. Fedorowicz, E. Demchuk, S. Li and H. Singh. Nearest-neighbor nonparametric method for estimating the configurational entropy of complex molecules. *J. Comput. Chem.* **28**, 655–668 (2007).

26. V. Hnizdo, J. Tan, B. J. Killian and M. K. Gilson. Efficient calculation of configurational entropy from molecular simulations by combining the mutual-information expansion and nearest-neighbor methods. *J. Comput. Chem.* **29**, 1605–1614 (2008).

27. U. Hensen, O. F. Lange and H. Grubmüller. Estimating absolute configurational entropies of macromolecules: The minimally coupled subspace approach. *PLoS One* **5**, e9179–e9178 (2010).

28. U. Hensen, H. Grubmüller and O. F. Lange. Adaptive anisotropic kernels for nonparametric estimation of absolute configurational entropies in high-dimensional configuration spaces. *Phys. Rev. E* **80**, 011913–011918 (2009).

19

Calculation of the Absolute Entropy from a Single Monte Carlo Sample

As pointed out earlier (Section 10.2), calculation of the average energy from a *single* Monte Carlo (MC), molecular dynamics (MD), or any other Boltzmann sample is straightforward: one has to sum up the molecular interactions, E_i, over each of the sample's configurations, i, and average them. On the other hand, a similar direct calculation of the absolute entropy seems hopeless because one would have to calculate the value of $\ln(P_i^B)$, where the Boltzmann probability, P_i^B, depends on the whole ensemble (through Z) and not on system configuration, i, alone. In this chapter, we present an approach suggested by Meirovitch, which demonstrates that the entropy is actually "written" on the system configurations in terms of the frequency of occurrence of partial clusters and "reading" this information leads to the absolute entropy at least approximately; in other words, calculation of entropy is "raised" to the same level as the calculation of energy and other geometrical properties. As pointed out in Chapter 18, the absolute entropy is very useful in structural biology, and the quasi-harmonic (QH) method described there is an example for a method that extracts the entropy from a single sample; however, QH is based on a harmonic approximation and is limited to a single microstate.

The approach mentioned above is implemented in two main methods, one called "*local states*" (LS) and the other "*hypothetical scanning*" (HS). Both methods are general in the sense that they are applicable to a wide range of systems, lattice and continuum (fluid) models, as well as macromolecules (synthetic polymers, proteins); in particular, these methods can handle proteins residing in a structurally limited microstate, in the random coil state, or in conformations which are hybrids of these two different states. With LS, which is, in particular, effective for peptides and lattice systems (e.g., Ising models), the entropy is expressed in terms of local states, which are local arrangements of dihedral angles or local spin clusters, respectively. HS is based on the scanning method (Section 14.5) and has been found to be more efficient than LS for polymers, fluids, and peptides. Furthermore, important derivatives of HS—the HSMC and HSMD methods have been established, where HS is combined with MC or MD, respectively. Finally, HSMD and thermodynamic integration (TI) have led to the hybrid HSMD-TI method, which is an efficient tool for calculating the free energy of binding (Chapter 23). Because of its higher versatility, HS will be discussed first.

19.1 The Hypothetical Scanning (HS) Method for SAWs

19.1.1 An Exact HS Method

For simplicity, we describe HS as applied to self-avoiding walks (SAWs). First, we point out that large samples generated by different exact simulation techniques lead to the same averages and fluctuations (within the statistical errors); this equivalence means that a given sample does not carry a memory of the simulation method with which it has been created. Now, suppose that a sample of size, n, of N-step SAWs has been created by MC (e.g., by the pivot algorithm), and we wish to extract from this sample the absolute S; we call this Boltzmann sample—the basic sample of size, n. Based on the above equivalence, one can assume that the basic sample has been generated with the complete scanning method (Section 14.5.1) rather than with MC. Therefore, one can reconstruct for each chain configuration, i, of the basic sample, the probability with which it has hypothetically been constructed by the complete scanning method

(Section 14.5.1). The reconstruction is carried out step-by-step by calculating for each (existing) step, $v_k(i)$, $k = 1,...,N$, the exact transition probability (TP) by counting all of the possible continuations of the chain in $f = N - k + 1$ future steps (f is called the future scanning parameter, which is maximal in this case); this is the essence of the exact HS method. The TP (denoted also by p_k) is [Equation (14.21)],

$$p_k = p(v_k \mid v_{(k-1)},...,v_1, f = N - k + 1) = Z_k^v \Big/ \sum_{v=1}^{4} Z_k^v \tag{19.1}$$

where the "past" steps, $v_m(i)$, $m = 1,...,k - 1$ of treated chain, i, are held fixed; notice that in the complete scanning method, the probabilities of all the possible directions, v, at step k are calculated, and one direction is accepted with a random number. Here (with HS), however, the "accepted" direction is already defined by the given chain (i), and a random number is not needed. Z_k^v is an *exact* future partition function, that is, the sum of all the future chains starting from direction, v, at step k (see Section 14.5.1). The product, Πp_k, leads to the Boltzmann probability, P_i^B [Equation (14.22)], and thus to the absolute entropy, $-k_B \ln P_i^B$. Because P_i^B is equal for all i ($P_i^B = 1/Z$), S can be obtained by reconstructing *any single* chain. As pointed out earlier, $P_i^B = 1/Z$ is a manifestation of the zero fluctuation of the free energy [Equation (5.5)], since for SAWs, $F = -TS$.

19.1.2 Approximate HS Method

Clearly, reconstructing the TPs with the exact HS method is limited to short chains due to the exponential growth of the number of the future chains. Therefore, in practice, the method can only be applied with a limited scanning parameter, $f < N - k + 1$ [1–3]. Notice that there is no attrition in this approximate process (unlike with the scanning method), since *existing* N-step chains are analyzed; however, the probability of SAW i, $P_i^0(f > 1)$ [Equation (14.18) with $N - k + 1 > f > 1$] is approximate and is normalized not only over SAWs, but over a larger group of chains that also contains self-intersecting walks. One defines an entropy functional, S^A:

$$S^A(f) = -k_B \sum_{\text{SAWs}(i)} P_i^B \ln P_i^0(f) > S \tag{19.2}$$

where to SAW i, which, in principle, will be selected with the correct P_i^B (by MC or any other correct technique), we assign the random variable, $\ln P_i^0(f)$. $S^A(f)$ is an upper bound of S, which can be deduced from the following arguments: In Equations (14.26) and (14.27), we have defined the probability, $P_i(f) = P_i^0(f)/A$, which is normalized over the group of SAWs alone, where $A < 1$. Replacing $\ln P_i^0(f)$ in Equation (19.2) by $\ln P_i(f)$ and using Jensen's inequality (Appendix, Section A4) would lead to $S^{A'}(f)$, which is a *rigorous* upper bound for S. While A can be estimated within the framework of the scanning method, it *cannot* be calculated from the given MC sample. However, since $A < 1$, $S^A(f)$ [Equation (19.2)] should also be an upper bound of S, where $S^A > S^{A'} > S$. S^A can be estimated by $\overline{S^A}$ from the basic sample of size, n,

$$\overline{S^A}(f) = -\frac{k_B}{n} \sum_{t=1}^{n} \ln P_{i(t)}^0(f) \tag{19.3}$$

$\overline{S^A}(f)$ is a statistical average with a fluctuation, $\sigma_A(f)$, that decreases with increasing, n, as $\sigma_A(f)/n^{1/2}$. Moreover, as f is increased, $\sigma_A(f)$ is expected to decrease, becoming zero for a complete scanning (an exact HS method); therefore, the required sample size, n, decreases with increasing f as well. Since a larger f leads to both smaller S^A and $\sigma_A(f)$, it is more "economical" to invest the computer time in increasing f than n. Still, for a long chain (where $f << N$), $S^A(f)$ will always remain an upper bound. However, as discussed later, approximate HS is also equipped with a lower bound functional, S^B (Section 19.6), which enables one improving further the estimation of the correct S, which is bound from both sides. Finally, for chain models with attractions, $S^A(f)$ leads to a lower bound for the free energy functional, $F^A(f) = <E> - TS^A(f)$. From now on, in most cases, we refer to *approximate* HS just by HS.

19.2 The HS Monte Carlo (HSMC) Method

The need to increase f can be achieved by the HSMC method, which is based on a version of the scanning method described in Section 14.5.12. There, the TPs are obtained, not by exact enumeration, but from a sample of future SAWs generated by an additional MC run. One can use this (stochastic) procedure to analyze a basic (given) sample of SAWs along the same lines described above for HS. More specifically, assume that the basic sample has been created on a square lattice, where, at the reconstruction of step k of the chain configuration, i, the current direction is, $v_k(i)$; in this case, one carries out an MC simulation of, n_f, future MC steps (sweeps) starting from an arbitrary future chain of f bonds, where the already determined steps $1,2,...,k-1$ are held fixed. During this run, one counts the number of times, $m_{v(k)}$, the actual direction, $v_k(i)$, appears in the MC sample, which leads to TP(f, n_f) [compare with Equation (14.45)],

$$\text{TP}_k(f, n_f) = p_k(f, n_f) = p(v_k | v_{(k-1)}, ..., v_1, f, n_f) = \frac{m_{v(k)}}{n_f} \tag{19.4}$$

Because MC is an exact method, for very large n_f, the calculated TP(f, n_f) will approach the TP that would be calculated by exact enumeration (approximate HS). As for HS, HSMC can be divided into two methods: a *partial* HSMC, where $f < N - k + 1$ and a *complete* HSMC, where the whole future is scanned, that is, $f = N - k + 1$. For partial HSMC, one obtains $P_i^0(f, n_f) = \Pi p_k(f, n_f)$, where $P_i^0(f, n_f)$ is defined over an ensemble that also includes self-intersecting walks; thus, $P_i^0(f, n_f)$ is always approximate even for $n_f \to \infty$. For complete HSMC, f is maximal, and the probability of SAW i, $P_i(n_f) = \Pi p_k(n_f)$ is defined *only* over the ensemble of SAWs (Figure 19.1).

It should be emphasized that unlike the deterministic probabilities of HS, for finite n_f, $p_k(n_f)$ [hence, $P_i(n_f)$] are stochastic, depending on the specific random number sequences used in the future MC simulations; therefore, the normalization condition, $\Sigma P_i(n_f) = 1$, might be somewhat violated. However, as n_f is increased, this noise decreases, where for $n_f \to \infty$ $P_i(n_f) \to P_i^B$, meaning that the method becomes deterministic, that is, equals to the exact HS. Thus, the approximation of the complete HSMC is determined solely by n_f, the higher is n_f, the better is the approximation. To be on the rigorous side, we first convert these stochastic functionals to deterministic ones by averaging over the MC-based probability distributions, $P_{i(j)}(n_f)$; $S^A(n_f)$ is changed to,

$$<S^A>_{P(n_f)} = -k_B \sum_{\text{SAWs}(i)} P_i^B \frac{1}{n_j} \sum_j^{n_j} \ln P_{i(j)}(n_f) \tag{19.5}$$

future chain is moved by MC

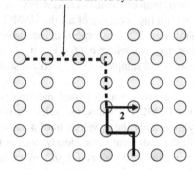

FIGURE 19.1 Reconstruction of the 4th step (an arrow in direction 2) of a SAW of $N = 20$ steps on a square lattice (for clarity, steps (bonds) 5–20 are not shown, as they do not take part in the reconstruction of the 4th step). The reconstruction is based on future SAWs of 5 steps. Thus, the 5-step dashed future chain shown constitutes a starting point for a long Monte Carlo simulation, which generates a sample of size n_f (steps 1–3 remain constant). Then, the number of visits, $m_{2(4)}$, of the first step of the future chain to direction, 2 is counted and the probability of step 4 is $m_{2(4)}/n_f$ [see Equation (19.4)] (compare with Figure 14.1); for a SAW of $N = 8$ steps, this procedure becomes the *complete* HSMC.

where n_j runs over all the possible probability distributions that can be defined by MC for a given n_f; a similar averaging can be defined for $S^A(f, n_f)$. However, for large enough n_f, the values of $\ln P_{i(j)}(n_f)$ for different j become very similar, and we approximate Equation (19.5) by a single probability distribution, $P_i(n_f)$, which to emphasize again, is defined only over the ensemble of SAWs [and $P_i(f, n_f)$ is defined over an ensemble, which includes all SAWs and some self-intersecting walks]. Thus, one can define the (upper bound) entropy functionals, $S^A(n_f)$ and $S^A(f, n_f)$, which are expected to decrease as n_f is increased; their relation to S is,

$$S^A(f,n_f) > S^A(n_f) = -k_B \sum_{\text{SAWs}(i)} P_i^B \ln P_i(n_f) > S \quad \text{large } n_f \tag{19.6}$$

$S^A(n_f)$ and $S^A(f, n_f)$ are statistical averages relying on both, the probability space and the experimental world through n_f, (see Section 1.4); therefore, they appear without a bar; however, the bar appears (as needed) on their estimations (from basic samples of size n),

$$\overline{S^A}(n_f) = -\frac{k_B}{n} \sum_{t=1}^{n} \ln P_{i(t)}(n_f) \tag{19.7}$$

For SAWs with attractions, E_i, the corresponding (lower bound) free energy functionals, $F^A(f,n_f)$ and $F^A(n_f)$ are expected to increase with n_f,

$$F^A(f,n_f) < F^A(n_f) = \sum_{\text{SAWs}(i)} P_i^B[E_i + k_B \ln P_i(n_f)] < F \tag{19.8}$$

For a SAW of N steps, one can distinguish among three types of errors, (1) a systematic error defined by $S^A(n_f, n \to \infty) - S$, which is also expressed by the fluctuation:

$$\sigma_A(n_f) = \left(\sum_{\text{SAWs}(i)} P_i^B[S^A(n_f) + k_B \ln P_i(n_f)]^2 \right)^{1/2} \tag{19.9}$$

As $n_f \to \infty$, $S^A(n_f) \to S$, $\sigma_A(n_f) \to 0$, and $S^A(n_f)$ can be obtained from any *single* chain configuration (a manifestation of the zero fluctuation of the free energy [Equation (5.5)]). Furthermore, for a given n, one would expect a *monotonic* decrease of $\sigma_A(n_f)$ to zero and $S^A(n_f)$ to S as n_f is increased. However, in practice, n is finite and, for a given n_f, there is also (2) a statistical error, $\sim \sigma_A(n_f)/n^{1/2}$. Clearly, since $\sigma_A(n_f)$ decreases with n_f, the required sample size, n, decreases as well; a third source for an error is the *noise* discussed earlier. However, within the accuracy of all the HSMC and HS studies carried out thus far, the effect of noise has never shown up. Thus, as expected, increasing f or n_f has always resulted in a monotonic decrease of S^A, $\sigma_A(n_f)$, or $\sigma_A(f, n_f)$, where $\sigma_A(f, n_f) \to \sigma_A(f)$ as n_f is increased, and $\sigma_A(f) \neq 0$. Thus, for both methods, the fluctuation and sample size, n, decrease simultaneously as n_f is increased—an important efficiency factor.

It is of interest to point out the alternative way for implementing the complete HSMC method (or HSMD – these are denoted together HSMC(D)). Thus, one selects a *single* chain from the n-size basic MC sample and reconstructs it by increasing the n_f values until a satisfying convergence of the results for $S^A(n_f)$ has been detected; here, the statistical error clearly does not exist. Also, since in a typical HSMC(D) study several decreasing results for $S^A(n_f)$ and $\sigma_A(n_f)$ are calculated for increasing n_f, one can extrapolate the function $S^A[\sigma_A(n_f)]$ to $\sigma_A = 0$, which should lead to the correct S.

In practice, one seeks to increase n_f (for a given n) until $S^A(n_f)$ does not change within a desired error. In this respect, *complete* HSMC(D) can be considered as an exact method. However, this convergence is not always achievable for a large molecule because the number of future simulations is $\sim N$, while the required computer time grows at least by N^2; also, the total error is accumulated over the N simulation steps; thus, F^A will remain a lower bound. This limitation is mitigated by the existence of several procedures (not restricted to SAWs), which lead to upper bounds and exact values for the free energy. Also, for proteins, one is typically interested in entropy (or free energy) *differences*, $\Delta S_{m,n}$, between two microstates, m and n; in this case, $\Delta S_{m,n}^{A}(n_f) = S_m^A(n_f) - S_n^A(n_f)$ usually converges to the correct, $\Delta S_{m,n}$, for relatively small values of n_f since the errors in S^A are typically comparable for both microstates, and they thus get cancelled in $\Delta S_{m,n}^A(n_f)$.

Summary: First, we emphasize that the HS/HSMC(D) approach presented above is based on averaging the logarithm values of the calculated TPs, which leads to the absolute entropy (or the free energy); this process is more efficient than taking the logarithm of $<\exp - (E/k_BT)>$ (where E fluctuates as $N^{1/2}$), which can occur in free energy perturbation (Section 17.7), umbrella sampling, and other techniques developed in Chapters 20 and 21. The HS/HSMC methods presented above for SAWs are quite general and can be extended (with some specific modifications) to more complex systems, such as continuum fluids and proteins. However, before treating these systems, we enrich our methodology further by also defining an upper bound for F (denoted F^B) and providing several procedures for its evaluation. We also define functionals, such as F^D, that lead to the exact F. As has been noted above, these functionals increase dramatically the accuracy of HS, LS, and the HSMC(D) techniques.

19.3 Upper Bounds and Exact Functionals for the Free Energy

As stated above, all the methods discussed in this section are applicable to complex systems, where the corresponding approximate probability is a function, not only of f and n_f, but of additional parameters. Thus, the corresponding (approximate) probability is denoted by a unified symbol, P^{HS}. Also, for simplicity (but without the loss of generality), the procedures of this section are developed for a discreet system. While in most cases, P^{HS} is stochastic, we assume that for large enough n_f, P^{HS} can be considered as deterministic (see Section 19.2). The theory is developed for P^{HS} (and P_i^B), which is defined over the entire system of interest, while for SAWs, $P_i^0(f)$ and $P_i^0(f, n_f)$ are normalized over a space that also includes self-intersecting walks. Therefore, to make the theory applicable also to polymers, these probabilities should be divided by the factor A [see Equation (14.26) and Section 19.1.2]. Finally, while the simulations can be performed by both MC and MD, for brevity, we consider in this Section (19.3) only MC, and because our main interest is in *complete* HSMC(D), for brevity, we refer in most cases to this method as HSMC. Finally, for generality, we shall consider the free energy rather than the entropy.

We start by writing the fluctuation (standard deviation) of F^A (denoted by σ_A), which is a generalization of the fluctuation for the approximate entropy, S^A [Equation (19.9)]:

$$\sigma_A = \left[\sum_i P_i^B \left[F^A - F_i^{HS} \right]^2 \right]^{1/2} = \left[\sum_i P_i^B \left[F^A - E_i - k_BT \ln P_i^{HS} \right]^2 \right]^{1/2} \tag{19.10}$$

where $F_i^{HS} = E_i + k_BT \ln P_i^{HS}$.

19.3.1 The Upper Bound F^B

In Equation (19.8), we have defined a lower bound for the free energy, F^A. One can also define another free energy functional denoted F^B [3], which is an upper bound of F,

$$F^B = \sum_i P_i^{HS} \left[E_i + k_BT \ln P_i^{HS} \right] \geq F \tag{19.11}$$

This relation is based on the free energy minimum principle [4], which states that $F^B(P_i^{HS}) > F$ unless $P_i^{HS} = P_i^B$, where $F^B = F$. Thus, F^B is an upper bound, which is expected to approach the correct free energy, F, when $P_i^{HS} \to P_i^B$. It is necessary to rewrite Equation (19.11), such that F^B can be estimated by importance sampling from a (Boltzmann) sample generated with P_i^B (rather than P_i^{HS}). Applying the identities $\sum_i P_i^{HS} = 1$ and $P_i^B / (\exp[-E_i / k_B T] / Z) = P_i^B / P_i^B = 1$ leads to:

$$F^B = \frac{\sum\limits_i P_i^B \left[P_i^{HS} \exp(E_i / k_B T)(E_i + k_B T \ln P_i^{HS}) \right]}{\sum\limits_i P_i^B \left[P_i^{HS} \exp(E_i / k_B T) \right]} = \frac{\sum\limits_i P_i^B \left[P_i^{HS} \exp(E_i / k_B T) F_i^{HS} \right]}{\sum\limits_i P_i^B \left[P_i^{HS} \exp(E_i / k_B T) \right]} \quad (19.12)$$

F^B is estimated by $\overline{F}^B$ from a Boltzmann sample of size n generated by MC:

$$\overline{F}^B = \frac{\sum\limits_{t=1}^n P_{i(t)}^{HS} \exp(E_{i(t)} / k_B T) F_{i(t)}^{HS}}{\sum\limits_{t=1}^n P_{i(t)}^{HS} \exp(E_{i(t)} / k_B T)} \quad (19.13)$$

It should be noted, however, that the statistical reliability of this estimation (unlike the estimation of F^A) decreases sharply with increasing system size, because the overlap between the probability distributions P_i^B and P_i^{HS} decreases exponentially (see discussion in [6]). For SAWs, it is easy to see [using the factor A, in Equation (14.26)] that F^B remains an upper bound if $P_i^0(f)$ or $P_i^0(f, n_f)$ replace P_i^{HS} in the equations above.

19.3.2 F^B Calculated by the Reversed Schmidt Procedure

The Schmidt procedure enables one to extract from a *biased* sample an effectively smaller *unbiased* sample generated with P_i^B [for details see Section 14.5.6]. The opposite way is also available, that is, to estimate F^B by using a "reversed-Schmidt procedure" [3,5], which enables one to extract from the given *unbiased* sample of size n (the basic sample) generated with P_i^B, an effectively smaller *biased* sample generated with P_i^{HS}. Thus, the configurations of the unbiased sample are treated consecutively. If a configuration, i, was accepted to the biased sample, the next configuration, j, would be accepted with a transition probability A_{ij}:

$$A_{ij} = \min\left\{ 1, \exp\left[(E_j - E_i) / k_B T \right] P_j^{HS} / P_i^{HS} \right\}. \quad (19.14)$$

Equation (19.14) is a generalized MC procedure, which satisfies the detailed balance condition and is carried out with random numbers. The acceptance rate, R_{biased}, provides a measure for the *effective size* of the accepted biased sample,

$$R_{biased} = n_{biased} / n \quad (19.15)$$

where n_{biased} is the number of accepted biased configurations. The effectiveness of this procedure is again limited by the overlap of the distributions, P_i^B and P_i^{HS}. The acceptance rate, R_{biased}, is a useful gauge of the reliability of the F^B value, the closer is R_{biased} to 1, the better is the overlap between P_i^B and P_i^{HS}, the closer is F^B to F, and the smaller is the sample size required to estimate F^B reliably. Using the biased sample, F^B can be calculated directly from Equation (19.11). With values for both F^A and F^B [Equations (19.11) or (19.13)], their average, F^M, defined by:

$$F^M = (F^A + F^B) / 2 \quad (19.16)$$

often becomes a better approximation than either of them individually. This is provided that their deviations from F (in magnitude) are approximately equal, and that the error in F^B is not too large. Typically, several improving approximations for F^A, F^B, and F^M are calculated as a function of n_f or f, and other parameters and their convergence enable one to determine the correct free energy with high accuracy. In the case of SAWs, $P_i^0(f)$ or $P_i^0(f, n_f)$ replace P_i^{HS} in Equation (19.14).

19.3.3 A Gaussian Estimation of F^B

We now describe an efficient method to estimate the free energy upper bound F^B [Equations (19.11) through (19.13)], which can effectively overcome the statistical limitations associated with the standard evaluations of F^B described in the previous sub-sections [6]. Notice, however, that the applicability of this Gaussian-based method depends on the form of the HSMC implementation. Thus, for argon and water treated by complete HSMC, the Gaussian assumption has been found to be well suited (see Sections 19.3.6 and 19.5.3). We start by rearranging Equation (19.12),

$$F^B = \frac{\sum_i P_i^B \exp[F_i^{HS}/k_B T][F_i^{HS}]}{\sum_i P_i^B \exp[F_i^{HS}/k_B T]} \tag{19.17}$$

where to recall $F_i^{HS} = [E_i + k_B T \ln P_i^{HS}]$. Equation (19.17) emphasizes an *explicit* dependence of F^B on the variable, F_i^{HS}, a quantity that is directly related to the average, F^A [Equation (19.8)], and the fluctuation, σ_A [Equation (19.10)]. Let us now assume that when configurations (i) are sampled from the Boltzmann distribution (that is with P_i^B), their corresponding F_i^{HS} values occur with the Gaussian distribution,

$$\rho(F_i^{HS}) = \rho(F') = \frac{1}{\sqrt{2\pi}\sigma_A} \exp\left[-(F' - F^A)^2 / 2(\sigma_A)^2\right] \tag{19.18}$$

which is thus determined solely by the two parameters, F^A (the mean) and its standard deviation, σ_A. Now, rather than summing over the configurations i with their weights, P_i^B, as in Equation (19.17), we can integrate over all values of F_i^{HS} weighted with $\rho(F_i^{HS})$. The numerator in Equation (19.17) becomes:

$$\sum_i P_i^B \exp[F_i^{HS}/k_B T][F_i^{HS}] \approx \frac{1}{\sqrt{2\pi}\sigma_A} \int (\exp[F'/k_B T][F']) \exp\left[-(F' - F^A)^2 / 2(\sigma_A)^2\right] dF'$$

$$= \left(\frac{(\sigma_A)^2}{k_B T} + F^A\right) \exp\left[\frac{1}{2}\left(\frac{\sigma_A}{k_B T}\right)^2 + \frac{F^A}{k_B T}\right] \tag{19.19}$$

where we used the integrals of Equations (A1.3) and (A1.4) in the Appendix; the denominator is:

$$\sum_i P_i^B \exp[F_i^{HS}/k_B T] \approx \frac{1}{\sqrt{2\pi}\sigma_A} \int (\exp[F'/k_B T]) \exp\left[-(F' - F^A)^2 / 2(\sigma_A)^2\right] dF' =$$

$$= \exp\left[\frac{1}{2}\left(\frac{\sigma_A}{k_B T}\right)^2 + \frac{F^A}{k_B T}\right]. \tag{19.20}$$

The ratio of the results in Equations (19.19) and (19.20) is the new (Gaussian) estimation of F^B, denoted F_G^B,

$$F_G^B = \frac{(\sigma_A)^2}{k_B T} + F^A \tag{19.21}$$

Thus, F_G^B depends only on F^A and σ^A, which is an advantage of F_G^B because these quantities are typically easier to estimate than F^B from Equation (19.13) or (19.14). Provided that the Boltzmann sample of F_i^{HS} values is approximately Gaussian, then $F_G^B \approx F^B$. Similar to Equation (19.16), one defines the average,

$$F_G^M = (F^A + F_G^B)/2 = F^A + \frac{1}{2}\frac{(\sigma_A)^2}{k_B T} \qquad (19.22)$$

where again, several approximations for F^A, F_G^B, and F_G^M can be calculated as a function of n_f (and other parameters), and their convergence leads to highly accurate free energy determination. Indeed, as mentioned above, results for argon and water (Section 19.5.3) and for SAWs (Section 19.3.6) have shown that this Gaussian distribution is a very good approximation, as there is an excellent agreement of F_G^B with F^B for cases where F^B is well converged.

19.3.4 Exact Expression for the Free Energy

The denominator of F^B in Equation (19.12) defines an exact expression for the partition function,

$$\frac{1}{Z} = \frac{1}{Z}\sum_i P_i^B (P_i^{HS}/P_i^B) = \sum_i P_i^B\left(P_i^{HS}\exp[E_i/k_B T]\right) = \sum_i P_i^B \exp[F_i^{HS}/k_B T] \qquad (19.23)$$

which is based on $\sum_i P_i^B(P_i/P_i^B) = 1$ [6]. Therefore, Equation (19.23) will hold for any approximation where P_i^{HS} is normalized over the same space as P_i^B, meaning that it will apply *only* to SAWs treated by complete HSMC. An *exact* expression for the correct free energy F, denoted by F^D, is:

$$F^D = k_B T \ln\left(\frac{1}{Z}\right) = k_B T \ln\left[\sum_i P_i^B \exp\left[F_i^{HS}/k_B T\right]\right] \qquad (19.24)$$

Note that while the (Boltzmann) average of the approximate F_i^{HS} values gives a bound for the free energy $\left(\sum_i P_i^B F_i^{HS} = F^A\right)$, the average of $\exp[F_i^{HS}/k_B T]$ leads to the free energy exactly. In fact, using Equation (19.24), it is easy to verify that F^A defines a lower bound (that is, $F^A \le F$), where we must have:

$$\exp[F^A/k_B T] = \exp\left[\sum_i P_i^B [F_i^{HS}/k_B T]\right] \le \sum_i P_i^B \exp[F_i^{HS}/k_B T] = \exp[F/k_B T] \qquad (19.25)$$

in accordance with Jensen's Inequality (Section A.4 in Appendix).

In practice, the efficiency of estimating F by F^D depends on the fluctuation of this statistical average, which is determined by the fluctuation of F_i^{HS} exponentiated. That is, if the fluctuations in F_i^{HS} are small, then the values for $\exp[F_i^{HS}/k_B T]$ do not vary drastically, and the averages for F^D (and F^B) can be estimated reliably. Still (as for F^B), the direct calculation of F through F^D will not be as statistically reliable as the corresponding calculation for the lower bound estimate, F^A. Obviously, as $F_i^{HS} \to F$ (that is, as $P_i^{HS} \to P_i^B$), all fluctuations become zero and F can be obtained from a single configuration. It should be pointed out that Equation (19.23) with $P_i^{HS} = 1/V^N$ was suggested for a lattice gas long ago by Salsburg et al. [see Equation (18.2) and [1] there].

19.3.5 The Correlation Between σ_A and F^A

The zero fluctuation property of the correct free energy can be exploited directly through the extrapolation of a series of improving approximations, $F^A(\beta)$ defined by β (β includes n_f and other parameters). The corresponding fluctuations, $\sigma_A(\beta)$:

$$\sigma_A^2(\beta) = \sum_i P_i^B \left\{F^A(\beta) - [E_i + k_B T \ln P_i^{HS}(\beta)]\right\}^2 \qquad (19.26)$$

are expected to decrease systematically as the approximation improves. It has been suggested [7] to express the correlation between $F^A(\beta)$ and $\sigma_A(\beta)$ by the approximate function,

$$F^A(\beta) = F^{\text{extp}} + C[\sigma_A(\beta)]^\delta, \tag{19.27}$$

where F^{extp} is the extrapolated value of the free energy and C and δ are parameters to be optimized by best-fitting results for $F^A(\beta)$ and $\sigma_A(\beta)$ for different approximations of β. One can also calculate the tangent to the function at $\sigma_A(\beta_{\text{best}})$, which is the lowest value for the fluctuation obtained from the best approximation, β_{best}. If Equation (19.27) defines a concave-down function, and this trend of $F^A(\sigma_A)$ is assumed to hold for better (uncalculated) approximations of F^A, the intersection of the tangent with the vertical axis [$\sigma_A = 0$] defines an upper bound for F, which is denoted F^{up}. This upper bound, along with the best value for the lower bound, $F^A(\beta_{\text{best}})$, can be used to define the average, F^{M2} (Figure 19.2):

$$F^{\text{M2}} = [F^A(\beta_{\text{best}}) + F^{\text{up}}]/2 \tag{19.28}$$

19.3.6 Entropy Results for SAWs on a Square Lattice

In Table 19.1, we present results for the entropy of SAWs of increasing length, N, on a square lattice [8]. The results for the various HSMC functionals are compared to those obtained with approximate HS [Equation (19.3)] denoted S_{HS} (based on scanning parameter, $f = 8$), the scanning method (S_{scan}, $f = 6$ [Equation (14.23)]), thermodynamic integration (S_{TI}), and exact enumeration (S_{series}), where the last two sets of results are taken from Table 17.1. The HSMC results were obtained from n reconstructions of a *straight* SAW chain of N bonds. S^A [Equation (19.7)] is an upper bound, and because $F = -TS$, S^B and S_G^B are lower bounds (as compared to F^B [Equation (19.13)] and F_G^B [Equation (19.21)], which constitute upper bounds). As expected, the upper bound, S^A, and its fluctuation σ_A, decrease monotonically with increasing the parameter, n_f, which is related to the number of future MC steps per bond; correspondingly, the lower bounds, S^B, and its Gaussian approximation S_G^B, monotonically increase. Their averages with S^A, which are denoted S^M and S_G^M, respectively, and the exact entropy functional S^D [see Equation (19.24)] are expected to lead to the correct entropy results. Indeed, they are always equal to the corresponding S_{TI} and S_{series} values, but within larger statistical errors, which are denoted by parentheses, e.g., $1.00(3) = 1.00 \pm 0.03$. The (deterministic) results of S_{HS}, which are based on a limited future($f = 8$) are inferior to the HSMC ones. The results of S_{scan} are equal to those of S_{TI} and S_{series} up to $N = 399$, with larger statistical errors. More details about the simulations and results based on reconstructing of a sample of SAWs (where results for R_{biased} [Equation (19.15)] are also provided) appear in [8].

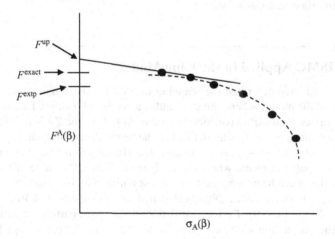

FIGURE 19.2 Schematic illustration of the free energy functional, $F^A(\beta)$ [Equation (19.8)] versus its fluctuation, $\sigma_A(\beta)$ [Equation (19.9)] as a function of the approximation, β. These results are fitted [Equation (19.27)] to a concave down function, where F^{extp} is the extrapolated value of the best fitted function to $\sigma_A(\beta) = 0$. The tangent to the function at the smallest $\sigma_A(\beta)$ value calculated is plotted, and its intersection with the vertical axis (at $\sigma_A(\beta) = 0$) defines the upper bound, F^{up} for F^A.

TABLE 19.1

Entropy Results for SAWs of N Steps on a Square Lattice Obtained by the complete HSMC method by Reconstructing a *single* Straight SAW

n_f	S^A/k_B	σ_A	S^B/k_B	S_G^B/k_B	S^M/k_B	S_G^M/k_B	S^D/k_B	n
			$N = 249$		$S_{SCAN} = 0.97836\,(2)$			
500	0.98391 (3)	0.00669 (4)	0.9727 (3)	0.9728 (1)	0.9783 (2)	0.97833 (7)	0.9783 (2)	63,000
5000	0.97889 (2)	0.00208 (4)	0.97782 (8)	0.97782 (5)	0.97836 (4)	0.97836 (3)	0.97836 (5)	9100
50,000	0.97840 (2)	0.00066 (4)	0.97829 (5)	0.97829 (2)	0.97835 (3)	0.97835 (2)	0.97835 (3)	930
S_{HS}	0.98306 (1)	0.00401 (1)	0.9745 (5)	0.9791 (3)	0.9788 (3)	0.9811 (2)	0.9799 (1)	176,000
S_{TI}	0.978358 (4)		0.978358 (4)	0.978358 (4)	0.978358 (4)	0.978358 (4)	0.978358 (4)	
Series	0.9783609 (1)		0.978360 (1)	0.978360 (1)	0.978360 (1)	0.978360 (1)	0.978360 (1)	
			$N = 399$		$S_{SCAN} = 0.97567\,(4)$			
500	0.98138 (6)	0.00540 (5)	0.9710 (5)	0.9697 (2)	0.9762 (3)	0.9756 (1)	0.9759 (3)	9500
5000	0.97625 (4)	0.00170 (5)	0.9751 (1)	0.97509 (8)	0.97567 (5)	0.97567 (5)	0.97567 (5)	2000
50,000	0.97568 (4)	0.00053 (5)	0.97557 (7)	0.97557 (5)	0.97563 (4)	0.97563 (4)	0.97563 (5)	225
S_{HS}	0.98141 (5)	0.00335 (5)	0.9743 (5)	0.9769 (3)	0.9779 (3)	0.9792 (2)	0.9782 (2)	5500
S_{TI}	0.975655 (8)		0.975655 (8)	0.975655 (8)	0.975655 (8)	0.975655 (8)	0.975655 (8)	
Sseries	0.975652 (1)		0.975652 (1)	0.975652 (1)	0.975652 (1)	0.975652 (1)	0.975652 (1)	
			$N = 599$		$S_{SCAN} = 0.97395\,(5)$			
500	0.98003 (8)	0.00445 (7)	0.9706 (8)	0.9682 (4)	0.9753 (4)	0.9741 (2)	0.9748 (5)	3000
5000	0.97466 (7)	0.00139 (7)	0.9736 (2)	0.9735 (1)	0.9741 (1)	0.97408 (9)	0.9741 (1)	450
50,000	0.97413 (5)	0.00036 (7)	0.9741 (1)	0.97405 (6)	0.97409 (6)	0.97409 (5)	0.97409 (5)	45
S_{TI}	0.97404 (1)		0.97404 (1)	0.97404 (1)	0.97404 (1)	0.97404 (1)	0.97404 (1)	
Series	0.974025 (1)		0.974025 (1)	0.974025 (1)	0.974025 (1)	0.974025 (1)	0.974025 (1)	

Source: Reprinted with permission from White, R.P. and Meirovitch, H., *J. Chem. Phys.*, 123, 214908–214911, 2005. Copyright 2005 by the American Institute of Physics.

Note: The MC procedure is based on the pivot algorithm with local moves. S^A is an upper bound for the entropy, and σ_A is its fluctuation. S^B and its Gaussian approximation, S_G^B are lower bounds, and their averages with S^A are denoted S^M and S_G^M, respectively. S^D is an exact entropy functional. n_f is related to the number of MC steps per bond (see text). S_{TI}, S_{scan}, S_{series}, and S_{HS} were obtained by thermodynamic integration, the scanning method, a series expansion formula, and the HS method, respectively. n is the number of reconstructions. The statistical error is defined by parentheses: 1.00 (3) = 1.00 ± 0.03. For more details, see text.

19.4 HS and HSMC Applied to the Ising Model

In Section 14.5.13, we have described the scanning method for the square Ising lattice based on a linear construction of the model, where spins are added to an initially empty lattice line-by-line, step-by-step. The TPs are based on partial partition functions $Z_k(+, f, T)$ and $Z_k(-, f, T)$ calculated by exact enumeration over a future square of f sites defined on the "empty" part of the lattice, while the already determined spins, at sites $1,\dots,k-1$ are held fixed. The HS picture for this model is similar to that described above for a polymer chain, where one analyzes a given MC sample of Ising configurations i by assuming that they have been generated with the scanning method. Thus, the scanning TP, $p(\sigma_k, f, T)$, of the *existing* spin σ_k, $(k = 1,\dots,N)$, is calculated using Equation (14.46), and the product of $p(\sigma_k, f, T)$, $k = 1,\dots,N$, leads to $P_i(f, T)$. The calculation of the above partition functions by exact enumeration leads to the HS method and to an upper bound $S^A[P_i(f, T)]$ for the entropy [Equation (19.2)]. This method was applied to the Ising model on a cubic lattice [9], where the results were compared to those obtained by series expansion and the LS method (see later Section 19.8). At that time the upper bounds S^B and S^D were not known.

One can also apply *partial* HSMC, where the spins of the future square on the lattice are simulated by MC. In this case, one obtains the TP, $p(\sigma_k, f, n_f, T)$, of the existing spin, σ_k, which depends on both—the size of the future box, f, and the *future* sample size, n_f. $p(\sigma_k, f, n_f, T)$ is proportional to the number of times, $n_{visit}(\sigma_k)$—the *existing* spin, σ_k, appears in the future MC sample,

$$p(\sigma_k, f, n_f, T) = \frac{n_{visit}(\sigma_k)}{n_f} \tag{19.29}$$

Also, if the whole future (that is, $f = N - k + 1$ future spins) is treated, one obtains the *complete* HSMC method, where Equation (19.29) leads to the TP values, $p(\sigma_k, n_f, T)$, where their product defines the corresponding $P_i(n_f, T)$ (for simplicity, "HS" is omitted). For these three methods, one can define an upper bound entropy functional, S^A, which depending on all of the parameters is:

$$S^A(f, n_f, T) > S^A(n_f, T) = -k_B \sum_i P_i^B \ln P_i(n_f, T) > S \tag{19.30}$$

In Equation (19.30), $S^A(f, n_f, T)$ is defined for partial HSMC and $S^A(n_f, T)$ for complete HSMC. Notice that unlike the SAWs case, all these probabilities and that for HS are defined on the configurational space of the Ising model. The HS and HSMC techniques are also applicable to the Ising model with an external magnetic field and to various lattice gas models.

19.5 The HS and HSMC Methods for a Continuum Fluid

Application of the HS method to a continuum fluid is similar to the procedure described above for the Ising model, with some modifications. For simplicity, we shall treat N argon atoms in a box of volume, V (*NVT* ensemble), interacting via a Lennard-Jones potential [Equation (17.23)]. In practice, the box is divided into $L^3 = L \times L \times L$ cubic cells with a maximal size that still guarantees that no more than one center of a spherical argon atom occupies a cell. During the analysis of configuration i, of a given MC (or MD) sample, the cells are visited orderly line-by-line layer-by-layer starting from one corner of the box until all of them have been treated. At step k of the process, N_k atoms (that is, N_k occupied cells) and $k - 1 - N_k$ vacant cells have already been treated. These N_k atoms are now positioned at their coordinates of configuration i, and together with the already visited vacant cells, they define the (frozen) "past;" the $L^3 - (k - 1)$ as yet unvisited cells (including target cell k) define the "future volume." In the process, one might consider the entire future of $f = L^3 - (k - 1)$ cubic cells or a partial future defined by a box of size f.

19.5.1 The HS Method

As for the Ising model, one defines two *future* canonical partition functions, $Z_k(-, f)$ and $Z_k(+, f)$, for vacant and occupied cell k, respectively (for brevity, we omit the temperature, T, from Z_k). $Z_k(+, f)$ and $Z_k(-, f)$ are obtained by scanning all of the possible configurations $\mathbf{x}^{N-N_k}$ of the remaining $N - N_k$ (future) atoms in the future volume, while the past volume is excluded, and for $Z_k(-, f)$, the target cell k is excluded as well. It is stressed that while the previously treated N_k atoms are fixed, their interactions with the future atoms are included in the calculation of $Z_k(+, f)$ and $Z_k(-, f)$. Now, if cell k is vacant, the TP$_k$ is, $p(-, f) = Z_k(-, f)/[Z_k(+, f) + Z_k(-, f)]$. If on the other hand, cell k is occupied, the future partition function, $Z_k(+, \mathbf{x}', f)$ is calculated, where one of the future atoms is fixed at the position, $\mathbf{x}'$—the exact location (inside the target cell k) at which an atom was exhibited in configuration i. $Z_k(+, \mathbf{x}', f)$ thus covers a portion of the total configurational volume spanned by $Z_k(+, f)$. TP$_k$ for an occupied cell is the probability density, $Z_k(+, \mathbf{x}', f)/[Z_k(+, f) + Z_k(-, f)]$. The TPs can be obtained by an exact enumeration [$\int \exp(-E/k_B T)$] over a partial box of size f, a process, defining the HS method, which has been found to be of a limited efficiency, since only relatively small future boxes (f) can be used [10].

19.5.2 The HSMC Method

One can also apply the partial HSMC(D) [11], or consider the complete HSMC(D) [6,12], where the entire future is treated, that is $f = L^3 - (k - 1)$. In what follows, we describe the latter case applied with HSMC, where, for brevity, the word *complete* will be omitted.

The process of obtaining the TPs is the following: At step k, the previously defined N_k atoms, as well as their associated images, are held fixed in their assigned positions (in configuration i), while all the remaining $N - N_k$ future atoms are allowed to move. An MC trajectory is generated for the $N - N_k$ future atoms, and the TP is determined from atom counts in the target cell k. The simulation is performed under standard periodic rules, with the exception that regions inside previously defined cells are excluded. Any trial move that would place a future atom into this previously assigned volume is rejected. A two-dimensional representation of the main simulation box is given in Figure 19.3.

The TPs are calculated (from the counts) in the following way. We denote by n_f the total number of attempted moves in the MC simulation for any reconstruction step k. n_{cell} is the number of counts for which an atom was observed in the target cell k. The probability for the target cell to be occupied (unoccupied) by an atom is thus given by p_{occ} (p_{unocc}):

$$p_{occ} = \frac{n_{cell}}{n_f} \quad \text{and} \quad p_{unocc} = 1 - p_{occ} \tag{19.31}$$

For the case where the target cell k is vacant in configuration i, the transition probability is $TP_k = P_{unocc}$. For an occupied target cell, one has to calculate the probability density, ρ_{occ}, for an atom to be located at the precise location (inside cell k) at which it is found in configuration i. For this, we define a much smaller volume, V_{cube} (inside cell k), termed a "cube," which is centered at the exact atom position, $\mathbf{x}'$ (in configuration i). We count the number of visitations of atoms within this cube during the future MC simulation, n_{cube}, and thus using Equation (19.31) estimate the probability density as:

$$\rho_{occ} = p_{occ}\left(\frac{n_{cube}}{n_{cell}}\right)\left(\frac{1}{V_{cube}}\right) = \left(\frac{n_{cube}}{n_f}\right)\left(\frac{1}{V_{cube}}\right) \tag{19.32}$$

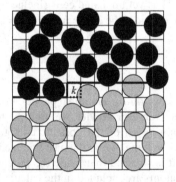

FIGURE 19.3 A two-dimensional (2d) illustration of the main simulation box at the kth step of the HSMC reconstruction. The 2d "volume" is divided into cells, where $k - 1$ of them have already been considered in previous steps (starting from the upper left corner). These $k - 1$ cells comprise the "past volume" (the region above the heavy lines), which contains previously treated fixed atoms that are denoted by full black circles, defined by the van der Waals radius. This region is excluded from the moveable future atoms (denoted by full gray circles), which are thus simulated in the "future volume" below the heavy lines, while in the presence of the fixed atoms. The future atoms can visit the target cell k (depicted by dotted lines), and their counts in this cell lead to the transition probability of an empty cell or the transition probability density of an occupied one. Note that for the case of an occupied target cell, counts are actually accumulated for visitations to a smaller region, V_{cube} (see text), located inside the target cell, but not shown in the figure.

Still, this probability density is uniform over the cube, while the particle is located at position, $\mathbf{x}'$. Therefore, we seek to improve ρ_{occ} by considering the probability density for the exact location, $\mathbf{x}'$; for that, we replace $1/V_{cube}$ by,

$$\frac{1}{V_{cube}} \rightarrow \frac{Z_{\mathbf{x}'}}{Z_{cube}} \tag{19.33}$$

At step k, there are $N_f = N - N_k$ future particles. Omitting for brevity, the variable k, $Z_{\mathbf{x}'}$ is a future partition function consisting of one future atom located at $\mathbf{x}'$, while the remaining $N_f - 1$ atoms are free to move in the future volume, while Z_{cube} is the cube's partition function obtained by integrating $Z_{\mathbf{x}}$ over the cube, $\int_{cube} Z_{\mathbf{x}} d\mathbf{x}$. Thus,

$$\frac{Z_{\mathbf{x}'}}{Z_{cube}} = \frac{\int\limits_{future} [\exp- E(\mathbf{x}^{N_f-1}, \mathbf{x}') / k_B T] d\mathbf{x}^{N_f-1}}{\int\limits_{\mathbf{x} \in cube} \int\limits_{future} [\exp- E(\mathbf{x}^{N_f-1}, \mathbf{x}) / k_B T] d\mathbf{x}^{N_f-1} d\mathbf{x}} \tag{19.34}$$

Multiplying the exponent in the numerator by:

$$[\exp- E(\mathbf{x}^{N_f-1}, \mathbf{x}) / k_B T][\exp+ E(\mathbf{x}^{N_f-1}, \mathbf{x}) / k_B T] = 1 \tag{19.35}$$

and performing an additional integration over the cube leads to V_{cube}, which thus should divide the numerator; we also define:

$$\Delta E(\mathbf{x}, \mathbf{x}', \mathbf{x}^{N_f-1}) = E(\mathbf{x}^{N_f-1}, \mathbf{x}) - E(\mathbf{x}^{N_f-1}, \mathbf{x}') \tag{19.36}$$

which is the difference between the interaction energy of the atom at $\mathbf{x}$ and the surroundings ($\mathbf{x}^{N_f-1}$) and the interaction energy of the atom at $\mathbf{x}'$ with the surroundings. Finally, we define the probability density $\rho(\mathbf{x}^{N_f-1}, \mathbf{x})$ for an atom to be at place $\mathbf{x}$ in the cube. One obtains,

$$\frac{Z_{\mathbf{x}'}}{Z_{cube}} = \frac{1}{V_{cube}} \int\limits_{\mathbf{x} \in cube} \int\limits_{future} \rho(\mathbf{x}^{N_f-1}, \mathbf{x}) \exp \Delta E(\mathbf{x}, \mathbf{x}', \mathbf{x}^{N_f-1}) / k_B T] d\mathbf{x}^{N_f-1} d\mathbf{x}$$

$$= \frac{1}{V_{cube}} \left\langle \exp \Delta E(\mathbf{x}, \mathbf{x}', \mathbf{x}^{N_f-1}) / k_B T \right\rangle_{future, cube} \tag{19.37}$$

This average is computed during the HSMC simulation in the following way. Every time an atom is found in the cube, one calculates the resulting (hypothetical) potential energy for this atom to be repositioned at $\mathbf{x}'$, this is done keeping all other ($N_f - 1$) atoms fixed. Thus, $\Delta E(\mathbf{x}, \mathbf{x}', \mathbf{x}^{N_f-1})/k_B T$ is calculated and accumulated, leading to the sum,

$$\frac{1}{n_{cube}} \sum_{t=1}^{n_{cube}} [\exp \Delta E(\mathbf{x}_{i(t)}, \mathbf{x}', \mathbf{x}_{i(t)}^{N_f-1})] \tag{19.38}$$

The *estimated* transition probability density is then:

$$\overline{\rho_{occ}} = \left(\frac{1}{n_f}\right)\left(\frac{1}{V_{cube}}\right) \sum_{t=1}^{n_{cube}} [\exp \Delta E(\mathbf{x}_{i(t)}, \mathbf{x}', \mathbf{x}_{i(t)}^{N_f-1})] \tag{19.39}$$

(where $\rho_{occ} \equiv \rho^{HS}$). Typical values for the ensemble average (in brackets) are on the order of 1. Nevertheless, the above correction has been found to improve the overall results significantly.

The total product of TP_k over all L^3 cells—a product of N transition probability densities ρ_{occ} and $L^3 - N$ transition probabilities for empty cells, gives rise to the estimate, $\rho^{HS}(\mathbf{x}^N)$ for the Boltzmann probability density, $\rho^B(\mathbf{x}^N)$,

$$\prod_k TP_k = N!\rho^{HS}(\mathbf{x}^N) \approx N!\frac{\exp[-E(\mathbf{x}^N)/k_BT]}{Z_N} \tag{19.40}$$

Notice that the counting procedure (for the TP_k) does not distinguish between labeled atoms, while in the integration leading to Z_N, all the labeled arrangements contribute, hence, $\rho(\mathbf{x}^N)$ is a labeled probability density, and $N!$ is required in Equation (19.40). It should be pointed out that for water, the procedure is more complicated because there are also rotational degrees of freedom. As has been discussed for HSMC(SAWs), ρ^{HS} is a *stochastic* probability density, due to the future MC scanning. However, if n_f is large, the related noise is reduced, and ρ^{HS} is expected to lead to an entropy functional S^A, which according to Jensen' Inequality (Section A4), constitutes an upper bound to S,

$$S^A = -k_B \int \rho^B \ln \rho^{HS} d\mathbf{x}^N \tag{19.41}$$

This is the basic implementation of HSMC to argon. More enhancements of the method can be found in [6,10,11]. Notice that the calculated free energy denoted A_c is:

$$A_c = -k_BT \ln\left(\frac{Z_N}{N!\sigma^{3N}}\right) \tag{19.42}$$

where s is the Lennard-Jones distance parameter.

19.5.3 Results for Argon and Water

19.5.3.1 Results for Argon

They are presented in Table 19.2 at $T = 96.53$K, at (reduced) density, $\rho^* = N\sigma^3/V = 0.846$ (to be distinguished from the probability density, ρ^{HS}), where V is the volume, N is the number of particles, and σ and ε are the standard Lennard-Jones energy and distance parameters, respectively [Equation 17.23)] [6]. All the free energy values in the table are given as $A_c/\varepsilon N$, where A_c is the configurational free energy [Equation (19.42)]. σ_A -the standard deviation of F^A, is also per atom. n is the sample size, and $\overline{n}_f$ is the average number of MC steps per cell. F^A is a lower bound, while F^B and F_G^B are upper bound functionals for the free energy; F^M, F_G^M, and F^D are expected to lead to the correct free energy. R_{biased} is defined in Equation (19.15). For definitions, see Section 19.3.

The table shows the expected decrease of $-F^A$ and σ_A and the corresponding increase of $-F^B$ and $-F_G^B$ as n_f is increased. However, for $N = 125$, the increase of $-F^B$ is less pronounced than that of $_-F_G^B$, which demonstrates that the latter functional is the more robust among the two. For $N = 64$, the results for F^M, F_G^M, and F^D are equal to the TI values with even somewhat smaller errors, while those errors exceed the TI error for $N = 125$ (besides for F_G^M). Also, R_{biasd}—the acceptance rate for the reversed-Schmidt procedure [Equation (19.15)], as expected, increases monotonically with increasing $\overline{n}_f$. Figure 19.4 demonstrates the converges of the upper bound and lower bound results of the free energy functional to the correct free energy value.

Table 19.3 presents HSMC free energy results ($A_c/(N)$ obtained from the correlation between F^A and its fluctuation σ_A (Section 19.3.5). The function, $F^A(\sigma_A)$, is approximated from multiple fits (data sets) created from various combinations of (σ_A, F^A) points (approximations) taken from Table 19.2, all for the same N; the results given are averages over all of the data sets. F^{exp} [Equation (19.27)] is the extrapolated free energy, F^{up} is an upper bound defined by the tangent of the fit at the best approximation, and F^{M2} is the average of F^{up} with the corresponding F^A value [Equation (19.28)]. The statistical error is defined in the caption of Table 19.1.

TABLE 19.2

Free Energy Results for Argon Obtained by HSMC

$\overline{n}_f$	$-F^A$	σ_A	$-F^B$	$-F_G^B$	$-F^M$	$-F_G^M$	$-F^D$	R_{biased}	n
				$N = 64$					
720,000	4.132 (1)	0.0330 (5)	4.064 (4)	4.046 (3)	4.098 (4)	4.089 (2)	4.096 (3)	0.11	581
1,440,000	4.117 (1)	0.0224 (5)	4.079 (4)	4.077 (2)	4.098 (4)	4.097 (1)	4.098 (3)	0.19	495
2,880,000	4.1085 (8)	0.0167 (5)	4.087 (3)	4.086 (2)	4.098 (3)	4.097 (1)	4.097 (2)	0.38	459
7,200,000	4.1046 (5)	0.0105 (5)	4.096 (2)	4.096 (1)	4.100 (2)	4.1002 (7)	4.100 (1)	0.53	371
14,400,000	4.1025 (5)	0.0078 (3)	4.097 (1)	4.0976 (6)	4.100 (1)	4.1001 (5)	4.1000 (8)	0.60	244
28,800,000	4.1019 (4)	0.0053 (5)	4.099 (1)	4.0997 (6)	4.100 (1)	4.1008 (5)	4.1007 (8)	0.76	174
TI	4.100 (1)		4.100 (1)	4.100 (1)	4.100 (1)	4.100 (1)	4.100 (1)		
				$N = 125$					
1,000,000	4.139 (1)	0.0246 (5)	4.08 (2)	4.045 (4)	4.11 (2)	4.092 (2)	4.10 (1)	0.08	362
2,000,000	4.124 (1)	0.0175 (6)	4.06 (2)	4.077 (4)	4.09 (2)	4.100 (2)	4.09 (1)	0.06	179
4,000,000	4.116 (1)	0.0110 (9)	4.10 (1)	4.097 (3)	4.11 (1)	4.107 (2)	4.108 (7)	0.31	125
10,000,000	4.1124 (6)	0.0083 (5)	4.10 (1)	4.102 (1)	4.10 (1)	4.1070 (9)	4.105 (6)	0.36	170
20,000,000	4.1102 (6)	0.0060 (5)	4.10 (1)	4.105 (1)	4.11 (1)	4.1074 (8)	4.107 (4)	0.43	99
TI	4.108 (1)		4.108 (1)	4.108 (1)	4.108 (1)	4.108 (1)	4.108 (1)		

Source: Reprinted with permission from White, R.P. and Meirovitch, H., *J. Chem. Phys.*, 121, 10889–10904, 2004. Copyright 2004 by the American Institute of Physics.

Note: Free energy values are given as $A_c/\varepsilon N$, where A_c is the configurational free energy, ε is the standard Lennard-Jones energy parameter, and N is the number of atoms. F^A is a lower bound of the free energy and σ_A is its fluctuation. F^B is an upper bound and F_G^B is the corresponding Gaussian approximation. F^M and F_G^M are the averages of F^A with F^B and F_G^B, respectively. F^D is the direct estimate for the free energy. R_{biased} is the acceptance rate for the reversed-Schmidt procedure, and R_G is the corresponding Gaussian result. $\overline{n}_f$ is the average number of MC steps per cell. n is the number of configurations analyzed (the sample size), where a single HSMC reconstruction was performed on each configuration. Results obtained by thermodynamic integration are denoted as TI. The statistical error appears in parenthesis, see Table 19.1.

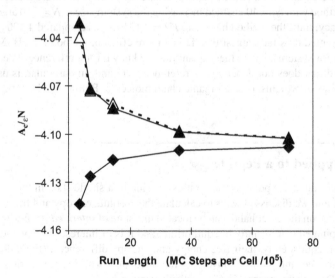

FIGURE 19.4 Free energy bounds as a function of HSMC run length for argon, $N = 32$ atoms. The HSMC run length on the horizontal axis is given as $\overline{M}_{tor}$, the average number of MC steps per cell. Shown are the free energy lower bound F^A (diamonds and solid lines), the upper bound F^B (open triangles and dashed lines), and the Gaussian upper bound F_G^B (solid triangles and solid lines). Free energies are given as $A_c/\varepsilon N$, where A_c is the configurational free energy defined in Equation (19.42), ε is the standard Lennard-Jones energy parameter, and N is the number of atoms. (Reprinted with permission from White, R.P. and Meirovitch, H., *J. Chem. Phys.*, 121, 10889–10904, 2004. Copyright 2004 by the American Institute of Physics.)

TABLE 19.3

Extrapolated Free Energy Results for Argon.

N	$-F^{up}$	$-F^{M2}$	$-F^{extp}$	$F(TI)$	Data Sets
64	4.0996 (3)	4.1010 (1)	4.1009 (2)	4.100 (1)	28
125	4.1036 (7)	4.1075 (6)	4.1065 (10)	4.108 (1)	10

Source: Reprinted with permission from White, R.P. and Meirovitch, H., *J. Chem. Phys.*, 121, 10889–10904, 2004. Copyright 2004 by the American Institute of Physics.

Note: Free energy values are given as $A_c/\varepsilon N$, where A_c is the configurational free energy, ε is the Lennard-Jones energy parameter, and N is the number of atoms. The function, $F^A(\sigma_A)$, is lower bound, F^{extp} is the extrapolated free energy, F^{up} is an upper bound defined by the tangent of the fit at the best approximation, and F^{M2} is the average of F^{up} and F^A. Results obtained by thermodynamic integration are denoted as TI. The statistical error is defined in the caption of Table 19.1. For more explanations see section 19.3.5 and the paragraph above Table 19.3.

The table reveals that the results for F^{extp} and F^{M2} are equal to the TI values within the error bars, which are larger for $N = 125$ than for $N = 64$.

19.5.3.2 Results for Water

HSMC results for the free energy [6] were obtained also for 64 TIP3P water molecules (for the definition of this water model see Section 16.5), where the trends shown in Tables 19.2 and 19.3 have been demonstrated as well—the accuracy is increased with increasing n_f—the amount of future MC simulations; the best result obtained is $F_G^M = 5.596$ (8), as compared to 5.599 (2) kcal/mol obtained by TI. These results and those of Tables 19.1 through 19.3 demonstrate that HSMC is an exact method in the sense that its accuracy can be increased at will by increasing a single parameter, n_f. However, HSMC cannot handle large systems because the time required for a future MC step, t_{step}, increases with N, the corresponding n_f should also be increased, and the total computer time is larger than $N^2 n_f$, where N is the number of molecules.

Still, one can resort to the partial HSMC method, where the number of future molecules, N_f, is limited. Using the latter method with $N_f = 40$ enabled increasing the system size to $N = 216$ argons, with the price of a reduced accuracy; thus, the method has led to $F^M = 4.131$ (3), as compared 4.120 (1) obtained by TI. All the results presented thus far suggest that TI is a more efficient method than HSMC for calculating the absolute S or F for systems, where these quantities are known for a reference state. However, HSMC is a general method that does not depend on a reference state and, in particular, is useful for peptides, side chains, and loops in proteins, or any organic chain molecule. Below, we describe HSMC as applied to a peptide.

19.6 HSMD Applied to a Peptide

The systems studied thus far—polymers and fluids—reside in a single state in thermodynamic equilibrium (Section 5.2), and we discussed ways to calculate the absolute entropy and free energy of this state. Peptides and proteins, on the other hand, can be located in a state of *intermediate flexibility*, where several microstates are populated significantly in equilibrium, and it is of interest to know their relative populations (stabilities) or, equivalently, their free energy and entropy differences. (For a discussion about the definition of a microstate, and problems involving intermediate flexibility, see Section 16.10). Therefore, extending HSMC(D) to these systems is of a particular importance.

A peptide in explicit water is commonly simulated by MD, therefore, instead of HSMC, we shall mostly refer below to the *complete* HSMD, which will be called just HSMD; application of HSMC(D) to peptides was published in [13]. As pointed out in Section 16.2, a $3d$ structure of a peptide or a protein is generally called "*conformation*" rather than "*configuration*," used for a state of a fluid. For simplicity, HSMD is described as applied to a peptide consisting of three glycine residues, depicted in Figure 19.4, where, for further simplicity, the oxygens and most of the hydrogens are discarded. As a first step, we order the heavy atoms along the chain and number them by k', $k' = 1,..,10$. The reconstruction of a peptide configuration, i, taken from an MD sample is performed in internal coordinates, where the bond lengths are assumed to be constant (for details, see [14]).

Thus, at step k' of the reconstruction process, the TPs of atoms $1,...,k' - 1$ have already been determined, and these atoms remain fixed at their positions in conformation, i. The position of atom k' is defined solely by two angles, a bond angle and a dihedral angle. These angles are ordered along the chain and are denoted by α_k, $k = 2k'$ ($k = 1, 2, ...,20$). Thus, the TP for atom k' is determined by α_{k-1} and α_k ($\alpha_{2k'-1}$ and $\alpha_{2k'}$), as described below. After the TP of atom k' has been determined, the atom is placed in its position in conformation, i, and the next atom is treated, and the process continues until TPs have been calculated for all the atoms of the peptide.

The figure is concentrated on step (or atom) $k' = 6$. At this stage, atoms $k' = 1...5$ (depicted by full spheres) define the fixed "past." At this step (6), one calculates the TPs related to bond angle α_k (defined by C'–C$_\alpha$–N) and dihedral angle α_{k-1} (defined by C'–C$_\alpha$–N–C'), which both define the position of atom C' ($k = 12$). For that, one defines the "future" atoms, which are the as yet untreated atoms, $k' = 6,...,10$; these atoms are depicted by empty spheres connected by dashed lines. Now, one carries out an MD simulation of the future atoms, while the past atoms $k' = 1...5$ are kept fixed in their positions. Every, say, 10 fs, the Cartesian coordinates of the future atoms of the peptide (and the coordinates of the corresponding "past" fixed atoms) are stored in files for a later analysis. Thus, altogether, n_f conformations are stored for step $k' = 6$ and a total of $10\,n_f$ conformations are stored for the molecule depicted in Figure 19.5. This defines stage 1 of the process. It should be noted that if the peptide is immersed in a box of water, the water molecules are also considered to be future molecules, and they are also moved in the MD simulation; however, their coordinates, in general, are not stored in files. Notice that the calculations of this stage 1 can be carried out with any of the available programs (AMBER, CHARMM, TINKER, NAMD, etc. discussed in Section 16.3).

TP($k' = 6$) is a conditional probability density, $\rho(\alpha_{11}\alpha_{12}|\alpha_{10},...,\alpha_1)$ (depending on the previous angles, $\alpha_{10},...,\alpha_1$), which is calculated (together with the other TPs) in stage 2 by a program which uses the stored data. Thus, two small segments (bins) $\delta\alpha_{k-1}$ and $\delta\alpha_k$ are centered at $\alpha_{k-1}(i)$ and $\alpha_k(i)$, respectively, and the number of *simultaneous* visits, n_{visit}, of the (stored) future chains to these two bins during the simulation is counted; for the general case, one obtains:

$$\rho(\alpha_{k-1}\alpha_k|\alpha_{k-2},\cdots,\alpha_1) \approx \rho^{HS}(\alpha_{k-1}\alpha_k|\alpha_{k-2},\cdots,\alpha_1) = \frac{n_{visit}}{n_f\delta\alpha_{k-1}\delta\alpha_k} \qquad (19.43)$$

In this equation, $\delta\alpha_k$ should appear with a Jacobian ($\sin \theta$), which, for simplicity, has been omitted. While the correct $\delta\cos(\alpha_k)$ is considered in the analysis of stage 2, we note, that for the models studied thus far,

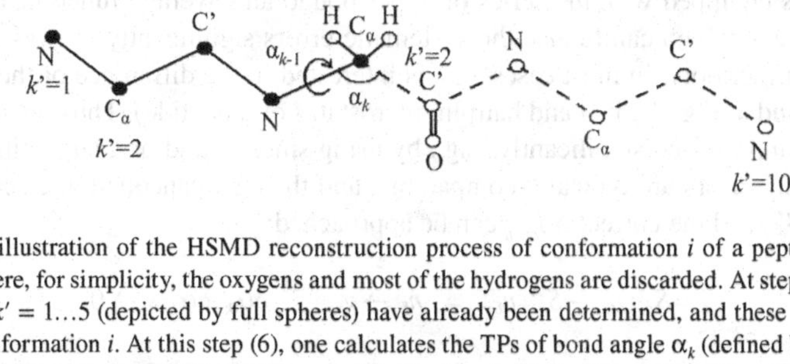

FIGURE 19.5 An illustration of the HSMD reconstruction process of conformation i of a peptide consisting of three glycine residues, where, for simplicity, the oxygens and most of the hydrogens are discarded. At step $k' = 6$, the TPs related to the "past" atoms $k' = 1...5$ (depicted by full spheres) have already been determined, and these atoms are kept fixed in their positions at conformation i. At this step (6), one calculates the TPs of bond angle α_k (defined by C'–C$_\alpha$–N) and dihedral angle α_{k-1} (defined by C'–C$_\alpha$–N–C'), which are related to C', where $k = 2k'$, and thus $k = 12$. The TPs are obtained from an MD simulation, where the as yet unreconstructed atoms $k' = 6...10$ (the "future" atoms) are moved (depicted by empty spheres connected by dashed lines), while the past atoms $k' = 1...5$ are kept fixed; notice that the future part should remain within the limits of the microstate and future-past interactions are taken into account. Small bins $\delta\alpha_{k-1}$ and $\delta\alpha_k$ are centered at the values α_{k-1} and α_k in i. The TP is calculated from the number of visits [n_{visit}, Equation (19.43)] of the future part to $\delta\alpha_{k-1}$ and $\delta\alpha_k$ *simultaneously* during the simulation. After $\rho^{HS}(\alpha_{11}\alpha_{12}|\alpha_{10},\cdots,\alpha_1)$ [Equation (19.44)] has been determined, the coordinates of C' (and O) are fixed at their positions in i, i.e., they become "past" atoms and the process continues.

the contribution of the Jacobians has been cancelled out in entropy *differences*, and the results with $\delta\alpha_k$ and $\delta\cos(\alpha_k)$ were found to be actually the same.

Clearly, the longer the (reconstruction) MD simulation (in stage 1), the larger is n_f, and the better is the approximation of $\rho^{HS}(\alpha_{k-1}\alpha_k | \alpha_{k-2}, \cdots, \alpha_1)$; this approximation, which will also be improved for smaller bins (provided that the statistics are adequate, that is, that n_f is sufficiently large.); the TPs become exact for $n_f \to \infty$ and $\delta\alpha_{k-1}$, $\delta\alpha_k \to 0$. Thus, in practice, the TPs will be somewhat approximate due to insufficient future sampling (finite n_f) and relatively large bin sizes (in [15], it is shown that $\delta\alpha_k$ and $\delta\alpha_{k+1}$ can be optimized). Therefore, the probability density of conformation i, ρ_i, will also be approximate, and its normalization might be somewhat violated. Notice that, typically, a peptide (loop) is simulated within the limits of a microstate, m, and therefore, the *future* peptide's conformations generated by MD at each step k' should remain within m—a condition which, in general, is satisfied. The corresponding probability density related to the peptide of K backbone atoms ($K = 10$ in the figure) is:

$$\rho^{HS}(\alpha_K, \cdots, \alpha_1) = \rho^{HS}([\alpha_k]) = \prod_{k=2,2}^{K} \rho^{HS}(\alpha_{k-1}\alpha_k | \alpha_{k-2}, \cdots, \alpha_1) \tag{19.44}$$

where the index, i, is omitted since it is defined by $\alpha_K, \cdots, \alpha_1$. $\rho^{HS}([\alpha_k])$ will define an approximate entropy functional S_m^A, for microstate, m. It can be shown (using Jensen's Inequality) that S_m^A constitutes a *rigorous* upper bound for the correct S (defined by replacing $\rho^{HS}([\alpha_k])$ by $\rho_m^B([\alpha_k])$ in Equation (19.45) below, see [3,6]):

$$S_m^A = -k_B \int_m \rho_m^B([\alpha_k]) \ln \rho^{HS}([\alpha_k]) d[\alpha_K] \tag{19.45}$$

ρ_m^B [Equation (19.45)] is the Boltzmann probability density of $[\alpha_K]$, with which the conformations were generated by MD in m ($S_m^A \geq S_m$ is also known as the Gibbs' inequality). We note again that $\rho^{HS}(\alpha_K, \cdots, \alpha_1)$ is a stochastic probability density, and therefore its normalization might be somewhat violated—a problem which in practice is expected to disappear with increasing n_f. However, as discussed in detail for SAWs, we shall consider $\rho^{HS}(\alpha_K, \cdots, \alpha_1)$ and S_m^A as "deterministic" entities, as long as in simulations, S_m^A will decrease as the approximation improves. S_m^A [Equation (19.45)], which is defined up to an additive constant, can be estimated from a Boltzmann MD sample of size n by $\overline{S}_m^A$ [see Equation (19.3)], where the corresponding free energy is estimated by $\overline{F}_m^A$.

S_m^A constitutes a measure of a pure geometrical character for the peptide flexibility, that is, with no *direct* dependence on the interaction energy. However, as discussed earlier, HSMC(D) is not an efficient method because the number of build-up steps increases with system size, and the step errors are accumulated. Still, one is equipped with the series of upper bound and average functionals F^B, F^M, F^D, etc. defined in Section 19.3, which can reduce the systematic errors significantly, as was shown for SAWs, argon, and water. Furthermore, in most cases, one is interested in the difference of the entropy between two microstates, m and n (e.g., helical and hairpin microstates of a peptide). Thus, while the accuracy of S_m^A and S_n^A can be compromised significantly [e.g., by using small n_f and/or a large bin $\delta\alpha_k$], it turns out that the corresponding errors are typically comparable, and thus get cancelled in calculations of differences $\Delta S_{m,n}^A = S_m^A - S_n^A$, and the correct $\Delta S_{m,n}$ can be approached:

$$\Delta S = S_m^A(n_f) - S_n^A(n_f) \qquad n_f \to \infty, \qquad \delta\alpha_k, \delta\alpha_{k-1} \to 0 \tag{19.46}$$

Thus, in practice, one calculates several approximations for S_m^A and S_n^A for increasing values of n_f and decreasing bins, verifying that both entropies decrease monotonically as the approximation improves, that is, both approach the correct values from above. Typically, the convergence of ΔS^A (to ΔS) is much faster than that of the individual entropies. This process is continued until ΔS^A does not change within a desired error, and thus it is considered to be exact within this error. Notice that the error also depends on the sample size, n, and other simulation parameters (see discussion on errors in Section 19.2). It should be emphasized that because the *whole* future is scanned at each reconstruction step, one has the freedom to determine the order with which these steps are carried out. Thus, the HSMD process can be started from any of the chain's ends or from the middle of the chain, treating first one half and then the second half of the chain, etc. While for large n_f, the entropy will be the same, in some cases, changing the reconstruction order can lead to more structural information [see the discussion following Equation (23.68)]. Finally, we point out that in stage 2, a program exists, which enables one using the raw data of stage 1 to produce results for $S^A(n_f,\alpha_k,\alpha_{k-1},n)$ for various sets of these parameters, where the best set of parameters minimizes $S^A(n_f,\alpha_k,\alpha_{k-1},n)$.

19.6.1 Applications

HSMD was applied to decaglycin, $(Gly)_{10}$, and the peptide, $(Val)_2(Gly)_6(Val)_2$, in vacuum, modeled by the AMBER6 force field in the helical, extended, and hairpin microstates [15]. The expected relations among, F^A, F^B, F^M, and F^D have always been satisfied by reconstructing both sample of chains or a single chain. These results are always larger (that is better) than results obtained by the QH and the LS method (described below—see also [16]). Thus, the free energy difference (ΔF^A) between *significantly different* microstates can be obtained from two simulations only, without the need to resort to a thermodynamic integration from microstate m to n.

HSMD was also applied to the flexible 7-residue surface loop, 304–310 (Gly-His-Gly-Ala-Gly-Gly-Ser) of the enzyme porcine pancreatic α-amylase [17]. As the ligand binds to the protein, the loop undergoes a structural change, and the interest has been to calculate from two separate MD samples of the free and bound microstates, the corresponding entropy and free energy differences. The calculation is based on the AMBER force field and AMBER with the implicit solvation, GB/SA (discussed in Sections 16.3 and 16,4, respectively). Again, the HSMD results for F^A are better (higher) than results obtained by QH and LS.

In Summary: HSMC(D) can treat any chain molecule (not necessarily a peptide) with and without side chains. One has to order the atoms along the chain and define the corresponding "dihedral" and bond angles. Also, as stated earlier, bond stretching is ignored because its contribution to *entropy differences* is expected to get cancelled. However, if necessary, this effect can be included easily in the framework of HSMD by defining a bond length bin $\delta(r^3/2)$. Obviously, this will require increasing n_f to get statistically reasonable values for n_{visit}, where n_{visit}, in this case, is the number of *simultaneous* visits of the future chain to the three bins $\delta\alpha_{k-1}$, $\delta\cos(\alpha_k)$, and $\delta(r^3/2)$. Finally, it should be emphasized that for a peptide in explicit water, the conformational entropy of the peptide is obtained without the entropy of water, which is taken into account by HSMD-TI discussed below.

19.7 The HSMD-TI Method

HSMD described thus far leads only to the conformational entropy of the peptide, even if it is solvated by a box of water. To calculate the entropy of the peptide/water system one would need to apply HSMD to the entire system, peptide + water, by reconstructing both components in any order desired. However, because one is also interested in the *conformational* entropy of the peptide (in water), the peptide should be reconstructed first. (Note that in such peptide reconstruction the *future* coordinates include the coordinates $\mathbf{x}^N$ of the N water molecules in the box that are moved by MD as well). Next, the configuration of water, $\mathbf{x}_i^N$, should be reconstructed in the presence of the *fixed* peptide structure in system configuration, i; as we have seen, this reconstruction of water is feasible (Section 19.5.3 and [6]), but it would be time consuming [6,18]. However, since one is interested mainly in the free energy difference between two microstates m and n of the same peptide, the contribution of water to the free energy, $F_{water}(i)$ (in the presence of a fixed peptide structure), can be obtained more efficiently by TI. Thus, one writes $F_{water}(i)$ as a sum of two contributions: (1) F_w—the free energy of free water in the box (that is, with zero interactions with the fixed peptide structure) and (2) F_i^{TI}, which is obtained by switching on gradually the peptide-water interactions from zero to their full values using a TI (or Free Energy Perturbation (FEP)) process. Thus, one obtains for the free energy difference (for the peptide-water system), ΔF_{mn},

$$\Delta F_{mn} = [E_{\text{intra-peptide}}(m) - TS_m^A + F_W + F^{TI}(m)] - [E_{\text{intra-peptide}}(n) - TS_n^A + F_W + F^{TI}(n)]$$

$$= \Delta E_{\text{intra-peptide}} - T\Delta S_{\text{peptide}} + \Delta F^{TI} \tag{19.47}$$

where F_W is cancelled, $E_{\text{intra-peptide}}(m)$ is the intra-peptide energy, S_m^A is the peptide's conformational entropy, and $\Delta S_{\text{peptide}}$ is the *converged* value of the differences $S_m^A - S_n^A$ [see Equation (19.46)]; all the quantities are averaged over the samples of size, n (to be distinguished from microstate n). Notice that for practical reasons, TI is performed in an opposite direction, where the peptide water interactions are gradually eliminated. This method, called HSMD-TI, has been suggested in [18], where the 7-residue surface loop of porcine pancreatic α-amylase is studied in TIP3P explicit water rather than in implicit solvation [17]. The intermediate flexibility of other mobile loops has been studied as well, among them—the 287–290 mobile loop of Acetylcholineesterase [19] and the 8-residue mobile loop, 45–52 of streptavidin [14]; the free energy of binding of ligand-protein complexes in explicit water has also been studied (see Chapter 23). These systems were modeled mainly by the AMBER force field and TIP3P water [Sections 16.3 and 16.5]. It should be emphasized that the functionals F^B, F^D, etc. are, *in practice*, *not* applicable within the scope of HSMD-TI.

19.8 The LS Method

The LS method [20] is a general technique, which enables one to extract the entropy from an MC or MD sample. The method is efficient for lattice models, such as the Ising model, and for continuum models of peptides, loops in proteins, etc.; LS is less efficient for random coil polymers and fluids. Therefore, we discuss below the application of LS to the Ising lattice and to peptides.

19.8.1 The LS Method Applied to the Ising Model

The LS method is described below for the Ising model on a square lattice of $N = L \times L$ spins. The local states method is based on the same philosophy described for HS and HSMC in Section 19.1. The difference lies in the definition of the TPs. Thus, one seeks to analyze a given MC sample of Ising lattices (generated at temperature, T), by assuming that each lattice has been filled with spins step-by-step, line-by-line, starting from an initially empty lattice using TPs. At step k of the reconstruction of lattice configuration, i, the TPs of the $k - 1$ "past" spins have already been determined, and these spins appear on the lattice with their signs defined on the reconstructed configuration i; the $N - k$ as yet untreated "future" sites are assumed to be empty, and the last L spins added to the lattice at sites $k - 1$, $k - 2$, ... $k - L$, which are "exposed" to the empty (future) part of the lattice, are called the *uncovered* spins. Because of the nearest neighbor interaction, the TP depends only on these L uncovered spins, where their effect is expected to decrease as the distance from site k increases. Thus, the most influential spins for determining TP_k are its nearest neighbor spins at sites $k - 1$ and $k - L$. The next influential spin is at site $k - L + 1$, which has the next shortest distance from site k in terms of lattice bonds on the empty lattice (2 bonds). The next effective uncovered spins are at sites $k - 2$ and $k - L + 2$, which are distant from k by three (empty) lattice bonds, etc. The lowest approximation is to assume that only the nearest neighbor spins affect TP_k. These sites can have $2^2 = 4$ spin configurations, which are called local states, and are denoted by $j, j = 1,...,4$. The calculation of TP_k is done in two rounds. First, the whole sample is visited, and for every site k of configuration, i, the current local state, j, and the spin at site k (σ_k) are identified and recorded. This leads to, $n(\sigma_k | j)$—the number of times the combination (σ_k, j) appear in the whole sample and to the transition probability $p(\sigma | j)$,

$$p(\sigma | j) = \frac{n(\sigma | j)}{Nn} \tag{19.48}$$

where nN is the total number of sites in the sample. Because the Ising model is homogeneous, the local states and TPs (unlike for SAWs) are site independent, and the subscript k has therefore been omitted in the above equation. Thus, the sample size, nN, of the local states increases with increasing both the lattice size, N, and the number of configurations, n. At stage 2, the sample is visited again, where the probability, $p(\sigma|j)$, for each site k is assigned according to the current σ_k and j [Equation 19.48]. The product of these TPs leads to an approximate probability, P_i, of configuration, i, which we write in the general case for b (back) uncovered spins,

$$P_i(b) = \prod_{k=1}^{N} p[\sigma_k | j(b,k)] \tag{19.49}$$

where the local state, $j(b, k)$, is a function of b and k. Clearly, as b is increased, the approximation improves, where for $b = L$ (that is, the entire set of uncovered spins is considered), LS becomes an exact method. However, the number of local states increases as 2^b, which puts a practical limit on the size of b. The above probability defines an entropy functional S^A, which constitutes an upper bound for the correct S [see discussion for Equation (19.2)],

$$S^A = -k_B \sum_i P_i^{B} \ln P_i(b) \tag{19.50}$$

It should be pointed out that the functionals introduced earlier, S^B, S^D, S^M, etc. (see Section 19.3) are also applicable to LS. Thus, LS is a convenient tool for calculating S and F due to its pure statistical nature, which does not require (in contrast to HS and HSMC) calculating the interaction energy. A graphical explanation of LS is given in Figure 19.6.

To demonstrate the high efficiency of LS, Table 19.4 below presents the excellent results obtained for the Ising model on a square lattice [7]. The results were calculated at three Ising temperatures, $J/k_BT = K$ [J is the Ising interaction constant (Equation (7.1))], in the hot region at $K = 0.40$, in the cold region at $K = 0.46$, and at the transition temperature, $K_c = 0.44068...$ The calculations are based on $b = 12$, where S^M and F^M are the averages of the upper and lower bounds of the entropy and free energy, respectively [Equation (19.16)]; E is the energy.

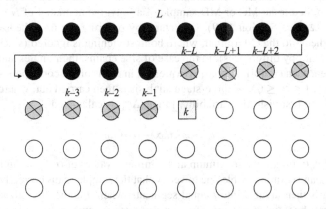

FIGURE 19.6 A diagram illustrating the philosophy of the LS method applied to the square Ising lattice of $L \times L$ spins. The LS method assumes that the Ising lattice can be constructed by adding spins step-by-step, line-by-line to an initially empty lattice using TPs, which depend on the already determined spins. The full circles in the figure denote lattice sites already filled with spins (± 1) in previous steps of process (steps $1,2,...,k − 1$). The TPs depend only on the L "uncovered" spins added in the last L steps (grey crossed circles). The figure presents an approximation based on the six uncovered spins closest to site k ($k − 1 \rightarrow k − 3$ and $k − L \rightarrow k − L + 2$); thus, there are altogether $2^6 = 64$ local states, j and their corresponding TPs, $p(+,j) = 1 − p(−, j)$, to determine spin k. These TPs are determined from the population of the corresponding local states in a large MC sample of the Ising model. Then, the probability of each member of the sample is obtained by the product of its L^2 specific TPs.

TABLE 19.4

Results for the Square Ising Model of $N = L^2$ Spins at Ising Temperature, K

$K = J/k_B T$	L	$-E/JN$	$-E_{exact}/JN$	$S^M/k_B N$	$S_{exact}/k_B N$	$-F^M/k_B TN$	$-F_{exact}/k_B TN$
0.40	100	1.1060 (2)	1.10608	0.43698 (9)	0.43693	0.879365 (4)	0.879364
0.46	80	1.581 (1)	1.5812	0.2317 (5)	0.23153	0.958877 (2)	0.958871
$0.44068 = K_c$	40	1.430 (2)	1.4298	0.29995 (7)	0.30013	0.92982 (9)	0.930095

Source: Reprinted with permission from Meirovitch, H., *J. Chem. Phys.*, 111, 7215–7224, 1999. Copyright 1999 by the American Institute of Physics.

Note: E is the energy, S is the entropy, and F is the free energy. The local-states results, based on correlation parameter, $b = 12$, are compared to the exact results of Onsager for $K = 0.40$ and 0.46 [21], and to those of Ferdinand and Fisher [22] for $K_c = 0.44068...$ S^M and F^M are defined in Equation (19.16).

The LS method was applied to several models simulated by MC. One is a lattice gas on a square lattice [23], where for the first time, the pressure was calculated from the free energy, and results in the two-phase region were obtained. In a subsequent study of a system of hard squares on a square lattice, very good results were obtained for the entropy, pressure, chemical potential, and several critical exponents [24]. Other studies of an Ising model on a cubic lattice [9], and an Ising antiferromagnet on the Face Centered Cubic (FCC) lattice [25], have led to better results for the entropy and the free energy than results obtained by thermodynamic integration in [26,27], respectively. It should be emphasized again that unlike LS, HS, and HSMC, thermodynamic integration is problematic when a phase transition occurs on the integration path. The very good results obtained in these studies demonstrate the suitability of LS to handle lattice models of that type. LS was also extended to problems in fluid dynamics by Chorin [28–30] and used by others [31,32]. However, for SAWs, LS (with TPs that are based on the "past") was found to be less efficient than HS, which due to its future "feelers," can handle efficiently the long-range interactions [33].

19.8.2 The LS Method Applied to a Peptide

Application of LS to a peptide is more complicated than to the Ising model because a peptide is not a homogenous system, and its internal coordinates are continuous variables. We describe this method, for simplicity, as applied to an MC or MD sample of a polyglycine model of N residues, say, in the helical microstate, Ω_h (see Section 16.9). One starts by ordering the $6N$ dihedral and bond angles sequentially from the N- to the C-terminal, where bond stretching is ignored (since it has been found to be ineffective in entropy differences); in the case of side chains, their angles should be arranged in this set as well. Then, each conformation, i, is expressed in terms of the corresponding set of (ordered) angles denoted $\{\alpha_k\}$, $1 \leq \alpha_k \leq 6N$. A three-stage analysis is then carried out, where the sample is visited three times. In the first visit, the variability range, $\Delta\alpha_k$, is calculated,

$$\Delta\alpha_k = \alpha_k(max) - \alpha_k(min) \tag{19.51}$$

where $\alpha_k(max)$ and $\alpha_k(min)$ are the maximum and minimum values of α_k found in the sample, respectively. $\Delta\alpha_k$, $\alpha_k(max)$, and $\alpha_k(min)$ enable one to verify that the sample spans correctly the Ω_h microstate.

Each range, $\Delta\alpha_k$, is then divided into l equal segments, where l is the discretization parameter. We denote these segments by $\nu_k(m)$ $m = 1,...,l$. Thus, an angle, α_k, is now represented by the segment, $\nu_k(m)$, to which it belongs, and a conformation, i, is expressed by the corresponding vector of segments $[\nu_1(m_1), \nu_2(m_2), ..., \nu_K(m_K)]$; however, for simplicity, we omit the letter m, and denote this vector by $(\nu_1, \nu_2, ..., \nu_K)$, where ν_2, for example, is representing, $\nu_2(m_2)$ of conformation, i. Under this discretization, a set of TP densities, $\rho(\alpha_k | \alpha_{k-1} \cdots \alpha_1)$, can, *in principle*, be estimated by:

$$\rho(\alpha_k | \alpha_{k-1} \cdots \alpha_1) \approx \frac{p(\alpha_k | \alpha_{k-1} \cdots \alpha_1)}{(\Delta\alpha_k / l)} = \frac{n(\nu_k, \cdots, \nu_1)}{n(\nu_{k-1}, \cdots, \nu_1)(\Delta\alpha_k / l)} \tag{19.52}$$

where $n(v_k, \cdots, v_1)$ is the number of times the *local state* [that is, the partial vector $(v_k, \cdots, v_1)$ representing $(\alpha_k, \cdots, \alpha_1)$] appears in the sample. Because the number of local states grows exponentially with k, one resorts to approximations based on smaller local states consisting of v_k and the b angles preceding it along the chain, that is, the vector $(v_k, v_{k-1}, \dots, v_{k-b})$, is considered; b (defined also for the Ising model) is called the correlation parameter.

Therefore, the sample is visited for the second time and, for a given b, one calculates the number of occurrences $n(v_k, v_{k-1} \dots, v_{k-b})$ of all the local states from which a set of TPs, $p(v_k | v_{k-1}, \dots, v_{k-b})$ are defined using Equation (19.52) for $n(v_k, v_{k-1}, \dots, v_{k-b})$. The sample is then visited for the third time, and for each member i of the sample, one determines the $6N$ local states and the corresponding TPs, whose product defines an *approximate* probability density $\rho_i(b, l)$ for conformation, i:

$$\rho_i(b,l) = \prod_{k=1}^{6N} p(v_k | v_{k-1}, \dots, v_{k-b}) / (\Delta \alpha_k / l), \qquad (19.53)$$

the larger are b and l, the better the approximation (for enough statistics). $\rho_i(b, l)$ allows one to define for the entropy and free energy, respectively, a rigorous upper bound, S^A, and a lower bound, F^A [Equations (19.2) and (19.8)]. Thus, with LS, the past is treated partially (through b), while the entire future is taken into account, in contrast to HS and partial HSMC, where the whole past is considered, but only a limited part of the future is taken into account through the scanning parameter, f. To improve LS, the parameters (b, l) should be increased, which require very large samples (meaning, a large disk space) to get the adequate statistics. On the other hand, to improve the HS and HSMC methods, one should increase computer time for increasing the future scanning parameter (f) and the MC simulations of the future chains, respectively.

LS has become an essential ingredient of a methodology for treating intermediate flexibility in cyclic peptides, protein loops, or bound ligands, which populate significantly several structurally different microstates in the thermodynamic equilibrium (see Section 16.9). Part of this methodology is a procedure for optimizing parameters of implicit solvation models (see Section 16.4) [34], which is carried out with an extensive conformational search using the *local torsional deformation* method for cyclic molecules [35]. Thus, using local torsional deformation, a set of significantly different energy minimized structures with energies close to the global energy minimum are identified. These structures become "seeds" for MC(MD) simulations, which span the conformational space of the corresponding microstates, i; the free energies, F_i, calculated by LS from these samples, lead to the relative populations, p_i, of the different microstates.

In other words, one can predict the most stable *solution* microstates and their populations ab initio, that is, without relying on experimental data. This methodology was introduced initially for the linear pentapeptide Leu enkephalin [36,37], for calculating nuclear magnetic resonance parameters—the nuclear Overhauser effect intensities and ^3J coupling constants, as averages of contributions from different microstates, i, weighted by p_i. Subsequently, nuclear magnetic resonance parameters of several cyclic peptides—a pentapeptide [38], a hexapeptide [39,40], and two heptapeptides [41] in dimethyl sulfoxide—have been calculated with the enhanced methodology (equipped with local torsional deformation and the optimization of the ASPs). The local states method was also applied to the cyclic hexapeptide cyclo-(Ala-Pro-D-Phe)$_2$ [42], where the reduction in entropy in going from vacuum to the crystal environment has been calculated. A much larger molecule studied is the gonadotropin releasing hormone (188 atoms), which is composed of a cyclic "base" and a relatively long tail [5]. The "tail-down" state of the entire molecule was found by LS to be more stable than "tail-up" in accord with previous predictions. Application of LS to the tail itself has shown that the tail-up state is of higher entropy than the tail-down, that is, it is the more flexible.

The local states method was also used for calculating the difference in the *backbone* conformational entropies of three loops of ten residues of the Ras protein, as a measure of their relative backbone flexibilities [43]. The correlation of these entropies with B-factors obtained from X-ray crystallography has been shown as well. The effect of entropy can be demonstrated by comparing energy differences between two microstates to the corresponding free energy differences. Using LS, such differences were obtained for three surface loops of 8, 9, and 12 residues of ribonuclease A [44]. Finally, LS was also applied to the linear peptide elastin [45].

Summary: As pointed out earlier, LS (unlike HS) is of a "geometrical" character, where no energy is involved in the determination of S. In this respect, the LS and the QH methods belong to the same category of techniques. However, while LS can handle samples consisting of several microstates and even the random coil state, QH leads to reliable results only for a single microstate. Still, for a single microstate, QH, which considers the (quadratic) correlations among *all* variables, is expected to lead to better results than LS. Indeed, for peptides [15,16] and a surface loop of the protein α-amylase [17,18], the entropy results of QH were found to be better (that is, smaller) than the LS results based on $b = 2$ and $l = 10$. However, LS has the potential for future improvements (by increasing b and l) with the constant enhancements in computer capabilities. Still, the corresponding results obtained with complete HSMC(D) have always been found to be the lowest because HSMC(D) considers *all* the (true) correlations among the variables, ν_k.

We note again that QH, MIE, NN, and MIE-NN discussed earlier (Sections 18.4 through 18.8) also consist of distributions of certain local states (which are not based on transition probabilities); however, unlike QH, the methods, LS, HSMC(D), MIE, NN, MIE-NN, and the mining minima technique (Section 18.3) are all applicable to samples spanning several microstates. Finally, it should be emphasized that the advantage of LS, HS, and HSMC(D) over the methods mentioned above, is that they provide both an upper and a lower bound for the entropy and the free energy.

REFERENCES

1. H. Meirovitch. Method for estimating the entropy of macromolecules with computer simulation. Chains with excluded volume. *Macromolecules* **16**, 249–252 (1983).
2. H. Meirovitch. Improved computer simulation method for estimating the entropy of macromolecules with hard-core potential. *Macromolecules* **16**, 1628–1631 (1983).
3. H. Meirovitch. Computer simulation of the free energy of polymer chains with excluded volume and with finite interactions. *Phys. Rev. A* **32**, 3709–3715 (1985).
4. T. L. Hill. *Statistical Mechanics Principles and Selected Applications.* (Dover, New York, 1987).
5. H. Meirovitch, S. C. Koerber, J. Rivier and A. T. Hagler. Computer simulation of the free energy of peptides with the local states method: Analogues of gonadotropin releasing hormone in the random coil and stable states. *Biopolymers* **34**, 815–839 (1994).
6. R. P. White and H. Meirovitch. Lower and upper bounds for the absolute free energy by the hypothetical scanning Monte Carlo method: Application to liquid argon and water. *J. Chem. Phys.* **121**, 10889–10904 (2004).
7. H. Meirovitch. Simulation of a free energy upper bound, based on the anti-correlation between an approximate free energy functional and its fluctuation. *J. Chem. Phys.* **111**, 7215–7224 (1999).
8. R. P. White and H. Meirovitch. Calculation of the entropy of random coil polymers with the hypothetical scanning Monte Carlo method. *J. Chem. Phys.* **123**, 214908–214911 (2005).
9. H. Meirovitch. Methods for estimating the entropy with computer simulation. The simple cubic Ising lattice. *J. Phys. A* **16**, 839–846 (1983).
10. A. Szarecka, R. P. White and H. Meirovitch. Absolute entropy and free energy of fluids using the hypothetical scanning method. I. Calculation of transition probabilities from local grand canonical partition functions. *J. Chem. Phys.* **119**, 12084–12095 (2003).
11. R. P. White and H. Meirovitch. Absolute entropy and free energy of fluids using the hypothetical scanning method. II. Transition probabilities from canonical Monte Carlo simulations of partial systems. *J. Chem. Phys.* **119**, 12096–12105 (2003).
12. R. P. White and H. Meirovitch. A simulation method for calculating the absolute entropy and free energy of fluids: Application to liquid argon and water. *Proc. Natl. Acad. Sci. USA* **101**, 9235–9240 (2004).
13. S. Cheluvaraja and H. Meirovitch. Simulation method for calculating the entropy and free energy of peptides and proteins. *Proc. Natl. Acad. Sci. USA* **101**, 9241–9246 (2004).
14. I. J. General and H. Meirovitch. Relative stability of the open and closed conformations of the active site loop of streptavidin. *J. Chem. Phys.* **134**, 025104–025117 (2011).
15. S. Cheluvaraja and H. Meirovitch. Calculation of the entropy and free energy of peptides by molecular dynamics simulations using the hypothetical scanning molecular dynamics method. *J. Chem. Phys.* **125**, 024905–024913 (2006).

16. S. Cheluvaraja and H. Meirovitch. Calculation of the entropy and free energy by the hypothetical scanning Monte Carlo Method: Application to peptides. *J. Chem. Phys.* **122**, 054903–054914 (2005).

17. S. Cheluvaraja and H. Meirovitch. Stability of the free and bound microstates of a mobile loop of α-amylase obtained from the absolute entropy and free energy. *J. Chem. Theory Comput.* **4**, 192–208 (2008).

18. S. Cheluvaraja, M. Mihailescu and H. Meirovitch. Entropy and free energy of a mobile loop in explicit water. *J. Phys. Chem. B* **112**, 9512–9522 (2008).

19. M. Mihailescu and H. Meirovitch. Absolute free energy and entropy of a mobile loop of the enzyme Acetyl choline esterase. *J. Phys. Chem. B* **113**, 7950–7964 (2009).

20. H. Meirovitch. Calculation of entropy with computer simulation methods. *Chem. Phys. Lett.* **45**, 389–392 (1977).

21. L. Onsager. Crystal statistics. I. A two-dimensional model with an order-disorder transition. *Phys. Rev.* **65**, 117–149 (1944).

22. A. E. Ferdinand and M. E. Fisher. Bounded and inhomogeneous Ising models. I. Specific-heat anomaly of a finite lattice. *Phys. Rev.* **185**, 832–846 (1969).

23. H. Meirovitch and Z. Alexandrowicz. Estimation of the pressure with computer simulation. The lattice-gas model. *Mol. Phys.* **34**, 1027–1035 (1977).

24. H. Meirovitch. A Monte-Carlo study of the entropy, the pressure and the critical behavior of the hard-square lattice gas. *J. Stat. Phys.* **30**, 681–698 (1983).

25. H. Meirovitch. Computer simulation study of hysteresis and free energy in the FCC Ising antiferromagnet. *Phys. Rev. B* **30**, 2866–2874 (1984).

26. K. Binder. Statistical mechanics of finite three-dimensional Ising models. *Physica* **62**, 508–526 (1972).

27. K. binder. Monte Carlo study of entropy for face-centered cubic Ising antiferromagnets. *Z. Phys. B* **45**, 61–69 (1981).

28. A. J. Chorin. Equilibrium statistics of a vortex filament with applications. *Commun. Math. Phys.* **141**, 619–631 (1991).

29. A. J. Chorin. Partition functions and equilibrium measures in two-dimensional and quasi-three-dimensional turbulence. *Phys. Fluids* **8**, 2656 (1996).

30. A. J. Chorin. *Vorticity and Turbulence.* (Springer-Verlag, Berlin, Germany, 1994).

31. S. K. Ma. *Statistical Mechanics.* (World Scientific, Singapore, 1985).

32. A. J. Schilijper and B. Smit. Two-sided bounds on the free energy from local states in Monte Carlo simulations. *J. Stat. Phys.* **56**, 247–260 (1989).

33. H. Meirovitch. On the simulation of the entropy of macromolecules with different flexibilities. *J. Chem. Phys.* **114**, 3859–3867 (2001).

34. C. Baysal and H. Meirovitch. Novel procedure for developing implicit solvation models for peptides and proteins. *J. Phys. Chem. B* **101**, 7368–7370 (1997).

35. C. Baysal and H. Meirovitch. New conformational search method based on local torsional deformations for cyclic molecules, loops in proteins, and dense polymer systems. *J. Chem. Phys.* **105**, 7868–7871 (1996).

36. E. Meirovitch and H. Meirovitch. New theoretical methodology for elucidating the solution structure of peptides from NMR data. II. Free energy of the dominant microstates of Leu-enkephalin and population-weighted average NOE intensities. *Biopolymers* **38**, 69–87 (1996).

37. H. Meirovitch and E. Meirovitch. New theoretical methodology for elucidating the solution structure of peptides from NMR data. III. Solvation effects. *J. Phys. Chem.* **100**, 5123–5133 (1996).

38. C. Baysal and H. Meirovitch. Ab initio structure prediction of a cyclic pentapeptide in DMSO based on an implicit solvation model. *Biopolymers* **53**, 423–433 (2000).

39. C. Baysal and H. Meirovitch. Determination of the stable microstates of a peptide from NOE distance constraints and optimization of atomic solvation parameters. *J. Am. Chem. Soc.* **120**, 800–812 (1998).

40. C. Baysal and H. Meirovitch. Free energy based populations of interconverting microstates of a cyclic peptide lead to the experimental NMR data. *Biopolymers* **50**, 329–344 (1999).

41. C. Baysal and H. Meirovitch. On the transferability of atomic solvation parameters: *Ab initio* Structural prediction of cyclic heptapeptides in DMSO. *Biopolymers* **54**, 416–428 (2000).

42. H. Meirovitch, D. H. Kitson and A. T. Hagler. Computer simulation of the entropy of polypeptides using the local states method: Application to Cyclo-(Ala-Pro-D-Phe)$_2$ in vacuum and the crystal. *J. Am. Chem. Soc.* **114**, 5386–5399 (1992).

43. H. Meirovitch and T. F. Hendrickson. The backbone entropy of loops as a measure of their flexibility. Application to a Ras protein simulated by molecular dynamics. *Proteins* **29**, 127–140 (1997).

44. B. Das and H. Meirovitch. Optimization of solvation models for predicting the structure of surface loops in proteins. *Proteins* **43**, 303–314 (2001).

45. Z. R. Wasserman and F. R. Salemme. A molecular dynamics investigation of the elastomeric restoring force in elastin. *Biopolymers* **29**, 1613–1631 (1990).

20

The Potential of Mean Force, Umbrella Sampling, and Related Techniques

In previous sections, we have described the free energy perturbation (FEP) and thermodynamic integration as fundamental tools for calculating free energy differences in complex systems. In this section, we introduce methods, designed in particular to calculate the potential of mean force (PMF), which lead to free energy differences along a reaction coordinate. We first discuss *"umbrella sampling"* and *"Bennett's acceptance ratio"*, and then present histogram techniques, such as the *"weighted histogram analysis method"* (WHAM).

20.1 Umbrella Sampling

We have seen [Equation (17.7)] that in a FEP step, the free energy difference, ΔF, is obtained by:

$$F_2 - F_1 = \Delta F = -k_B T \ln < \exp[-\Delta E(\mathbf{x}^N)/k_B T] >_1 . \tag{20.1}$$

The limitation of this approach for handling large systems or large windows has been emphasized. Thus, while a Monte Carlo (MC) simulation will select configurations from a narrow region around the typical energy $E_1^*(T)$, the largest contribution of the random variable $\exp[-\Delta E/k_B T]$ is obtained for $\mathbf{x}^N$ values for which $E_2(\mathbf{x}^N) << E_1(\mathbf{x}^N)$, and this region might not be reached in a practical simulation; therefore, small samples would provide wrong results. The idea of Torrie and Valleau [1,2] has been to apply importance sampling, that is, to select conformations with a non-Boltzmann *biased* probability distribution, $P^w(\mathbf{x}^N)$, which is wide enough to include the important energy regions; $P^w(\mathbf{x}^N)$ is defined with the help of a known function of the coordinates, $w(\mathbf{x}^N)$,

$$P^w = \frac{w(\mathbf{x}^N)\exp[-E_1(\mathbf{x}^N)/k_B T]}{\int w(\mathbf{x}^N)\exp[-E_1(\mathbf{x}^N)/k_B T]} = \frac{w(\mathbf{x}^N)\exp[-E_1(\mathbf{x}^N)/k_B T]}{Z_w} \tag{20.2}$$

The calculation of Z_w might be as difficult as the calculation of Z itself. However, each of the integrals in the numerator and denominator of ΔF [Equation (20.1)] can be converted into a statistical average defined with P^w,

$$\Delta F = -k_B T \ln \frac{1/Z_w \int w(\mathbf{x}^N)\exp[-E_1(\mathbf{x}^N)/k_B T]\{\exp[-\Delta E(\mathbf{x}^N)/k_B T]/w(\mathbf{x}^N)\}d\mathbf{x}^N}{1/Z_w \int \{w(\mathbf{x}^N)\exp[-E_1(\mathbf{x}^N)/k_B T]/w(\mathbf{x}^N)\}d\mathbf{x}^N}$$

$$= -k_B T \ln \frac{< \exp[-\Delta E/k_B T]/w >_w}{< 1/w >_w} . \tag{20.3}$$

$<>_w$ means an average defined with P^w. To evaluate these averages, one has to sample with P^w, while the *value* of P^w is not required. Because the value of $w(\mathbf{x}^N)$ is known, one can define, for example, an (asymmetric) MC procedure [Equation (10.11)] based on the probability $P^w(\mathbf{x}^N)$ [Equation (20.2)], and thus the transition probabilities are:

$$p_{ij} = \min\{1, (w_j / w_i)\exp[-(E_j - E_i)/k_B T]\}, \tag{20.4}$$

reminiscent in form to the Schmidt's procedure, Equation (14.36). Note that the normalization factor Z_w is cancelled. This procedure is the basis of umbrella sampling. The problem is to find a function w that induces minimal fluctuations for both averages. Obviously, for:

$$w = \frac{\exp[-\Delta E / k_B T]}{<\exp[-\Delta E / k_B T]/ w >_w} \tag{20.5}$$

the fluctuation of the random variable in the numerator of Equation (20.3) is nullified, whereas $w = 1/<1/w>$ nullifies the fluctuation in the denominator. However, these functions are unknown, and for large $\Delta E/k_B T$, the fluctuations of both random variables will still be large even for the most optimal w, meaning that the expected increase in the window size will be limited. In this context, we point out that the function, w, defined by Bennett can be optimized (see Section 20.2).

Torrie and Valleau used a tabular function for w that was used to calculate ΔF between a reference fluid and a Lennard-Jones system of 32 particles. Several attempts were made later to define a better function, w, for the umbrella sampling of fluids in general [3,4]. These methods led only to moderate improvements in efficiency, meaning that for *macroscopic* perturbations, several windows should be used between the initial and target Hamiltonians, and their contributions to the free energy should be combined. However, Valleau [5,6] developed a systematic way for computing, w, which enables one calculating system properties over a wide range of temperatures and densities. This problem was already considered in 1967 by McDonald and Singer [7,8], who developed a precursor of umbrella sampling. Umbrella sampling has been found useful for calculating the potential of mean forces of reaction coordinates, as discussed in Section 20.3.

20.2 Bennett's Acceptance Ratio

Another significant contribution, which overcomes limitations in the Zwanzig's FEP procedure and umbrella sampling, is Bennett's *"acceptance ratio procedure"* [9], already mentioned above. Thus, with FEP, the difference in the free energy, $\Delta F_{1,2}$, between two potential energies E_1 and E_2 (defined on the same volume, V) is obtained from a single simulation based on E_1 or E_2 [Equation (20.1)]. Bennett suggested expressing $\Delta F_{1,2}$ as a ratio of two averages, which requires one to carry out two simulations with respect to both E_1 and E_2,

$$\frac{Z_1}{Z_2} = \frac{Z_1 \int w(\mathbf{x}^N)\exp\{-[E_2(\mathbf{x}^N) + E_1(\mathbf{x}^N)]/k_B T\}d\mathbf{x}^N}{Z_2 \int w(\mathbf{x}^N)\exp\{-[E_2(\mathbf{x}^N) + E_1(\mathbf{x}^N)]/k_B T\}d\mathbf{x}^N} = \frac{\langle w\exp[-E_1 / k_B T]\rangle_2}{\langle w\exp[-E_2 / k_B T]\rangle_1} \tag{20.6}$$

where $w(\mathbf{x}^N)$ is a function, which is chosen to minimize the variances of both the numerator and denominator. The free energy difference is:

$$\frac{\Delta F}{k_B T} = \ln\langle w\exp[-E_1/k_B T]\rangle_2 - \ln\langle w\exp[-E_2 / k_B T]\rangle_1 \tag{20.7}$$

The averages of the random variables, $w\exp{-E_1/k_B T}$ and $w\exp{-E_2/k_B T}$, can be estimated from MC [or molecular dynamics (MD)] *uncorrelated* samples of sizes n_1 and n_2, respectively. Clearly, the accuracy of this estimation improves as the overlap between the two probability densities grows and as n_1 and n_2 increase; therefore, one seeks to find a function, w, that minimizes the variance of ΔF. The assumption is made that the difference, $(\Delta F_{est} - \Delta F)^2$, between the correct ΔF and its estimation, ΔF_{est}, is approximately Gaussian [$\ln(x_1) - \ln(x_2) = \ln(x_1/x_2) = \ln(x_1/(x_1 + \Delta x) = \ln(1/(1 + \Delta x/x_1) \approx \Delta x/x$; $(\Delta x)^2 = <x^2> - <x>^2$]; thus, treating each term in Equation (20.7)] separately, the combined n-dependent variance is,

$$(\Delta \bar{F}_{est} - \Delta F)^2 \approx \sigma^2 = \frac{\left\langle [w \exp(-E_2/k_BT)]^2 \right\rangle_1 - [\left\langle w \exp(-E_2/k_BT) \right\rangle_1]^2}{n_1[< w \exp(-E_2/k_BT) >_1]^2}$$

$$+ \frac{\left\langle [w \exp(-E_1/k_BT)]^2 \right\rangle_2 - [\left\langle w \exp(-E_1/k_BT) \right\rangle_2]^2}{n_2[\left\langle w \exp(-E_1/k_BT) \right\rangle_2]^2} \tag{20.8}$$

$$= \frac{\int \left\{ \left[\frac{Z_1}{n_1} \exp(-E_2/k_BT) + \frac{Z_2}{n_2} \exp(-E_1/k_BT) \right] w^2 \exp[-(E_1 + E_2)/k_BT] \right\} d\mathbf{x}^N}{\left\{ \int w \exp[-(E_1 + E_2)/k_BT] d\mathbf{x}^N \right\}^2} - \frac{1}{n_1} - \frac{1}{n_2}$$

Before minimization, it should be noted that multiplying w by a constant does not change the expression of Equation (20.8); therefore, without loss of generality, the integral in the denominator can be made stationary with respect to variation of w, and thus the integral can be fixed to a constant,

$$\int w \exp[-(E_1 + E_2)/k_BT] d\mathbf{x}^N - \text{const.} = 0 \tag{20.9}$$

Next, one defines a Lagrange function based on the numerator of Equation (20.8) and the left hand side of Equation (20.9), where the latter is multiplied by a Lagrange multiplier, λ. Derivation of this function with respect to w and equating the *integrands* to zero lead to:

$$\left\{ \left[\frac{Z_1}{n_1} \exp(-E_2/k_BT) + \frac{Z_2}{n_2} \exp(-E_1/k_BT) \right] \exp[-(E_1 + E_2)/k_BT] \right\} w \delta w$$

$$+ \lambda \exp[-(E_1 + E_2)/k_BT] \delta w = 0 \tag{20.10}$$

where $w(\mathbf{x}^N)$ is:

$$w = \frac{\text{const.}}{\frac{Z_1}{n_1} \exp(-E_2/k_BT) + \frac{Z_2}{n_2} \exp(-E_1/k_BT)}. \tag{20.11}$$

Inserting w in the right hand side of Equation (20.6) leads to:

$$\frac{Z_1}{Z_2} = \frac{\left\langle \left[1 + \exp(E_1 - E_2 + C)/k_BT) \right]^{-1} \right\rangle_2}{\left\langle \left[1 + \exp(E_2 - E_1 - C)/k_BT) \right]^{-1} \right\rangle_1} \exp(C/k_BT), \tag{20.12}$$

where C is defined by,

$$\exp(C/k_BT) = \frac{Z_1 n_2}{Z_2 n_1}. \tag{20.13}$$

Equation (20.12) can be expressed in terms of the Fermi-Dirac function $f(x) = 1/[1 + \exp(x/k_BT)]$,

$$\frac{Z_1}{Z_2} = \frac{\left\langle f(E_1 - E_2 + C) \right\rangle_2}{\left\langle f(E_2 - E_1 - C) \right\rangle_1} \exp(C/k_BT). \tag{20.14}$$

Equation (20.14) is valid for *any* value of C, while the optimal C [defined by Equation (20.13)] is a priori unknown since it depends on Z_1/Z_2 – our target of calculation. Still, the optimal C can be obtained by a self-consistent procedure as follows. Suppose that $\langle f(E_2 - E_1 - C)\rangle_1$ and $\langle f(E_1 - E_2 + C)\rangle_2$ have been estimated from two MC samples of sizes, n_1 and n_2, respectively; using an arbitrary C, one obtains,

$$\langle f(E_2 - E_1 - C)\rangle_1 \rightarrow \frac{1}{n_1} \sum_{k=1}^{n_1} f_k(E_2 - E_1 - C)$$

$$\langle f(E_1 - E_2 + C)\rangle_2 \rightarrow \frac{1}{n_2} \sum_{k=1}^{n_2} f_k(E_1 - E_2 + C) \tag{20.15}$$

Inserting Equation (20.15) in Equation (20.14) and taking the logarithm leads to:

$$\frac{\Delta F}{k_B T} = \ln \frac{\sum_{k=1}^{n_2} f_k(E_1 - E_2 + C)}{\sum_{k=1}^{n_1} f_k(E_2 - E_1 - C)} - \ln \frac{n_2}{n_1} + \frac{C}{k_B T} \tag{20.16}$$

while the optimal C, obtained from Equation (20.13) is:

$$\frac{\Delta F}{k_B T} = -\ln \frac{n_2}{n_1} + \frac{C}{k_B T}. \tag{20.17}$$

Thus, Equation (20.16) becomes consistent to Equation (20.17) when:

$$\sum_{k=1}^{n_2} f_k(E_1 - E_2 + C) = \sum_{k=1}^{n_1} f_k(E_2 - E_1 - C). \tag{20.18}$$

In practice, one calculates these summations for different values of C until Equation (20.18) is satisfied; then ΔF is calculated by Equation (20.17).

The minimal variance, σ^2, is obtained by substituting w in Equation (20.8) with the optimized w [Equation (20.11)]; one obtains,

$$\sigma^2 = \frac{1}{\int \frac{n_1 n_2 \rho_1 \rho_2}{n_1 \rho_1 + n_2 \rho_2} d\mathbf{x}^N} - \frac{n_1 + n_2}{n_1 n_2} \tag{20.19}$$

where $\rho_i = \exp(-E_i/k_B T)/Z_i$ $(i = 1,2)$. Since σ^2 is a decreasing function of n_1 and n_2, one can find a number $\bar{n}$ between them where:

$$\sigma^2 = \frac{2}{\bar{n}} \left[\frac{1}{\int \frac{2\rho_1 \rho_2}{\rho_1 + \rho_2} d\mathbf{x}^N} - 1 \right] \tag{20.20}$$

The integral in Equation (20.20) is a measure of the overlap in space between the two densities (for $\rho_1 = \rho_2$, $\sigma^2 = 0$; when ρ_1 and ρ_2 are completely separated in space $\sigma^2 = \infty$). Equation (20.20) says that estimation of Z_1/Z_2 with a desired accuracy requires sample sizes that are greater than the reciprocal of the overlap between ρ_1 and ρ_2. For the method to work, ρ_1 and ρ_2 should overlap to some extent, where the transition between them is relatively smooth. Additional components of the acceptance ratio method and a more detailed discussion about its performance appear in Bennett's original paper.

The acceptance ratio method is characterized by unique ingredients, which do not appear in the methods studied earlier—umbrella sampling and Zwanzig's perturbation theory. Thus, (1) while Bennett's theory is derived mainly within the framework of probability space, it also involves components of the simulation world—the sample sizes, n_1 and n_2 (see Section 1.4). (2) The variance is minimized, and (3) the difference in the free energy is determined self-consistently; also, as pointed out at the outset, Bennett's method is based on simulating two Hamiltonians. Most of these points have been borrowed by methods developed later by Swendsen's group (see Sections 20.4–20.5 below), and partially by "*the multicanonical method*", and "*the method of expanded ensembles*" described in Chapter 21. Indeed, the sophisticated acceptance ratio method has been found to perform better than other techniques in comparison studies conducted during the years; however, these studies are mostly based on simplified models (see Section 21.6.1).

Finally, note that in Chapter 22 (Section 22.5), both Bennett's method and umbrella sampling are used for obtaining the chemical potential through calculation of Z_N and Z_{N+1}. In this case, unlike in Equations (20.1, 3, and 6), the velocities parts of the partition functions are not cancelled and should be evaluated separately.

20.3 The Potential of Mean Force

The potential of mean force, W, is the free energy calculated under the restriction that some of the coordinates are constant. For a simple liquid, Ben Naim [10] defines,

$$W(\mathbf{x}_1, \mathbf{x}_2) = -k_B T \ln g(\mathbf{x}_1, \mathbf{x}_2)$$

$$= -k_B T \ln \left\{ \frac{N(N-1)}{\rho^2 Z} \int \exp[-E(\mathbf{x}^{(N-2)} \,|\, (\mathbf{x}_1, \mathbf{x}_2)) / k_B T] d\mathbf{x}^{(N-2)} \right\} \tag{20.21}$$

Here, $\mathbf{x}_1$ and $\mathbf{x}_2$ are held fixed and the integration is over the $(N - 2)$ remaining coordinates. Because $N(N - 1)$ pairs of molecules can occupy these positions, the integral is multiplied by this factor. The integral divided by Z is the probability density, $\rho^2(\mathbf{x}_1, \mathbf{x}_2)$, to find two particles at these positions. $g(\mathbf{x}_1, \mathbf{x}_2) = \rho^2(\mathbf{x}_1, \mathbf{x}_2)/\rho^2$ ($\rho = N/V$) is the pair correlation function.

For a simple fluid, $W = W(r)$, that is, W depends only on the distance r between the particles. When $r \to \infty$, $\rho^2(\mathbf{x}_1, \mathbf{x}_2) = \rho(\mathbf{x}_1)\rho(\mathbf{x}_2) = \rho^2 \to g(\mathbf{x}_1, \mathbf{x}_2) = 1$ and $W = 0$. $W(r)$ is the work required to bring two particles from infinite separation to distance r; W is called the PMF. However, in practice, one is not interested in the absolute W, but in the shape of W, from which relative free energies can be obtained.

In general, one defines a reaction coordinate, ξ, and calculates the PMF along this coordinate; an example is two hydrophobic particles in water, where the PMF is calculated as a function of their distance $r = \xi$. Another example is the angle $\chi = \xi$ of a side chain, where of interest is the PMF for changing χ from one torsional potential well to another. For a more complex ξ, defining a probability function [Equation (20.21)] might not be straightforward because of the need to calculate the Jacobian of the transformation of the coordinates. In practice, a standard (unbiased) long simulation will not cover the full range of ξ of interest. Therefore, one can carry out several simulations along ξ using umbrella sampling.

Thus, one selects several points $\xi_0{}^i$ along the reaction coordinate which constitute "seeds" around which simulations will be performed. To keep the system around ξ_0 (for simplicity, we omit i), a harmonic restoring potential $U(\xi_0, \xi)$ is applied.

$$U(\xi_0, \xi) = K(\xi - \xi_0)^2 \tag{20.22}$$

where K is a parameter and the biasing function w is,

$$w(\mathbf{x}^N) = \exp\{-K[\hat{\xi}(\mathbf{x}^N) - \xi_0(\mathbf{x}^N)]^2 / k_B T\} \tag{20.23}$$

In practice, one must estimate the probability [Equation (20.21)] in several values of ξ around ξ_0. $P(\xi)$ can be written:

$$P(\xi) = \frac{1}{Z} \int \delta[\xi - \hat{\xi}(\mathbf{x}^N)] \exp[-E(\mathbf{x}^N) / k_B T] d\mathbf{x}^N \qquad (20.24)$$

where δ is Dirac's delta function and the "hat" above $\hat{\xi}(\mathbf{x}^N)$ emphasizes that this is a function, as compared to the single value ξ. To evaluate this probability density (for various values $\xi = \xi'$), we multiply the integrand above and that in Z (below) by w/w. Also, we divide the integral and Z both by $\int w \exp[-E(\mathbf{x}^N)/k_B T]$ (nothing is changed),

$$P(\xi) = \frac{1}{Z} \int \delta[\xi - \hat{\xi}(\mathbf{x}^N)] \exp[-E(\mathbf{x}^N) / k_B T] d\mathbf{x}^N$$

$$= \frac{\dfrac{\displaystyle\int w(\mathbf{x}^N) \exp[-E(\mathbf{x}^N) / k_B T]\{\delta[\xi - \hat{\xi}(\mathbf{x}^N)] / w(\mathbf{x}^N)\} d\mathbf{x}^N}{\displaystyle\int w(\mathbf{x}^N) \exp[-E(\mathbf{x}^N) / k_B T] d\mathbf{x}^N}}{\dfrac{\displaystyle\int w(\mathbf{x}^N) \exp[-E(\mathbf{x}^N) / k_B T] / \{w(\mathbf{x}^N)\} d\mathbf{x}^N}{\displaystyle\int w(\mathbf{x}^N) \exp[-E(\mathbf{x}^N) / k_B T] d\mathbf{x}^N}}$$

$$= \frac{\dfrac{\displaystyle\int w(\mathbf{x}^N) \exp[-E(\mathbf{x}^N) / k_B T] \delta[\xi - \hat{\xi}(\mathbf{x}^N)] d\mathbf{x}^N}{\displaystyle\int w(\mathbf{x}^N) \exp[-E(\mathbf{x}^N) / k_B T] d\mathbf{x}^N} \exp[+U(\xi) / k_B T]}{\dfrac{\displaystyle\int w(\mathbf{x}^N) \exp[-E(\mathbf{x}^N) / k_B T] \exp[+U(\hat{\xi}(\mathbf{x}^N) / k_B T] d\mathbf{x}^N}{\displaystyle\int w(\mathbf{x}^N) \exp[-E(\mathbf{x}^N) / k_B T] d\mathbf{x}^N}} \qquad (20.25)$$

where in the last expression, $w(\mathbf{x}^N)$ has been substituted by its definition [Equation (20.23)] and the δ-function has been operated, leading to $\exp[+U(\xi) / k_B T]$. Thus,

$$P(\xi) = P^w(\xi) \frac{\exp[+U(\xi) / k_B T]}{< \exp[+U / k_B T] >_w} \qquad (20.26)$$

where $P^w(\xi)$ is the biased probability distribution defined in the upper fraction of Equation (20.25). $P(\xi)$ can be estimated by Metropolis MC using the biased probability [Equation (20.4)].

$$p_{ij} = \min\{1, (w_j/w_i) \exp[-(E_j - E_i)/k_B T] = \min\{1, \exp[-(E_j - E_i + U_j - U_i)/k_B T] \qquad (20.27)$$

That is, a trial structure is selected at random (within some small region); it is accepted according to the above p_{ij}. This simulation is used for estimating both the numerator and denominator; thus, one obtains,

$$\overline{P^w}(\xi) = \frac{1}{n_{\text{trial}}} \sum_{t=1}^{n_\xi} 1_t = \frac{n_\xi}{n_{\text{trial}}}$$

$$\qquad (20.28)$$

$$\overline{< \exp[U / k_B T] >_w} = \frac{1}{n_{\text{trial}}} \sum_{t=1}^{n_{\text{trial}}} \exp[U_t / k_B T]$$

where n_{trial} is the number of MC snapshots and n_ξ is the number of times ξ was observed. Notice that the bias introduced in the simulation is eliminated in the calculation of averages (similarly, in principle, to the Rosenbluth and Rosenbluth method, Section 14.4). For example, if $\xi = \xi_0$, $P^w(\xi)$ is expected to be too large and thus will be decreased by the factor, $\exp[+U(\xi)/k_BT]/<\exp[+U/k_BT]>_w = 1/<\exp[+U/k_BT]>_w < 1$. On the other hand, if ξ is distant from ξ_0, $P^w(\xi)$ will be too small and thus will be increased by the factor, $\exp[+U(\xi)/k_BT]/<\exp[+U/k_BT]>_w > 1$; clearly, in this case of $P^w(\xi) = 0$ or a statistically unreliable simulation, this compensation process will not work. Replacing MC by MD, the system will be simulated with the restoring potentials where the parameters in Equation (20.28) are estimated directly from the biased sample.

$P(\xi)$ is calculated around each of the seeds, $\xi_0{}^i$, where the constant K determines the range of ξ around each $\xi_0{}^i$ that will be sampled effectively. However, the normalization of the probability density of different windows $(\xi_0{}^i)$ is not the same because only small (different) parts of space are visited at each window. Therefore, the distance between neighbor seeds should not be too large because one wants to allow overlap of $P(\xi)$ results of neighbor seeds. The results for different windows can be tailored together as follows:

The match between $P_1(\xi)$ and $P_2(\xi)$ (around neighbor seeds $\xi_0{}^1$ and $\xi_0{}^2$) can be obtained by selecting ξ_a from the overlap region, calculating the ratio $P_1(\xi_a)/P_2(\xi_a)$, and changing any probability $P_2(\xi)$ from the second window to the tailored probability, $P(\xi)$,

$$P(\xi) = \frac{P_2(\xi)P_1(\xi_a)}{P_2(\xi_a)}.$$

(20.29)

The PMF is:

$$PMF = -k_BT \ln P(\xi).$$

(20.30)

The fact that the energies at different windows are not significantly different (unless near a transition state) and that U is a local interaction (meaning that $\exp[U/k_BT]$ is relatively small) contributes to the success of this umbrella sampling procedure. Still, the system should be equilibrated for each window, which might require long simulations.

This procedure was first suggested by Pangali, Rao, and Berne [11]. They studied various aspects of the hydrophobic effect using MC and calculating the PMF of two non-polar particles approaching each other in a bath of water molecules. If the hydrophobic interaction exists, one would expect to see the lowest PMF when the two particles become close to each other. The PMF taken from their paper presented in Figure 20.1 shows indeed such a decrease.

Many PMF studies for a wide range of problems have been carried out during the years, using the method described above and other techniques. However, more robust methodologies developed in the 1990s of the previous century [e.g., WHAM discussed in the following sections] have become the methods of choice. Still, it is beneficial to outline some of the earlier work which demonstrates the richness of this field with respect to both methods and systems. An early study in this area is that of the Karplus' group [12], who studied the PMF of rotation of a tyrosine ring in the protein Bovine Pancreatic Trypsin Inhibitor (BPTI). With methods developed by Voter [13] and Tobias and Brooks III [14], the PMF is not obtained by umbrella sampling, but by calculating differences in free energy for small deviations of $\pm\Delta\xi$.

Selecting the reaction coordinate and choosing an efficient umbrella sampling function is not always straightforward for complex systems and several attempts have been made to design better bias functions, w, and improved procedures. Mezei [15] suggested an iterative procedure in which the PMF results obtained from one set of simulations is used to define a better w for the next set. He found this "*adaptive umbrella sampling*" to be better than the usual umbrella sampling procedure in his application of this strategy to two conformational states of the alanine dipeptide in water. Hooft, van Eijck, and Kroon [16] suggested another method in this spirit in which, unlike Mezei's method, information from all the sets of simulations is taken into account. Their method was used for calculating the PMF of the central torsional

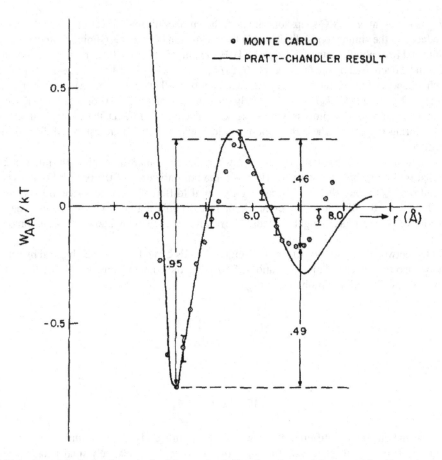

FIGURE 20.1 Potential of mean force obtained from a simulation of two Lennard-Jones particles (shown in circles) dissolved in 214 ST2 water molecules. Solid line shows the Pratt-Chandler result. (Reprinted with permission from Pangali, C. et al., *J. Chem. Phys.*, 71, 2975, 1979. Copyright 1979 by the American Institute of Physics.)

angle of glycol in vacuum, in water, and in CCl_4. For these systems, the efficiency of the method was found to be comparable to that of another method they had used earlier [17]. However, the first method in this category was suggested by Paine and Scheraga, who applied it to the 5-residue peptide Met-enkephalin [18].

Efficiency studies concerning umbrella sampling were also carried out by Beutler and van Gunsteren, who applied various procedures to 1,2-dichloroethane in water [19] and to glycine dipeptide in vacuum and in water [20]. An important part of these studies was to check the performance of two-dimensional bias functions. In a subsequent paper, Beutler et al. [21] further applied such functions to study PMFs of side chain rotations in the peptide antamanide and compared their results to those obtained from nuclear magnetic resonance (NMR) experiments. Finally, methods for better matching the results of different widows to form a single probability density function have also been suggested [22].

20.3.1 Applications

These methodology-based investigations were tested on relatively simple systems, and simple systems have been also the target of a vast number of applications involving PMF calculations. Thus, different conformational states of the alanine dipeptide in vacuum and water were obtained from PMF calculations [22,23]. PMFs of this system solvated by explicit water and modeled by the Poisson-Boltzmann equation and a hydrophobic term were found by Marone et al. to be comparable [24]; PMFs of the glycine dipeptide in water for two reaction coordinates were studied by Fraternali and van Gunsteren [25].

A large amount of work based on MC and MD simulations has been devoted to explaining the association of NaCl in water, using different models for water and the solute and varying computational parameters, such as cut-off distances of the interactions and the boundary conditions [26–34]. Chandrasekhar et al. studied the PMF of an S_N2 reaction between a chloride ion and methyl chloride in aqueous solution and demonstrated the large influence of the solvent on the PMF, as compared to the corresponding vacuum energy profile [35,36]. Another reaction, studied by Rosky's group, is of a sodium-dimethyl phosphate ion pairing [37,38].

Larger systems have also been studied. Thus, Tobias and Brooks III used their own technique to calculate the PMF of the central torsional angle of *n*-butane in vacuum [39], water, and CCl_4 [40] (see also [41,42]). They also applied umbrella sampling to evaluate the free energy of folding/unfolding of one turn of α helix of alanine and valine in water [43], along with the conformational equilibria of two blocked dipeptides as models of reverse turns [44]. We also mention the work of Roux and Karplus [45], who calculated the PMF of Na^+ and K^+ ions as a function of position in a model of gramicidin A, and the work of Dang and Kollman who calculated the free energy of association of 9-methyladenine and 1-methylthymine bases in water [46]. Finally, Warshel and collaborators have used umbrella sampling for calculating energy profiles in chemical reactions in solutions and proteins, as well as for ligand binding to proteins [47–49].

Increasing system size further would require a significant increase in computer time. In this context, it is of interest to mention the work of Wallqvist and Covell [50] (see Section 17.5.2), who studied by MD the free energy profiles of the conformational change of dodecane ($CH_3(CH_2)_{10}CH_3$) from an extended to a hairpin state in vacuum and in a box of 717 water molecules. Clearly, a hairpin to helix transformation would be much more complex. Bozko and Brooks III studied the PMF of a three-helix bundle, where the reaction coordinate was the radius of gyration [51]. Another study in which the PMF of a global conformation change was calculated is that of Wang et al. [52], who investigated twisting of β sheets in polypeptides containing up to 15 residues. More references for the early studies can be found in [53].

In summary, umbrella sampling and related methods have been found to be useful tools for calculating PMFs (hence, free energy differences) of relatively simple systems, where the reaction coordinate can be defined efficiently. In the sections below, more efficient approaches for calculating PMFs are described, based on ideas developed by Bennett and Swendsen's group.

20.4 The Self-Consistent Histogram Method

In Section 14.5.9 [Equation (14.44)], we have described a method suggested by Ferrenberg and Swendsen (based on a histogram of the energy values), where a *single* simulation carried out at a given temperature, T, leads to results for the probabilities, $P_{T_i}^B(E)$ at different *neighbor* temperatures T_i in terms of $P_T^B(E)$ (see [27], Chapter 14). In a subsequent paper [54], these authors have extended this idea to an *optimized* method—the self-consistent histogram method (SCHM), which combines data from an arbitrary number of simulations to obtain information over a wide range of parameter values. For example, the free energy can be obtained for a wide range of temperatures. Following Bennett's ideas, SCHM enables one to minimize the errors of the simulations. This is in contrast to Zwanzig's perturbation method and umbrella sampling, where the individual errors generated in n steps ("windows") are accumulated to a total error, which increases with n.

For simplicity, we introduce HCSM as applied to the ferromagnetic Ising model on a square lattice of N spins, where nearest neighbor spins σ_i and σ_j interact via $\varepsilon_{ij} = -J\sigma_i\sigma_j$; J is a constant energy parameter [Equation (7.1)]. Since the number of nearest neighbor spins is four, the energy E_i of system configuration i is ($i = 1,2,\ldots,2^N$):

$$E_i = -\frac{1}{2}\sum_{k=1}^{N}\sum_{j=1}^{4} J\sigma_{k(i)}\sigma_{j(i)} \qquad (20.31)$$

and the partition function, Z, is:

$$Z = \sum_{i=1}^{2^N} \exp(-E_i / k_B T) = \sum_E \Omega(E) \exp(-\beta E) \qquad (20.32)$$

where $\Omega(E)$ is the density of states and, for brevity, we define $\beta = 1/k_B T$. We consider R MC simulations, where the kth simulation is performed at inverse temperatures β_k ($k = 1,...,R$); the corresponding kth sample size consists of n_k snapshots, which are arranged in energy bins (histograms), $\{N_k(E)\}$, that is, the population of the bin at E is $N_k(E)$. The goal is to calculate the free energies related to the temperatures, β_k. Assuming that $N_k(E)$ is determined by the Poisson distribution [see Equations (1.38) and (1.47)]; the *estimated* variance is:

$$\delta^2[N_k(E)] = g_k \overline{N_k(E)} \qquad (20.33)$$

where the bar means an average based on many MC simulations of length n_k. If successive MC configurations are independent, then $g_k = 1$, otherwise:

$$g_k = 1 + 2\tau_k \qquad (20.34)$$

where τ_k is the correlation time [55].

20.4.1 Free Energy from a Single Simulation

We show now that a simulation at temperature, β_k, enables one to calculate the difference in free energy, $\beta_k F(\beta_k) - \beta F(\beta)$, for a neighbor temperature, β. For that, we define the function $z_k(\beta)$, which formally is a mixture of ensemble and simulation components,

$$\frac{\overline{z_k(\beta)}}{n_k} = \sum_E \frac{\overline{N_k(E)}}{n_k} \exp[(\beta_k - \beta)E] = \sum_E \overline{P_k(E)} \exp[(\beta_k - \beta)E] \approx$$

$$\approx \frac{\sum_E \Omega(E)\exp(-\beta_k + \beta_k - \beta)E}{\sum_E \Omega(E)\exp(-\beta_k E)} = \frac{Z(\beta)}{Z(\beta_k)} \qquad (20.35)$$

Defining the free energy function, $f(\beta) = \beta F(\beta) = -\ln Z(\beta)$ leads to the difference, Δf:

$$\Delta f = f(\beta_k) - f(\beta) = \ln \overline{z_k(\beta)} - \ln n_k. \qquad (20.36)$$

Clearly, this method will not work if the probability distributions $P_k(E)$ and $P_\beta(E)$ do not overlap; therefore, as the system size increases, β should be closer to β_k, $\beta \to \beta_k$. Notice also that not ΔF, but only Δf is calculated; therefore, only if the absolute value of $F(\beta_k)$ is known, the absolute value of $F(\beta)$ can be obtained as well. This method, based on a single sample, is extended below to include the effect of several simulations at different temperatures, β_k.

20.4.2 Multiple Simulations and The Self-Consistent Procedure

We write again the probability $P_k(E)$ of bin E,

$$P_k(E) = \frac{\Omega(E)\exp(-\beta_k E)}{Z(\beta_k)} \approx \frac{\overline{N_k(E)}}{n_k} \qquad (20.37)$$

where the *ensemble* expression in the middle term of Equation (20.37) is equated approximately to the *estimated* expression on the right hand side. Therefore, $\Omega(E)$, the density of states is:

$$\Omega(E) = \overline{N_k(E)}n_k^{-1}\exp(\beta_k E - f_k) \tag{20.38}$$

[Notice that $\Omega(E)$ and other quantities defined below are *estimated*, and according to our convention, should appear with a bar; however, for simplicity, in most estimated quantities in this section and in the following one, the bar is omitted]. Contributions of results obtained from the R simulations will be combined now [with a priori unknown probabilities $p_k(E)$] to a general expression, which will be shown to lead to an improved estimation of $\Omega(E)$,

$$\Omega(E) = \sum_{k=1}^{R} p_k(E)\overline{N_k(E)}n_k^{-1}\exp(\beta_k E - f_k); \quad \sum_{k=1}^{R} p_k(E) = 1 \tag{20.39}$$

Based on Equation (20.33) [compare also with Bennett's Equation (20.8)], the error, $\delta^2\Omega(E)$, is

$$\delta^2\Omega(E) = \sum_{k=1}^{R} p_k(E)^2 g_k \overline{N_k(E)}n_k^{-2}\exp 2(\beta_k E - f_k). \tag{20.40}$$

Substituting $\overline{N_k(E)}$ [Equation (20.38)] in Equation (20.40) leads to:

$$\delta^2\Omega(E) = \sum_{k=1}^{R} p_k(E)^2 g_k \Omega(E)n_k^{-1}\exp(\beta_k E - f_k). \tag{20.41}$$

Now, $\delta^2\Omega(E) + \lambda(\Sigma p_k(E) -1)$ is minimized with respect to the $p_k(E)$'s, where λ is a Lagrange multiplier; since the summands are independent, each can be treated independently; one obtains,

$$2p_k(E)g_k\Omega(E)n_k^{-1}\exp(\beta_k E - f_k) + \lambda = 0 \tag{20.42}$$

and due to normalization,

$$p_k(E) = \frac{g_k^{-1}n_k \exp(f_k - \beta_k E)}{\displaystyle\sum_{k=1}^{R} n_k g_k^{-1}\exp(f_k - \beta_k E)}. \tag{20.43}$$

Thus,

$$\Omega(E) = \frac{\displaystyle\sum_{k=1}^{R} g_k^{-1}N_k(E)}{\displaystyle\sum_{k=1}^{R} n_k g_k^{-1}\exp(f_k - \beta_k E)}. \tag{20.44}$$

Now, we define the *unnormalized* probability, $P(E,\beta)$ [compare with Equation (20.37)],

$$P(E,\beta) = \Omega(E)\exp(-\beta E). \tag{20.45}$$

Substituting $\Omega(E)$ by its value in Equation (20.44) leads to the multiple-histogram equation,

$$P(E,\beta) = \frac{\left[\sum_{k=1}^{R} g_k^{-1} N_k(E)\right] \exp(-\beta E)}{\sum_{k=1}^{R} n_k g_k^{-1} \exp(f_k - \beta_k E)}. \tag{20.46}$$

Also, summing over E, both sides of Equation (20.45) for a given k lead on the right side to $Z(\beta_k)$, thus,

$$\exp- f_k = \sum_{E} P(E,\beta_k) \tag{20.47}$$

The values of f_k are found self-consistently by iterating Equations (20.46) and (20.47). It should be noted that adding an arbitrary constant to these two equations does not affect the solution, meaning that only free energy differences, Δf, can be calculated. To obtain the absolute free energies, one should evaluate the free energy at a temperature for which its absolute value (F_k) is known; for the Ising model, such temperatures are the absolute zero or infinite T. Clearly, a Δf value for any temperature not appearing in the set can be obtained by Equation (20.36) applied to a neighbor temperature from the set. Having in hand a set of f_k enables one to calculate $P(E,\beta)$ [Equation (20.45)] for other β values, and thus to evaluate the average value of an operator $A(E)$ at β:

$$\overline{A(E,\beta)} = \sum_{E} A(E)P(E,\beta) \Big/ \sum_{E} P(E,\beta) \tag{20.48}$$

To calculate the statistical error in $P(E,\beta)$, we insert in $\delta^2\Omega$ [Equation (20.40)], the expressions for the probabilities [Equation (20.43)], which lead to:

$$\delta^2\Omega = \frac{\sum_{k=1}^{R} g_k^{-1} N_k(E)}{\left[\sum_{k=1}^{R} n_k g_k^{-1} \exp(f_k - \beta_k E)\right]^2}. \tag{20.49}$$

Dividing $\delta^2\Omega$ by Ω^2 [Equation (20.44)] and taking the square root gives,

$$\frac{\delta\Omega(E)}{\Omega(E)} = \frac{\delta P(E,\beta)}{P(E,\beta)} = \left(\sum_{k=1}^{R} g_k^{-1} N_k(E)\right)^{-1/2} \tag{20.50}$$

Equation (20.50) shows that the method always reduces the statistical errors for longer simulations and when additional simulations (samples) are added to the analysis. It should be noted, however, that Equation (20.46) is based on exponentials of extensive variables, which suggests that SCHM would be limited to relatively small systems. Thus, for a large system, the energies that contribute to the partition function concentrate in a very narrow range around the typical energy, $E(T)^*$ [see Equation (5.11)], and to obtain overlapping distributions, the range of treated temperatures is decreased and R should increase. The authors show that for the square Ising model of 256 sites, the method works well.

However, we note that for the Ising model, the local states method (Section 19.8.1) appears to have some advantages over SCHM. Thus, local states provide the *absolute* entropy from a single simulation at a temperature of interest. Also, the estimation of the local states transition probabilities improves with increasing both sample size and system size, N; furthermore, as the size of the local states increases, the fluctuation of the free energy decreases to zero [Equations (5.5–5.6)]. However, the ideas of SCHM are very useful in application to biological systems for calculating not only free energy differences, but also potentials of mean force along reaction coordinates. This widened method called *"the weighted histogram analysis"* is described below.

20.5 The Weighted Histogram Analysis Method

To describe WHAM, we mostly follow the description and notations of the original reference [56]. The Hamiltonian used conventionally in structural biology is the force field described earlier [Equation (16.1)], which will be denoted here by $\hat{H}_0(\mathbf{x})$, where $\mathbf{x}^N$, which stands for the $3N$ coordinates is denoted here for brevity by $\mathbf{x}$. As in umbrella sampling, where harmonic potentials are added to the force field, we introduce L restraining potentials, $\hat{V}_i(\mathbf{x})$, which are multiplied by the corresponding coupling parameters, λ_i. The general Hamiltonian $\hat{H}_{\{\lambda\}}(\mathbf{x})$ is thus,

$$\hat{H}_{\{\lambda\}}(\mathbf{x}) = \hat{H}_0(\mathbf{x}) + \sum_{i=1}^{L} \lambda_i \hat{V}_i(\mathbf{x}) = \sum_{i=0}^{L} \lambda_i \hat{V}_i(\mathbf{x}) \qquad (20.51)$$

where circumflexes over the symbols denote functions. The symbol in braces, $\{\lambda\}$, stands for the set of values λ_i, $i = 1,2,...,L$, where $\{0\}$ indicates that all λ_i, $i = 1,2,...,L$ are zero; λ_0 is always unity, and $\hat{V}_0(\mathbf{x}) = \hat{H}_0(\mathbf{x})$. The restraining potentials are designed to shift the sampling distribution along a coordinate of interest, such as a reaction coordinate, $\hat{\xi}(\mathbf{x})$. Several restraining potentials enable one to sample different regions of a long reaction path, and by adjusting the values of λ_i, any region of interest along, $\hat{\xi}(\mathbf{x})$, can be preferentially sampled. Thermodynamically, the λ_i can be considered as intensive parameters multiplying the energies, $V_i(\mathbf{x})$, which typically are *not* extensive. Thus, the partition function is a function of $\{\lambda\}$ and β,

$$Z_{\{\lambda\},\beta} = \sum_{\{V\},\xi} \Omega(\{V\},\xi) \prod_{i=0}^{L} \exp(-\lambda_i \beta V_i) \qquad (20.52)$$

where $\Omega(\{V\},\xi)$ is a generalized density of states,

$$\Omega(\{V\},\xi) = \int d\mathbf{x} \delta[\xi - \hat{\xi}(\mathbf{x})] \prod_{i=0}^{L} \delta[V_i - \hat{V}_i(\mathbf{x})]. \qquad (20.53)$$

$\Omega(\{V\},\xi)$ is independent of $\{\lambda\}$ and β.

Free energies and PMF can be obtained after corrections for the bias introduced by the restraining potentials. We write now the probability density, $P_{\{\lambda\},\beta}(\xi)$, obtained from a simulation with the Hamiltonian $\hat{H}_{\{\lambda\}}(\mathbf{x})$ at inverse temperature, $\beta = 1/k_B T$:

$$P_{\{\lambda\},\beta}(\xi) = \frac{1}{Z_{\{\lambda\},\beta}} \int \delta[\xi - \hat{\xi}(\mathbf{x})] \exp[-\beta \hat{H}_{\{\lambda\}}(\mathbf{x})] d\mathbf{x} = \left\langle \delta[\xi - \hat{\xi}(\mathbf{x})] \right\rangle_{\{\lambda\},\beta} = \exp[-\beta W_{\{\lambda\},\beta}(\xi)] \qquad (20.54)$$

where $W_{\{\lambda\},\beta}(\xi)$ is the (biased) potential of mean force,

$$W_{\{\lambda\},\beta}(\xi) = -k_B T \ln P_{\{\lambda\},\beta}(\xi). \tag{20.55}$$

Equation (20.54) is equivalent to Equation (20.24). While $P_{\{\lambda\},\beta}(\xi)$ is obtained directly from the (biased) simulation, one is interested in $P_{\{0\},\beta}(\xi)$ – the probability density that would be obtained from an unbiased sampling, where $\hat{H}_{\{\lambda\}}(\mathbf{x}) = \hat{H}_0(\mathbf{x})$, that is, $\lambda_i = 0$, $i = 1,...,L$, and $\lambda_0 = 1$ in Equation (20.51),

$$P_{\{0\},\beta}(\xi) = \exp[-\beta W_{\{0\},\beta}(\xi)]. \tag{20.56}$$

With Equations (20.52–20.56) above, $P_{\{0\},\beta}(\xi)$ can be expressed as,

$$P_{\{0\},\beta}(\xi) = \frac{Z_{\{\lambda\},\beta}}{Z_{\{0\},\beta}} \left\langle \delta[\xi - \hat{\xi}(\mathbf{x})] \prod_{i=1}^{L} \exp \beta \lambda_i \hat{V}_i(\mathbf{x}) \right\rangle_{\{\lambda\},\beta} \tag{20.57}$$

Restricting the restraining potentials to be functions of $\hat{\xi}(\mathbf{x})$ only (a common case), that is:

$$\hat{V}_i(\mathbf{x}) \equiv \hat{V}_i(\hat{\xi}(\mathbf{x})) \tag{20.58}$$

leads to the desired relations between the biased $P_{\{\lambda\},\beta}(\xi)$ and $W_{\{\lambda\},\beta}(\xi)$, and the corresponding unbiased $P_{\{0\},\beta}(\xi)$ and $W_{\{0\},\beta}(\xi)$; thus:

$$P_{\{0\},\beta}(\xi) = P_{\{\lambda\},\beta}(\xi) \frac{Z_{\{\lambda\},\beta}}{Z_{\{0\},\beta}} \exp\left[\sum_{i=1}^{L} \beta \lambda_i \hat{V}_i(\xi) \right] \tag{20.59}$$

and

$$W_{\{0\},\beta}(\xi) = W_{\{\lambda\},\beta}(\xi) - \beta^{-1} \ln \frac{Z_{\{\lambda\},\beta}}{Z_{\{0\},\beta}} - \sum_{i=1}^{L} \lambda_i \hat{V}_i(\xi) \tag{20.60}$$

These equations, which constitute a generalization of those discussed earlier in Section 20.3, provide the basis for two theories, *the single and multiple histogram equations*; these theories are extensions for biomolecules of the single and multiple SCHM procedures described in Sections 20.4.1 and 20.4.2, respectively.

20.5.1 The Single Histogram Equations

The objective is to generate the PMF profile of the coordinate ξ from a *single* simulation. Now, assume that to enhance sampling, the simulation is carried out at a relatively high temperature $T_1 = 1/k_B\beta_1$, where restraining potentials with coupling parameters $\lambda_1, \lambda_2,...,\lambda_L$ are imposed as well (still, $\lambda_0 = 1$); the data for ξ are divided into B bins denoted, $\xi_0, \xi_1,..., \xi_{B-1}, \xi_B$. However, we are interested in the probability to find ξ at the mth bin when the biasing potentials are removed and (in general) at a lower temperature $T_2 = 1/k_B\beta_2$. This is obtained by applying the ideas of Section 20.4.1 for the present much more complex biomolecule model. Thus, we first write the desired probability density, $P_{\beta_2,\{0\}}[\xi \in (\xi_m, \xi_{m+1})]$, in the *unbiased* ensemble, and then build upon it the probability density in the *biased* ensemble. Again, $\mathbf{x}$ is the $3N$ Cartesian coordinates of the system, and Z is the partition function,

$$P_{\beta_2,\{0\}}[\xi \in (\xi_m, \xi_{m+1})] = \frac{\int \exp(-\beta_2 \hat{V}_0(\mathbf{x})) \int_{\xi_m}^{\xi_{m+1}} \delta(\hat{\xi}(\mathbf{x}) - \xi) d\mathbf{x} d\xi}{\int \exp(-\beta_2 \hat{V}_0(\mathbf{x})) \int_{\xi_0}^{\xi_B} \delta(\hat{\xi}(\mathbf{x}) - \xi) d\mathbf{x} d\xi}$$

$$= \frac{\frac{1}{Z(\beta_1,\{\lambda\})} \int \exp\left[-\beta_2\hat{V}_0(\mathbf{x}) + \beta_1\hat{V}_0(\mathbf{x}) - \beta_1\hat{V}_0(\mathbf{x}) - \sum_{i=1}^{L}\beta_1\lambda_i\hat{V}_i(\mathbf{x}) + \sum_{i=1}^{L}\beta_1\lambda_i\hat{V}_i(\mathbf{x})\right] \int_{\xi_m}^{\xi_{m+1}} \delta(\hat{\xi}(x) - \xi) d\mathbf{x} d\xi}{\frac{1}{Z(\beta_1,\{\lambda\})} \int \exp\left[-\beta_2\hat{V}_0(\mathbf{x}) + \beta_1\hat{V}_0(\mathbf{x}) - \beta_1\hat{V}_0(\mathbf{x}) - \sum_{i=1}^{L}\beta_1\lambda_i\hat{V}_i(\mathbf{x}) + \sum_{i=1}^{L}\beta_1\lambda_i\hat{V}_i(\mathbf{x})\right] \int_{\xi_0}^{\xi_B} \delta(\hat{\xi}(\mathbf{x}) - \xi) d\mathbf{x} d\xi}$$

$$= \frac{\int P_{\beta_1,\{\lambda\}}(\mathbf{x}) \exp\left[\hat{V}_0(\mathbf{x})(\beta_1 - \beta_2) + \sum_{i=1}^{L}\beta_1\lambda_i\hat{V}_i(\mathbf{x})\right] \int_{\xi_m}^{\xi_{m+1}} \delta(\hat{\xi}(x) - \xi) d\mathbf{x} d\xi}{\int P_{\beta_1\{,\lambda\}}(\mathbf{x}) \exp\left[\hat{V}_0(\mathbf{x})(\beta_1 - \beta_2) + \sum_{i=1}^{L}\beta_1\lambda_i\hat{V}_i(\mathbf{x})\right] \int_{\xi_0}^{\xi_B} \delta(\hat{\xi}(\mathbf{x}) - \xi) d\mathbf{x} d\xi}$$

(20.61)

An estimation of the two integrals in the last line is obtained from a (biased) MD run (based on β_1 and $\lambda_1, \lambda_2, \ldots, \lambda_L$), which leads to:

$$\bar{P}_{\beta_2}[\xi \in (\xi_m, \xi_{m+1})] = \frac{\sum_{j=1}^{\eta(m)} \exp\left[(\beta_1 - \beta_2)V_{0,j}^m + \sum_{i=1}^{L}\beta_1\lambda_i V_{i,j}^m\right]}{\sum_{k=1}^{B}\sum_{j=1}^{\eta(k)} \exp\left[(\beta_1 - \beta_2)V_{0,j}^k + \sum_{i=1}^{L}\beta_1\lambda_i V_{i,j}^k\right]}$$

(20.62)

where the number of snapshots at the kth bin is $\eta(k)$, $k = 1, \ldots, B$ and $V_{i,j}^k$ is the value of the restraining potential i of the jth snapshot of the kth bin. It is convenient to express $\bar{P}_{\beta_2}$ in a somewhat different way,

$$\bar{P}_{\beta_2}(\xi) = \frac{\sum_{\{V\}} N_{\{\lambda\},\beta_1}(\{V\},\xi)\exp\left[(\beta_1 - \beta_2)V_0 + \sum_{i=1}^{L}\beta_1\lambda_i V_i\right]}{\sum_{\{V\},\xi} N_{\{\lambda\},\beta_1}(\{V\},\xi)\exp\left[(\beta_1 - \beta_2)V_0 + \sum_{i=1}^{L}\beta_1\lambda_i V_i\right]}$$

(20.63)

where $N_{\{\lambda\},\beta_1}(\{V\},\xi)$ is the value taken by the histogram at the set $\{V\}$ (which includes V_0) and ξ, and the bar over the probability density emphasizes that it is an estimated P.

20.5.2 The WHAM Equations

The WHAM equations have been already developed in Section 20.4 within SCHM for the square Ising model; in this case, histograms of MC snapshots are accumulated for a given energy value, E, based on several MC runs at different temperatures. In the previous Section 20.5.1, a PMF profile of the reaction coordinate ξ has been generated based on a *single* simulation; in this section, a PMF profile will be obtained from *several* simulations. Thus, consider R MD (or MC) simulations based on the Hamiltonian, $\hat{H}_{\{\lambda\}}$ [Equation (20.51)], where the kth simulation is carried out at inverse temperature $\beta_k = 1/k_B T_k$ with the corresponding set of L coupling parameters $\{\lambda\}_k \equiv \lambda_{i,k}$ $i = 1, \ldots, L$. The starting point is the equation for the density of states, $\Omega_k(\{V\},\xi)$, obtained by the kth simulation, which is equivalent to the expression developed previously in Equations (20.37–20.38),

$$\Omega_k(\{V\},\xi) = N_k(\{V\},\xi)n_k^{-1}\left[\exp\left(\sum_{i=0}^{L}\beta_k\lambda_{i,k}V_i\right) - A_k\right] \quad (k=1,2,...,R) \tag{20.64}$$

Thus, $N_k(\{V\},\xi)$ is the value taken by the histogram at $\{V\}$ and ξ during the kth simulation, n_k is the total number of snapshots generated in the kth simulation, and $\Sigma\beta_k\lambda_{i,k}V_i$ is the energy term; these quantities replace $N_k(E)$, n_k, and the energy, $\beta_k E$ in Equation (20.38), respectively. $f_k = \beta_k A_k$ is a dimensionless free energy of the system defined by the Hamiltonian of Equation (20.51) with coupling parameters, $\{\lambda\}_k$; A_k is the (Helmholtz) free energy of the kth system.

The objective of WHAM is to obtain the best estimate for the probability density $P_{\{\lambda\},\beta}(\{V\},\xi)$. As before [Equation (20.39)], this is achieved by expressing $\Omega(\{V\},\xi)$ as a weighted sum of the R estimates $\Omega_k(\{V\},\xi)$, $(k=1,2,...,R)$,

$$\Omega(\{V\},\xi) = \sum_{k=1}^{R}p_k(\{V\})\Omega_k(\{V\},\xi) \qquad \sum_{k=1}^{R}p_k(\{V\})=1 \tag{20.65}$$

The error, $\delta^2\Omega(\{V\},\xi)$, is then minimized using the same procedure described in Section 20.4.2, which leads to the optimized $\Omega(\{V\},\xi)$ [compare with $\Omega(E)$, Equation (20.44)]:

$$\Omega(\{V\},\xi) = \frac{\displaystyle\sum_{k=1}^{R}N_k(\{V\},\xi)}{\displaystyle\sum_{k=1}^{R}n_k\exp\left(f_k-\beta_k\sum_{i=0}^{L}\lambda_{i,k}V_i\right)} \tag{20.66}$$

From $\Omega(\{V\},\xi)$, the WHAM equations are obtained. Thus, one calculates the *unnormalized biased* probability distribution, $P_{\{\lambda\},\beta}(\{V\},\xi)$, for temperature β and a set of coupling parameters $\{\lambda\} \equiv \lambda_j, j = 1,...,L$.

$$\bar{P}_{\{\lambda\},\beta}(\{V\},\xi) = \frac{\left(\displaystyle\sum_{k=1}^{R}N_k(\{V\},\xi)\right)\exp\left(-\beta\sum_{j=0}^{L}\lambda_j V_j\right)}{\displaystyle\sum_{k=1}^{R}n_k\exp\left(f_k-\beta_k\sum_{i=0}^{L}\lambda_{i,k}V_i\right)} \tag{20.67}$$

from which $\exp(-f_k)$ is obtained:

$$\exp(-f_k) = \sum_{\{V\},\xi}\bar{P}_{\{\lambda\}_k\beta_k}(\{V\},\xi). \tag{20.68}$$

Equations (20.67–20.68) constitute an extension of the previous equations (20.46–20.47), respectively. However, the factors g_i appearing in Equations (20.46–20.47) have been omitted here because Kumar et al. [56] have commented that the effect of g_i in biomolecular systems is negligible; however, these factors should be considered close to phase transition, for example. By iterating Equations (20.67–20.68), the f_k values can be determined self-consistently. In practice, one selects initial values (say, $f_k = 0$, $k = 1,2,...,L$) for Equation (20.67), where the resulted $P_{\{\lambda\},\beta}(\{V\},\xi)$ lead to improved f_k values through Equation (20.68), and the process continues; experience has shown that the convergence is relatively fast [56]. After the f_k's have been determined and substituted in Equation (20.67), the unbiased probability density $P_{\{0\},\beta}(\{V\},\xi)$ can be determined, where the effect of the restraining potentials (the bias) is removed; thus,

$$\bar{P}_{\{0\},\beta}(\{V\},\xi) = \frac{\left(\sum_{k=1}^{R} N_k(\{V\},\xi)\right)\exp(-\beta\lambda_0 V_0)}{\sum_{k=1}^{R} n_k \exp\left(f_k - \beta_k \sum_{i=0}^{L} \lambda_{i,k} V_i\right)} \tag{20.69}$$

Equation (20.69), in its general form, requires generating $(L + 1 + D_\xi)$ dimensional histograms, where D_ξ is the dimension of ξ, which can be larger than one. Now, if all the restraining potentials are explicit functions of the reaction coordinate(s), the dimensionality of the histograms can be reduced to $1 + D_\xi$, and Equations (20.67–20.68) become:

$$\bar{P}_{\{\lambda\},\beta}(V_0,\xi) = \frac{\left(\sum_{k=1}^{R} N_k(V_0,\xi)\right)\exp\left(-\beta\lambda_0 V_0 - \beta\sum_{i=1}^{L}\lambda_i \hat{V}_i(\xi)\right)}{\sum_{k=1}^{R} n_k \exp\left(f_k - \beta_k \lambda_0 V_0 - \beta_k \sum_{i=1}^{L}\lambda_{i,k}\hat{V}_i(\xi)\right)} \tag{20.70}$$

$$\exp(-f_k) = \sum_{V_0,\xi} \bar{P}_{\{\lambda\}_k,\beta_k}(V_0,\xi) \tag{20.71}$$

These equations can be simplified further in the case where all simulations are carried out at the same temperature β; then $\exp(-\beta\lambda_0 V_0)$ is cancelled in Equation (20.70), and inserting $\lambda_i = 0$, for $i = 1,2,...,L$ leads (after carrying out the self-consistent procedure) to $P_\beta(\xi)$:

$$\bar{P}_\beta(\xi) = \frac{\left(\sum_{k=1}^{R} N_k(V_0,\xi)\right)}{\sum_{k=1}^{R} n_k \exp\left(f_k - \beta\sum_{i=1}^{L}\lambda_{i,k}\hat{V}_i(\xi)\right)} \tag{20.72}$$

In [56], Kumar et al. studied the PMF of the pseudorotation phase angle of the sugar ring in deoxy-adenosine—a molecule consisting of 31 atoms; this system was modeled by the all-atom AMBER force field in vacuum (see Section 16.3); in their 1995 paper [57], the two-dimensional PMF (of the dihedral angles, γ and χ) of the same molecule was studied by a Monte Carlo method. In 1993, Boczko and Brooks, III [58] found WHAM to be better in some respects than a conventional technique, as applied to the blocked tripeptide $(Ala)_3$. In a 1995 paper [59], Roux studied, by WHAM, two-dimensional PMFs of the alanine dipeptide. All these (early) studies were conducted in vacuum, and to mimic solvent effects, in most cases a distance-dependent dielectric constant has been used (see Section 16.4).

20.5.3 Enhancements of WHAM

As described above, WHAM is, in general, a very versatile methodology for calculating the potential of mean force and free energy differences, in particular, the free energy of binding. WHAM has become a standard tool for analyzing trajectories generated (e.g., at different temperatures) by enhanced-sampling methods, such as umbrella sampling, replica exchange (parallel tempering), simulated tempering, or the method of expanded ensembles; these methods are described in detail in the next Chapter 21, in

Sections 21.1–21.5, respectively. The inherent versatility of WHAM has led to extensive studies of its performance for systems of increasing complexity; attempts have also been made to enhance its efficiency in various ways, in particular, by improving the iteration procedure that can become extremely inefficient as the number of trajectories, R, is increased [60]. A detailed discussion of this literature is beyond the scope of this book; still, we present below a (mostly) descriptive summary of the relevant methodologies, concentrating on the complexity of the systems studied. A short review of this specific field with an extensive list of references can be found in [61].

In 1997, Bartels and Karplus [62] suggested a new multidimensional adaptive umbrella sampling technique for MD simulations, which enables a uniform sampling of the conformational space spanned by several degrees of freedom. The efficiency of the method is achieved by using WHAM to combine the results from different simulations; the method was applied to the alanine and threonine dipeptides. A similar method (for MC simulations) was suggested a year earlier by Kumar et al. [63], where PMFs of (φ,ψ) of the blocked alanine dipeptide, and of the end-to-end distance of $(Ala)_{10}$ and $(Gly)_{10}$ were studied; these systems were modeled by the ECEPP and AMBER force fields in vacuum (Section 16.4).

In 2001, Souaille and Roux [64] used WHAM to combine FEP with umbrella sampling calculations. They studied (by MD) the free energy of solvation of a K^+ ion, a water molecule, and an argon atom, immersed in spheres of up to 100 explicit water molecules. These authors also studied a system of a point charge and a point dipole. While the WHAM and FEP results were found to be comparable, WHAM has shown some advantages compared to the standard FEP and thermodynamic integration.

Latter, in 2005, Gallicchio et al. [65] developed a novel Bayesian statistical uncertainty estimation method for any quantity derived from WHAM and used it to validate the calculated potential of mean force. They studied the 16-residue G peptide, which is the C-terminal-hairpin of the B1 domain of protein G (41-GEWTYDDATKTFTVTE-56). This system, modeled by the OPLS all-atom force field (see Section 16.3) and the analytical generalized Born implicit solvent model (see Section 16.4), was simulated by replica exchange MD. Applying WHAM to the set of samples generated at different temperatures (T-WHAM) made it possible to determine the PMF (at room temperature) in the transition region between a minor α-helical population and the major β-hairpin population. They identified a possible transition path between these structures along which the peptide retains a significant amount of secondary structure.

In 2009, Bereau and Swendsen [60] proposed a new algorithm for solving the WHAM equations (for estimating free energies differences). The algorithm (applied to the Ising model) provides a more natural way of approaching the problem and improves convergence compared to the widely used direct iteration method. The authors also study how parameters (temperature, pressure, etc.) of the independent simulations should be chosen to optimize the accuracy of the set of free energies. A year later Hub et al. [66] presented a new WHAM implementation, termed g-WHAM, which estimates statistical errors using the technique of bootstrap analysis. Three bootstrap methods were tested and their efficiencies discussed by computing the PMF along the distance between two methanol molecules in vacuum.

In a 2012 paper, Zhu and Hummer [67] developed methodologies which address the following challenges: (1) how to obtain fast and accurate solutions of the coupled non-linear WHAM equations, (2) how to quantify the statistical errors of the resulting free energies, (3) how to diagnose possible systematic errors, and (4) how to optimally allocate computational resources. Their method was applied for studying the free energy of the passage of an N_a ion through the transmembrane pore of the ligand-gated ion channel using the CHARMM force field in vacuum (see Section 16.3).

In 2008, Shirts and Chodera [68] presented a new estimator alternative to WHAM called "*the multistate Bennett acceptance ratio*" (MBAR) because it reduces to the Bennett acceptance ratio estimator, BAR, when only two states are considered. MBAR has significant advantages over WHAM-based procedures, as it does not require the sampled energy range to be discretized to produce histograms; thus, the bias due to energy binning is eliminated, and in many cases, the computing time is reduced significantly. Additionally, an estimate of the statistical uncertainty is provided for all estimated quantities. In the large sample limit, MBAR is unbiased and has the lowest variance of any known estimator for making use of equilibrium data collected from multiple states. In their study, Shirts and Chodera calculated the PMF for the folding transition of a DNA hairpin system of 20 base pairs, combining data from multiple optical tweezer measurements under different external biasing potentials.

In a later (2012) paper, Tan et al. [69] demonstrate the validity and value of MBAR. They discuss two statistical arguments to derive the MBAR equations, showing that this binless method can be used, not only for estimating free energies and equilibrium expectations, but also to estimate equilibrium distributions. A number of useful results from the statistical literature are provided, including the determination of the MBAR estimators by the minimization of a convex function. This leads to an approach to the computation of MBAR free energies by optimization algorithms, which can be more effective than existing algorithms. These ideas were tested by studying the binding of several ligands to the 103-residue protein FKBP12. These systems were modeled by the OPLS force field and an implicit solvent model.

Another non-iterative alternative to WHAM is *"the statistical-temperature WHAM"* (ST-WHAM) [70,71], which determines the density of states through its logarithmic derivative, or the statistical temperature. In this way, the method estimates the density of states non-iteratively with minimal approximation. ST-WHAM can be regarded as a refinement of the more approximate *"umbrella integration method"* (UIM) [72,73]. However, the extension to multidimensional ensembles, e.g., the isothermal-isobaric ensemble, can be numerically challenging [70]. These studies are based on simple systems, such as the Ising model, the alanine dipeptide, or the polypeptide (Ala)$_8$.

Derivations of WHAM, ST-WHAM, and UIM are described below. On the other hand, the details of a numerical improvement of the implementation of WHAM and MBAR based on the method of *"direct inversion in the iterative subspace"* (DIIS) [74] is more complex and will not be discussed here; the theory of DIIS, which is iterative in nature, can be found in [61], as applied to the Ising model, a Lennard-Jones fluid, and the villin headpiece (35 residues) immersed in a box of 1989 TIP3P water molecules (see Section 16.5). In this study, the efficiencies of DIIS, WHAM, ST-WHAM, and UIM are compared.

20.5.4 The Basic MBAR Equation

Omitting the g_k factors, Equations (20.46–20.47) can be written as a single expression:

$$f_i = -\ln \int \frac{\left[\sum_{k=1}^{R} N_k(E)\right]\exp(-\beta_i E)}{\sum_{k=1}^{R} n_k \exp(f_k - \beta_k E)} dE = -\ln \Xi_i(\mathbf{f}) \tag{20.73}$$

where $\Xi_i(\mathbf{f})$ denotes the integral on the right hand side as a function of $\mathbf{f} = (f_1, ...,f_R)$. The converged set, $\mathbf{f}$, can be obtained by the iterations [Equation (20.47)],

$$f_i^{\text{new}} = -\ln \Xi_i(\mathbf{f}^{\text{old}}). \tag{20.74}$$

The histogram-free form of f_i is based on Equation (20.73). The first step is to avoid the histogram dependency of WHAM [through $N_k(E)$] by defining [64]:

$$N_k(E) = \sum_{\mathbf{x}}^{(k)} \delta[E(\mathbf{x}) - E)] \tag{20.75}$$

where $E(\mathbf{x})$ is the energy function and $\sum_{\mathbf{x}}^{(k)}$ denotes a sum over trajectory frames $\mathbf{x}$ of simulation k. Inserting Equation (20.75) in Equation (20.73) leads to the histogram-free (or binless) MBAR,

$$f_i = -\ln \sum_{j=1}^{R} \sum_{\mathbf{x}}^{j} \frac{q_i(\mathbf{x})}{\sum_{k=1}^{R} n_k q_k(\mathbf{x})\exp(f_k)} \tag{20.76}$$

where $q_i(\mathbf{x}) \equiv \exp[-\beta_i E(\mathbf{x})]$. Equation (20.75) can be extended to umbrella sampling.

20.5.5 ST-WHAM and UIM

ST-WHAM—the non-iterative alternatives to WHAM [72,73] are obtained by taking the logarithmic derivative of $d_k(E)$—the kth summand of the denominator of Equation (20.44) (again, omitting the factors g_k),

$$d_k(E) = n_k \exp(-\beta_k E + f_k). \tag{20.77}$$

The logarithmic derivative is:

$$\frac{d}{dE} \ln \sum_{k=1}^{R} d_k(E) = \frac{\sum_{k=1}^{R} d_k(E)\beta_k}{\sum_{k=1}^{R} d_k(E)} = -\frac{\sum_{k=1}^{R} N_k(E)\beta_k}{\sum_{k=1}^{R} N_k(E)} \tag{20.78}$$

The right hand side of Equation (20.78) has been obtained from the definition of the density of states, $\Omega_1(E)$, for a *single* histogram [Equation (20.44)]:

$$\Omega_1(E) = \frac{N_k(E)}{d_k(E)} \tag{20.79}$$

$\Omega(E)$ is obtained by integration of Equation (20.78),

$$\Omega(E) = \left[\sum_{k=1}^{R} N_k(E) \right] \exp \left[\int^{E} \frac{\sum_{k=1}^{R} N_k(E')\beta_k}{\sum_{k=1}^{R} N_k(E')} dE' \right] \tag{20.80}$$

Equations (20.79–20.80) constitute the ST-WHAM method. Notice that in evaluating the integral, one may encounter an empty bin. To decrease the effect of the resulting error, one can let the integrand borrow the value from the nearest non-empty bin. ST-WHAM is most convenient in one dimension, and its results usually differ only slightly from those of WHAM [72].

In UIM [72,73], the distribution $N_k(E)$ is further approximated as a Gaussian. As mentioned earlier, application of ST-WHAM becomes difficult when more thermodynamic variables exist in the ensemble of interest; thus, extension of the method, for example, to the isothermal-isobaric ensemble, can be numerically challenging [72].

20.5.6 Summary

A great deal of work has been devoted to studying the effectiveness of WHAM, to the improvement of its efficiency, and to the development of alternative techniques. The above review shows that in the early works, the methodologies have been tested on relatively simple systems (e.g., the alanine dipeptide); however, these systems became more complex in the course of time, with the advent of stronger computers. Still, the modeling used has been limited, since none of the systems reviewed (besides the villin headpiece treated with DIIS [74]) has been studied with explicit solvation. While this enables one checking the performance of competing techniques, the comparison with the experiment might be problematic. In fact, the calculated free energy of binding of ligands to the protein FKBP12 of Tan et al. [69] has been found to be *significantly* larger than the experimental values. Finally, many methods studied in this chapter and in the next one are discussed extensively in the book "*Understanding Molecular Simulations*" by Frenkel and Smit [75].

REFERENCES

1. G. M. Torrie and J. P. Valleau. Monte Carlo free energy estimates using non-Boltzmann sampling: Application to the sub-critical Lennard-Jones fluid. *Chem. Phys. Lett.* **28**, 578–581 (1974).
2. G. M. Torrie and J. P. Valleau. Nonphysical sampling distributions in Monte Carlo free-energy estimation – umbrella sampling. *J. Comput. Phys.* **23**, 187–199 (1977).
3. H. L. Scott and C. Y. Lee. The surface tension of water: A Monte Carlo calculation using an umbrella sampling algorithm. *J. Chem. Phys.* **73**, 4591–4596 (1980).
4. M. Mezei. Excess free energy of different water models computed by Monte Carlo methods. *Mol. Phys.* **47**, 1307–1315 (1982).
5. J. P. Valleau. Density-scaling Monte Carlo study of subcritical Lennard-Jones. *J. Chem. Phys.* **99**, 4718–4728 (1977).
6. J. P. Valleau. Density-scaling: A new Monte Carlo technique in statistical mechanics. *J. Comp. Phys.* **96**, 193–216 (1991).
7. I. R. McDonald and K. Singer. Calculation of the thermodynamic properties of liquid argon from Lennard-Jones parameters by a Monte Carlo method. *Discuss. Faraday Soc.* **43**, 40 (1967).
8. I. R. McDonald and K. Singer. Examination of the adequacy of the 12–6 potential for liquid argon by means of Monte Carlo calculations. *J. Chem. Phys.* **50**, 2308–2315 (1969).
9. C. H. Bennett. Efficient estimation of free energy differences from Monte Carlo data. *J. Comp. Phys.* **22**, 245–268 (1976).
10. A. Ben Naim. *Statistical Thermodynamics for Chemists and Biochemists* (Plenum Press, New York, 1992).
11. C. Pangali, M. Rao and B. J. Berne. A Monte Carlo simulation of the hydrophobic interaction. *J. Chem. Phys.* **71**, 2975 (1979).
12. S. H. Northrup, M. R. Pear, C. Y. Lee, J. A. McCammon and M. Karplus. Dynamical theory of activated processes in globular proteins. *Proc. Natl. Acad. Sci. USA* **79**, 4035–4039 (1982).
13. A. F. Voter. A Monte Carlo method for determining free energy difference and transition state theory rate constants. *J. Chem. Phys.* **82**, 1890–1899 (1985).
14. D. J. Tobias and C. L. Brooks, III. Calculation of free energy surfaces using the methods of thermodynamic perturbation theory. *Chem. Phys. Lett.* **142**, 472–476 (1987).
15. M. Mezei. Adaptive umbrella sampling: Self-consistent determination of the non-Boltzmann bias. *J. Comp. Phys.* **68**, 237–248 (1987).
16. R. W. W. Hooft, B. P. van Eijck and J. Kroon. An adaptive umbrella sampling procedure in conformational analysis using molecular dynamics and its application to glycol. *J. Chem. Phys.* **97**, 6690–6693 (1992).
17. R. W. W. Hooft, B. P. van Eijck and J. Kroon. Use of molecular dynamics methods in conformational analysis. Glycol. A model study. *J. Chem. Phys.* **97**, 3639–3646 (1992).
18. G. H. Paine and H. A. Scheraga. Prediction of the native conformation of a polypeptide by a statistical-mechanical procedure. I. Backbone structure of enkephalin. *Biopolymers* **24**, 1391–1436 (1985).
19. T. C. Beutler and W. F. van Gunsteren. The computation of a potential of mean force: Choice of the biasing potential in the umbrella sampling technique. *J. Chem. Phys.* **100**, 1492–1497 (1994).
20. T. C. Beutler and W. F. van Gunsteren. Umbrella sampling along linear combinations of generalized coordinates. Theory and application to a glycine dipeptide. *Chem. Phys. Lett.* **237**, 308–316 (1995).
21. T. C. Beutler, T. Bremi, R. R. Ernst and W. F. van Gunsteren. Motion and conformation of side chains in peptides. A comparison of 2D umbrella-sampling molecular dynamics and NMR results. *J. Phys. Chem.* **100**, 2637–2645 (1996).
22. M. Mezei, P. K. Mehrotra and D. L. Beveridge. Monte Carlo determination of the free energy and internal energy of hydration for the Ala dipeptide at 25°C. *J. Am. Chem. Soc.* **107**, 2239–2245 (1985).
23. H. Resat, P. V. Maye and M. Mezei. The sensitivity of conformational free energies of the alanine dipeptide to atomic site charges. *Biopolymers* **41**, 73–81 (1997).
24. T. J. Marrone, M. K. Gilson and J. A. McCammon. Comparison of continuum and explicit models of solvation: Potentials of mean force for alanine dipeptide. *J. Phys. Chem.* **100**, 1439–1441 (1996).
25. F. Fraternali and W. F. Van Gunsteren. Conformational transitions of a dipeptide in water: Effects of imposed pathways using umbrella sampling techniques. *Biopolymers* **34**, 347–355 (1994).
26. M. Berkowitz, O. A. Karim, J. A. McCammon and P. J. Rossky. Sodium chloride ion pair interaction in water: Computer simulation. *Chem. Phys. Lett.* **105**, 577–580 (1984).

27. A. C. Belch, M. Berkowitz and J. A. McCammon. Solvation structure of a sodium chloride ion pair in water. *J. Am. Chem. Soc.* **108**, 1755–1761 (1986).
28. E. Guardia, R. Rey and J. A. Padro. Potential of mean force by constrained molecular dynamics: A sodium chloride ion-pair in water. *Chem. Phys.* **155**, 187–195 (1991).
29. D. E. Smith and L. X. Dang. Computer simulations of NaCl association in polarizable water. *J. Chem. Phys.* **100**, 3757–3766 (1994).
30. J. Gao. Simulation of the Na+CL$^-$ ion pair in supercritical water. *J. Phys. Chem.* **98**, 6049–6053 (1994).
31. R. A. Friedman and M. Mezei. The potentials of mean force of sodium chloride and sodium dimethyl-phosphate in water: An application of adaptive umbrella sampling. *J. Chem. Phys.* **102**, 419–426 (1995).
32. L. X. Dang, J. E. Rice and P. A. Kollman. The effect of water models on the interaction of the sodium–chloride ion pair in water: Molecular dynamics simulations. *J. Chem. Phys.* **93**, 7528–7529 (1990).
33. J. V. Eerden, W. J. Briels, S. Harkema and D. Feil. Potential of mean force by thermodynamic integration: Molecular dynamics simulation of decomplexation. *Chem. Phys. Lett.* **164**, 370–376 (1989).
34. H. Resat, M. Mezei and J. A. McCammon. Use of the grand canonical ensemble in potential of mean force calculations. *J. Phys. Chem.* **100**, 1426–1433 (1996).
35. J. Chandrasekhar, S. F. Smith and W. L. Jorgensen. S_N2 reaction profiles in the gas phase and aqueous solution. *J. Am. Chem. Soc.* **106**, 3049–3050 (1984).
36. J. Chandrasekhar, S. F. Smith and W. L. Jorgensen. Theoretical examination of the S_N2 reaction involving chloride ion and methyl chloride in the gas phase and aqueous solution. *J. Am. Chem. Soc.* **107**, 154–163 (1985).
37. S. E. Huston and P. J. Rossky. Free energies of association for the sodium-dimethyl phosphate ion pair in aqueous solution. *J. Phys. Chem.* **93**, 7888–7895 (1989).
38. S.-W. Chen and P. J. Rossky. Potential of mean force for a sodium dimethyl phosphate ion pair in aqueous solution: A further test of the extended RISM theory. *J. Phys. Chem.* **97**, 6078–6082 (1993).
39. D. J. Tobias and C. L. Brooks, III. Molecular dynamics with internal coordinate constraints. *J. Chem. Phys.* **89**, 5115–5127 (1988).
40. D. J. Tobias and C. L. Brooks, III. The thermodynamics of solvophobic effects: A molecular-dynamics study of *n*-butane in carbon tetrachloride and water. *J. Chem. Phys.* **92**, 2582–2592 (1990).
41. C. D. Bell and S. C. Harvey. Comparison of free energy surfaces for extended-atom and all-atom models of *n*-butane. *J. Phys. Chem.* **90**, 6595–6597 (1986).
42. W. L. Jorgensen and J. K. Buckner. Use of statistical perturbation theory for computing solvent effects on molecular conformation: Butane in water. *J. Phys. Chem.* **91**, 6083–6085 (1987).
43. D. J. Tobias and C. L. Brooks, III. Thermodynamics and mechanism of α helix initiation in alanine and valine peptides. *Biochemistry* **30**, 6059–6070 (1991).
44. D. J. Tobias, S. F. Sneddon and C. L. Brooks, III. Reverse turns in blocked dipeptides are intrinsically unstable in water. *J. Mol. Biol.* **216**, 783 -796 (1990).
45. B. Roux and M. Karplus. Ion transport in a model gramicidin channel structure and thermodynamics. *Biophys. J.* **59**, 961–981 (1991).
46. L. X. Dang and P. A. Kollman. Molecular dynamics simulations study of the free energy of association of 9-methyladenine and 1-methylthymine bases in water. *J. Am. Chem. Soc.* **112**, 503–507 (1990).
47. A. Warshel. Dynamics of reactions in polar solvents. Semiclassical trajectory studies of electron-transfer and proton-transfer reactions. *J. Phys. Chem.* **86**, 2218–2224 (1982).
48. J. Aqvist and A. Warshel. Energetics of ion permeation through membrane channels solvation of Na+ by gramicidin A. *Biophys. J.* **56**, 171–182 (1989).
49. A. Warshel, F. Sussman and J.-K. Hwang. Evaluation of catalytic free energies in genetically modified proteins. *J. Mol. Biol.* **201**, 139–159 (1988).
50. A. Wallqvist and D. G. Covell. Free-energy cost of bending n-dodecane in aqueous solution. Influence of the hydrophobic effect and solvent exposed area. *J. Phys. Chem.* **99**, 13118–13125 (1995).
51. E. M. Boczko and C. L. Brooks, III. First-principles calculation of the folding free energy of a three-helix bundle protein. *Science* **269**, 393–396 (1995).
52. L. Wang, T. O'Connell, A. Tropsha and J. Hermans. Molecular simulations of β-sheet twisting. *J. Mol. Biol.* **262**, 283–293 (1996).
53. H. Meirovitch. Calculation of the free energy and entropy of macromolecular systems by computer simulation. In *Reviews in Computational Chemistry*, K.B. Lipkowitz and D.B. Boyed (Eds.), Vol. 12, 1–74. Wiley-VCH, New York, 1998.

54. A. M. Ferrenberg and R. H. Swendsen. Optimized Monte Carlo data analysis. *Phys. Rev. Lett.* **63**, 1195–1198 (1989).
55. H. Muller-Krumbhaar and K. Binder. Dynamic properties of the Monte Carlo method in statistical mechanics. *J. Stat. Phys.* **8**. 1–24 (1973).
56. S. Kumar, D. Bouzida, R. H. Swendsen, P. A. Kollman and J. M. Rosenberg. The weighted histogram analysis method for free-energy calculations on biomolecules. I. The method. *J. Comput. Chem.* **13**, 1011–1021 (1992).
57. S. Kumar, J. M. Rosenberg, D. Bouzida, R. H. Swendsen and P. A. Kollman. Multidimensional free energy calculations using the weighted histogram analysis method. *J. Comput. Chem.* **16**, 1339–1350 (1995).
58. E. M. Boczko and C. L. Brooks, III. Constant-temperature free energy surfaces for physical and chemical processes. *J. Phys. Chem.* **97**, 4509–4513 (1993).
59. B. Roux. The calculation of the potential of mean force using computer simulations. *Comput. Phys. Commun.* **91**, 275–282 (1995).
60. T. Bereau and R. H. Swendsen. Optimized convergence for multiple histogram analysis. *J. Comp. Phys.* **228**, 6119–6129 (2009).
61. C. Zhang, C.-L. Lai and B. M. Pettitt. Accelerating the weighted histogram analysis method by direct inversion in the iterative subspace. *Mol. Simul.* **42**, 1079–1089 (2016).
62. C. Bartels and M. Karplus. Multidimensional adaptive umbrella sampling: Applications to main chain and side chain peptide conformations *J. Comput. Chem.* **18**, 1450–1462 (1997).
63. S. Kumar, P. W. Payne and M. Vásquez. Method for free-energy calculations using iterative technique. *J. Comput. Chem.* 17, 1269–1275 (1996).
64. M. Souaille and B. Roux. Extension to the weighted histogram analysis method: Combining umbrella sampling with free energy calculations. *Comput. Phys. Commun.* **135**, 40–57 (2001).
65. E. Gallicchio, M. Andrec, A. K. Felts and R. M. Levy. Constant-temperature free energy surfaces for physical and chemical processes. *J. Phys. Chem. B* **109**, 6722–6731 (2005).
66. J. S. Hub, B. L. de Groot and D. van der Spoel. g-wham – a free weighted histogram analysis implementation including robust error and autocorrelation estimates. *J. Chem. Theory Comput.* **6**, 3713–3720 (2010).
67. F. Zhu and G. Hummer. Convergence and error estimation in free energy calculations using the weighted histogram analysis method. *J. Comput. Chem.* **33**, 453–465 (2012).
68. M. R. Shirts and J. D. Chodera. Statistically optimal analysis of samples from multiple equilibrium states. *J. Chem. Phys.* **129**, 124105–124110 (2008).
69. Z. Tan, E. Gallicchio, M. Lapelosa and R. M. Levy Theory of binless multi-state free energy estimation with applications to protein-ligand binding. *J. Chem. Phys.* **136**, 144102–144114 (2012).
70. J. Kim, T. Keyes and J. E. Straub. Communication: Iteration-free, weighted histogram analysis method in terms of intensive variables. *J. Chem. Phys.* **135**, 061103–061117 (2011).
71. M. K. Fenwick. A direct multiple histogram reweighting method for optimal computation of the density of states. *J. Chem. Phys.* **129**, 125106–125107 (2008).
72. J. Kästner and W. Thiel. Bridging the gap between thermodynamic integration and umbrella sampling provides a novel analysis method: Umbrella integration. *J. Chem. Phys.* **123**, 144104–144105 (2005).
73. J. Kästner. Umbrella integration in two or more reaction coordinates. *J. Chem. Phys.* **131**, 034109–034118 (2009).
74. P. Pulay. Convergence acceleration of iterative sequences. The case of SCF iteration. *Chem. Phys. Lett.* **73**, 393–398 (1980).
75. D. Frenkel and B. Smit. *Understanding Molecular Simulations: From Algorithms to Applications* (Academic Press, New York, 2002).

21

Advanced Simulation Methods and Free Energy Techniques

An essential problem with dynamical simulation methods, such as the standard molecular dynamics (MD) and Monte Carlo (MC) techniques, is their inefficiency to induce crossing of energy barriers, in particular, in protein and fluid systems. In this chapter, we discuss several sophisticated techniques, based on MC and MD, which can overcome this problem at least partially; they are: the "*replica exchange method*" (REM), "*the multicanonical*" (MUCA) method, the related "*Wang and Landau technique*," the "*method of expanded ensembles*" (EXE, known also as "*simulated tempering*"), and "*the adaptive integration procedure*" (AIM). Like WHAM (Section 20.5) and Bennett's method (Section 20.2), most of these techniques are based on a self-consistent procedure that also leads directly to free energy differences, ΔF; replica exchange, in this respect, is an outlier, which, however, can lead to ΔF if combined with WHAM, for example. Finally, we discuss "*Jarzynski's non-equilibrium method*" for calculating free energy differences, which, together with AIM, constitute an alternative to the free energy perturbation (FEP) and thermodynamic integrations (TI) methods discussed earlier in Chapter 17. In this chapter, the word "configuration" is mostly used for describing the spatial structure of a system ($\mathbf{x}^N$), while "conformation" is mostly used for the structure of a peptide or a protein.

21.1 Replica-Exchange

The REM (also called parallel tempering) was suggested by Swendsen and Wang [1], as applied originally to a spin-glass system. The method was further developed for spin glasses (e.g., [2–4]) and for polymers [5], where its potential for structure determination of biomolecules was demonstrated first by Hansmann [6], who applied it to the pentapeptide Met enkephalin (see Section 16.8.2); REM has been further developed in the course of the years and, due to its relative simplicity, has become a popular tool in computational structural biology and other areas. We shall describe two versions of replica exchange, temperature-based REM (T-REM) or simply just REM and Hamiltonian-based REM denoted H-REM.

21.1.1 Temperature-Based REM

With T-REM, several MC simulation runs of the same system are carried out *independently* in parallel at different temperatures, $T_0 < T_1 < T_2 < \ldots < T_M$, where T_0 is the temperature of interest (typically $T_0 = 300$ K) (Figure 21.1).

During simulation, every k MC steps, two *neighbor* replicas (copies) are chosen at random and their current configurations, denoted for simplicity by $\mathbf{x}_i$ ($\mathbf{x}_i \equiv \mathbf{x}_i^N$) and $\mathbf{x}_j$ [with energies, $E(\mathbf{x}_i)$ and $E(\mathbf{x}_j)$] are exchanged with an "asymmetric" Metropolis transition probability (Equation 10.11):

$$p\left(\begin{array}{c} x_i \rightarrow x_j \\ x_j \rightarrow x_i \end{array}\right) = \begin{cases} 1 & \text{for } \Delta \leq 0 \\ \\ \exp(-\Delta) & \text{for } \Delta > 0 \end{cases} \tag{21.1}$$

where Δ is:

$$\Delta = \left[\beta_i - \beta_j\right]\left[E(\mathbf{x}_j) - E(\mathbf{x}_i)\right]. \quad \left(\beta_i = 1/k_\mathrm{B}T_i\right) \tag{21.2}$$

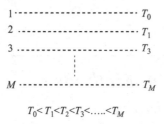

$$T_0 < T_1 < T_2 < T_3 < \ldots < T_M$$

FIGURE 21.1 A set of M temperatures $T_0 < T_1 < T_2 < T_3 < \ldots < T_M$ for MC/MD simulations. The exchange among these replicas increases the chance for typical high energy configurations to visit the lowest temperature replica, T_0, thus leading to crossing of energy barriers at T_0.

To obtain Δ [Equation (21.2)], one should recall that in asymmetric MC, Δ is the difference between the final energy factor $\left(\beta_2 E_i + \beta_1 E_j\right)$ and the initial one $\left(\beta_2 E_j + \beta_1 E_i\right)$ (for simplicity, we treat two replicas, 1 and 2, where $E(\mathbf{x}_i)$ is denoted by E_i); this leads to Δ [Equation (21.2)],

$$\Delta_{12} = \beta_2 E_i + \beta_1 E_j - \beta_2 E_j - \beta_1 E_i = \beta_1 (E_j - E_i) + \beta_2 (E_i - E_j) = (\beta_1 - \beta_2)(E_j - E_i) \tag{21.3}$$

From the point of view of probability theory, we have the following picture: if the simulations were carried out within each replica m only, the mth sample would be related to the canonical ensemble based on β_m, and on the entire set of configurations (elementary random variables, denoted by X_m), where the corresponding partition function is $Z_m \equiv Z(\beta_m)$. However, when exchange between replicas is allowed, the theoretical framework becomes an extended system (to be distinguished from an extended ensemble) defined by the product space, $(X_0 \cdot X_1 \cdot \ldots \cdot X_M)$, where vectors of configurations, $(\mathbf{x}_i^{m=0}, \mathbf{x}_j^{m=1}, \ldots, \mathbf{x}_k^M)$, constitute its elementary random variables ($\mathbf{x}_i^{m=0}$ is configuration $\mathbf{x}_i$ of the $m = 0$ ensemble, etc.). The partition function of the extended system is:

$$Z = \prod_{m=0}^{M} Z_m \tag{21.4}$$

and the probability of the configuration, $(\mathbf{x}_i^0, \mathbf{x}_j^1, \ldots, \mathbf{x}_k^M)$, in the extended system is:

$$P(\mathbf{x}_i^0, \mathbf{x}_j^1, \ldots, \mathbf{x}_k^M) = \frac{\exp-\left[\beta_0 E(\mathbf{x}_i^0) + \beta_1 E(\mathbf{x}_j^1) \ldots + \beta_M E(\mathbf{x}_k^M)\right]}{Z} \tag{21.5}$$

The MC transition probability for exchanging configurations of different ensembles is obtained from the detailed balance condition using Equation (21.5); thus, by considering (for simplicity, as above) only replicas 1 and 2, and denoting $\mathbf{x}_j^1 \equiv \mathbf{x}_j(1)$, the result of Equation (21.2) is recovered:

$$\frac{p(\mathbf{x}_i(1) \to \mathbf{x}_i(2); \mathbf{x}_j(2) \to \mathbf{x}_j(1))}{p(\mathbf{x}_j(1) \to \mathbf{x}_j(2); \mathbf{x}_i(2) \to \mathbf{x}_i(1))} = \frac{\exp-\left[\beta_2 E(\mathbf{x}_i) + \beta_1 E(\mathbf{x}_j)\right]}{\exp-\left[\beta_2 E(\mathbf{x}_j) + \beta_1 E(\mathbf{x}_i)\right]} \tag{21.6}$$

$$= \exp-[\beta_2 - \beta_1][E(\mathbf{x}_i) - E(\mathbf{x}_j)] = \exp-\Delta$$

According to Equation (21.5), a configuration in the extended system has a *non-canonical ensemble* weight. Alternatively, one can adopt the point of view that a configuration in replica m is distributed according to the (canonical) Boltzmann probability density defined by β_m, where the exchange of configurations in REM is considered as a global MC update, which leads to decreased correlations, hence, to a widened configurational range for the mth simulation. Therefore, after a REM simulation has been relaxed, one can estimate the canonical ensemble averages of a quantity $A(\beta_m)$ for each temperature, β_m, based on its sample size, n_m,

$$\overline{A(\beta_m)} = \frac{1}{n_m} \sum_{t=1}^{n_m} A_{i(t)}(\beta_m) \quad m = 0,1,...,M \tag{21.7}$$

where $A_{i(t)}(\beta_m)$ is the value of A for the tth sampled configuration in the mth replica. One can also use the self-consistent histogram method (Section 20.4) to estimate the averages, $A(\beta)$ at temperatures, β, differing (but not considerably) from the $M + 1$ REM temperatures. For that, one calculates first the free energies, f_m, self-consistently using Equations (20.46) and (20.47), and then uses Equation (20.48); thus, free energy differences are calculated as well.

The above theory suggests that with REM, the simulated system at the temperature of interest (T_0) has a better chance than with standard MC to visit high energy configurations and thus to cross energy barriers (leading potentially to protein folding, for example). However, this can happen only if the number of replicas is sufficiently large. More specifically, let us consider again the two neighbor replicas 1 and 2 with *ensemble* average energies, $<E_1>$ and $<E_2>$, respectively; these energies are extensive–proportional to the number of particles, N, while their fluctuations grow only as $N^{1/2}$.

Therefore, the energy difference between systems of energies, mN and nN ($m > n$ are small numbers) is $N(m - n)$, while the sum of fluctuations is only $N^{1/2}(m^{1/2} + n^{1/2})$ and thus for large enough N, no energetic overlap will practically occur between these replicas. In other words, the *extensive* difference, $<E_2> - <E_1>$, will lead to a negligible probability exp $- \Delta$, unless $\beta_1 - \beta_2$ is reduced, or generally, the number of replicas is increased by a factor of $N^{1/2}$. Indeed, it has been shown [4,7] that to cover a temperature range adequately, the number of required replicas should grow as the square root of the number of particles in the system. Practically, the exchange rate between neighbor replicas should be large enough, say above 10%, and approximately equal for all pairs. This might require, not only to increase the number of temperatures, but also to distribute them optimally, which might require deviating from the equally spaced arrangement. For example, the exchange rate between neighbor replicas i and j is expected to grow with T, hence, the corresponding difference, ΔT_{ij}, can be increased appropriately. Another limitation of T-REM is that efficiency gains are not guaranteed if the barriers to sampling do not depend on the temperature, e.g., in the case of entropic barriers [8,9].

The dependence of the number of replicas on system size is demonstrated, for example, in two early MD studies of Berne's group of the 16-residue peptide, GEWTYDDATKTFTVTE, which is the C-terminus β-hairpin of protein G. Thus, 18 equally spaced replicas (at the range, 270–690 K) were needed for a study based on an implicit solvent [10], as compared to 64 replicas required for the larger system with explicit water [11]. It should be pointed out that implementing REM is suitable for parallel simulations, where each replica resides on a different node and configurations are transferred between nodes. However, because this transfer might become time consuming for a large system, one can instead exchange the corresponding temperatures, leaving the configurations at their replicas. The number of visits of say, T_0, to different replicas during the simulation constitutes a measure for the efficiency of the REM process [6].

The16-residue peptide mentioned above [10,11] was simulated by MD. It should be noted that implementing MD within the framework of REM requires some modifications, which were introduced first by Sugita and Okamoto [8], who called the method REMD. In this case, one has to also consider the kinetic energy, $E(\mathbf{p}_i)$, which depends on the momenta, $\mathbf{p}_i$ ($\mathbf{p}_i \equiv p_i^N$). For a large system, the total energy related to replica i is $E_{tot}(i) = E(\mathbf{x}_i) + E(\mathbf{p}_i) = E(\mathbf{x}_i) + 3Nk_B T_i$. Thus, to take the kinetic energy into account, the momenta, which are transferred to replica 1 from 2, should be multiplied by the factor $(T_2/T_1)^{1/2}$, while the momenta transferred to replica 1 is multiplied by $(T_1/T_2)^{1/2}$; it is easy to realize that these multiplications leave Equation (21.3) intact, that is, Δ depends only on the potential energies. Alternatively, due to the relation, $3k_B T = m<\mathbf{v}^2>$, one can scale the masses, m by T_2/T_1 and T_1/T_2, respectively. A specific mass scaling (defined prior to the simulation) might be important for handling efficiently the effect of viscosity. In this case, however, an appropriate velocity ($\mathbf{v}$) and momentum scaling should be defined to keep the system within the framework of the canonical ensemble [12,13].

Because MD is the method of choice in structural biology, most of the studies in this field have been carried out with REMD. Therefore, from now on, we shall refer to REM (or equivalently to T-REM) introduced above as T-REMD and sometimes for emphasis, as *standard* T-REMD. For the same reason, Hamiltonian REM discussed in the next sub-section will be called H-REMD.

A comparison between the efficiencies of REMD and MD was carried out by Nymeyer [14]. A great deal of work has been devoted to improving the efficiency of T-REMD for larger systems. For example, relying on ideas of Fukunishi et al. [7] and Sugita et al. [15] for H-REMD (see next sub-section), Liu et al. [16] suggested procedure where the solute is heated, while the water stays cold. Thus, the exchange probability for this *replica exchange method with solute tempering* (REST) scales only with the number of degrees of freedom of the solute (e.g., a protein), but not with the (typically) large number of water molecules. More specifically, the total energy of the system, E_0, is composed of three components, E_p, E_{pw}, and E_{ww}, which are respectively, the internal energy of the protein, the protein-water interaction energy, and the water-water interactions,

$$E_0(\mathbf{x}) = E_p(\mathbf{x}) + E_{pw}(\mathbf{x}) + E_{ww}(\mathbf{x}) \tag{21.8}$$

where again, for simplicity, we denote $\mathbf{x}^N$ by $\mathbf{x}$. $E_0(\mathbf{x})$ is the energy of the lowest replica, which is labeled by the index 0, and its temperature is T_0. As one climbs the replica ladder, the potential energy is rescaled (for replica i) as follows,

$$E_i(\mathbf{x}) = E_p(\mathbf{x}) + \left[\frac{\beta_0 + \beta_i}{2\beta_i}\right] E_{pw}(\mathbf{x}) + \left[\frac{\beta_0}{\beta_i}\right] E_{ww}(\mathbf{x}). \tag{21.9}$$

Clearly, when $\beta_i = \beta_0$, $E_i(\mathbf{x}) = E_0(\mathbf{x})$, and if the protein becomes the entire system (no solvent) REST reduces to T-REMD. Following Equation (21.3), for two neighbor replicas denoted 1 and 2, the energies defined by Equation (21.9) lead to the following Δ,

$$\Delta_{12} = (\beta_1 - \beta_2)[E_p(\mathbf{x}_2) - E_p(\mathbf{x}_1) + 1/2E_{pw}(\mathbf{x}_2) - 1/2E_{pw}(\mathbf{x}_1)] \tag{21.10}$$

where E_{ww}, which causes the poor scaling with system size in standard T-REMD, cancels out and does not appear in this expression. The acceptance probability (for exchange) is thus much larger for the scaled potentials. It is important to recognize that because the potential energies of all of the replicas other than the lowest one are deformed, this method only samples the correct distribution of replica 0.

The performance of REST [16] was compared to that of standard T-REMD for the small molecule, alanine dipeptide dissolved in 512 water molecules. Three replicas at 300, 420, and 600K (23%–29% exchange rate) were needed with REST, as compared to 22 replicas required with standard T-REMD (22%–29%). Also, between these methods, REST was found to be the most efficient (in Central Processing Unit (CPU) time) with respect to calculating thermodynamic averages and for the ab initio folding of the molecule. However, in a later study of larger molecules by the same group [17], REST has not performed better than standard T-REMD. The systems studied there are: (1) the 16-residue α-helix, AAAAKAAAAKAAAAKA immersed in a box of 2522 water molecules, (2) the 16-residue peptide GEWTYDDATKTFTVTE (discussed above) in a box of 1361 water molecules (see also [10,11]), and (3) TrpCage – the 20-residue miniprotein, NLYIQWLKDGGPSSGRPPPS with up to 1000 randomly placed water molecules.

A remedy for this inefficiency of REST is a method called REST2 (developed by the same group) based on the modified scaling [18],

$$E_i(\mathbf{x}) = \frac{\beta_i}{\beta_0} E_p(\mathbf{x}) + \left[\frac{\beta_i}{\beta_0}\right]^{1/2} E_{pw}(\mathbf{x}) + E_{ww}(\mathbf{x}). \tag{21.11}$$

We shall not describe REST2 further (see [18]) and will not elaborate on the reasons for the weakness of REST and the strength of REST2. We only comment that for the folding of the 20-residue TrpCage in explicit water, REST2 required ten replicas between 300 and 512K. This system size and previously mentioned sizes of protein-water systems are important as a measure of performance for the various methods. In this context, we specify below three more T-REMD studies, which give an idea about the system's complexity that can be handled by T-REMD.

PROBLEM 21.1

Describe REST2 and discuss its efficiency

(1) García and Onuchic [19] applied standard T-REMD to the 46-residue fragment B of the staphylococcal protein immersed in a cubic box containing 5107 explicit water molecules (16,055 atoms). Using 82 replicas (277–548 K) of the water-protein system, they were able to reveal the folding mechanism of this three-helix bundle protein. (2) Paschek and García [20] studied the folding-unfolding of the C-terminal (41–56) fragment of protein G contained in a box of 1424 explicit water molecules. Using a T-REMD procedure, which also depends on the density, they employed a total of 253 replicas between 320 and 515 K within the density range 0:96 to 1:16 g/cm³. (3) Roe et al. [21] investigated the folding cooperativity in a three-stranded β-sheet of the 20-residue peptide, VFITSdPGKTYTEVdPGOKILQ. However, they used the generalized Born implicit solvent, which required only 12 replicas within the range, 260–570 K.

In Summary: The T-REMD studies involving peptide or protein folding represent a typical type of problem treated in structural biology. Generally, free energy landscapes, as functions of suitable reaction coordinates, can also be obtained with the help of WHAM, for example. As pointed out above, the size and complexity of the systems studied constitute measures of efficiency.

We have discussed in some detail an approach (REST, REST2) for enhancing the efficiency of standard T-REMD; these papers [16–18] are part of a rich body of literature devoted to the same aim. For example, in [22–26], ideas are presented for optimizing the number and allocation of the different temperatures and ways are discussed for generating samples (replicas) of *effectively* equal size (as has been already said, higher temperatures lead to effectively larger samples due to reduced correlations). The molecules involved in these studies are the 46-residue protein A [23], the 67-residue protein GSα₃W [25], and other peptides. While these proteins may be considered relatively large, we emphasize that they are treated in vacuum and in implicit solvents. On the other hand, the theory developed in [26] was tested on a more realistic system consisting of 15,500 explicit water molecules. It is beyond the scope of this book to discuss all the methodological and application papers of the T-REM(D) field; the interested reader is advised to check articles [1–26], the early reviews [27,28], and references cited therein.

21.1.2 Hamiltonian-Dependent Replica Exchange

T-REM(D) is a straightforward technique, which does not require any a-priori knowledge of the investigated problem and can often be used blindly. On the other hand, the user cannot embed in the simulation set-up any knowledge about the problem and only relatively small systems can be handled (still, temperature-based manipulations, as those implemented in REST can lead to enhanced efficiency).

A way to alleviate the limitations of T-REM(D) is provided by Hamiltonian REM(D) [H-REM(D)], in which the different replicas are simulated (in most cases) at a constant temperature, while the Hamiltonian (force field) is used as a replica-coordinate. This approach is based on the recognition that it is not strictly necessary to use the same Hamiltonian for all simulations since they are parallel and non-interacting. Thus, due to the same formal role that $1/k_B T$ and H play in $\exp - (H/k_B T)$, procedures can be devised, where the entire force field, or only a subset of degrees of freedom (e.g., the Lennard-Jones parameters of some atoms) are scaled by a parameter, λ, to define various replicas. Thus, by treating only atoms that mostly effect the molecular stability ("hot" atoms), larger systems can be handled, because the number of required replicas grows as the square root of the number of the *treated* atoms. As pointed out above, it is not strictly necessary to use the same Hamiltonian for all replicas – a freedom that allows biasing the upper ladder ("heated") Hamiltonians in arbitrary ways to accelerate sampling. In what follows, we refer only to H-REMD, as all the applications are based on this method. Notice that scaling the velocities in H-REMD is not required since T is kept constant.

In general, the Hamiltonian (configurational energy) can be divided into an unperturbed part, $E_{up}(\mathbf{x})$ and L perturbed parts, $V_k(\mathbf{x})$, $k = 1,...,L$, which are scaled by the parameters, λ_k, respectively. For replica m the total energy is:

$$E_m(\mathbf{x}) = E_{up}(\mathbf{x}) + \sum_{k=1}^{L} \lambda_{k,m} V_k(\mathbf{x}) \qquad m = 0,...,M \qquad \lambda_{k,m} \leq 1 \tag{21.12}$$

where for $m = 0$ all $\lambda_k = 1$. We shall treat now the general case, where in addition to changes in the parts of the Hamiltonian, the temperature, β_m, depends on replica m as well. Considering again, neighbor replicas 1 and 2 with the corresponding configurations $\mathbf{x}_1$ and $\mathbf{x}_2$, lead to Δ_{12} [see Equation (21.3)],

$$\Delta_{12} = \beta_1\left[E_{up}(\mathbf{x}_2) + \sum_{k=1}^{L} \lambda_{k,1} V_k(\mathbf{x}_2)\right] + \beta_2\left[E_{up}(\mathbf{x}_1) + \sum_{k=1}^{L} \lambda_{k,2} V_k(\mathbf{x}_1)\right]$$

$$-\beta_1\left[E_{up}(\mathbf{x}_1) + \sum_{k=1}^{L} \lambda_{k,1} V_k(\mathbf{x}_1)\right] - \beta_2\left[E_{up}(\mathbf{x}_2) + \sum_{k=1}^{L} \lambda_{k,2} V_k(\mathbf{x}_2)\right] \tag{21.13}$$

$$= \sum_{k=1}^{L} [\beta_1\lambda_{k,1} - \beta_2\lambda_{k,2}][V_k(\mathbf{x}_2) - V_k(\mathbf{x}_1)] + \beta_1\left[E_{up}(\mathbf{x}_2) - E_{up}(\mathbf{x}_1)\right] + \beta_2\left[E_{up}(\mathbf{x}_1) - E_{up}(\mathbf{x}_2)\right]$$

In the case of constant temperature, β, the unperturbed terms, E_{up}, cancel each other and Δ_{12} reduces to,

$$\Delta_{12} = \beta \sum_{k=1}^{L} [\lambda_{k,1} - \lambda_{k,2}][V_k(\mathbf{x}_2) - V_k(\mathbf{x}_1)]. \tag{21.14}$$

It should be pointed out that replicas can be defined with other combinations of $\beta_m/\lambda_{k,m}$. For example, $\lambda_{k,m}$ is defined differently for each m, while β_m is changed only every l replicas, that is, $m = nl$, $n = 1,2,...$ and $m \leq M$. In the more general case, where λ_k is a function of $V_k(\mathbf{x})$ (that is, $\lambda_k V_k(\mathbf{x})$ is replaced by $V_k(\mathbf{x},\lambda_k)$) Equation (21.13) for Δ_{12} with constant β becomes,

$$\Delta_{12} = \beta \sum_{k=1}^{L} V_k(\mathbf{x}_2,\lambda_{k,1}) + V_k(\mathbf{x}_1,\lambda_{k,2}) - V_k(\mathbf{x}_1,\lambda_{k,1}) - V_k(\mathbf{x}_2,\lambda_{k,2}) \tag{21.15}$$

As mentioned earlier, the first Hamiltonian-based REMD was introduced by Sugita et al. [15], who studied a blocked alanine trimer in implicit solvent. However, instead of using the procedure delineated above, they developed a *replica-exchange umbrella sampling method*, where a (*biased*) harmonic restraint potential, $V_m(\mathbf{x})$, defined for replica, m, is added to the force field; $V_m(\mathbf{x}) = k_m(\xi(\mathbf{x}) - d_m)^2$, where ξ is a reaction coordinate (the end-to-end distance, in this case), d_m is a midpoint value, and k_m is the strength of the restraining potential. This system was studied by T-REMD (16 replicas; all $V_m(\mathbf{x}) = 0$), by H-REMD (14 replicas, $T = 300$ K), and by combinations of four temperatures and four restraints, $V_m(\mathbf{x})$ (16 replicas). The potential of mean force as a function of ξ was calculated by WHAM, and the advantages of these techniques over regular umbrella sampling have been demonstrated.

The Hamiltonian REMD procedure described in Equations (21.13–21.15) was introduced by Fukunishi et al. [7]. They treated a coarse-grained model of the peptide $(Ala)_{16}$ with a standard type force field, $E_{up}(\mathbf{x})$, and an implicit solvation term, $V(\mathbf{x})$, which takes into account the effect of hydrophobicity; $V(\mathbf{x})$ is similar to the model defined in Equation (16.2). In this H-REMD calculation, only the hydrophobic term is scaled by the parameters, λ_m,

$$E_m(\mathbf{x}) = E_{up}(\mathbf{x}) + \lambda_m V(\mathbf{x}) \tag{21.16}$$

where four replicas were used in the calculations.

Affentranger et al. [29] applied H-REMD to the 18-residue tip of the P domain of calreticulin that was immersed in 2808 explicit water molecules, by scaling part of the charges and the Lennard–Jones (LJ) parameters. More specifically, the interactions were divided into three groups, meaning that Equation (21.16) is defined with $L = 2$,

$$E_m(\mathbf{x}) = E_{up}(\mathbf{x}) + \lambda_m V_1(\mathbf{x}) + \lambda_m^2 V_2(\mathbf{x}) \tag{21.17}$$

where V_1 represents the sum of: (1) all LJ protein-solvent interactions, (2) electrostatic interactions between side-chain and main-chain atoms, and (3) electrostatic interactions between side-chain atoms and the solvent. V_2 accounts for the sum of: (1) protein-protein LJ interactions and (2) electrostatic interactions between side-chain atoms; solvent-solvent interactions were left unchanged. Eight replicas were found to be enough for handling the above system. To calculate the potential of mean force by WHAM, Equation (21.17) was reformulated to be compatible with Equation (20.51),

$$E_m(\mathbf{x}) = E_0(\mathbf{x}) + (\lambda_m - 1)V_1(\mathbf{x}) + (\lambda_m^2 - 1)V_2(\mathbf{x}) \tag{21.18}$$

where, as before, $E_0(\mathbf{x})$ is the force-field energy of the whole system. The scaling procedure defined above for H-REMD is similar in essence to that used for the temperature-based REST [Equation (21.8)]. A version of this H-REMD was implemented later in the GROMACS 4 program [30] (see section 16.3), where it is applied to the trans-activation 1 peptide (Ace-ETFSDLWKLLPENNH2) of p53 N-terminal domain, solvated by 3387 water molecules. This calculation required 10 replicas as compared to 62 replicas needed with standard T-REMD. A similar method was used by Meli and Colombo [31] to study the folding of villin headpiece (35 residues) and protein A (62 residues) in explicit water, which required 10 and 12 replicas, respectively.

Equations (21.13–21.15), which epitomize the basic theory of H-REMD, can be applied to a wide range of systems by scaling (for each system) the most effective parts of the Hamiltonian. For example, in [32] and [33], scaling is applied to the (φ,ψ) backbone angles and to side chain angles of peptides, respectively. In [34] and [35], the softcore parameters of softcore potentials (LJ and electrostatic) are changed, and in [36], the LJ hardcore parameter (van der Waals radius) is scaled. [37] presents a two-stage procedure in which: (1) a distance-dependent penalty potential is derived from a coarse-grained elastic network model, and (2) this potential (which can drive a structure away from its current conformation) is embedded in a new H-REMD procedure, which leads to an efficient conformational search of peptides modeled by a conventional force field. In another study [38], the scaling targets are hydrogen bonds, and *adapted* hydrogen bond potentials are defined within the framework of a H-REMD scheme. In [39], an attempt has been made to combine *self-guided Langevin dynamics* with H-REMD. In [40], a constant pH-REMD method has been proposed. Finally, in [41], it is suggested to increase the efficiency of the combined method – umbrella sampling and H-REMD – by optimizing the positions of the sampling windows; this is achieved by maximizing a suitable scoring function. Finally, it should be noted, that in accord with results presented earlier, many of the H-REMD procedures were shown to be superior to standard T-REMD. However, efficiency comparisons among the various H-REMD techniques themselves are not available.

In Summary: Replica exchange is a simple, but powerful idea for inducing conformational changes in peptides, proteins, and other frustrated systems (e.g., spin glasses). While standard T-REMD is a straightforward well defined procedure, it is limited to relatively small systems since the temperature changes are applied to the entire (extensive) Hamiltonian. H-REMD, on the other hand, operates only on a limited number, N_h, of "hot" *solute* atoms, therefore, it can handle much larger systems, as the required number of replicas, M, is proportional to $N_h^{1/2}$ (rather than to $N^{1/2}$). Also, H-REMD can be designed in various ways, which are tailored to the specific properties of the Hamiltonian of interest; this flexibility improves the performance further. Indeed, applications discussed in this sub-section show that for H-REMD, the required value of M is smaller than 20, while for T-REMD, M (in some cases) can be an order of magnitude larger. Clearly, procedures that are hybrids of T- and H-REMD are expected to enhance efficiency even further (e.g., [42]).

As mentioned earlier, an important performance criterion for REMD is the maximal protein size (in explicit water) that can be handled. However, because most of the H-REMD studies have been methodological in nature, they have been tested on relatively small systems, and only few of them have challenged the size issue. Among the largest proteins studied, we have mentioned the 46-residue fragment B of the staphylococcal protein, and protein A (62 residues). To this list, one can add the helical proteins α3D (73 residues), α3W (67 residues), Fap1-NRα (81 residues), S-836 (102 residues), and some mutants, which were folded in explicit water by Zhang and Ma [42]; these authors used a single-trajectory-based tempering method with a high temperature dihedral bias. One would expect the treated protein length to grow further in the future as computers become more powerful. Finally, the literature for H-REMD is extensive (larger than for T-REMD) and only a limited number of selected papers could be discussed here. For a more complete coverage, see [1–42] and references cited therein.

21.2 The Multicanonical Method

The usual Metropolis MC method is typically applied at constant T; thus, for proteins at 300K, MC becomes an inefficient procedure for searching large regions of conformational space, due to high energy barriers. This limitation can be removed to a large extent by replica exchange, discussed in the previous section. Another method in this category is the multicanonical (MUCA) method of Berg and Neuhaus [43,44]. MUCA was originally introduced for spin systems, but its potential for the structure determination of peptides and proteins was recognized by Hansmann and Okamoto [45], who extended the method to such problems. Lee [46] has found an alternative implementation of the multicanonical algorithm – the *"entropic sampling method"* that is simpler for some applications than the original procedure [43]; MUCA was also extended to MD simulations [47,48]. We shall describe MUCA below using Lee's formulation [46].

As said above, the usual Metropolis method is performed at constant T and the probability to visit different energies is very uneven. On the other hand, with MUCA, all of the system's energies are visited with the same probability, which enables the system to move from energy to energy with actually no barrier.

Assume a simple lattice chain model with attractive interactions (Section 9.5.3), where L is the total number of different energies in the system. Let the degeneracy of energy, E, be denoted by $v(E)$. The multicanonical probability, P_i^M, of chain configuration i (with energy E_i) is:

$$P_i^M = \frac{1}{L}\frac{1}{v(E_i)} = \frac{1}{L}\exp[-S(E_i)] = P^M(E_i)\exp[-S(E_i)] \qquad (21.19)$$

Therefore, each energy is visited with the same probability $1/L$, and then all configurations with energy E_i are chosen with equal probability $1/v(E_i)$ or $\exp[-S(E_i)]$, where $S(E_i)$ is the *microcanonical entropy* of E_i divided by k_B. To recall, with the T-based MC, the (Boltzmann) probability of E is not $1/L$ but:

$$P^B(E) = \frac{v(E)\exp[-E/k_B T]}{Z} \qquad (21.20)$$

One can simulate the system with the Metropolis method according to P_i^M [Equation (21.19)] (instead of P_i^B) using transition probabilities that satisfy detailed balance,

$$p_{ij} = \min(1, \exp[-S(E_j) + S(E_i)]) = \min[1, v(E_i)/v(E_j)] \qquad (21.21)$$

The simulation is performed as follows: Assume that structure i has been chosen at step t of the process, and we seek to generate a *trial* structure j for step $t + 1$. j is selected at random with probability T_{ij} (Equation 10.9), within a predefined region around the current structure i (for peptides, this is best done in internal coordinates), and then Equation (21.21) is applied. Clearly, with such T_{ij}, chain conformations of energies with high degeneracy will be selected more frequently as trial configurations. However,

p_{ij} [Equation (21.21)], favors structures with low degenerate energies, and thus the net result is that different energies will be visited democratically – with the same probability. This way, the simulation can reach the low energy structures (which contribute mostly to Z) without becoming trapped.

The problem is that the entropies are not known a priori. The entropic sampling version [46] provides a simple prescription for building in a recursive way a function $J(E)$ that is proportional to $S(E)$. The process consists of several separate simulation runs carried out with

$$p_{ij} = \min\{1, \exp[J(E_i) - J(E_j)]\} \tag{21.22}$$

where the improved values of $J(E)$ obtained at simulation k are used to perform the $(k + 1)$th simulation, and the process continues until all energies are visited approximately with the same probability. First, the energy range is divided into bins. The process can be started with a usual MC run at very high T, where the Boltzmann probability is approximately random; the number of visits (MC steps) $H(E)$ to each bin is calculated. Now, each of the bins is assigned an initial value $J(E) = 1$, which is corrected after the initial run in the following way:

$$J_{new}(E) = \begin{cases} J_{old}(E) + \ln H(E) & \text{if} \quad H(E) > 0 \\ J_{old}(E) & \text{if} \quad H(E) = 0 \end{cases} \tag{21.23}$$

The next simulation is carried out with the new set of $J(E)$ using Equation (21.21) [$J(E)$ replacing $S(E)$], where $H(E)$ is put to 0 and is calculated again during the simulation; subsequently, the set of $J(E)$ is updated and the process continues.

This procedure can better be explained as applied to a system of two energies with different degeneracies, $v(E_1) \gg v(E_2)$. Because states are selected at random ($T_{12} = T_{21}$), after one cycle, $H(E_1) > H(E_2)$, and for a *very long* cycle,

$$\frac{H(E_1)}{H(E_2)} \approx \frac{v(E_1)}{v(E_2)} \tag{21.24}$$

Thus, a production run can be performed with $H(E_1)$ and $H(E_2)$ using Equation (21.21), where the two energies will be visited (approximately) an equal number of times. However, starting from a short cycle, one might encounter the situation, $H(E_1)/H(E_2) = Av(E_1)/v(E_2)$ for $A > 1$. Thus, in the next cycle, the somewhat *distorted* $H(E_1)/H(E_2)$ [now $J(E_1)/J(E_2)$] is expected to lead to $H'(E_1)/H'(E_2) \approx 1/A$ rather than 1. In this case, the correction machinery of the method is being activated, that is:

$$\frac{H(E_1)}{H(E_2)} \frac{H'(E_1)}{H'(E_2)} = \frac{v(E_1)A}{v(E_2)} \frac{1}{A} \approx \frac{v(E_1)}{v(E_2)} \tag{21.25}$$

While the idea is interesting, it should be stressed that the simulations for building the density of states [$J(E)$] are computationally expensive and practically always approximate, that is, a fully flat histogram is not achievable (a typical time series of the energy is shown in Figure 21.2); in this respect, REM has an advantage over MUCA. For a system with many different energies, one has to confirm that the low energies have been visited; in particular, for proteins, reaching the low energy (small) region in conformational space (protein folding) might not be trivial even with MUCA.

The final set of $J(E)$'s can be used to calculate statistical averages $\overline{X}(T)$ at *significantly different* temperatures, T, based on a single MUCA run using the reweighting formula [see Equation (14.44)]:

$$\overline{X}(T) = \frac{\sum\limits_{E} X(E)\exp[J(E) - E/k_B T]}{\sum\limits_{E} \exp[J(E) - E/k_B T]}. \tag{21.26}$$

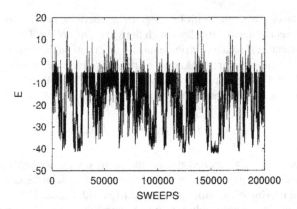

FIGURE 21.2 Typical time series of the energy visited using MUCA. The calculation was done for the hepta-peptide deltorphin modeled by ECEPP/2 with dielectric constant, $\varepsilon = 2$ in vacuum; the number of variables (dihedral angles) is 36. The figure shows the last 1/5 of the total production run. The system visits an energy range of ~50 kcal/mol. (From Yasar, F. et al.: Efficiency of the multicanonical simulation method as applied to peptides of increasing Size: The heptapeptide deltorphin. *Journal of Computational Chemistry.* 2002. 23. 1127–1134. Copyright Wiley-VCH Verlag GmbH & Co. KGaA. Reproduced with permission.)

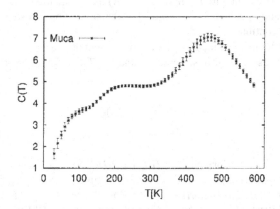

FIGURE 21.3 Specific heat calculated for the hepta-peptide deltorphin in vacuum (see Figure 21.2) over a wide range of temperatures by the reweighting formula, Equation (21.26). For the significance of the peaks see [54]. (From Yasar, F. et al.: Efficiency of the multicanonical simulation method as applied to peptides of increasing Size: The heptapeptide deltorphin. *Journal of Computational Chemistry.* 2002. 23. 1127–1134. Copyright Wiley-VCH Verlag GmbH & Co. KGaA. Reproduced with permission.)

This formula demonstrates an important advantage of the multicanonical method over conventional MC, which requires a separate simulation for each temperature. The specific heat as a function of temperate is shown in Figure 21.3. (Notice that results for different temperatures, T_i, based on a single MC run at T can be obtained with the method of Ferrenberg and Swendsen [Equation (14.44)], but only for T_i that are close enough to T.). For a large system, most of the contribution to $<X(T)>$ (e.g., X an extensive variable) comes from a very narrow region around a typical energy, $E^*(T)$ (Section 5.2); therefore, the bin's size should be decreased, which requires longer simulations to gain sufficient accuracy. In this case, one can limit the energy region studied, while decreasing correspondingly the range of temperatures considered. The entropy difference between two temperatures T_1 and T_2 can be calculate (at least approximately) from, $J[E^*(T_1)] - J[E^*(T_2)]$, where $E^*(T_1)$ and $E^*(T_2)$ are the typical energies for the corresponding temperatures. Thus, MUCA is a convenient tool for identifying first-order transitions, where two different energies [with different $J(E)$ values] have the same free energies.

Related to the above discussion, it should be pointed out that the ability of MUCA to cover the whole energy range is limited to relatively small systems, due to the orders of magnitude differences among $v(E)$ values, which for an N-spin Ising model, for example, range from 1 to 2^N. Therefore, in practice (as pointed out above), the simulation is performed within the energy range of interest, which for Ising models is typically around phase transition. On the other hand, for peptides and proteins, the interest generally is to find the most stable microstate (preferably of lowest free energy, at least of lowest energy), which is extremely small and might not be reached with the *random* transition probabilities, T_{ij}; clearly, in this case, MUCA's machinery will not be operative at all. Therefore, the efficiency of the simulation depends on the specific system, its size, and the MUCA procedure used.

Implementation of MUCA requires calculating the a priori unknown weights by an iterative procedure that is not straightforward and can be a stumbling block for newcomers. A recursive procedure has been suggested that takes into account the statistical errors of all previous iterations, and thus enables an *automatic* calculation of the weights; thus, the need for human intervention after each iteration is avoided. This problem was addressed first by Berg's group, where the procedure was tested for magnetic lattice models [49–52]. Later, a detailed translation of this recursion procedure for continuum models of peptides has been carried out by Yasar et al., as applied to Leu enkephalin [53] and, subsequently, to the heptapeptide with bulky side chains, deltorphin (H-Tyr[1]-DMet[2]-Phe[3]-His[4]-Leu[5]-Met[6]-Asp[7]-NH2) [54]. Both systems were modeled in vacuum by ECEPP/2 (see Section 16.3), with dielectric constant, $\varepsilon = 2$, where the number of variables (dihedral angles) is 24 and 19 for Leu enkephalin defined with variable ω and with constant ω (=180°), respectively. For deltorphin with $\omega = 180°$, the number of variables is 36. Energy minimization of conformations of the MUCA trajectories were found to provide good conformational coverage of the low energy regions of the molecules. However, these results are somewhat inferior to those obtained with the Monte Carlo minimization method (Section 16.8.2).

21.2.1 Applications

To examine the behavior of MUCA, the early studies were based on simplified force fields, such as ECEPP [Section 16.3], which consists of constant bond lengths and bond angles, hence, on significantly reduced number of variables – the backbone dihedral angles (φ, ψ) and the side chain angles, χ. Also, solvent effects were described only implicitly or were not considered at all (a peptide in vacuum). Furthermore, popular molecules for performance studies have been the pentapeptides Met- and Leu-enkephalin, not only due their relatively small size, but because their ground state energies have been obtained by other techniques.

Using MUCA, Hansmann and Okamoto were able to find the microstate containing the global energy minimum of Met enkephalin [45]. In another study, they compared the performances of MUCA, $1/k$-sampling, and simulated tempering (Section 21.4) to fold Met-enkephalin and have found these methods to have a comparable efficiency [55]. Hansmann and Okamoto also showed that a few variants of MUCA outperform simulated annealing as applied to Met-enkephalin [56]. Using ECEPP (with dielectric constant, $\varepsilon = 2$), Hansmann and Okamoto investigated the helix-coil transition in 10-residue oligomers, of valine, glycine, and alanine. Their results were compatible with the experimental assertions that alanine is a helix-forming, valine is a helix-indifferent, and glycine is a helix-breaker [57].

It is important to mention the early work of Hao and Scheraga. Thus, to increase efficiency, they implemented in MUCA a biased set of transition probabilities, T_{ij} (rather than the random set, $T_{ji} = T_{ji}$) and "*jump walking techniques*" [58]. Treating simplified lattice models of proteins (of 27–57 residues), their objective has been to identify highly foldable amino acid sequences [59] and to optimize force field parameters for folding [60, 61, see also 62]. These tasks were carried out by folding experiments with their improved MUCA procedure, which was also used to study an entropy driven coil-globule first-order transition [58,63].

In later years, MUCA has been applied to larger systems. For example, Hansmann's group [64] compared the efficiency of MUCA and replica exchange MD by folding an all-atom Gō model of the 75-residue protein, MNK6, with an end-to-end β-sheet [A Gō model (due to Nobuhiro Gō) is a simplified model, where the lowest energy structure is known by construction.]. They found MUCA to be the more efficient method, as it produced folding/unfolding events by a factor of 30 more than replica exchange.

Then, they studied by MUCA the folding landscape of the 36-residue DS119 with a physical all-atom force field and an implicit solvent, to find that the folding mechanism is greatly affected by the central helix. To understand whether and how this folding mechanism depends on the energy function, in a subsequent paper [65], Uyaver and Hansmann compared MD and MC implementations of MUCA based on different force fields. While the native structure was obtained with a similar frequency, the folding mechanism was found to be different.

Still, in all the above examples, MUCA has been tested as applied to relatively simplified models in implicit solvent. Nakajima, Nakamura, and Kidera of the Osaka group (who extended MUCA in 1997 to MD simulations [47,48]) have taken the challenge to apply MUCA to realistic all-atom macromolecular models in *explicit solvent*. Thus, in the following years, the Osaka group has carried out extensive work (tens of papers) based on MUCA-MD and enhancements introduced to it [66–69]. We shall only present several examples which demonstrate the efficiency of MUCA-MD to handle large systems. Thus, an early paper [70] investigates the stability of the 8-residue β-hairpin peptide, Ala-Ile-Val-Asn-Gly-Lys-Thr-Tyr in explicit water. A later paper is a study in explicit water of the relative stability of α and β secondary structures (and other microstates) of a 40-residue protein (the C-terminal domain of histone-like nucleoid- structuring (H-NS) protein) [71]. The MUCA-MD results were in accord with those obtained by nuclear magnetic resonance (NMR), and the importance of using a reliable force field has been emphasized. Another study is the ab initio folding of a 57-residue protein – the first repeat of human glutamylprolyl-tRNA synthetase (EPRS-R1) in explicit solvent [72]. The fact that the simulation started from a fully extended structure means that no knowledge about the native structure has been introduced. The authors have found that the lowest free energy cluster (which is also the largest cluster) carries the highest similarity to the NMR structure.

21.2.2 MUCA-Summary

The above discussion demonstrates the importance of MUCA as a conformational search tool, in particular, in structural biology. Clearly, we could discuss only a limited number of publications selected from the vast body of literature related to this technique. Still, the tedious calculation of weights (in particular, with increasing system size) has led to attempts to circumvent this problem by combining MUCA with other techniques. Thus, Sugita and Okamoto have suggested a three-stage procedure, where: (1) one performs a T-REMD study consisting of a set of temperatures which cover the energy range of interest, (2) the density of states, $\nu(E)$, is obtained by applying WHAM to the T-REMD results, and (3) the $\nu(E)$ values are used in a MUCA production run [73–75]. This method, called, the "*replica-exchange multicanonical algorithm*," was developed together with a related technique – "*multicanonical replica-exchange*." A "*multicanonical parallel tempering technique*" was also developed by Faller, Yan, and de Pablo [76], while Escobedo's group suggested determining the MUCA weights based on Bennett's optimized acceptance ratio method [77]. To this category, one should add the "*virtual-system coupled multicanonical molecular dynamics simulation*" of Higo, Umezawa, and Nakamura [69]. For a more complete discussion about the relations of MUCA with other techniques, see [78] and references cited therein; reference [79] is an early review about MUCA and other advanced conformational search techniques for protein folding.

21.3 The Method of Wang and Landau

In the previous paragraph, we have pointed out the difficulty in calculating sufficiently accurate results for the density of states using the self-consistent procedure of MUCA, and some solutions to this problem have been mentioned. An additional important contribution in this direction is the method of Wang and Landau (WL) [80,81]. The algorithm consists again on the basic observation of MUCA, that if one performs a random walk in energy space with a probability proportional to $\nu(E)^{-1}$, a flat histogram is generated for the energy distribution. For the Ising model, for example, the procedure is carried out as follows:

At the very beginning of the random walk, the density of states is a priori unknown, so one sets the densities of states $v(E)$ of all energies, E to $v(E) = 1$. Then, one begins the random walk in energy space by flipping spins randomly. In general, if E_1 and E_2 are energies before and after a (trial) spin has been flipped, the transition probability is:

$$p(E_1 \to E_2) = \min\left[1, \frac{v(E_1)}{v(E_2)}\right]. \tag{21.27}$$

But unlike with MUCA, one updates the corresponding density of states by multiplying the existing value by a modification factor $f > 1$, that is, $v(E) \to f v(E)$; more specifically, acceptance or rejection of E_2 leads to the multiplications, $f v(E_2)$ or $f v(E_1)$, respectively. The initial modification factor WL chose was $f = f_0 = e^1 = 2.718228\ldots$, while other values at this range are acceptable as well. However, with f_0 too small, the random walk will spend an extremely long time to reach all possible energies, while a too large f_0 will lead to high statistical errors.

One keeps walking randomly in energy space, modifying the density of states until the accumulated histogram, $H(E)$ – the number of visits to each energy E, becomes approximately "flat". At this point, the logarithm of the density of states converges to the true value with an accuracy proportional to $\ln(f)$. Then one reduces the modification factor to a finer one according to some recipe, such as $f_1 = (f_0)^{1/2}$ (any function that monotonically decreases to 1 will do) and resets the histogram to $H(E) = 0$. Then one begins the next level random walk (based on the $v(E)$ values obtained in the previous round) with a finer modification factor $f = f_1$; this is continued until the histogram again becomes "flat"; then one stops and reduces the modification factor as before, that is, $f_{i+1} = (f_i)^{1/2}$. The MC steps needed for a given f generally increase as the modification factor is refined.

The simulation process is stopped when the modification factor is smaller than some predefined final value [such as $f_{\text{final}} = \exp(10^{-8}) \approx 1.000, 000, 01$. It is impossible to obtain a perfectly flat histogram, and the phrase "flat histogram" means that histogram $H(E)$ for all possible E is not less than $x\%$ of the average histogram $H(E)$, where $x\%$ is chosen according to the size, complexity of the system, and the desired accuracy of the density of states. For the 32×32 Ising model with nearest neighbor interactions, WL find that $x\%$ can be chosen as high as 95%; however, if one chooses a too high $x\%$ for a large system, the criterion for "flatness" may never be satisfied, and the program may run forever. They also point out that $\ln(f_{\text{final}})$ cannot be chosen arbitrarily small, since the modified $\ln[v(E)]$ will not differ from the unmodified one to within the number of digits in the double precision accuracy of the calculation. If this happens, the algorithm will not converge to the true value, and the program may run forever. Even if f_{final} is within range, but too small, the calculation might take excessively long to finish. While the accuracy of the density of states depends strongly on f_{final}, it is also affected by many other factors, such as the complexity and size of the system, criterion of the flat histogram, and other details of the implementation of the algorithm.

As for MUCA, for a large system, it would be difficult to calculate all the $v(E)$'s correctly from a single simulation. Therefore, multiple random walks can be performed, each for a different energy range, where MC moves beyond the range are rejected. The resultant pieces of the density of states can then be joined together and used to produce canonical averages for the calculation of thermodynamic quantities at any temperature [see Equation (21.26)].

With the conventional versions of MUCA, the density of states is updated from the histogram, $H(E)$, directly only after enough histogram entries have been accumulated. Because of the exponential growth of the density of states in energy space, this process is relatively inefficient due to the *linear* accumulation of the histogram. On the other hand, in the WL algorithm, the density of states is modified at each step of the random walk, which allows approaching the true values much faster than with the conventional methods, especially for large systems. $H(E)$ is only used to check whether the histogram is flat enough to go to the next level random walk with a finer modification factor. Clearly, by the end of the simulation, only a relative density of states is obtained. The absolute values can be calculated if the exact value is known for some energy (e.g., for the square Ising lattice $v(E) = 2$ for $E = 2N$).

During the random walk (especially for the early stage of iteration), the algorithm does not exactly satisfy the detailed balance condition, since the density of states is modified constantly; however, after many iterations, the density of states converges to the true value very quickly as the modification factor approaches 1. From Equation (21.27), we have:

$$\frac{1}{v(E_1)} p(E_1 \rightarrow E_2) = \frac{1}{v(E_2)} p(E_2 \rightarrow E_1) \tag{21.28}$$

where $1/v(E_1)$ is the probability at the energy level E_1 and $p(E_1 \rightarrow E_2)$ is the transition probability from E_1 to E_2 for the random walk. It can thus be concluded that the detailed balance condition is satisfied to within the accuracy proportion to $\ln(f)$.

The WL procedure looks relatively simple and results with high accuracy were obtained by them for the square Ising model (largest system of 256 × 256 sites) and a 2d 10-state Potts model (200 × 200 sites). Notice that like with MUCA, the *relative* free energy (and entropy) can be calculated for different temperatures:

$$F(T) = -k_B T \ln \sum_E v(E) \exp[-E/k_B T] \tag{21.29}$$

Only if the absolute value of $v(E)$ (for any E) is known, the absolute F and S can be obtained by matching the absolute and MC results of $v(E)$.

All thermodynamic quantities discussed by WL for the Ising model (energy, entropy, and specific heat) are directly related to the energy. However, it is possible to calculate any quantity which may not directly relate to energy, that is, the random walk is defined for a more complex parameter space. For that, one needs to know the Hamiltonian, and the algorithm can be optimized to estimate the relevant density of states to the property and temperature range of interest. An example would be the end-to-end distance, R, of a polymer, where values of $v(E,R)$ rather than $v(E)$ are calculated from which $\overline{R(T)}$ can be obtained at different temperatures. This comment applies also to MUCA.

21.3.1 The Wang and Landau Method-Applications

The WL method has been extended to simulations in continuum and to different ensembles, where an extensive work has been carried out by de Pablo's group. In an early work [82], this group examined the effect of amino acid sequence and chain length on the random coil–helix and random coil–beta sheet transitions in model proteins based on a cubic lattice and a knowledge-based force field. This group also treated a system of up to 110 Lennard–Jones particles in the (μVT) ensemble based on the calculation of $v(N, E)$ [83].

A way to enhance efficiency is by carrying out several independent WL runs in parallel, each within a relatively narrow window of energies, where the energies of neighbor windows have some overlap; configurations are swapped between windows within the framework of parallel tempering (REM). Thus, by superimposing the overlapping energy windows, $v(E)$ values are obtained over a relatively large energy range [84]. In this way, folding transitions in deca-alanine and Met-enkephalin have been studied with and without an implicit solvent. The WL method with parallel tempering has also been applied in a solid-liquid equilibria study based on a Lennard–Jones fluid and NaCl; this study consists of a two-dimensional density of states $v(E, V)$ [85]. In another paper of this type, the effects of chaperonin-like cage-induced confinement on protein stability were studied for proteins of varying sizes and topologies described by minimalist Gō models [86].

Another method for calculating the density of states is based on the thermodynamic equation [compare with Equation (2.20)]:

$$\ln v(N,E,V) = \int \frac{1}{k_B T(E)} dE \tag{21.30}$$

where the *configurational temperature*, $T(E)$, can be obtained from the forces $\mathbf{F}_i$ exerted on particles i, during an MC simulation [87]:

$$\frac{1}{k_B T(E)} = \frac{\left\langle \sum_i -\nabla \cdot \mathbf{F}_i \right\rangle}{\left\langle \sum_i |\mathbf{F}_i|^2 \right\rangle} \tag{21.31}$$

Yan and de Pablo [88,89] suggested obtaining $T(E)$ from the averages of these forces and their derivatives calculated during a WL process. In particular, results for $T(E)$ obtained for $f \to 1$ (where detailed balance is satisfied) are expected to lead to unbiased estimates for $\ln[v(N, V, E)]$ (unlike the bias introduced to $\ln[v(N, V, E)]$ in the early stages of the histogram-based WL procedure). This method called, *"the configurational temperature density of states"* (CTDOS) also provided the free energy. The advantage of CTDOS (equipped with parallel tempering) over the usual WL method (WLDOS) has been demonstrated in [89] by applying both methods to the 16-residue β-hairpin taken from the C-terminal fragment (41–56) of protein G. CTDOS was also used for analyzing the entropic and energetic contributions to the stability of proteins near surfaces by studying four proteins described by Gō models [90]. Finally, an additional method called CTDOSH has been suggested and has been found to be more efficient than CTDOS as applied to the Lennard-Jones fluid [91].

In [92], the WL method was combined with an expanded ensemble (see next sub-section). The combined method was used for calculating the potential of mean force between a spherical particle suspended in nematic liquid crystal and the surface of the container. This method called, the *"expanded ensemble density of states"* (EXEDOS) was used later for studying the mechanical stretching of proteins [93]. More specifically, the expanded ensemble has been expressed in terms of the stretched end-to-end distance of a 15-mer alanine in an implicit solvent, where the corresponding potentials of mean force were calculated. EXEDOS was also applied to the study of crystallization [94].

Finally, molecular simulations were used to quantify the mechanical stability of four helical β-peptides (of ~10 residues) by subjecting them to an unfolding process and calculating the potential of mean force. These molecules, modeled with an implicit solvent, were treated efficiently by EXEDOS. However, EXEDOS was found inadequate to handle the more complex systems based on explicit solvent, which, thus, were treated instead by another method [95].

21.4 The Method of Expanded Ensembles

Lyubartsev et al. [96] developed a method which enables crossing energy barriers from a *single* MC run, where differences in free energy are provided as a by-product of the simulation. This method, which is called EXE, was developed independently by Marinari and Parisi [2], who called it *"simulated tempering."*

Consider an *NVT* ensemble with the Hamiltonian $H(\mathbf{x}^N)$ and the reciprocal temperature $\beta = 1/k_B T$. A set of reciprocal temperatures, β_m is defined.

$$0 = \beta_0 < \beta_1 < \beta_2 < \ldots < \beta_M = \beta \qquad (\infty = T_0 > T_1 > T_1 > \ldots T_M = T). \tag{21.32}$$

For each β_m, we define a canonical ensemble with the same Hamiltonian, fixed N and V, which lead to the partition function, Z_m,

$$Z_m = \frac{1}{N!} \int d\mathbf{x}^N \exp[-\beta_m H(\{\mathbf{x}^N\})]. \tag{21.33}$$

Using Z_m, we define an expanded ensemble with the partition function, Z:

$$Z = \sum_{m=0}^{M} Z_m \exp(\eta_m) \tag{21.34}$$

where η_m are constants to be determined. The expansion of the *NVT* ensemble implies the transfer from a single fixed reciprocal temperature β to a set of temperatures $\{\beta_m\}$ using the factors $\exp(\eta_m)$. This situation is reminiscent of the grand-canonical partition function [Equation (6.37)], which constitutes a summation over canonical partition functions of different *N:*

$$Z(\mu,T,V) = \sum_N Z_N \exp(\mu N/k_\mathrm{B}T). \tag{21.35}$$

Correspondingly, each canonical ensemble, *m*, becomes a sub-ensemble of the expanded ensemble [Equation (21.34)]. The probability to find the *m*th system in the expanded ensemble is:

$$P_m^\mathrm{e} = \frac{Z_m \exp(\eta_m)}{Z} \tag{21.36}$$

and the probability (in the expanded system) of configuration $\mathbf{x}^N$ at *m* is:

$$P_m(\mathbf{x}^N) = P_m^\mathrm{e} p(\mathbf{x}^N|m) = \frac{Z_m \exp(\eta_m)}{Z} \frac{\exp[-\beta_m H(\mathbf{x}^N)]}{Z_m} = \frac{\exp[-\beta_m H(\mathbf{x}^N) + \eta_m]}{Z} \tag{21.37}$$

One can apply the Metropolis MC procedure to the expanded ensemble using two types of moves: (1) the usual displacement of particles (coordinates) at a fixed temperature and (2) a transfer of configuration from β_m to β_n (typically, $\beta_{m\pm1}$); in this case,

$$\frac{p_{mn}}{p_{nm}} = \frac{P_m(\mathbf{x}^N)}{P_n(\mathbf{x}^N)} = \frac{\exp[-H(\mathbf{x}^N)\beta_n + \eta_n]}{\exp[-H(\mathbf{x}^N)\beta_m + \eta_m]} = \exp[-H(\mathbf{x}^N)(\beta_n - \beta_m) + \eta_n - \eta_m] \tag{21.38}$$

and the corresponding (asymmetric MC) transition probability density, p_{mn} is:

$$p_{mn} = \min\{1, \exp[-H(\mathbf{x}^N)(\beta_n - \beta_m) + \eta_n - \eta_m]\}. \tag{21.39}$$

Now:

$$\frac{P_m^\mathrm{e}}{P_n^\mathrm{e}} = \frac{Z_m}{Z_n} \exp(\eta_m - \eta_n) = \exp(-\beta_m F_m + \beta_n F_n + \eta_m - \eta_n) \tag{21.40}$$

and free energy differences can be obtained for any pair of temperatures. For a fluid, the case $\beta_0 = 0$ corresponds to an ideal gas, where the partition function is known exactly, $Z_0 = V^N/N!$ Therefore, one can obtain the *absolute* free energy,

$$\beta_m F_m = -\ln(Z_m) = -\ln(P_m^\mathrm{e}/P_0^\mathrm{e}) + \eta_m - \eta_0 - \ln(Z_0) \tag{21.41}$$

provided that η_m and η_0 are known (Unlike replica exchange, only a *single* replica is treated in the simulation, which is sought to visit the higher temperatures.). A perfect simulation, where the single replica spends equal time in all temperatures, would be achieved if for all *m:*

$$\eta_m = \beta_m F_m \quad \text{or} \quad P_m^\mathrm{e}/P_n^\mathrm{e} = 1 \tag{21.42}$$

Thus, η_m is based on the (a priori) *unknown* F_m – the aim of the study! Clearly, applying, $\eta_n = 1$ for all *n* will not reach this goal of democracy, as the system will remain in the lowest temperature.

A similar problem has been encountered with respect to the entropies in MUCA, and the remedy here is the same: one starts the simulation with a set of $v_n = 1$ (or $v_n = e$) for all n ($\eta_n = -\ln v_n$); the v_n values are then built up recursively in several separate runs using the procedure described for MUCA (or that of WL with a factor f). When $\ln(P_m^e)$ ($m = 1,...,M$) become approximately equal, one can carry out a production run (typically the η_n are negative). Notice that for a complex system (e.g., a protein), the $T = \infty$ reference state is not applicable, and, in practice, only *differences* in free energy, $\beta_m F_m - \beta_n F_n$ can be calculated by EXE. EXE can also be used to calculate other quantities, such as the potential of mean force along a reaction coordinate, ξ, such as the distance between two specific molecules in solvent or the end-to-end distance of a polymer. In this case, Equation (21.39) becomes (see Equation (21.55) and [84]):

$$p_{mn} = \min\{1, \exp\{-\beta[H(\xi_n) - H(\xi_m)] + \eta_n - \eta_m\} \tag{21.43}$$

Like replica exchange, EXE enables the system to cross energy barriers by inducing exchange of conformations among different temperatures. However, this method shares the same difficulties encountered with MUCA – the need to carry out many simulations prior to the production run for building up the set of η_n. This set is never perfect, and the computer time required for its calculation grows strongly with system size. On the other hand, for calculating free energy differences, the set of η_n can be approximate, as long as enough visits occur to the temperatures of interest.

21.4.1 The Method of Expanded Ensembles-Applications

In general, EXE has not been used by itself, but has been "teamed" with other methods to enhance efficiency. We have already mentioned its combination with the WL technique within the framework of EXEDOS. Fenwick and. Escobedo [97] combined EXE with replica exchange and used their protocol to study protein folding by treating a single linear heteropolymer chain having 27 monomers on a simple cubic lattice. The combined method was found to outperform the individual methods at low temperatures. In another study of this group [98], EXE has been combined with Bennett's acceptance-ratio method for calculating the chemical potential of block copolymer chains in melt. This method was further developed in [99], as applied to the solvation of large hard spheres into a fluid of small spheres and the mesophase formation of a block copolymer-homopolymer mixture. In [100], this method was extended for chemical potential calculations in cases where the number of molecules fluctuate, as in the grand canonical ensemble. The method was applied to the solvation of large hard spheres in a fluid of small and large spheres, and to the vapor-liquid equilibrium of a chain system. Finally, notice that further discussions about calculating the chemical potential based on EXE and other techniques appear in Section 22.7.

21.5 The Adaptive Integration Method

An interesting development in the TI category is the AIM for computing free energies due to Fasnacht et al. [101]. However, AIM can treat other quantities, such as radial distribution functions and potentials of mean force. A general TI process [Equation (17.12)] is based on the integral:

$$\Delta F = \int_0^1 \left\langle \frac{dH(\lambda, \mathbf{x}^N)}{d\lambda} \right\rangle_\lambda d\lambda, \tag{21.44}$$

where $0 \le \lambda \le 1$ defines a hybrid Hamiltonian, $H(\lambda) = (1 - \lambda)E_0 + \lambda E_1$, that is varied between two energy functions E_0 and E_1 [$H(\lambda)$ can also be defined by more general non-linear scalings.]. This integral is commonly evaluated by carrying out l *separate* MD (or MC) simulations [based on $H(\lambda)$] at l intermediate λ values; the corresponding l averages (in conformational space, $\mathbf{x}^N$) of the derivative $<dH(\lambda, \mathbf{x}^N)/d\lambda>_\lambda = <E_1(\mathbf{x}^N) - E_0(\mathbf{x}^N)>_\lambda$ are then calculated. With AIM, on the other hand, the sampling is performed within an MC procedure that allows transitions between coordinates as well as between different λ values. Therefore, the parameter λ is treated as an additional coordinate, which leads to an expanded $(\lambda, \mathbf{x}^N)$ ensemble. The (a priori unknown) partition function, Z_λ, for a given λ is,

$$Z_\lambda = \int \exp[-H(\lambda, \mathbf{x}^N)/k_BT]d\mathbf{x}^N, \tag{21.45}$$

and the Boltzmann probability density to find a conformation, $\mathbf{x}^N$, in the ensemble based on $H(\lambda, \mathbf{x}^N)$ is:

$$P^B(\mathbf{x}^N/\lambda) = \exp[-H(\lambda, \mathbf{x}^N)/k_BT]/Z_\lambda. \tag{21.46}$$

In accord with the definition of the expanded ensemble for the temperature discussed in the previous section, we define here an expanded ensemble with respect to λ at constant T (β) based on the expanded partition function, Z:

$$Z = \sum_{\lambda=0}^{M} Z_\lambda \exp(\eta_\lambda) \text{ and } P^e(\lambda) = \frac{Z_\lambda \exp(\eta_\lambda)}{Z} \tag{21.47}$$

where $P^e(\lambda)$ is the probability to find λ in the expanded ensemble. A normalized probability for $(\lambda, \mathbf{x}^N)$ in the expanded ensemble is [see Equation (21.37)]:

$$P(\mathbf{x}^N, \lambda) = P^e(\lambda)P(\mathbf{x}^N/\lambda) = \frac{Z_\lambda \exp(\eta_\lambda)}{Z} \frac{\exp[-H(\mathbf{x}^N, \lambda)\beta]}{Z_\lambda} = \frac{\exp[-H(\mathbf{x}^N, \lambda)\beta + \eta_\lambda]}{Z} \tag{21.48}$$

which leads to the MC transition probability,

$$p(\lambda_1 \rightarrow \lambda_2) = \min\left[1, \exp\{-[H(\lambda_2, \mathbf{x}^N) - H(\lambda_1, \mathbf{x}^N) + \hat{F}_{\lambda_1} - \hat{F}_{\lambda_2}]/k_BT\}\right] \tag{21.49}$$

In accord with Equation (21.42), a "democratic" sampling requires $\eta_\lambda = \beta F_\lambda = -\ln Z_\lambda$, and because F_λ is not known a priori, an *approximate* free energy, $\hat{F}_\lambda = -k_BT \ln \hat{Z}_\lambda$ (appears with a hat), replaces the correct F_λ in Equation (21.49). The values of $\hat{F}_\lambda(\hat{Z}_\lambda)$ are built up with an adaptive procedure *different* from that used in the methods described previously. More specifically, during the simulation, the averages $<dH(\lambda, \mathbf{x}^N)/d\lambda>_\lambda = <E_1(\mathbf{x}^N) - E_0(\mathbf{x}^N)>_\lambda$ for each λ visited are calculated; these values are numerically integrated, leading to new values of $\hat{F}_{\lambda_i} - \hat{F}_{\lambda_j}$ for all pairs i, j [Equation (21.44)]. The new $\hat{F}_{\lambda_i}$ values are inserted in Equation (21.49) for the next round of the simulation, etc. As the simulation continues, the running averages of the free energy derivatives become more accurate, making the estimated differences, $\hat{F}_{\lambda_i} - \hat{F}_{\lambda_j}$, increasingly accurate as well. The detailed balance condition will be satisfied (albeit asymptotically) with all λ values (bins) being visited an equal number of times.

It should be noted the one can carry out also the more traditional procedure (e.g., of MUCA and WL), starting from $\hat{Z}_\lambda = 1$ for all λ, where $\hat{Z}_\lambda$ is increased by 1 for each visit to the λ bin. Thus, asymptotically (that is, for a very long run) the ratios of the $\hat{Z}_\lambda$ values will attain stability, $\hat{Z}_{\lambda,i}/\hat{Z}_{\lambda,j} \approx Z_{\lambda,i}/Z_{\lambda,j}$. Notice also that for each λ, the transitions between the coordinates ($\mathbf{x}^N$) can be carried out by any canonical simulation technique (MC, MD, etc.).

The authors claim that a larger number of bins (λ values) can be treated with AIM than with TI (for the same amount of computer time), which leads to a much finer resolution. Another potential advantage of AIM lies in the fact that a bin might be visited many times during the simulation, each visit starts from a different structure (seed), leading to an adequate sampling of the contributing microstate(s) for this λ. With TI, on the other hand, only a single simulation (starting from one seed) is typically performed, and the coverage of the contributing microstates is expected to be more limited.

AIM draws from the previously developed methods, MUCA, EXE, and WL, and it is related to the λ-dynamics approach, where λ is treated as a thermodynamic variable [102–104]. However, AIM also provides a prescription for simulating other quantities, such as radial distribution functions, potentials of mean force, or chemical potentials. For example, the radial distribution function for the separation $r_{ij}(\mathbf{x}^N)$ between particles i and j is:

$$P_{ij}(r) = \frac{\int \exp[-\beta H(\mathbf{x}^N)]\delta[r - r_{ij}(\mathbf{x}^N)]d\mathbf{x}^N}{Z} = \frac{Z_r}{Z} \tag{21.50}$$

where, for simplicity, the partition function $Z_{r_{ij}}$ is denoted by Z_r. Also, we denote by $\mathbf{x}_r^N$ a configuration within the partial ensemble, where $r_{ij} = r$. The (conditional) probability density to find $\mathbf{x}_r^N$ within this ensemble is:

$$P(\mathbf{x}_r^N/r_{ij} = r) = \frac{\exp[-\beta H(\mathbf{x}_r^N)]}{Z_r} \tag{21.51}$$

In the present discussion [related to $P_{ij}(r)$], the role of λ (for the free energy) is replaced by r. Thus, one can define an expanded partition function Z^e and probability, $P^e(r)$, [see Equations (21.44–21.48)], where:

$$Z^e = \sum_r Z_r \exp(\eta_r) \text{ and } P^e(r) = \frac{Z_r \exp(\eta_r)}{Z^e} \tag{21.52}$$

and the probability of $\mathbf{x}_r^N$ in this ensemble is:

$$P(\mathbf{x}_r^N, r) = P^e(r)P(\mathbf{x}_r^N/r) = \frac{Z_r \exp(\eta_r)}{Z^e} \frac{\exp[-\beta H(\mathbf{x}_r^N)]}{Z_r} = \frac{\exp[-\beta H(\mathbf{x}_r^N) + \eta_r]}{Z^e}. \tag{21.53}$$

For a "democratic" sampling, $P^e(r_1)/P^e(r_2) = 1$, one obtains:

$$\exp(\eta_{r_2} - \eta_{r_1}) = \frac{P_{ij}(r_1)}{P_{ij}(r_2)} \tag{21.54}$$

and the MC transition probability is:

$$p(\mathbf{x}_{r_1}^N \to \mathbf{x}_{r_2}^N) = \min\left[1, \exp\left\{-\beta[H(\mathbf{x}_{r_2}^N) - H(\mathbf{x}_{r_1}^N)]\right\} \frac{\hat{P}_{ij}(r_1)}{\hat{P}_{ij}(r_2)}\right] \tag{21.55}$$

Thus, as for the free energy, the $P_{ij}(r)$ values are unknown a priori, and therefore they appear with a hat in Equation (21.55). Their correct values can be obtained self-consistently, as described earlier. Fasnacht et al. have tested AIM extensively as applied to a dense $2d$ Lennard–Jones fluid of $N = 36$ particles and have found its performance to be somewhat better than those of the Wang-Landau method and a related method of Yan and de Pablo [88]. See also Section 21.6.1.

21.6 Methods Based on Jarzynski's Identity

Equation (17.7) defines the FEP method, which enables one to calculate the difference in free energy, $\Delta F_{1,0}$, between two systems characterized by force fields (Hamiltonians) E_0 and E_1; however, if E_0 and E_1 differ significantly, FEP becomes inefficient. A standard remedy is defining a set of intermediate force fields, E_λ, using a parameter, λ ($0 \leq \lambda \leq 1$), where $E_\lambda = (1 - \lambda) E_0 + \lambda E_1$; thus, $\Delta F_{0,1}$ is obtained as a sum of free energies, ΔF_i, calculated between pairs of neighbor (hence, very similar) force fields, which can lead to high accuracy and precision. In practice, the E_λ are defined by a set of k discrete λ values $0 = \lambda_0 < \lambda_1, \ldots < \lambda_i \ldots < \lambda_{k-1} = 1$ and $\Delta F_{1,0}$ reads,

$$\Delta F_{1,0} = \sum_{i=0}^{i=k-1} \Delta F_i = -k_B T \sum_{i=0}^{k-1} \ln < \exp-[E(\lambda_{i+1}) - E(\lambda_i)] / k_B T >_{\lambda_i}. \tag{21.56}$$

Clearly, the larger is k and the longer is the time devoted for simulating ΔF_i, the better is their estimation and thus the estimation of their sum, $\Delta F_{1,0}$. However, in practice, carrying out a reversible transformation is unfeasible and $\Delta F_{1,0}$ is expected to be an upper bound, as discussed in Section 17.2. There we have pointed out that $\Delta F_{1,0}$ can be interpreted as the amount of work, W, applied to the system; thus,

$$W \geq \Delta F_{1,0} = -k_B T \sum_{i=0}^{k-1} \ln \langle \exp-W(\lambda_i) / k_B T \rangle_{\lambda_i} \tag{21.57}$$

where W is equal to $\Delta F_{1,0}$ for a reversible process and is larger than $\Delta F_{1,0}$ for an irreversible process. Because λ is an external parameter affecting the Hamiltonian, the term, $E(\lambda_{i+1}) - E(\lambda_i)$ can be considered as the instantaneous work, $W(\lambda_{i+1})$, applied to the system (see Equation 17.10). Before proceeding, it should be emphasized that the interpretation of work presented above and previously in Section 17.2 is "by no means an accepted textbook definition of work and may be a subject to dispute. The related results, however, do not depend on the validity of this interpretation" (Hendrix [105]).

A more recent approach for calculating $\Delta F_{1,0}$ (related to the above discussion) is due to Jarzynski, who suggested a *fast* (irreversible) switching between E_0 and E_1 based on the set of the λ-Hamiltonians defined above; he arrived at the following identity [106,107],

$$\Delta F_{1,0} = -k_B T \ln \langle \exp[-W_f / k_B T] \rangle_0 \tag{21.58}$$

where W_f is the total work accumulated during a specific switch; $\langle \cdots \rangle_0$ represents an average of $\exp[-W_f / k_B T]$ obtained from an ensemble of non-reversible switches, that all start from an equilibrium *NVT* ensemble defined by the Hamiltonian, E_0. This method is sometimes called *fast growth*, as compared to the quasi-reversible method defined in Equation (21.56) above.

We present a derivation of Jarzynski's identity due to Crooks [108], because it is based on theoretical considerations developed previously in this book. Thus, assume a canonical *NVT* system, which is kept initially at equilibrium with a heat bath at temperature, T. A parameter λ is assigned to the Hamiltonian (as described above), which is switched, and the system relaxes to a new equilibrium state; the interest is to calculate the difference in free energy, ΔF, between these two states. Clearly, as discussed above, this objective can typically be achieved with FEP (or TI), where the Hamiltonian is changed very slowly (quasi-reversibly). However, in the procedure suggested by Jarzynski, the Hamiltonian is changed *irreversibly*, where λ is switched between initial and final values over a *finite* length of time.

For simplicity, we consider a discrete time and phase space, where the internal state of the system at time, t, is denoted by i_t, and the corresponding external parameter is λ_t; the energy of the system at time, t, is denoted by $E(i_t, \lambda_t)$. We now consider the evolution of this system through time as λ is moved through a fixed sequence $\{\lambda_0, \lambda_1, \ldots, \lambda_\tau\}$ leading to the particular path in phase space,

$$i_0 \xrightarrow{\lambda_1} i_1 \xrightarrow{\lambda_2} i_2 \xrightarrow{\lambda_3} \cdots \xrightarrow{\lambda_\tau} i_\tau \tag{21.59}$$

At time $t = 0$, the system is at state i_0 and the control parameter is λ_0, where the time evolution of the system is considered to occur in two sub-steps. First, the parameter λ is moved to a new value λ_1, which takes the amount of work $W_1 = E(i_0, \lambda_1) - E(i_0, \lambda_0)$. Then the state of the system, at constant λ_1, is relaxed to state i_1, where during this evolution, the system exchanges a quantity $Q_1 = E(i_1, \lambda_1) - E(i_0, \lambda_1)$ of heat with the heat bath. This evolution through phase space is repeated τ times, leading to the corresponding W_t and Q_t. The total work performed *on the system*, W (see definition in Section 2.9), the total heat exchanged with the heat bath, Q, and the total change of energy, ΔE, are,

$$W = \sum_{t=1}^{\tau-1} W_t = \sum_{t=0}^{\tau-1} [E(i_t, \lambda_{t+1}) - E(i_t, \lambda_t)]$$

$$Q = \sum_{t=1}^{\tau-1} Q_t = \sum_{t=1}^{\tau} [E(i_t, \lambda_t) - E(i_{t-1}, \lambda_t)] \tag{21.60}$$

$$\Delta E = Q + W = E(i_\tau, \lambda_\tau) - E(i_0, \lambda_0)]$$

The distinction between W_t and Q_t is important. If, for example, i_0 and i_1 represent two equilibria ensembles, the related reversible work, $W_{r,1}$, in going from i_0 to i_1, is defined as the free energy difference, $W_{r,1} = \Delta F = F(T, \lambda_1) - F(T, \lambda_0)$ (Section 2.9). Therefore, the difference $W_{d,1} = W_1 - W_{r,1}$ can be defined as the related dissipative work, where $W_{d,1}/k_B T$ is the change in the entropy of the system and heat bath (universe) in units of k_B due to the dissipative work. It should be pointed out that the work and the dissipative work depend on the path followed through phase space, while the reversible work depends only on the initial and final (equilibrium) ensembles.

It is useful to consider reversing the direction of time, where states are visited in the opposite direction to that described in Equation (21.59), that is, starting at state i_τ and ending at state i_0,

$$i_0 \xleftarrow{\lambda_1} i_1 \xleftarrow{\lambda_2} i_2 \xleftarrow{\lambda_3} \cdots \xleftarrow{\lambda_\tau} i_\tau \tag{21.61}$$

It should be noted that the forward path begins with a change in λ, whereas the reverse path begins with a change in the internal state of the system. The definition of the corresponding W, Q, and ΔE can be obtained with Equation (21.59) by reversing the direction of time. Unless said otherwise, these quantities are in the forward direction.

The present derivation is based on the assumption that the transition between two neighbor states, i_t and i_{t+1}, is Markovian, meaning that the probability of this transition depends only on i_t, but not on the previously visited states (history). Therefore, the probability of the entire path described in Equation (21.59) can be split into a product of its constituent single time step (transition) probabilities $p(i_{t-1} \xrightarrow{\lambda_t} i_t)$:

$$P(i_0 \xrightarrow{\lambda_1} i_1 \xrightarrow{\lambda_2} i_2 \xrightarrow{\lambda_3} \cdots i_{\tau-1} \xrightarrow{\lambda_\tau} i_\tau)$$
$$= p(i_0 \xrightarrow{\lambda_1} i_1) p(i_1 \xrightarrow{\lambda_2} i_2) \cdots p(i_{\tau-1} \xrightarrow{\lambda_\tau} i_\tau) \tag{21.62}$$

If one further assumes that the single time steps are microscopically reversible [see Section 11.7 and discussion prior to Equation (1.75)], they satisfy the detailed balance condition for all fixed values $\{\lambda_0, \lambda_1, \ldots, \lambda_\tau\}$ of the parameter λ.

$$\frac{p(A \xrightarrow{\lambda} B)}{p(A \xleftarrow{\lambda} B)} = \frac{P(B|\lambda)}{P(A|\lambda)} = \frac{\exp{-[E(B,\lambda)/k_B T]}}{\exp{-[E(A,\lambda)/k_B T]}} \tag{21.63}$$

We now write the ratio of the probabilities for the forward and reverse paths; using the Markovian and detailed balance conditions for the single time steps; one obtains:

$$\frac{P(i_0 \xrightarrow{\lambda_1} i_1 \xrightarrow{\lambda_2} i_2 \xrightarrow{\lambda_3} \cdots i_{\tau-1} \xrightarrow{\lambda_\tau} i_\tau)}{P(i_0 \xleftarrow{\lambda_1} i_1 \xleftarrow{\lambda_2} i_2 \xleftarrow{\lambda_3} \cdots i_{\tau-1} \xleftarrow{\lambda_\tau} i_\tau)} = \frac{p(i_0 \xrightarrow{\lambda_1} i_1) p(i_1 \xrightarrow{\lambda_2} i_2) \cdots p(i_{\tau-1} \xrightarrow{\lambda_\tau} i_\tau)}{p(i_0 \xleftarrow{\lambda_1} i_1) p(i_1 \xleftarrow{\lambda_2} i_2) \cdots p(i_{\tau-1} \xleftarrow{\lambda_\tau} i_\tau)}$$

$$= \frac{\exp-[E(i_1,\lambda_1)/k_BT]\ \exp-[E(i_2,\lambda_2)/k_BT]\cdots}{\exp-[E(i_0,\lambda_1)/k_BT]\ \exp-[E(i_1,\lambda_2)/k_BT]\cdots} \quad \frac{\exp-[E(i_\tau,\lambda_\tau)/k_BT]}{\exp-[E(i_{\tau-1},\lambda_\tau)/k_BT]} = \exp-\left(\frac{Q}{k_BT}\right) \quad (21.64)$$

The last equality stems from the fact that:

$$\frac{\exp-[E(i_t,\lambda_t)/k_BT]}{\exp-[E(i_{t-1},\lambda_t)/k_BT]} = \exp-[E(i_t,\lambda_t)-E(i_{t-1},\lambda_t)]/k_BT) = \exp-\frac{Q_t}{k_BT} \quad (21.65)$$

where Q_t and Q are the amounts of heat exchanged with the heat bath during step t and along the entire path of the *forward* process, respectively [see Equation (21.60)]; $-Q_t/k_BT$ is the corresponding change in the entropy of the heat bath. Equations (21.64–21.65) essentially constitute a statement that the detailed balance continues to hold for Markovian microscopically reversible systems irrespective of how much work is performed on the system [see Equation (1.70)]. We now add the condition that both the forward and reverse paths start from the equilibrium distributions, $P(i_0,\lambda_0)$ and $P(i_\tau,\lambda_\tau)$, which leads to:

$$\frac{P(i_0,\lambda_0)(P(i_0 \xrightarrow{\lambda_1} i_1 \cdots i_{\tau-1} \xrightarrow{\lambda_\tau} i_\tau)}{P(i_\tau,\lambda_\tau)P(i_0 \xleftarrow{\lambda_1} i_1 \cdots i_{\tau-1} \xleftarrow{\lambda_\tau} i_\tau)} = \exp\frac{\Delta E - \Delta F - Q}{k_BT} = \exp\frac{W_d}{k_BT} \quad (21.66)$$

To explain this result, notice that the effect of the two equilibrium distributions is:

$$\frac{P(i_0,\lambda_0)}{P(i_\tau,\lambda_\tau)} = \frac{Z_{i_\tau,\lambda_\tau}\exp-E(i_0,\lambda_0)}{Z_{i_0,\lambda_0}\exp-E(i_\tau,\lambda_\tau)} = \exp\frac{[E(i_\tau,\lambda_\tau)-E(i_0,\lambda_0)]-[F_{i_\tau,\lambda_\tau}-F_{i_0,\lambda_0}]}{k_BT} = \exp\frac{\Delta E - \Delta F}{k_BT} \quad (21.67)$$

Since $\Delta F = W_r$ and $\Delta E - Q = W$ [Equation (21.60)], one obtains, $\Delta E - Q - \Delta F = W - W_r = W_d$, which is the exponent (divided by k_BT) appearing on the right hand side of Equation (21.66). Now we have reached the stage which enables proving Jarzynski's identity,

$$< \exp(-W/k_BT) >_0 = \exp-(\Delta F/k_BT). \quad (21.68)$$

The $<\ldots>_0$ is an average over all phase space paths starting from a canonical equilibrium state denoted by 0, where the sequence of the external control parameter $\{\lambda_t\}$ is kept fixed; this average over all forward paths is:

$$< \exp(-W/k_BT) >_0 = \sum_{i_0, i_1, \cdots i_\tau} P(i_0,\lambda_0)(P(i_0 \xrightarrow{\lambda_1} i_1 \cdots i_{\tau-1} \xrightarrow{\lambda_\tau} i_\tau) \exp\frac{-W}{k_BT} \quad (21.69)$$

This average over the forward time path can be expressed in terms of an average over the reversed path using Equation (21.66), and the desired Jarzynski's identity [Equation (21.68)] is obtained,

$$< \exp(-W/k_BT) >_0 = \sum_{i_0, i_1, \cdots i} P(i_\tau,\lambda_\tau)P(i_0 \xleftarrow{\lambda_1} i_1 \cdots i_{\tau-1} \xleftarrow{\lambda_\tau} i_\tau) \exp\frac{W_d - W}{k_BT} = \exp\frac{-\Delta F}{k_BT} \quad (21.70)$$

or

$$\int \rho(W)\exp(-W/k_BT)dW = \exp\frac{-\Delta F}{k_BT} \quad (21.71)$$

where $\rho(W)$ is the distribution of the work values. Crooks also shows that Jarzynski's identity applies to the NpT ensemble, where ΔG replaces ΔF in Equation (21.68). Notice that for $\tau \to 0$, Jarzynski's identity reduces to the free energy perturbation.

A direct implementation of Jarzynski's formula is straightforward following the above derivation. Thus, a set of λ values, $\lambda_0 = 0, \lambda_1,...,\lambda_\tau = 1$ is determined first, and an MD (or MC) equilibrium simulation is carried out at T, based on the potential energy, E_0. Then, a configuration $\mathbf{x}_1$ is selected from the sample and: (1) the instantaneous work $E(\mathbf{x}_1,\lambda_1) - E(\mathbf{x}_1,\lambda_0)$ is calculated and (2) applying MD with $E(\lambda_1)$, the system relaxes over a short period of time from $\mathbf{x}_1$ to $\mathbf{x}_2$, and the process is repeated τ times until the final relaxation is performed with, $\lambda = 1$. In this way, N-independent simulations (switches) are carried out (based on different sets of random numbers), where the accumulated non-equilibrium work, $W_{\tau,n}$, for the nth run ($n = 1,2,...,N$) is obtained with Equation (21.60). ΔF is estimated according to Equation (21.71),

$$\overline{\Delta F} = -k_B T \ln \left[\frac{1}{N} \sum_{n=1}^{N} \exp(-W_{\tau,n} / k_B T) \right] \qquad (21.72)$$

The crucial question is how efficient is this estimation, in particular, how is it compared to FEP (or TI). First notice that the integral [Equation (21.71)] is dominated by *improbable* low work values, W_l, for which $\rho(W_l)\exp(-W_l/k_B T)$ is large; however, the *estimation* $\overline{\Delta F}$ [Equation (21.72)] is based instead on the most probable work values, W_p, with low contribution to the integral [109]. Thus, practically, $\overline{\Delta F}$ is biased, the faster the switches, the larger is this bias (due to a larger separation between the contributions of W_l and W_p), which can be decreased only by increasing the sample size, N. Therefore, increasing efficiency has been an early central aim [110], e.g., by biasing the selection of rare, most-contributing "rapid" paths [109–115] or developing alternatives to Jarzynski's identity [116].

However, we shall not elaborate on these interesting procedures since they have been tested mostly on highly simplified models, and the conclusions might not fully apply to complex systems such as proteins. Also, Jarzynski's equality, in spite of its theoretical importance, has not led thus far to an improved efficiency over other techniques (see below), in particular, over FEP [105,111,113,117].

Furthermore, since $W_{\tau,n}$ [Equation (21.72)] is an extensive parameter, the large fluctuation of $\exp(-W_{\tau,n}/k_B T)$ might limit the system size that can be handled. On the other hand, if $E_1 - E_0$ is significant, a typical application of FEP will require several separate simulations based on the exponents of $(\lambda_{i+1}) - E(\lambda_i) = \Delta\lambda(E_1 - E_0)$, where, however, $\Delta\lambda$ can be decreased at will (by increasing the number of intermediate λ values). While the number of simulations increases, the reduced exponents enable one to treat larger systems.

21.6.1 Jarzynski's Identity versus Other Methods for Calculating ΔF

As pointed out above, only a few performance tests between Jarzynski's method and other techniques have been carried out. For example, Shirts and Pande [118] reviewed and developed theoretical estimates for the bias and variance of Jarzynski's identity, TI, and Bennett's method. They applied these procedures to toy models, but could not define a preferred method for calculating ΔF; however, in applications to simple atomistic models, the lowest variance and bias were obtained with Bennett's method. In another study [119], the accuracy and precision of nine free energy methods were compared among Jarzynski' method, TI, AIM, FEP, Bennett's method, and single-ensemble path sampling [119]. ΔF was calculated for growing a (neutral) Lennard-Jones sphere in water and for charging a Lennard-Jones sphere in water. The efficiency was found to depend on the system and the extent of accuracy sought, where overall, AIM was found to be the most efficient. Jarzynski's identity was also applied to realistic systems of proteins [120], where steered MD was used for calculating the potential of mean force for unbinding acetylcholine from the alpha7 nicotinic acetylcholine receptor ligand-binding domain; four different procedures were checked in this study (see also [121] and references cited therein).

In this context, it is of interest to mention a paper by Bruckner and Boresch [122], who investigated the relative efficiency of TI, three variants of FEP, and the Bennett's acceptance ratio for computing relative and absolute solvation free energy differences. Using the criterion of efficiency—"the minimal number of λ states required for gaining accurate results," they have found Bennett's method to be the best, followed by the "doublewide" variant of FEP. Interestingly, the performance of the commonly used TI was found to be rather poor due to the use of the trapezoidal rule as a method of numerical quadrature. Therefore, in a following paper [123], they tested various numerical quadrature methods within the framework of TI and found Simpson's rule and spline integration to be the best, leading to an equal performance to Bennett's method. However, TI was found to be more susceptible than Bennett's method to the details of the hybrid Hamiltonian. While, overall, Bennett's method was found to be the most robust, Bruckner and Boresch discuss scenarios when TI may become the preferable procedure.

This discussion demonstrates the sensitivity of the performance of a method to implementation details and to the size and complexity of the system tested. In fact, the above conclusions are based on studies of relatively small systems, such as the alchemical mutation of benzyl phosphonate into difluorobenzyl phosphonate, and the calculation of the absolute solvation free energy of blocked glycine (N-acetyl-Gly-methylamide). In this context, it should also be emphasized that an important component of the studies discussed above is the implementation of an efficient soft-core potential [Equation (17.24)], which handles the divergence of the alchemical TI procedure at the end points, $\lambda = 0$ or 1; see also [124–126].

21.7 Summary

MC and MD are relaxation type methods, which makes them poor conformational search tools at room temperature for systems with a rugged potential energy surface, such as proteins. Therefore, the advent of the sophisticated methods discussed in this chapter (e.g., REMD and MUCA), which lead to an efficient crossing of energy barriers, constitutes an important contribution to the field of structure determination. Still, protein folding remains a formidable task because of the difficulty to scan the huge conformational space for the tiny native microstate. Most of these methods also provide differences in free energy, hence, the potential of mean force (as a function of a reaction coordinate), and other related quantities.

We have discussed in detail only six methods; however, certain combinations among them (and additional techniques) have led to new procedures with increased performance and applicability, enriching thereby the field significantly. We have mentioned the "*replica-exchange multicanonical algorithm*," "*replica-exchange umbrella sampling method*," EXE with replica exchange and with Bennett method, and CTDOS equipped with parallel tempering. These methods are only a small representative set of a larger group of hybrid methods; however, describing them in detail is beyond the scope of this book.

As pointed out in the previous sub-section (21.6.1), comparing the efficiencies of different techniques is not straightforward. Still, to provide a crude criterion for evaluating performance, the systems in the application sections have been described in detail, where, for example, the size and sequence of proteins, protein segments, or peptides have been given together with the specific modeling used. We have encountered models of different quality, from a chain residing on a lattice, to its more faithful coarse-grained description in continuum with implicit solvation (or in vacuum). Clearly, with the most realistic model, a protein described by a conventional force field is immersed in a "box" of explicit solvent; therefore, it would be desirable to compare various methods, as applied to this model. Based on the data presented in this chapter one can cautiously conclude that for structure determination, replica exchange emerges as the most efficient method, as it has handled significantly larger protein systems than the other methods. Clearly, this stems partially from the fact that REM is a simple "direct" method, free of the self-consistent calculation of the density of states required by MUCA, WL, EXE, and AIM.

REFERENCES

1. R. H. Swendsen and J. S. Wang. Replica Monte Carlo simulation of spin-glasses. *Phys. Rev. Lett.* **57**, 2607–2609 (1986).
2. E. Marinari and G. Parisi. Simulated tempering: A new Monte Carlo scheme. *Europhys. Lett.* **19**, 451–458 (1992).
3. C. J. Geyer and E. A. Thompson. Annealing Markov chain Monte Carlo with applications to ancestral inference. *J. Amer. Stat. Assoc.* **90**, 909–920 (1995).

4. K. Hukushima and K. Nemoto. Exchange Monte Carlo method and application to spin glass simulations. *J. Phys. Soc. Jpn.* **65**, 1604–1608 (1996).
5. M. C. Tesi, E. J. J. van Rensburg, E. Orlandini and S. G. Whittington. Monte Carlo study of the interacting self-avoiding walk model in three dimensions. *J. Stat. Phys.* **82**, 155–181 (1996).
6. U. H. E. Hansmann. Parallel tempering algorithm for conformational studies of biological molecules. *Chem. Phys. Lett.* **281**, 140–150 (1997).
7. H. Fukunishi, O. Watanabe and S. Takada. On the Hamiltonian replica exchange method for efficient sampling of biomolecular systems: Application to protein structure prediction. *J. Chem. Phys.* **116**, 9058–9067 (2002).
8. Y. Sugita and Y. Okamoto. Replica-exchange molecular dynamics method for protein folding. *Chem. Phys. Lett.* **314**, 141–151 (1999).
9. J. Machta. Strengths and weaknesses of parallel tempering. *Phys. Rev. E* **80**, 056706-7 (2009).
10. R. Zhou, B. J. Berne and R. Germain. The free energy landscape for ß hairpin folding in explicit water. *Proc. Natl. Acad. Sci. USA* **98**, 14931–14936 (2001).
11. R. Zhou and B. J. Berne. Can a continuum solvent model reproduce the free energy landscape of a ß-hairpin folding in water? *Proc. Natl. Acad. Sci. USA* **99**, 12777–12782 (2002).
12. T. Nagai and T. Takahashi. Mass-scaling replica-exchange molecular dynamics optimizes computational resources with simpler algorithm. *J. Chem. Phys.* **141**, 114111-10 (2014).
13. T. Nagai and T. Takahashi. Momentum and velocity scaling rules in replica-exchange molecular dynamics simulations with mass manipulation. Proceedings of Computational Science Workshop 2014 (CSW2014). *JPS Conf. Proc.* **5**, 011009 (2015).
14. H. Nymeyer. How efficient is replica exchange molecular dynamics? An analytic approach. *J. Chem. Theory Comput.* **4**, 626–636 (2008).
15. Y. Sugita, A. Kitao and Y. Okamoto. Multidimensional replica exchange method for free-energy calculations. *J. Chem. Phys.* **113**, 6042–6051 (2000).
16. P. Liu, B. Kim, R. A. Friesner and B. J. Berne. Replica exchange with solute tempering: A method for sampling biological systems in explicit water. *Proc. Natl. Acad. Sci. USA* **102**, 13749–13754 (2005).
17. X. Huang, M. Hagen, B. Kim, R. A. Friesner, R. Zhou and B. J. Berne. Replica exchange with solute tempering: Efficiency in large scale systems. *J. Phys. Chem. B* **111**, 5405–5410 (2007).
18. L. Wang, R. A. Friesner and B. J. Berne. Replica exchange with solute scaling: A more efficient version of replica exchange with solute tempering (REST2). *J. Phys. Chem. B* **115**, 9431–9438 (2011).
19. A. E. García and J. N. Onuchic. Folding a protein in a computer: An atomic description of the folding/unfolding of protein A. *Proc. Natl. Acad. Sci. USA* **100**, 13898–13903 (2003).
20. D. Paschek and A. E. García. Reversible temperature and pressure denaturation of a protein fragment: A replica exchange molecular dynamics simulation study. *Phys. Rev. Lett.* **93**, 238105-4 (2004).
21. D. R. Roe, V. Hornak and C. Simmerling. Folding cooperativity in a three stranded β-sheet model. *J. Mol. Biol.* **352**, 370–381 (2005).
22. C. Predescu, M. Predescu and C. V. Ciobanu. On the efficiency of exchange in parallel tempering Monte Carlo simulations. *J. Phys. Chem. B* **109**, 4189–4196 (2005).
23. N. Rathore, M. Chopra and J. J. de Pablo. Optimal allocation of replicas in parallel tempering simulations. *J. Chem. Phys.* **122**, 24111–24118 (2005).
24. W. Nadler and U. H. E. Hansmann. Optimized explicit-solvent replica exchange molecular dynamics from scratch. *J. Phys. Chem. B* **112**, 10386–10387 (2008).
25. W. Nadler, J. H. Meinke and U. H. E. Hansmann. Folding proteins by first-passage times-optimized replica exchange. *Phys. Rev. E* **78**, 061905-4 (2008).
26. M. K. Prakash, A. Barducci and M. Parrinello. Replica temperatures for uniform exchange and efficient roundtrip times in explicit solvent parallel tempering simulations. *J. Chem. Theory Comput.* **7**, 2025–2027 (2011).
27. Y. Okamoto. Generalized-ensemble algorithms: Enhanced sampling techniques for Monte Carlo and molecular dynamics simulations. *J. Mol. Graph. Model.* **22**, 425–439 (2004).
28. D. J. Earl and M. W. Deem. Parallel tempering: Theory, applications, and new perspectives. *Phys. Chem. Chem. Phys.* **7**, 3910–3916 (2005).
29. R. Affentranger, I. Tavernelli and E. E. Di Iorio. A novel Hamiltonian replica exchange MD protocol to enhance protein conformational space sampling. *J. Chem. Theory Comput.* **2**, 1228–1234 (2006).

30. T. Terakawa, T. Kameda and S. Takada. On easy implementation of a variant of the replica exchange with solute tempering in GROMACS. *J. Comput. Chem.* **32**, 1211–1491 (2011).

31. M. Meli and G. Colombo. A Hamiltonian replica exchange molecular dynamics (MD) method for the study of folding, based on the analysis of the stabilization determinants of proteins. *Int. J. Mol. Sci.* **14**, 12157–12169 (2013).

32. S. Kannan and M. Zacharias. Enhanced sampling of peptide and protein conformations using replica exchange simulations with a peptide backbone biasing-potential. *Proteins* **66**, 697–706 (2007).

33. C. Xu, J. Wang and H. Liu. A Hamiltonian replica exchange approach and its application to the study of side-chain type and neighbor effects on peptide backbone conformations. *J. Chem. Theory Comput.* **4**, 1348–1359 (2008).

34. J. Hritz and C. Oostenbrink. Optimization of replica exchange molecular dynamics by fast mimicking. *J. Chem. Phys.* **127**, 204104-13 (2007).

35. J. Hritz and C. Oostenbrink. Hamiltonian replica exchange molecular dynamics using soft-core interactions. *J. Chem. Phys.* **128**, 144121-10 (2008).

36. S. G. Itoh, H. Okumura and Y. Okamoto. Replica-exchange method in van der Waals radius space: Overcoming steric restrictions for biomolecules. *J. Chem. Phys.* **132**, 134105-8 (2010).

37. M. Zacharias. Combining elastic network analysis and molecular dynamics simulations by Hamiltonian replica exchange. *J. Chem. Theory Comput.* **4**, 477–487 (2008).

38. J. Vreede, M. G. Wolf, S. W. de Leeuw and P. G. Bolhuis. Reordering hydrogen bonds using Hamiltonian replica exchange enhances sampling of conformational changes in biomolecular systems. *J. Phys. Chem. B* **113**, 6484–6494 (2009).

39. M. S. Lee and M. A. Olson. Protein folding simulations combining self-guided Langevin dynamics and temperature-based replica exchange. *J. Chem. Theory Comput.* **6**, 2477–2487 (2010).

40. Y. Meng and A. E. Roitberg. Constant pH replica exchange molecular dynamics in biomolecules using a discrete protonation model. *J. Chem. Theory Comput.* **6**, 1401–1412 (2010).

41. D. S. Dashti and A. E Roitberg. Optimization of umbrella sampling replica exchange molecular dynamics by replica positioning. *J. Chem. Theory Comput.* **9**, 4692–4699 (2013).

42. C. Zhang and J. Ma. Folding helical proteins in explicit solvent using dihedral-biased tempering. *Proc. Natl. Acad. Sci. USA* **109**, 8139–8144 (2012).

43. A. Berg and T. Neuhaus. Multicanonical algorithms for first order phase transition. *Phys. Lett. B* **267**, 249–253 (1991).

44. B. A. Berg and T. Neuhaus. Multicanonical ensemble: A new approach to simulate first-order phase transitions. *Phys. Rev. Lett.* **68**, 9–12 (1992).

45. U. H. E. Hansmann and Y. J. Okamoto. Prediction of peptide conformation by multicanonical algorithm: New approach to the multiple-minima problem. *J. Comput. Chem.* **14**, 1333–1338 (1993).

46. J. Lee. New Monte Carlo algorithm: Entropic sampling. *Phys. Rev. Lett.* **71**, 211–214 (1993).

47. U. H. E. Hansmann, Y. Okamoto and F. Eisenmenger. Molecular dynamics, Langevin and hybrid Monte Carlo simulations in a multicanonical ensemble. *Chem. Phys. Lett.* **259**, 321–330 (1996).

48. N. Nakajima, H. Nakamura and A. Kidera. Multicanonical ensemble generated by molecular dynamics simulation for enhanced conformational sampling of peptides. *J. Phys. Chem. B* **101**, 817–824 (1997).

49. B. A. Berg and T. Celik. New approach to spin-glass simulations. *Phys. Rev. Lett.* **69**, 2292–2295 (1992).

50. B. A. Berg. Multicanonical recursions. *J. Stat. Phys.* **82**, 323–342 (1996).

51. B. A. Berg. Algorithmic aspects of multicanonical simulations. *Nucl. Phys. B* **63** (Suppl.) A-C, 982–984 (1998).

52. A. Mitsutake and Y. Okamoto. α-Helix propensities of homo-oligomers in aqueous solution studied by multicanonical algorithm. *Chem. Phys. Lett.* **309**, 95–100 (1999).

53. F. Yasar, T. Çelik, B. A. Berg and H. Meirovitch. Multicanonical procedure for continuum peptide models *J. Comput. Chem.* **21**, 1251–1261 (2000).

54. F. Yasar, H. Arkin, T. Çelik, B. A. Berg and H. Meirovitch. Efficiency of the multicanonical simulation method as applied to peptides of increasing size: The heptapeptide deltorphin. *J. Comput. Chem.* **23**, 1127–1134 (2002).

55. U. H. E. Hansmann and Y. Okamoto. Numerical comparisons of three recently proposed algorithms in the protein folding problem. *J. Comput. Chem.* **18**, 920–933 (1997).

56. U. H. E. Hansmann and Y. Okamoto. Comparative study of multicanonical and simulated annealing algorithms in the protein folding problem. *Phys. A* **212**, 415–437 (1994).

57. Y. Okamoto and U. H. E. Hansmann. Thermodynamics of helix-coil transitions studied by multicanonical algorithms. *J. Phys. Chem.* **99**, 11276–11287 (1995).
58. M.-H. Hao, and H. A. Scheraga. Monte Carlo simulation of a first-order transition for protein folding. *J. Phys. Chem.* **98**, 4940–4948 (1994).
59. M.-H. Hao and H. A. Scheraga. Statistical thermodynamics of protein folding: Sequence dependence. *J. Phys. Chem.* **98**, 9882–9893 (1994).
60. M.-H. Hao and H. A. Scheraga. How optimization of potential functions affects protein folding. *Proc. Natl. Acad. Sci. USA* **93**, 4984–4989 (1996).
61. M.-H. Hao and H. A. Scheraga. Optimizing potential functions for protein folding. *J. Phys. Chem.* **100**, 14540–14548 (1996).
62. A. Kolinski, W. Galazka and J. Skolnick. On the origin of cooperativity of protein folding: Implications from model simulations. *Proteins* **26**, 271–287 (1996).
63. M.-H. Hao and H. A. Scheraga. Statistical thermodynamics of protein folding: Comparison of a mean-field theory with Monte Carlo simulations, *J. Chem. Phys.* **102**, 1334–1348 (1995).
64. P. Jiang, F. Yasar and U. H. E. Hansmann. Sampling of protein folding transitions: Multicanonical versus replica exchange molecular dynamics. *J. Chem. Theory Comput.* **9**, 3816–3825 (2013).
65. S. Uyaver and U. H. E. Hansmann. Multicanonical Monte Carlo simulations of a de novo designed protein with end-to-end β-sheet. *J. Chem. Phys.* **140**, 065101-4 (2014).
66. J. Higo, N. Kamiya, T. Sugihara, Y. Yonezawa and H. Nakamura. Verifying trivial parallelization of multicanonical molecular dynamics for conformational sampling of a polypeptide in explicit water. *Chem. Phys. Lett.* **473**, 326–329 (2009).
67. J. Higo and H. Nakamura. Virtual states introduced for overcoming entropic barriers in conformational space. *Biophysics* **8**, 139–144 (2012).
68. J. Higo, J. Ikebe, N. Kamiya and H. Nakamura. Enhanced and effective conformational sampling of protein molecular systems for their free energy landscapes. *Biophys. Rev.* **4**, 27–44 (2012).
69. J. Higo, K. Umezawa and H. Nakamura. A virtual-system coupled multicanonical molecular dynamics simulation: Principles and applications to free-energy landscape of protein–protein interaction with an all-atom model in explicit solvent. *J. Chem. Phys.* **138**, 184106-11 (2013).
70. J. Higo, O. V. Galzitskaya, S. Ono and H. Nakamura. Energy landscape of a β-hairpin peptide in explicit water studied by multicanonical molecular dynamics. *Chem. Phys. Lett.* **337**, 169–175 (2001).
71. J. Ikebe, N. Kamiya, H. Shindo, H. Nakamura and J. Higo. Conformational sampling of a 40-residue protein consisting of α and β secondary-structure elements in explicit solvent. *Chem. Phys. Lett.* **443**, 364–368 (2007).
72. J. Ikebe, D. M. Standley, H. Nakamura and J. Higo. Ab initio simulation of a 57-residue protein in explicit solvent reproduces the native conformation in the lowest free-energy cluster. *Protein Sci.* **20**, 187–196 (2011).
73. Y. Sugita and Y. Okamoto. Replica-exchange multicanonical algorithm and multicanonical replica-exchange method for simulating systems with rough energy landscape. *Chem. Phys. Lett.* **329**, 261–270 (2000).
74. A. Mitsutake, Y. Sugita and Y. Okamoto. Replica-exchange multicanonical and multicanonical replica-exchange Monte Carlo simulations of peptides. II. Application to a more complex system. *J. Chem. Phys.* **118**, 6676–6688 (2003).
75. T. Yoda, Y. Sugita and Y. Okamoto. Cooperative folding mechanism of a hairpin peptide studied by a multicanonical replica-exchange molecular dynamics simulation. *Proteins* **66**, 846–859 (2007).
76. R. Faller, Q. Yan and J. J. de Pablo. Multicanonical parallel tempering. *J. Chem. Phys.* **116**, 5419–5423 (2002).
77. M. K. Fenwick and F. A. Escobedo. On the use of Bennett's acceptance ratio method in multicanonical-type simulations. *J. Chem. Phys.* **120**, 3066–3074 (2004).
78. A. Mitsutakea, Y. Morib and Y. Okamoto. Multi-dimensional multicanonical algorithm, simulated tempering, replica-exchange method, and all that. *Phys. Procedia* **4**, 89–105 (2010).
79. U. H. Hansmann and Y. Okamoto. The generalized-ensemble approach for protein folding simulations. *Annu. Rev. Comp. Phys.* **6**, 129–157 (1999).
80. F. Wang and D. P. Landau. Efficient multiple-range random walk algorithm to calculate the density of states. *Phys. Rev. Lett.* **86**, 2050–2053 (2001).
81. F. Wang and D. P. Landau. Determining the density of states for classical statistical models: A random walk algorithm to produce a flat histogram. *Phys. Rev. E* **64**, 056101-16 (2001).

82. N. Rathore and J. J. de Pablo. Monte Carlo simulation of proteins through a random walk in energy space. *J. Chem. Phys.* **116**, 7225–7230 (2002).

83. Q. Yan, R. Faller and J. J. de Pablo. Density-of-states Monte Carlo method for simulation of fluids. *J. Chem. Phys.* **116**, 8745–8749 (2002).

84. N. Rathore, T. A. Knotts and J. J. de Pablo. Density of states simulations of proteins. *J. Chem. Phys.* **118**, 4285–4290 (2003).

85. E. A. Mastny and J. J. de Pablo. Direct calculation of solid-liquid equilibria from density-of-states Monte Carlo simulations. *J. Chem. Phys.* **122**, 124109-6 (2005).

86. N. Rathore, T. A. Knotts and J. J. de Pablo. Confinement effects on the thermodynamics of protein folding: Monte Carlo simulations. *Biophys. J.* **90**, 1767–1773 (2006).

87. O. G. Jepps, G. Ayton and D. J. Evans. Microscopic expressions for the thermodynamic temperature. *Phys. Rev. E.* **62**, 4757–4763 (2000).

88. Q. L. Yan and J. J. de Pablo. Fast calculation of the density of states of a fluid by Monte Carlo simulations. *Phys. Rev. Lett.* **90**, 035701–035704 (2003).

89. N. Rathore, T. A. Knotts, IV and J. J. de Pablo. Configurational temperature density of states simulations of proteins. *Biophys. J.* **85**, 3963–3968 (2003).

90. T. A. Knotts, N. Rathore and J. J. de Pablo. An entropic perspective of protein stability on surfaces. *Biophys. J.* **94**, 4473–4483 (2008).

91. M. Chopra and J. J. de Pablo. Improved density of states Monte Carlo method based on recycling of rejected states. *J. Chem. Phys.* **124**, 114102-6 (2006).

92. E. B. Kim, R. Faller, Q. Yan, N. L. Abbott and J. J. de Pablo. Potential of mean force between a spherical particle suspended in a nematic liquid crystal and a substrate. *J. Chem. Phys.* **117**, 7781–7787 (2002).

93. N. Rathore, Q. Yan and J. J. de Pablo. Molecular simulation of the reversible mechanical unfolding of proteins. *J. Chem. Phys.* **120**, 5781–5788 (2004).

94. M. Chopra, M. Müller and J. J. de Pablo. Order-parameter-based Monte Carlo simulation of crystallization. *J. Chem. Phys.* **124**, 134102-8 (2006).

95. C. A. Miller, S. H. Gellman, N. L. Abbott and J. J. de Pablo. Mechanical stability of helical *b*-peptides and a comparison of explicit and implicit solvent models. *Biophys. J.* **95**, 3123–3136 (2008).

96. A. P. Lyubartsev, A. A. Martsinovski, S. V. Shevkunov and P. N. Vorontsov-Velyaminov. New approach to Monte Carlo calculation of the free energy: Method of expanded ensembles. *J. Chem. Phys.* **96**, 1776–1783 (1992).

97. M. K. Fenwick and F. A. Escobedo. Expanded ensemble and replica exchange methods for simulation of protein-like systems. *J. Chem. Phys.* **119**, 11998–12010 (2003).

98. F. J. Martínez-Veracoechea and F. A. Escobedo. Simulation of the gyroid phase in off-lattice models of pure diblock copolymer melts. *J. Chem. Phys.* **125**, 104907-12 (2006).

99. F. A. Escobedo and F. J. Martínez-Veracoechea. Optimized expanded ensembles for simulations involving molecular insertions and deletions. I. Closed systems. *J. Chem. Phys.* **127**, 174103-10 (2007).

100. F. A. Escobedo. Optimized expanded ensembles for simulations involving molecular insertions and deletions. II. Open systems. *J. Chem. Phys.* **127**, 174104-12 (2007).

101. M. Fasnacht, R. H. Swendsen and J. M. Rosenberg. Adaptive integration method for Monte Carlo simulations. *Phys. Rev. E* **69**, 056704-15 (2004).

102. X. Kong and C. L. Brooks, III. λ-dynamics: A new approach to free energy calculations. *J. Chem. Phys.* **105**, 2414–2423 (1996).

103. J. L. Knight and C. L. Brooks, III. Multisite λ-dynamics for simulated structure activity relationship studies. *J. Chem. Theory Comput.* **7**, 2728–2739 (2011).

104. J. L. Knight and C. L. Brooks, III. Applying efficient implicit nongeometric constraints in alchemical free energy simulation. *J. Comput. Chem.* **32**, 3423–3432 (2011).

105. D. A. Hendrix and C. Jarzynski. A "fast growth" method for computing free energy differences. *J. Chem. Phys.* **114**, 5974–5981 (2001).

106. C. Jarzynski. Nonequilibrium equality for free energy differences. *Phys. Rev. Lett.* **78**, 2690–2693 (1997).

107. C. Jarzynski. Equilibrium free energy differences from nonequilibrium measurements: A master equation approach. *Phys. Rev. E* **56**, 5018–5035 (1997).

108. G. E. Crooks. Nonequilibrium measurements of tree energy differences for microscopically reversible Markovian systems. *J. Stat. Phys.* **90**, 1481–1487 (1998).

109. F. M. Ytreberg and D. M. Zuckerman. Efficient use of nonequilibrium measurement to estimate free energy differences for molecular systems. *J. Comput. Chem.* **25**, 1749–1759 (2004).
110. F. M. Ytreberg and D. M. Zuckerman. Single-ensemble nonequilibrium path-sampling estimates of free energy differences. *J. Chem. Phys.* **120**, 10876–10879 (2004).
111. C. Jarzynski. Rare events and convergence of exponentially averaged work values. *Phys. Rev. E* **73**, 046105-10 (2006).
112. D. Wu and D. A. Kofke. Rosenbluth-sampled nonequilibrium work method for calculation of free energies in molecular simulation. *J. Chem. Phys.* **122**, 204104-13 (2005).
113. H. Oberhofer, C. Dellago and P. Geissler. Biased sampling of nonequilibrium trajectories: Can fast switching simulations outperform conventional free energy calculations methods? *J. Phys. Chem.* **109**, 6902–6915 (2005).
114. W. Lechner, H. Oberhofer, C. Dellago and P. Geissler. Equilibrium free energies from fast switching trajectories with large time step. *J. Chem. Phys.* **124**, 044113-12 (2006).
115. J. MacFadyen and I. A. Andricioaei. Skewed-momenta method to efficiently generate conformational-transition trajectories. *J. Chem. Phys.* **123**, 074107-9 (2005).
116. A. B. Adib. Free energy surfaces from nonequilibrium processes without work measurement. *J. Chem. Phys.* **124**, 144111-5 (2006)
117. A. Pohorille, C. Jarzynski and C. Chipot. Good practices in free-energy calculations. *J. Phys, Chem. B* **114**, 10235–10253 (2010).
118. M. R. Shirts and V. S. Pande. Comparison of efficiency and bias of free energies computed by exponential averaging, the Bennett acceptance ratio, and thermodynamics integration. *J. Chem. Phys.* **122**, 144107-16 (2005).
119. F. M. Ytreberg, R. H. Swendsen and D. M. Zuckerman. Comparison of free energy methods for molecular simulations. *J. Chem. Phys.* **125**, 184114-11 (2006).
120. D. Zhang, J. Gullingsrud and J. A. McCammon. Potential of mean force for a acetylcholine unbinding from the alpha7 nicotinic acetylcholine receptor ligand-binding domain. *J. Am. Chem. Soc.* **128**, 3019–3026 (2006).
121. H. Xiong, A. Crespo, M. Marti, D. Estrin and A. E. Roitberg. Free energy calculations with non-equilibrium methods: Applications of the Jarzynski relationship. *Theor. Chem. Acc.* **116**, 338–346 (2006).
122. S. Bruckner and S. Boresch. Efficiency of alchemical free energy simulations. I. A practical comparison of the exponential formula, thermodynamic integration, and Bennett's acceptance ratio method. *J. Comput. Chem.* **32**, 1303–1319 (2011).
123. S. Bruckner and S. Boresch. Efficiency of alchemical free energy simulations. II. Improvements for thermodynamic integration. *J. Comput. Chem.* **32**, 1320–1333 (2011).
124. P. Floris, F. P. Buelens and H. Grubmüller. Linear-scaling soft-core scheme for alchemical free energy calculations. *J. Comput. Chem.* **33**, 25–33 (2012).
125. T. Steinbrecher, I. Joung and D. A. Case. Soft-core potentials in thermodynamic integration: Comparing one- and two-step transformations. *J. Comput. Chem.* **32**, 3253–3263 (2011).
126. P. Liu, F. Dehez, W. Cai and C. Chipot. Toolkit for the analysis of free-energy perturbation calculations. *J. Chem. Theory Comput.* **8**, 2606–2616 (2012).

22

Simulation of the Chemical Potential

The chemical potential plays an important role in chemical equilibrium, chemical reactions, and the calculation of the free energy of binding discussed in Chapter 23. In this chapter, we discuss the main simulation methods for calculating the chemical potential of fluid systems and polymers; a thorough review of this area can be found in the 1997 paper by Kofke and Cummings [1], more recent references appear in [2].

22.1 The Widom Insertion Method

For an ideal gas in the canonical ensemble, the chemical potential, μ, has been derived in Section 4.3,

$$\mu_{IG}(N,V,T) = -k_B T \ln \frac{1}{\Lambda^3} + k_B T \ln \rho = \mu^0(T) + k_B T \ln \rho \tag{22.1}$$

where $\rho = N/V$ is the density. For a non-ideal *large* system, the chemical potential can be written as:

$$\mu(N,V,T) = -k_B T \ln \frac{1}{\Lambda^3} - k_B T \ln \frac{Z_N}{N Z_{N-1}} \tag{22.2}$$

and the excess chemical potential above its ideal gas value is thus:

$$\mu_{ex}(N,V,T) = \mu(N,V,T) - \mu_{IG}(N,V,T) = -k_B T \ln \frac{Z_N}{V Z_{N-1}} \tag{22.3}$$

A central method for calculating μ_{ex} is the particle insertion method published first by Widom [3] and derived independently by Jackson and Klein [4]. Concentrating on a simple fluid described by the canonical ensemble, the excess chemical potential, μ_W, obtained by the Widom method is:

$$\frac{Z_N}{V Z_{N-1}} = \frac{\int \exp(-E_N / k_B T) d\mathbf{x}_1 ... d\mathbf{x}_N}{V \int \exp(-E_{N-1} / k_B T) d\mathbf{x}_1 ... d\mathbf{x}_{N-1}}$$

$$= \frac{\int \exp(-E_{N-1}/k_B T) \left[\int \exp(-e_N/k_B T) d\mathbf{x}_N \right] d\mathbf{x}_1 ... d\mathbf{x}_N}{V \int \exp(-E_{N-1}/k_B T) d\mathbf{x}_1 ... d\mathbf{x}_{N-1}} \tag{22.4}$$

$$= \int P_{N-1}^B [\int \frac{1}{V} \exp(-e_N/k_B T) d\mathbf{x}_N] d\mathbf{x}_1 ... d\mathbf{x}_{N-1}$$

and μ_W becomes:

$$\mu_W = -k_B T \ln \int d\mathbf{x}_N < \frac{1}{V} \exp[-e_N/k_B T] >_{N-1} \tag{22.5}$$

which we write as:

$$\mu_W = -k_{\mathrm{B}}T \ln < \exp[-e_N/k_{\mathrm{B}}T] >_{V^{-1},N-1}. \tag{22.6}$$

In this derivation, $F_N - F_{N-1}$ is expressed as a ratio of two partition functions, of N and $N-1$ particles. For brevity, E_N denotes the configurational energy based on N particles, that is, $E_N = E(\mathbf{x}_1,...,\mathbf{x}_N)$. We seek to write this ratio as a statistical average with a Boltzmann probability density, P_{N-1}^{B}, for a system of $N-1$ particles. For that, the integrand in the numerator (based on N particles) is divided into two parts, one based on E_{N-1}, and the other is the interaction energy e_N of a *single* [(N)th] particle with the rest of the $N-1$ particles. For a given set, $\mathbf{x}_1,...,\mathbf{x}_{N-1}$, this "ghost" particle moves over the volume at random [with probability density (V^{-1})], where the interaction of $\exp(-e_N/k_{\mathrm{B}}T)$ with the $(N-1)$-particles is integrated and averaged; these instantaneous averages are then averaged by the integration over P_{N-1}^{B}. These double averages are denoted in Equation (22.6) by $<>_{V^{-1},N-1}$. However, for simplicity, we shall omit the V^{-1} sign; thus:

$$\mu_W = -k_{\mathrm{B}}T \ln < \exp[-e_N/k_{\mathrm{B}}T] >_{N-1} \tag{22.7}$$

In practice, one carries out a standard Metropolis Monte Carlo (MC) or molecular dynamics (MD) simulation of the $(N-1)$-particle system; every predefined time interval the current configuration, $\mathbf{x}_1,...,\mathbf{x}_{N-1}$ (denoted i for brevity) is treated. Thus, to estimate the corresponding integral $g_i = \int V^{-1}\exp(-e_N/k_{\mathrm{B}}T)d\mathbf{x}_N$, one carries out k uniformly distributed attempted insertions (j) of the test particle to the volume (at configuration i), where its interaction with the $(N-1)$-particle template is calculated. g_i is estimated by the arithmetic average,

$$\overline{g_i} = \frac{1}{k}\sum_{t=1}^{k}\exp[-e_{N,j,t}/k_{\mathrm{B}}T]. \tag{22.8}$$

The estimated average value, $\overline{g}$ of g_i over the MC sample of size n, and the corresponding estimation of μ_W are:

$$\overline{g} = \frac{1}{n}\sum_{t=1}^{n}\overline{g}_{i(t)} \quad \rightarrow \quad \overline{\mu_w} = -k_{\mathrm{B}}T \ln\overline{g} \tag{22.9}$$

Similar to procedures discussed earlier (e.g., free energy perturbation (FEP) and thermodynamic integration, TI, see Chapter 17), particle insertion is an exact method with efficiency, that, however, depends on the density. Thus, the method works very well for relatively dilute systems, becoming extremely inefficient for highly dense systems since most of the attempts to insert a particle at random would fail. For a moderately dense system, the main contribution to $\exp[-e_N/k_{\mathrm{B}}T]$ comes from low probable configurations with large cavities, which would not appear in a usual MC sample; therefore, large samples are required to guarantee unbiased results. Several procedures have been suggested for improving the efficiency of the Widom insertion method (see [1]).

22.2 The Deletion Procedure

An expected remedy for the high density problem encountered with insertion would be to apply instead a *deletion* process by calculating $\mu_{\mathrm{SG}} = -(F_{N-1}-F_N) = k_{\mathrm{B}}T \ln Z_{N-1}/Z_N$, where Z_{N-1}/Z_N is expressed as a statistical average in P_N^{B}; this idea was suggested first by Shing and Gubbins [5]. Thus, according to our convention, $E_{N-1} = E_N + e_N$, and the integrand in the numerator, $\exp(-E_N/k_{\mathrm{B}}T)$ can be expressed by, $\exp(-E_N + e_N)/k_{\mathrm{B}}T)$,

$$\frac{VZ_{N-1}}{Z_N} = \frac{\int \exp(-E_{N-1}/k_BT)d\mathbf{x}_1...d\mathbf{x}_N}{\int \exp(-E_N/k_BT)d\mathbf{x}_1...d\mathbf{x}_N}$$

$$= \frac{\int \exp(-E_N/k_BT)\exp(+e_N/k_BT)d\mathbf{x}_1...d\mathbf{x}_N}{\int \exp(-E_N/k_BT)d\mathbf{x}_1...d\mathbf{x}_N} \tag{22.10}$$

$$= \int P_N^B \exp(+e_N/k_BT)d\mathbf{x}_1...d\mathbf{x}_N$$

Thus,

$$\mu_{SG} = k_BT \ln < \exp(e_N/k_BT) >_N \tag{22.11}$$

As for insertion, during an MC (or MD) simulation, a system configuration i is chosen, a ("deleted") particle is selected at random (probability $1/N$), its interaction energy, $+e_N$, is calculated, and $\exp(e_N/k_BT)$ is stored. In fact, to increase efficiency, the contribution of several randomly chosen particles can be calculated, where their average is the contribution of configuration i. μ_{SG} is estimated along the lines described above for the insertion process [Equations (22.8) and (22.9)]. While the deletion of a particle from a dense system is expected to be more efficient than insertion, notice that repulsive interactions are not bound from above and thus might be difficult to control (average) in a dense environment. Therefore, even for a moderately dense system, significant contributions to μ_{SG} come from low probable configurations with highly dense clusters, which would be missing from a realistic sample. Indeed, applications of the deletion process have generally led to wrong results for the chemical potential [1]. Finally, for a system of hard spheres, deletion is not applicable since it would lead to the wrong answer, $\exp[e_N/k_BT] = 1$ ($\mu_{SG} = 0$), for every experiment.

However, Shing and Gubbins have shown [5] that better results can be obtained by combining the insertion and deletion techniques. Thus, defining, for simplicity, $e = e_N$, we obtain,

$$< \exp(-e/k_BT) >_N = \int f_{N-1}(e)\exp(-e/k_BT)de \tag{22.12}$$

$$< \exp(e/k_BT) >_N = \int g_N(e)\exp(e/k_BT)de \tag{22.13}$$

where $f_{N-1}(e)de$ is the probability for finding the energy of the ghost particle (in a fluid of $N - 1$ particles) in the range e to $e + de$. Similarly, $g_N(e)de$ is the probability for finding the energy of the ghost particle (in a fluid of N particles) within e and $e + de$. From these definitions, one obtains the relation [5],

$$g_N(e) = \exp(\mu_{ex}/k_BT)f_{N-1}(e)\exp(-e/k_BT) \tag{22.14}$$

where μ_{ex} represent both μ_W and μ_{SG}, which are expected to be the same, and $\exp(\mu_{ex}/k_BT)$ is a constant independent of e. $g_N(e)$ and $f_{N-1}(e)$ are obtained by histograms generated from simulations related to Equations (22.7) and (22.11), respectively. The excess chemical potential can be obtained from Equation (22.14) if both g_N and f_{N-1} are known for the same value of e. Also, Equation (22.14) can be used to calculate f_{N-1} for regions of the ghost energy, e, where g_N can be evaluated accurately, but f_{N-1} cannot, such as in regions of negative e; the chemical potential can then be obtained reliably from Equation (22.12). In practice, Equation (22.14) is better than the insertion and deletion methods individually, but it fails at high densities, where g_N and f_{N-1} do not overlap.

22.3 Personage's Method for Treating Deletion

As has been pointed out above, simulation studies using deletion have been found generally inferior to those carried out by insertion; however, due to the potential of the deletion mechanism to handle dense systems, several attempts have been made to improve this process. Treating a Lennard-Jones (LJ) system, Parsonage [6–8] has realized that Equation (22.10) can be written as

$$\frac{VZ_{N-1}}{Z_N} = \frac{\int \exp(-E_{N-1}/k_BT)\exp(-e_N/k_BT)\exp(+e_N/k_BT)d\mathbf{x}_1...d\mathbf{x}_N}{\int \exp(-E_N/k_BT)d\mathbf{x}_1...d\mathbf{x}_N} \tag{22.15}$$

However, in the case of a hard-core potential, Equation (22.15) and a corresponding MC simulation are not compatible. This is demonstrated, from example, by the LJ potential, $E(r_{ij})$, used by Personage,

$$E(r_{ij}) = 4\varepsilon\left[\left(\frac{\sigma}{r}\right)^{12} - \left(\frac{\sigma}{r}\right)^6\right] \quad \begin{array}{ll} = \infty & \text{for} \quad r_{ij} < \alpha\sigma \\ & \text{for} \quad \beta\sigma > r_{ij} > \alpha\sigma \\ = 0 & \text{for} \quad r_{ij} > \beta\sigma \end{array} \tag{22.16}$$

where r_{ij} is the distance between particles i and j. Thus, to avoid overlaps during the MC process, a *trial* move of particle i is performed first, and the distances, r_{ij}, from i to the $N-1$ other particles are calculated. If r_{ij} is found smaller than the hard-core distance, $\alpha\sigma$, the energy is considered infinite and the move is rejected (σ is the LJ distance, where $\alpha < 1$ and β are cut-off distance factors). On the other hand, this trial move is accepted in Equation (22.15), as $\exp(-e_N/k_BT)\exp(+e_N/k_BT) = 1$. Therefore, the deletion process should be corrected for the case of hard-core potentials. The derivation starts from Widom's formula, where the particles' overlaps are taken into account by dividing both Z_N and Z_{N-1} by the same intermediate function,

$$\int w\exp(-E_{N-1}/k_BT)d\mathbf{x}_1...d\mathbf{x}_{N-1}d\mathbf{x}_N \tag{22.17}$$

where w is a function defined below. Thus,

$$< \exp[-\mu_w/k_BT] >_{N-1} = \frac{Z_N}{VZ_{N-1}} = \frac{\int \exp(-E_N/k_BT)d\mathbf{x}_1...d\mathbf{x}_N}{V\int \exp(-E_{N-1}/k_BT)d\mathbf{x}_1...d\mathbf{x}_{N-1}}$$

$$= \frac{\int w\exp(-E_{N-1}/k_BT)d\mathbf{x}_1...d\mathbf{x}_{N-1}d\mathbf{x}_N}{\int \exp(-E_{N-1}/k_BT)d\mathbf{x}_1...d\mathbf{x}_{N-1}d\mathbf{x}_N} \frac{\int w\exp(-E_N/k_BT)d\mathbf{x}_1...d\mathbf{x}_N}{\int w\exp(-E_{N-1}/k_BT)d\mathbf{x}_1...d\mathbf{x}_N}$$

$$= \frac{\int w\exp(-E_{N-1}/k_BT)d\mathbf{x}_1...d\mathbf{x}_{N-1}d\mathbf{x}_N}{\int \exp(-E_{N-1}/k_BT)d\mathbf{x}_1...d\mathbf{x}_{N-1}d\mathbf{x}_N} \frac{1}{\dfrac{\int w\exp(-E_{N-1}/k_BT)d\mathbf{x}_1...d\mathbf{x}_N}{\int w\exp(-E_N/k_BT)d\mathbf{x}_1...d\mathbf{x}_N}} \tag{22.18}$$

$$= <w>_{N-1} \frac{1}{\int wP_N^B \exp(+e_N/k_BT)d\mathbf{x}_1...d\mathbf{x}_N} = <w>_{N-1} \frac{1}{<w\exp[+e_N/k_BT]>_N}$$

In Equation (22.18), $w = 1$ means that there is no core overlap between the Nth and the remaining $N - 1$ particles, otherwise $w = 0$. Notice that in the derivation of Equation (22.18), we used

$$\int w\exp(-E_N/k_BT)d\mathbf{x}_1...d\mathbf{x}_N = \int \exp(-E_N/k_BT)d\mathbf{x}_1...d\mathbf{x}_N \qquad (22.19)$$

$<w>_{N-1}$ is the mean value of the fraction of successful attempts in trial insertions of a hard sphere with the appropriate core diameter into the system of $N - 1$ particles. $[<w\exp(+e_N/k_BT)>_N]^{-1}$ is the Shing and Gubbins average [Equation (22.11)] calculated according to Equation (22.18) [Notice again, the last line of Equation (22.18) is based on Equation (22.10).]. According to Equation (22.3), one obtains:

$$\mu(N,V,T) = \mu^{IG}+\mu^{ex} = \mu^0(T)+k_BT\ln\rho-k_BT\ln<w>_{N-1}+k_BT\ln<w\exp(+e_N/k_BT)>_N \qquad (22.20)$$

It should first be pointed out that for hard spheres, where $\mu'_{SG} \equiv k_BT\ln<w\exp(+e_N/k_BT)>_N = 0$, μ^{ex} is not zero, but $\mu^{ex} = -k_BT\ln<w>_{N-1}$, which is the same value which would be obtained by the Widom method [μ'_{SG} denotes $\mu_{SG}(w)$ calculated in Equation (22.18)]. Note that the calculation of $<>_N$ and $<w>_{N-1}$ are based on systems of N and $N - 1$ particles, respectively; however, for a large system, the results for N and $N - 1$ are expected to be approximately the same, and a single simulation would be adequate. Equation (22.20) can be equated with the Widom result, $\mu^0(T)+k_BT\ln\rho+\mu_W$, leading to:

$$\mu_W-\mu'_{SG} = -k_BT\ln<w>_{N-1}. \qquad (22.21)$$

This theory has been verified by MC simulations of N LJ particles [Equation (22.16)], from which μ_W and μ'_{SG} were estimated, while simulations of $N - 1$ particles were used for calculating $<w>_{N-1}$ [7]. Results are given for $\beta = 2.5\sigma$, for several values of α [Equation (22.16)], and for several temperatures and densities. Equation (22.21) is nicely satisfied for values of α close to $\alpha = 1$, within the range $\alpha = 0.875$–0.975, where the repulsions (overlaps) are moderate; however, when α is decreased further, the stronger LJ repulsions (which contribute significantly to μ'_{SG}) do not appear in a relatively short simulation, and the results for μ'_{SG} become significantly too small. While the highest density studied in this paper is $\rho^* = \rho\sigma^3 = 0.75$, in a subsequent article [8], the method shows relatively fast convergence with reliable results for densities up to $\rho^* = 0.95$ and temperatures, $T^* = k_BT/\varepsilon = 1$ and 1.2.

It should be pointed out that the (deletion-based) equation of Z_N/Z_{N-1} [Equation (22.18)] can be obtained from Bennett's formula [Equation (20.6)], which for the present problem reads:

$$\frac{Z_N}{VZ_{N-1}} = \frac{\dfrac{\int w\exp(-E_{N-1}/k_BT)d\mathbf{x}_1...d\mathbf{x}_{N-1}}{Z_{N-1}}}{\dfrac{\int w\exp(-E_{N-1}/k_BT)d\mathbf{x}_1...d\mathbf{x}_{N-1}d\mathbf{x}_N}{Z_N}}$$

$$= <w>_{N-1} \frac{1}{<w\exp[E_N-E_{N-1})/k_BT]>_N} = <w>_{N-1} \frac{1}{<w\exp[e/k_BT]>_N} \qquad (22.22)$$

Kofke and Cummings [1,9] pointed out that in addition to the derivation presented in Equations (22.18) and (22.22), one can define three more formulas for calculating Z_N/Z_{N-1}. For example, it is easy to see that the second line of Equation (22.18) leads to the insertion-based expression:

$$\frac{Z_N}{VZ_{N-1}} = <w>_{N-1} \left\langle \frac{\exp[-e/k_BT]}{w} \right\rangle_w \qquad (22.23)$$

which has been named "*staged insertion*" by Kofke and Cummings [1,9]. If the first term in the second line of Equation (22.18) (defined for simplicity as a/b) is treated as 1/(b/a), one obtains the insertion-based umbrella sampling definition for Z_N/Z_{N-1} [see Equation (20.2)]:

$$\frac{Z_N}{VZ_{N-1}} = \frac{1}{<1/w>_w} \left\langle \frac{\exp[-e/k_BT]}{w} \right\rangle_w . \tag{22.24}$$

Finally, one can obtain the "*staged deletion*" expression,

$$\frac{Z_N}{VZ_{N-1}} = \frac{1}{<1/w>_w} \frac{1}{<w\exp[e/k_BT]>_N} \tag{22.25}$$

22.4 Introduction of a Hard Sphere

In Parsonage's derivation, the hard-core effect is introduced within the (LJ) potential energy by the parameter α [Equation (22.16)]. Kofke and Cummings suggested to leave the potential energy intact, while replacing the effect of α by applying hard sphere conditions on the Nth particle. They carried out preliminary MC studies of the (modified) staged insertion [Equation (22.24)] and Bennett's derivation [Equation (22.21)].

Kofke and Cummings work was followed by a more extensive study by Boulougouris, Economou and Theodorou (BET) [10], who carried out a detailed derivation of the Bennett formula [Equation (22.21)] for the LJ potential with a hard sphere within the NpT ensemble (other ensembles were studied as well). BET pointed out that the deletion scheme of Shing and Gubbins [Equation (22.16)] is somewhat biased because a system of $N-1$ particles is not a typical equilibrated system due to the hole created by removing the Nth particle. BET have claimed that this flaw is corrected by their two-stage derivation, where: (1) the $N-1$ system with a hole is transferred to the same system, where a hard sphere occupies the hole and (2) the latter system is transferred to the N-particles system. For the NVT ensemble, one first defines for the Nth particle, a Heaviside step function, $H(x_i,N)$:

$$H(x_i,N) = \begin{array}{cc} 0 & \text{if } |\mathbf{x}_i - \mathbf{x}_N| < d_{\text{core}} \\ 1 & \text{if } |\mathbf{x}_i - \mathbf{x}_N| \geq d_{\text{core}} \end{array} \quad i = 1,\cdots,N-1 \tag{22.26}$$

where d_{core} is a hard-core diameter, which should be optimized for the force field used; in this respect, the BET formulation is more general than that of Parsonage. Now, the intermediate function [Equation (22.17)] is replaced by:

$$\int \prod_{i=1}^{N-1} H(x_i,N)\exp(-E_{N-1}/k_BT)dx_1\cdots dx_N \tag{22.27}$$

and following the same steps used to derive Equation (22.18) leads to the analog of the Bennett formula [Equation (22.22)],

$$\frac{Z_N}{VZ_{N-1}} = \frac{\left\langle \prod_{i-1}^{N-1} H(x_i,N) \right\rangle_{N\,1}}{\left\langle \prod_{i=1}^{N-1} H(x_i,N)\exp E^N(\mathbf{x}_1,\cdots,\mathbf{x}_N)/k_BT \right\rangle_N} \tag{22.28}$$

where

$$E^N(\mathbf{x}_1,\cdots,\mathbf{x}_N) = E_N(\mathbf{x}_1,\cdots,\mathbf{x}_N) - E_{N-1}(\mathbf{x}_1,\cdots,\mathbf{x}_{N-1}). \tag{22.29}$$

BET call this method the "*staged particle deletion*" (SPD).

Results for the excess chemical potential of a LJ system in the *NVT* ensemble were calculated for the pairs, $(T^*, \rho^*) = (1.0, 0.8), (1.0, 0.705), (1.35, 0.9)$, and $(0.75, 0.85)$ [10]. A very good agreement has been achieved between MC results based on the Widom and SPD methods and those obtained from the equation of state [11]. As for the Parsonage theory, implementation of SPD requires carrying out two MC runs based on $N - 1$ and N particles. However, for a large system, the difference between results obtained for $N - 1$ and N is expected to be negligible, which indeed has been verified by the BET calculations. Thus, a significant save in computer time was gained by estimating the averages in the numerator and denominator of Equation (22.28) from the same MC sample. In fact, SPD was found to be more efficient than Widom's procedure by a factor of 10 in terms of computer time. SPD was also found to be more efficient than staged insertion [Equation (22.23)] based on $H(x_i, N)$ [Equation (22.26)].

In a later paper [12], SPD was extended to chain molecules, where some enhancements to the method were introduced. Again, results in the *NVT* ensemble for ethane and for tangent sphere dimers, obtained by SPD, required less computer time than those obtained by Widom's test particle insertion method, by factors ranging from 3.5 to 1.5. The SPD technique was developed further to a method called the "*direct particle deletion*". In the direct particle deletion, the intermediate stage of SPD ($N - 1$ particles and a hard sphere) is replaced by a direct transition from the system of N particles to that of $N - 1$ with a hole. However, discussing this method in detail is beyond the scope of this book; see [13] and references cited therein.

22.5 The Ideal Gas Gauge Method

Another approach for calculating the chemical potential is the "*ideal gas gauge method*" (IGGC) due to Neimark and Vishnyakov [14]. We shall only present the main idea of this method as applied to a LJ fluid. Thus, the system of interest at a given T and volume, V, is considered to be at a chemical equilibrium with a system of ideal particles at the same T and volume, V_g, that serves as a gauge system. The total number of particles in the combined system, $N_{\text{tot}} = N + N_g$, is constant and expected to be relatively *large*, say, $N \geq 100$, where N and N_g are the number of particles in the system of interest and the gauge system, respectively. Because particles can move from system to system, N (and N_g) change in time.

A Metropolis MC procedure, which allows the exchange of particles, would lead to equilibration, where the *average* numbers of particles in these systems are $\bar{N}$ and $\bar{N}_g$, which, for a *large* N_{tot}, the following relation holds,

$$\mu(\bar{N}) = \mu(\bar{N}_g) = -k_B T \ln\left(\frac{V_g}{\Lambda^3 \bar{N}_g}\right) \tag{22.30}$$

Thus, $\mu(\bar{N})$ is obtained from the known $\mu(\bar{N}_g)$. The MC procedure is based on three different trial moves. The first is a (canonical type) particle displacement within the system,

$$p_{ij} = \min\left\{1, \exp(-\Delta E_{ij}/k_B T)\right\} \tag{22.31}$$

where the energy difference is $\Delta E_{ij} = E_j - E_i$. The second move is of a particle from V_g to V ($i \rightarrow j$); in this case [see Equations (14.33)–(14.36)], $T_{ij} = (1/N_g)(1/V)$, while $T_{ji} = 1/(N + 1)(1/V_g)$; thus, $T_{ij}/T_{ji} = V_g(N + 1)/VN_g$ and $A_{ij} = (T_{ij}/T_{ji})^{-1} [\exp - (E_j - 0)] = [V_g(N + 1)/VN_g]^{-1} \exp(-\Delta E_{ij})$, $(\Delta E_{ij} = E_j)$, which leads to,

$$p_{V_g \rightarrow V} = \min\{1, A_{ij}\} = \min\{1, \exp[-\Delta E_{ij}/k_B T - \ln(V_g(N+1)/(VN_g))]\} \tag{22.32}$$

In a similar way, the probability, $p_{V \to v_g}$, to move a particle from the system to the gauge system is

$$p_{V \to v_g} = \min\{1, \exp{-\Delta E_{ij}/k_B T} - \ln[V(N_g + 1)/(V_g N)]\} \tag{22.33}$$

Generally, one is interested in the chemical potential for a given density, $\rho_{given} = N/V$. Clearly, with IGGC described above, N and V (therefore, the density, ρ) are undetermined a priori. Thus, before performing a production MC run, one has to carry out several short trial runs at constant N_{tot}, where $\rho_{given} = \bar{N}/V$ is obtained by changing adequately V (or V and V_g). Notice also that in the transition $V_g \to V$, one tries to insert a particle from V_g into the occupied volume V by LJ particles; in this respect, IGGC is similar to the Widom insertion method.

IGGC is a manifestation of the *"mesoscopic canonical ensemble,"* which constitutes an intermediate between the canonical and the grand canonical ensembles. In the grand ensemble, the system under study exchanges particles with an infinite reservoir of ideal particles of a −well-defined chemical potential, while the number of particles in the system, N, fluctuates. In the canonical ensemble, the system is closed and N is constant. With the mesoscopic canonical ensemble, the system of interest (of volume V) is partially open, exchanging particles with the finite-volume reservoir (V_g) of ideal particles; thus, N fluctuates, and the level of its fluctuations is controlled by V_g. As V_g is increased or decreased, the mesoscopic canonical ensemble transforms into the grand ensemble or the canonical ensemble, respectively.

It should be pointed out that due to fluctuations, a single MC run can lead, in principle, to results for $\mu(\rho)$ for different N. While this is impractical for a large system, where the fluctuation of ρ decreases with increasing system size, this can be useful for a small system. In this case, Equation (22.30) becomes somewhat more complex depending also on the ratio of probabilities $\ln(P_{N_g}/P_{N_g-1})$ obtained from the histogram of distributions of particles in the gauge cell. Indeed, Neimark and Vishnyakov show [14] that for a small system, obtaining a set of results for μ for different N values at a given T, IGGC is more efficient than the Widom insertion method by a factor of 10 in terms of computer time.

22.6 Calculation of the Chemical Potential of a Polymer by the Scanning Method

Assume a system of N_p polymer chains, each of N bonds enclosed in a box of volume V. For simplicity, we consider self-avoiding walks (SAWs) on a square lattice, where $V = L^2$. To obtain the chemical potential, one has to calculate the free energy involved in adding a chain to the system. Naturally, this can be achieved by the Widom method, where the $(N_p + 1)$th chain (rather than a particle) is inserted into the system. Siepmann [15] suggested to insert the chain using the Rosenbluth & Rosenbluth method, and his related *configurational-bias Monte Carlo method* was also used to calculate μ of a polymer [16,17] (Section 14.4). However, we shall discuss this idea as implemented by the more efficient scanning method (Section 14.5), where f steps are scanned ahead instead of a single Rosenbluth & Rosenbluth step [18].

The partition function is the number of distinguishable different system configurations (that is, the different arrangements of the N_p SAWs in the box). Each system configuration i has a degeneracy, $N_p!2^{N_p}$, and the partition function is thus:

$$Z = \frac{1}{N_p!2^{N_p}} \sum_{\text{SAWs},i} 1_i. \tag{22.34}$$

The Boltzmann probability of system configuration i is:

$$P_i^B = \frac{1}{\sum_i 1_i} \tag{22.35}$$

and the chemical potential is thus:

$$\mu = F(N_p + 1) - F(N_p) = -\ln\frac{Z(N_p+1)}{Z(N_p)} = -\ln\frac{1}{2(N_p+1)}\sum_i P_i^{\mathrm{B}}(N_p)M_i \tag{22.36}$$

where M_i is the number of different ways to insert a test chain [the $(N_p + 1)$th chain] into the "frozen" configuration i of N_p SAWs; for simplicity, the factor $k_{\mathrm{B}}T$ has been omitted in Equation (22.36), which is the Widom insertion equation for SAWs (Equation (22.36) corrects an error in Equation (4) of [18]). To estimate M_i by importance sampling, it should first be expressed as a statistical average:

$$M_i = \sum_j 1_j = \sum_j G_j\left[\frac{1}{G_j}\right] \tag{22.37}$$

where G_j is a probability distribution normalized over *all* the test chain configurations, and the expression in the brackets is a random variable [see Equation (10.6)]. In practice, the N_v vacant sites in the system, $N_v = V - N_pN$, are numbered and a vacant site, s, is selected at random with the probability $p_s = 1/N_v$. s is a "seed" from which an attempt is made to generate a chain, j, step-by-step with the scanning method based on a future scanning parameter, f. The probability of a successfully generated chain is:

$$G_j(s,f) = \frac{1}{N_v}P_j(s,f) \tag{22.38}$$

where $P_j(s,f)$ is the probability of the chain constructed by the scanning method. Notice that the construction of a chain may fail, therefore, $P_j(s,f)$ is normalized over an ensemble, which also includes partial chains that cannot be completed due to a violation of the excluded volume interaction. Therefore, M_i, [Equation (22.37)] can be expressed as,

$$M_i = \sum_{s,j} 1_{s,j}\exp(-E_{s,j}) = \sum_{j,s} G_j(s,f)\left[\frac{\exp(-E_{s,j})}{G_j(s,f)}\right] \tag{22.39}$$

where the summation runs over all the vacant sites, s, and the corresponding j. In contrast to Equation (22.37), j in Equation (22.39) also runs over test chain configurations that do not satisfy the excluded volume condition; to each of these chains, an energy $E_{s,j} = \infty$ is assigned, where $E_{s,j} = 0$ is defined for chain configurations that satisfy the excluded volume condition [e.g., see Equation (14.23)]. M_i can be estimated by $\bar{M}_i$:

$$\bar{M}_i = \frac{1}{n_0}\sum_{t=1}^{n_{\mathrm{suc}}}\frac{1}{G_j(s,f,t)} \tag{22.40}$$

where n_0 is the number of chain constructions started and n_{suc} is the number of constructions succeeded; $G_j(s,f,t)$ is the probability of SAW j obtained at time, t, of the process. The variance of M_i decreases with increasing f, however, unlike the case of a single chain, even for a complete scanning ($f = N - k$ for construction step, k), the variance is not zero (because $P_j(s,f)$, and thus $G_j(s,f)$ depend on s). The system is simulated by MC and every predefined number of MC steps, M_i for the current configuration i is estimated from n_0 trial insertions, and the chemical potential is estimated according to Equation (22.36).

This method is efficient for moderate densities and chain lengths; for example, in [18], SAWs of $N = 30$ and density $N_pN/V = 0.555$ were successfully studied on a square lattice. Still, more efficiency can be gained by breaking the inserted chain into k sub-chains, which are appended gradually during the simulation, while their contributions to μ are calculated and added up. One can easily show that:

$$\mu = \mu_1 + \mu_2 + \dots \mu_k \tag{22.41}$$

where $\mu_1 = -[\ln Z(N_p, N, 1, N/k) - \ln Z(N_p, N)]$; $\mu_2 = [\ln Z(N_p, N, 1, 2N/k) - \ln Z(N_p, N, 1, N/k)]$, etc.

$Z(N_p, N, 1, N/k)$ is the partition function of a system consisting of N_p chains of N monomers and one chain of N/k monomers. To obtain μ_1, M_1 is calculated with Equations (22.40) by inserting the first sub-chain of N/k monomers into a system with N_p chains using an optimized scanning parameter, f_1. M_1 is estimated every n_{MC} Monte Carlo steps, and μ_1 is obtained with Equation (22.36). In the next stage, a chain of N/k monomers is appended to the existing sub-chain of N/k monomers with a scanning procedure based on optimized f_2. Notice that in this case, Equation (22.36) is used without the factor 1/2. The process is then continued until μ_k has been calculated. The chemical potential of short chains can be estimated reliably because a significant contribution to μ comes from system configurations with relatively small voids, which dominate the ensemble and thus appear frequently in an MC sample. As the density increases the sub-chain's length should be decreased, becoming a single monomer at a high density. In this case, n_{MC} simulations are carried out for each monomer, which might require a large computer time. The advantage of the scanning method lies in its flexibility: it can be adjusted to provide the most efficient simulation for a given density, chain length, chain complexity, and other system conditions.

22.7 The Incremental Chemical Potential Method for Polymers

The above discussion about constructing a chain monomer-by-monomer has been motivated by the "*incremental chemical potential*" (ICM) method of Kumar, Szleifer, and Panagiotopoulos [19]. ICM was developed for a chain model of beads connected by bonded interactions, where a LJ interaction is defined between any pair of non-bonded beads. As in the previous sub-section, the system consists of N_p chains, each of N beads and one partial chain of $k + 1$ beads, where our interest is in the residual chemical potential, μ_r, due to the *attached* $(k + 1)$th bead. The interaction energy of this bead with the $N_pN + k$ other beads is denoted by $E_{k+1}(\mathbf{x}_{k+1})$, where $\mathbf{x}_{k+1}$ is the Cartesian coordinates of the $(k + 1)$th bead. μ_r is thus given by:

$$-\frac{\mu_r}{k_BT} = \ln\left\langle \exp[-E_{k+1}(\mathbf{x}_{k+1})/k_BT] \right\rangle_{N_pN+k} \tag{22.42}$$

where the averaging is performed with a Boltzmann probability density based on the system of $N_pN + k$ beads. Notice the difference between this method and the method described in the previous section for inserting a single bond with the Rosenbluth & Rosenbluth method. While ICM can handle relatively dense systems, it is time consuming because the chemical potential of the entire chain requires N different simulations where the corresponding μ_r values are added up [Equation (22.40)]. However, Kumar, Szleifer, and Panagiotopoulos have assumed that after m initial steps ($m \sim 15$), μ_r becomes constant, and thus the contribution to μ from the rest of the chain can be obtained efficiently, simply by $(N - m)\mu_r$, where μ_r is calculated for a single arbitrary k value, $k > m$. The main interest in ICP has been in the context of this approximation, which, however, has been found to be limited to linear homopolymers at high temperature. This assumption has not been tested enough and should be taken with caution [20–23].

Another technique called "*expanded variable-length chain method*" (EVALENCH), which is close in spirit to ICP, has been suggested by Escobedo and de Pablo [24]. With EVALENCH, the system is described within the framework of an extended ensemble, EXE (Section 21.4). Thus, in the MC simulation, a monomer can be deleted from the "tagged" chain or appended to it, where these transitions are governed by incremental chemical potential parameters. EVALENCH and ICP are expected to handle relatively dense systems with comparable efficiency; however, EVALENCH requires optimizing the incremental chemical potential parameters [η_m, Equation (21.34)] and, therefore, application of ICP is the more straightforward for handling relatively dense systems. Later, Escobedo's group has suggested another method for calculating the chemical potential by combining EXE with Bennett's acceptance-ratio method, see Section 21.4 and [98–100] of Chapter 21. Finally, an extension of ICM has been suggested by Neimark's group [2], based on the *ideal gas gauge method* described above in Section 22.5.

22.8 Calculation of μ by Thermodynamic Integration

Thus far, we have presented methods for calculating the chemical potential based on insertion, deletion, or appending partial chains. However, for highly dense systems, it will become difficult to insert or append even a single monomer. In this case, one would have to resort to methods where the monomer-system interactions are gradually increased from zero to their full value using TI; an alternative and sometimes easier procedure is to gradually eliminate these interactions by TI. This approach, applied to LJ particles, was first suggested by Mon and Griffiths [25].

More specifically, for a canonical system of N particles, the chemical potential, $\mu(N,V,T)$, is obtained by decreasing slowly the interactions, $E(\mathbf{x}^{N-1}, \mathbf{x}_N)$ between the Nth molecule and the other $N - 1$ molecules, using, say, an MD simulation. As discussed in detail in Section 17.5, such elimination can be obtained by a parameter λ, $\lambda = 1 \rightarrow 0$, where:

$$E(\mathbf{x}^{N-1}, \mathbf{x}_N, \lambda) = \lambda E(\mathbf{x}^{N-1}, \mathbf{x}_N) \tag{22.43}$$

The free energy involved in this process is obtained by integrating the derivative of the interaction energy with respect to λ, which leads to [see Equation (17.12)]:

$$-\mu(N,V,T) = \Delta F = \int_{\lambda=1}^{\lambda=0} <E(\mathbf{x}^{N-1}, \mathbf{x}_N)>_\lambda d\lambda \tag{22.44}$$

One would expect that application of non-linear scaling (see [Equation (17.24)] for the LJ potential) would allow treating quite dense systems. Indeed, TI procedures are used successfully in structural biology for calculating the free energy of solvation, and the free energy of binding in complex protein systems, where molecules are created or annihilated.

Thermodynamic integration can also be used for calculating the chemical potential of polymer systems, as has been shown by Müller and Paul [26], who applied TI to the bond-fluctuating model, where the entire chain is inserted; Wilding and Müller [27] treated the same model with a modified TI procedure within the framework of the extended ensemble technique. The integration of the free energy from an ideal chain to a SAW described in Section 17.4.1 is another example for the potential of TI for polymers. Still, treating a long polymer at high density might require cutting the chain into segments. While TI calculations might sometimes be lengthy, the methodology is straightforward, and it thus seems strange that this direct technique has not been used more frequently for calculating the chemical potential.

REFERENCES

1. D. A. Kofke and P. T. Cummings. Quantitative comparison and optimization of methods for evaluating the chemical potential of molecular simulation. *Mol. Phys.* **92**, 973–996 (1997).
2. C. J. Rasmussen, A. Vishnyakov and A. V. Neimark. Calculation of chemical potentials of chain molecules by the incremental gauge cell method. *J. Chem. Phys.* **135**, 214109-14 (2011).
3. B. Widom. Some topics in the theory of fluids. *J. Chem. Phys.* **39**, 2808–2812 (1963).
4. J. L. Jackson and L. S. Klein. Potential distribution method in equilibrium statistical mechanics. *Phys. Fluids* **7**, 228–231 (1964).
5. K. S. Shing and K. E. Gubbins. The chemical potential in dense fluids and fluid mixtures via computer simulation. *Mol. Phys.* **46**, 1109–1128 (1982).

6. N. G. Parsonage. Determination of the chemical potential by the particle insertion method and by its inverse. *J. Chem. Soc., Faraday Trans.* **91**, 2971–2973 (1995).

7. N. G. Parsonage. Chemical-potential paradox in molecular simulation: Explanation and Monte Carlo results for a Lennard-Jones fluid. *J. Chem. Soc., Faraday Trans.* **92**, 1129–1134 (1996).

8. N. G. Parsonage. Computation of the chemical potential in high density fluids by Monte Carlo method. *Mol. Phys.* **89**, 1133–1144 (1996).

9. D. A. Kofke and P. T. Cummings. Precision and accuracy of staged free-energy perturbation methods for computing the chemical potential by molecular simulation. *Fluid Phase Equilib.* **150**, 41–49 (1998).

10. G. C. Boulougouris, I. G. Economou and D. N. Theodorou. On the calculation of the chemical potential using the particle deletion scheme. *Mol. Phys.* **96**, 905–913 (1999).

11. J. K. Johnson, J. A. Zollweg and K. E. Gubbins. The Lennard-Jones equation of state revisited. *Mol. Phys.* **78**, 591–618 (1993).

12. G. C. Boulougouris., I. G. Economou and D. N. Theodorou. Calculation of the chemical potential of chain molecules using the staged particle deletion scheme. *J. Chem. Phys.* **115**, 8231–8237 (2001).

13. M. G. De Angelis, G. C. Boulougouris and D. N. Theodorou. Prediction of infinite dilution benzene solubility in linear polyethylene melts via the direct particle deletion method. *J. Phys. Chem. B* **114**, 6233–6246 (2010).

14. A. V. Neimark and A. Vishnyakov. A simulation method for the calculation of chemical potentials in small inhomogeneous, and dense systems. *J. Chem. Phys.* **122**, 234108-10 (2005).

15. J. I. Siepmann. A method for the direct calculation of chemical potentials for dense chain systems. *Mol. Phys.* **70**, 1145–1158 (1990).

16. B. Smit. Grand canonical Monte Carlo simulation of chain molecules: Adsorption isotherms of alkanes in zeolites. *Mol. Phys.* **85**, 153–172 (1995).

17. G. C. A. M. Mooij and D. Frenkel. A systematic optimization scheme for configurational bias Monte Carlo. *Molec. Simul.* **17**, 41–55 (1996).

18. H. Meirovitch. Simulation of the chemical potential of polymers. *Comput. Theor. Polym. Sci.* **8** (1–2), 219–227 (1998).

19. S. K. Kumar, I. Szleifer and A. Z. Panagiotopoulos. Determination of the chemical potentials of polymeric systems from Monte Carlo simulations. *Phys. Rev. Lett.* **66**, 2935–2938 (1991).

20. B. Smit, G. C. A. M. Mooij and D. Frenkel. Comment on Determination of the chemical potential of polymeric systems from Monte Carlo simulations. *Phys. Rev. Lett.* **68**, 3657 (1992).

21. S. K. Kumar, I. Szleifer and A. Z. Panagiotopoulos. Kumar, Szleifer, and Panagiotopoulos Reply. *Phys. Rev. Lett.* **68**, 3658 (1992).

22. S. K. Kumar. The chain length dependence of the chemical potentials of macromolecular systems at zero density: Exact calculations and Monte Carlo simulations. *J. Chem. Phys.* **96**, 1490–1497 (1992).

23. S. K. Kumar. A modified real particle method for the calculation of the chemicals potentials of molecular systems. *J. Chem. Phys.* **97**, 3550–3556 (1992).

24. F. A. Escobedo and J. J. de Pablo. Monte Carlo simulation of the chemical potential of polymers in an expanded ensemble. *J. Chem. Phys.* **103**, 2703–2710 (1995).

25. K. K. Mon and R. B. Griffiths. Chemical potential by gradual insertion of a particle in Monte Carlo Simulation. *Phys. Rev. A* **31**, 956–959 (1985).

26. M. Müller and W. Paul. Measuring the chemical potential of polymer solutions band melts in computer simulations. *J. Chem. Phys.* **100**, 719–724 (1994).

27. N. B. Wilding and M. Müller. Accurate measurements of the chemical potential of polymeric systems by Monte Carlo simulation. *J. Chem. Phys.* **101**, 4324–4330 (1994).

23

The Absolute Free Energy of Binding

The non-covalent association of molecules is central to many biological processes, such as the action of hormones, the recognition of antigens by the immune system, the catalysis of chemical reactions by enzymes, and the action of drugs. Therefore, development of simulation methods for calculating the affinity (free energy) of molecular binding is important for academic and practical reasons, as such methods will help in elucidating the mechanisms of complex biological processes, which might be useful in rational drug design. A well established approach in this area is based on *simplified* computationally fast scoring functions for screening large data bases of ligands (small molecules). However, for the refinement stage, it is important to devise highly accurate methods for calculating the *absolute* (standard) free energy of binding (denoted ΔA^0) of a ligand to an enzyme, based on detailed molecular interactions and rigorous statistical mechanics. This whole topic will be discussed in the present chapter on several levels, experimental, (classical) thermodynamics, and statistical mechanics (see [1,2]), where advanced theories for calculating ΔA^0 will be reviewed, among them HSMD-TI (introduced in Section 19.7), which combines the hypothetical scanning (HS) method, molecular dynamics (MD), and thermodynamic integration (TI). Finally, a detailed application of HSMD-TI to a complex with experimentally known ΔA^0 will be discussed. The treatment of binding is based on the theory of chemical equilibrium, where the chemical potential plays a central role.

23.1 The Law of Mass Action

The law of mass action was discovered as an experimental law in 1864 by the Norwegian scientists, Guldberg and Waage, and rediscovered in 1879 (independently) by van't Hoff. To explain the law, we start with the chemical reaction:

$$a\mathrm{A} + b\mathrm{B} \rightleftharpoons c\mathrm{C} + d\mathrm{D} \qquad (23.1)$$

where A, B, C, and D, are chemical species, and a, b, c, and d, are their stoichiometric coefficients. The law states: The rate of a chemical reaction is proportional to the molar concentration of the reacting substances. Thus, *at equilibrium:*

$$K_c = \frac{\left[\mathrm{C}\right]^c \left[\mathrm{D}\right]^d}{\left[\mathrm{A}\right]^a \left[\mathrm{B}\right]^b} \qquad (23.2)$$

where [] denotes the molar concentration, M, which is the number of moles per liter solution, and K_c is the *equilibrium constant* in units of $M^{c+d-a-b}$; usually products appear in the nominator. If $K_c \gg 1$, the equilibrium tends to the right meaning that the products are favored. In the other case ($K_c \ll 1$), the equilibrium tends to the left favoring the reactants. The mass action law [Equation (23.2)] expresses a *kinetic equilibrium* because the number of left to right reactions per unit time is equal to the number of right to left reactions per unit time; more specifically: the left to right rate is proportional to $[\mathrm{A}]^a[\mathrm{B}]^b$, and it is equal to $K'[A]^a[B]^b$, while the right to left rate is proportional to $[C]^c[D]^d$, and it is equal to $K''[C]^c[D]^d$: K' and K'' are forward and reversed reaction constants. In a kinetic equilibrium, the rates are equal,

$$K'[A]^a[B]^b = K''[C]^c[D]^d \tag{23.3}$$

and defining $K_c = K'/K''$ leads to Equation (23.2); K_c is independent of the concentration, but depends on the temperature.

The thermodynamic derivation of the law of mass action is given in Section 23.5. However, to define the basis of the law, we discuss the following simple example. Assume the reaction $2A + B \rightarrow Q$, ($Q = BA_2$) and imagine that A, B, and Q move randomly on a large square lattice of $V = L \times L$ sites, where N_A, N_B, and N_Q - the numbers of A, B, and Q are large, while N_A, N_B, and $N_Q \ll V$. Thus, the probabilities for occupying a site by A, B and Q are approximately $P_A = N_A/V$, $P_B = N_B/V$, and $P_Q = N_Q/V$, respectively. If one further assumes that a B molecule has two nearest neighbor reaction sites for an A molecule, the probability for a BA_2 molecule is $P_A P_A P_B = P_B(P_A)^2$, where the probability of a Q molecule is P_Q. Clearly, $P_B(P_A)^2 \sim [B][A]^2$ and $P_Q \sim [Q]$. Since the rates are proportional to the corresponding concentrations, we obtain for the forward reaction (to the right)

$$(\text{rate} \rightarrow) \sim [B][A]^2. \tag{23.4}$$

and for the reversed reaction (to the left),

$$(\text{rate} \leftarrow) \sim [BA_2]. \tag{23.5}$$

Notice that Equation (23.1) describes a primary reaction. If intermediate reactions occur, one should apply Equation (23.1) to each one of them and calculate the resulting K_c.

23.2 Chemical Potential, Fugacity, and Activity of an Ideal Gas

In Section 4.3, we have calculated the chemical potential for an ideal gas using thermodynamics and statistical mechanics. Since these results are used in the coming sections, we summarize them here.

23.2.1 Thermodynamics

We start from the differential for the Gibbs free energy, $dG(N, p, T) = -SdT + VdP + \Sigma\mu_i dN_i$. Since for a chemical reaction in equilibrium, $\Sigma\mu_i dN_i = 0$ [see Equation (23.28)], at constant T, $dG = Vdp$, and we get Equation (4.21),

$$G - G^0 = \int_{p^0}^{p} Vdp = \int_{p^0}^{p} \frac{Nk_BT}{p} dp = Nk_BT \ln\frac{p}{p^0} \tag{23.6}$$

where G^0 is the standard free energy and p^0 is the standard pressure, usually 1 atmosphere. If N is taken to be 1 mole, G and G^0 become μ and μ^0, respectively. Therefore:

$$\mu(T, p) = \mu^0(T) + RT \ln(p / p^0) \tag{23.7}$$

and for $p = p^0$, $\mu(T, p^0) = \mu^0(T)$ – the standard chemical potential. $\mu^0(T)$ is not provided by thermodynamics, but can be obtained for a specific model by statistical mechanics as described below.

23.2.2 Canonical Ensemble

For an ideal gas, we obtained [Equation (4.17)]:

$$\mu(V, T) = -k_BT \ln\frac{1}{\Lambda^3} + k_BT \ln\rho = \mu^0(T) + k_BT \ln\rho = -k_BT \ln\frac{1}{\Lambda_c^3 c^0} + k_BT \ln(c / c^0) \tag{23.8}$$

$\mu^0(T) = -k_B T \ln \Lambda^{-3}$ is the standard chemical potential with a value depending on the units used for ρ. Using the density, c, in molar ($c^0 = 1$ molar $= 1$ mol/L solution), the unit of Λ should be changed to Λ_c, and the two terms on the right hand side of Equation (23.8) become dimensionless. In general, the choice of c^0 defines the units used and thus the value of μ^0, where for $c = c^0$, $\mu = \mu^0$.

23.2.3 NpT Ensemble

Dividing both terms in Equation (4.20) by p^0, replacing k_B by R, and expressing Λ according to the units of p^0 (denoted Λ_p) leads to,

$$\mu(p,T) = -RT \ln\left(\frac{1}{\Lambda_p^3 p^0} RT \right) + RT \ln \frac{p}{p^0} = \mu_p^0(T) + RT \ln \frac{p}{p^0}. \tag{23.9}$$

Notice that $k_B T \Lambda^{-3}$ has the dimension of pressure and p^0 defines the standard unit of pressure used, which typically is, $p^0 = 1$ atmosphere, therefore for $p = p^0$, $\mu = \mu^0$.

For *real gases*, where the ideal gas conditions are not satisfied (that is, p is not equal to RT/V), we define two additional parameters, the *fugacity*, f:

$$f = \exp(\mu / RT) \tag{23.10}$$

and the *activity*, a:

$$a = \exp[(\mu - \mu^0) / RT]. \tag{23.11}$$

Therefore,

$$\mu(p,T) = \mu_p^0(T) + RT \ln \frac{f}{f^0} = \mu_p^0(T) + RT \ln a. \tag{23.12}$$

where $\mu_p^0(T)$ is a function of f^0. For an ideal gas, $f/f^0 = a = p/p^0$, leading to Equation (23.9); for $a = 1$, $\mu = \mu^0$.

23.3 Chemical Potential in Ideal Solutions: Raoult's and Henry's Laws

A solution may contain several components. For simplicity, assume a solution of *large* amount of solvent, A and *small* amount of solute, B. Unlike an ideal gas, where the particles do not interact, in an ideal solution, forces exist, but they are the same for AA, BB, and AB (see [1]).

The partial vapor pressure p_A (p_B) is a good measure for the tendency of A (B) to escape from the solution into the vapor; thus, for the state of affairs in the solution at *equilibrium*,

$$\mu_A^{sol} = \mu_A^{vapor} \qquad \mu_B^{sol} = \mu_B^{vapor} \tag{23.13}$$

μ^{vapor} can be approximated by an ideal gas [Equation (23.7)]; assuming $p^0 = 1$ atm.

$$\mu_A^{sol} = \mu_A^{vapor} = \mu_A^0 + RT \ln p_A \tag{23.14}$$

where p_A is the partial pressure of A above the solution (this is the pressure of A, if only A would occupy the available volume, that is, no vapor of B). The correction for non-ideality is usually small.

A solution is said to be ideal if the escaping tendency of each component is proportional to the mole fraction of that component in the solution; this is due to the similar forces for AA, AB, and BB. Two laws related to ideal solutions—Raoult's law and Henry's law.

23.3.1 Raoult's Law

This law states that:

$$p_A = X_A p_A^0 \tag{23.15}$$

where p_A^0 is the vapor pressure of pure liquid, A (at the same temperature of p_A) and X_A is the mole fraction of A,

$$X_A = \frac{n_A}{n_A + n_B + n_C + ...} \tag{23.16}$$

n_A, n_B, etc. are the number of moles of A, B, etc. Raoult's law applies to a *large* amount of solvent, that is, to the limit, $X_A \to 1$.

23.3.2 Henry's Law

If X_B is sufficiently small (dilute), B is practically surrounded uniformly by A molecules (irrespective of the fact that A and B may form non-ideal solutions at higher concentrations of B) and thus the escaping tendency of B is proportional to its mole fraction,

$$p_B = kX_B \tag{23.17}$$

k is determined experimentally. Henry's law applies to the limit $X_B \to 0$.

23.4 Chemical Potential in Non-ideal Solutions

For non-ideal solutions, the equation for μ has the same structure as for an ideal gas, where non-ideal values of the fugacity and the activity, $f/f^0 = a$ replace the ideal gas values, p/p^0. We define two standard states, for the solvent (A) and the solute (B).

23.4.1 Solvent

For the solvent, we obtain [Equation (23.12)],

$$\mu_A = \mu_A^0 + RT \ln \frac{f_A}{f_A^0} = \mu_A^0 + RT \ln a_A \tag{23.18}$$

The standard state (for which $a_A = 1$) is a pure solvent with $p_A^0 = 1$ atm at the given T. Therefore, with this choice of a standard state (under ideal conditions, that is, $X_A \to 1$), Raoult's law becomes

$$a_A \approx \frac{p_A}{p_A^0} = \frac{X_A p_A^0}{p_A^0} = X_A \tag{23.19}$$

that is, $X_A \to 1$ and $a_A \to 1$. For non-ideal conditions, we define the *activity coefficient*, γ_A, where

$$a_A = \gamma_A X_A \tag{23.20}$$

Again, for an ideal solution, both $X_A \to 1$ and $\gamma_A \to 1$ ($p_A^0 = 1$ atm).

23.4.2 Solute

The standard state for the solute can be derived based on Henry's law; however, we shall define it in terms of molar concentration (molarity – # moles per liter solution), by defining the non-ideal activity a_B [based on Equation (23.7)]:

$$a_B = \gamma_B c_B \tag{23.21}$$

where c_B *is* the molar concentration of B and γ_B is the activity coefficient for B; thus:

$$\mu = \mu^0 + RT \ln \frac{c}{c^0} \quad \Rightarrow \quad \mu = \mu^0 + RT \ln \frac{\gamma_B c_B}{c^0}. \tag{23.22}$$

For a dilute (ideal solution), $\gamma_B = 1$; for a real solution, $\gamma_B \neq 1$, in other words:

$$\gamma_B \rightarrow 1 \quad \text{as} \quad c_B \rightarrow 0$$

or

$$a_B \rightarrow c_B \quad \text{as} \quad c_B \rightarrow 0 \tag{23.23}$$

Thus, the standard state ($a_B = 1$) is a *hypothetical* state, where the concentration of B is 1 molar and B possesses the properties of a very dilute solution (the word hypothetical is used because in reality one molar solution generally will not behave ideally—as a dilute solution).

23.5 Thermodynamic Treatment of Chemical Equilibrium

Consider a one-phase system at equilibrium at constant V and T; the system may be a gas, liquid, or solid. We write again the chemical reaction [Equation (23.1)] (see [2,3]),

$$\nu_A A + \nu_B B \underset{\leftarrow}{\overset{\rightarrow}{}} \nu_C C + \nu_D D \tag{23.24}$$

which can also be presented as:

$$\nu_A A + \nu_B B - \nu_C C - \nu_D D = 0 \tag{23.25}$$

Defining a variable λ, we now induce an infinitesimal change in the number of reactants, which leads to an interconversion of molecules, while keeping their total number intact (the system is closed),

$$dN_j = \nu_j d\lambda \qquad j = A, B, C, D \tag{23.26}$$

Our interest is in the change in the Helmholtz free energy of the reaction, dA. This infinitesimal change does not affect T and V, which remain constant, but only leads to a change in the number of particles, N_j, thus:

$$dA(T,V,N) = \frac{\partial A}{\partial T} dT + \frac{\partial A}{\partial V} dV + \sum_j \frac{\partial A}{\partial N_j} dN_j$$

$$= -SdT - pdV + \sum_j \mu_j dN_j = \sum_j \mu_j dN_j = \left(\sum_j \mu_j \nu_j \right) d\lambda \tag{23.27}$$

where $\mu = \mu(V,T)$. Because we deal with a closed system in *equilibrium*, the free energy is minimal with respect to all changes in λ, that is, $(\partial A/\partial \lambda)_{T,V} = 0$, or $(\Sigma \mu_j \nu_j)d\lambda = 0$, which is a manifestation of the second law [Equation (2.70)]; since $d\lambda$ is arbitrary, $\Sigma \mu_j \nu_j = 0$. Thus, the general *thermodynamic* equation of chemical equilibrium is,

$$\Delta A = \sum_j \nu_j \mu_j = \nu_C \mu_C + \nu_D \mu_D - \nu_A \mu_A - \nu_B \mu_B = 0 \qquad (23.28)$$

To derive a similar equation for the Gibbs free energy, G, one considers a system at constant p and T; Euler's theorem [Equation (2.75)] leads to:

$$G = \mu_A N_A + \mu_B N_B - \mu_C N_C - \mu_D N_D \qquad (23.29)$$

In an infinitesimal process, where (small) $\nu_i d\lambda$ molecules are changed in the reaction, the intensive parameters, p and T, remain approximately constant and the change of G is:

$$dG = \mu_A dN_A + \mu_B dN_B - \mu_C dN_C - \mu_D dN_D = (\mu_A \nu_A + \mu_B \nu_B - \mu_C \nu_C - \mu_D \nu_D)d\lambda \qquad (23.30)$$

Again, for a closed system in equilibrium, $dG = 0$, and because $d\lambda$ is arbitrary, one obtains,

$$dG = \nu_A \mu_A + \nu_B \mu_B - \nu_C \mu_C - \nu_D \mu_D = 0 \qquad (23.31)$$

where $\mu = \mu(p, T)$ and ΔG is per mole. Equations (23.28) and (23.31) are general—not limited to ideal gas mixtures, for example. We now express these equations in term of the chemical potential,

$$\mu = \mu^0 + k_B T \ln \frac{\gamma c}{c^0} \qquad (23.32)$$

assuming the *same* activity coefficient, γ, for all species and a standard concentration c^0 ($c^0 = 1$ molar); thus,

$$\Delta A = \nu_C \mu_C^0 + \nu_D \mu_D^0 - \nu_A \mu_A^0 - \nu_B \mu_B^0 + k_B T \ln \frac{c_C^{\nu_C} c_D^{\nu_D}}{c_A^{\nu_A} c_B^{\nu_B}} \left(\frac{c^0}{\gamma} \right)^{-\Delta \nu} = 0 \quad \Delta \nu = \nu_C + \nu_D - \nu_A - \nu_B \qquad (23.33)$$

which leads to:

$$\Delta A^0 = \nu_C \mu_C^0 + \nu_D \mu_D^0 - \nu_A \mu_A^0 - \nu_B \mu_B^0 = -k_B T \ln \frac{c_C^{\nu_C} c_D^{\nu_D}}{c_A^{\nu_A} c_B^{\nu_B}} \left(\frac{c^0}{\gamma} \right)^{-\Delta \nu} = -k_B T \ln K_c. \qquad (23.34)$$

This equation recovers the law of mass action [Equation (23.2)].

Thus, if $\Delta \nu \neq 0$, both ΔA^0 and K_c depend on the chosen standard state. ΔA^0 is called the *standard free energy change of the reaction*. Equation (23.34) is derived for the canonical ensemble. $\Delta G^0(p)$ can be derived in the same way (see [2.3]).

These derivations of the law of mass action consist of pure *thermodynamic* considerations. In the next sections, we show that application of statistical mechanics enables one to obtain the equilibrium constant based on partition functions.

23.6 Chemical Equilibrium in Ideal Gas Mixtures: Statistical Mechanics

Consider a simple reaction based on three ideal gases, A, B, and C; from Equation (23.28),

$$\nu_A \mu_A + \nu_B \mu_B = \nu_C \mu_C. \qquad (23.35)$$

The partition functions of these species (denoted by $i = 1, 2, 3$) are independent, thus:

$$Q_i(N,V,T) = \frac{q_i(V,T)^{N_i}}{N_i!} \quad \text{and} \quad Q(N_1,N_2,V) = \frac{q_1(V,T)^{N_1}}{N_1!} \frac{q_2(V,T)^{N_2}}{N_2!} \quad (23.36)$$

In Equation (4.1), we obtained, $q = V/\Lambda^3$, where in the case of diatomic or polyatomic molecules, for example, q also includes contributions from internal rotations and vibrations. The chemical potential is:

$$-\frac{\mu_i}{k_B T} = \left(\frac{\partial \ln Q_i}{\partial N_i}\right)_{N,V,T} = \ln \frac{q_i(V,T)}{N_i} \quad (23.37)$$

Substituting these expressions for μ in Equation (23.35) above leads to:

$$\frac{N_C^{v_C}}{N_A^{v_A} N_B^{v_B}} = \frac{q_C^{v_C}}{q_A^{v_A} q_B^{v_B}} \quad (23.38)$$

and multiplying both sides of Equation (23.38) by $V^{v_A + v_B - v_C}$ leads to:

$$\frac{\rho_C^{v_C}}{\rho_A^{v_A} \rho_B^{v_B}} = \frac{q_C'^{v_C}(T)}{q_A'^{v_A}(T) q_B'^{v_B}(T)} = K_c(T) \quad (23.39)$$

where $q' = q/V$ is independent of V, and $\rho = N/V$ is the density (when ρ is expressed in molar, it has been denoted by c). Equation (23.39) is a statistical mechanics formulation of the law of mass action for a mixture of ideal gases. K_c depends only on T (but not on N and V) and can be calculated from the partition functions q'. We have obtained for μ [Equation (4.17)],

$$\mu(T,V) = \mu^0(T) + k_B T \ln \rho \quad (23.40)$$

meaning that $\rho = \exp[(\mu - \mu^0)/k_B T]$. Therefore, inserting ρ in Equation (23.39) and using Equation (23.35), $v_C \mu_C - v_A \mu_A - v_B \mu_B = 0$: leads to

$$\frac{\rho_C^{v_C}}{\rho_A^{v_A} \rho_B^{v_B}} = \exp -[(v_C \mu_C^0 - v_A \mu_A^0 - v_B \mu_B^0)/k_B T] = \exp -[\Delta A^0/k_B T] = K_c(T) \quad (23.41)$$

where ΔA^0 is the standard free energy change of the reaction [Equation (23.34)],

$$\Delta A^0 \equiv v_C \mu_C^0(T) - v_A \mu_A^0(T) - v_B \mu_B^0(T) \quad (23.42)$$

Notice, that, for simplicity, the derivation is based on ρ rather than ρ/ρ^0, where ρ^0 is the standard concentration, considering ρ^0 would add the factor $(\rho^0)^{-\Delta v}$ to the left hand side of Equation (23.41), where $\Delta v = v_C - v_A - v_B$ [see Equation (23.34)].

23.7 Pressure-Dependent Equilibrium Constant of Ideal Gas Mixtures

For an ideal gas, the partial pressures of species i is, $p_i = \rho_i k_B T$; substituting it in Equation (23.39) leads to:

$$\frac{p_C^{v_C}}{p_A^{v_A} p_B^{v_B}} = \frac{[q_C'(T) k_B T]^{v_C}}{[q_A'(T) k_B T]^{v_A} [q_B'(T) k_B T]^{v_B}} = (k_B T)^{(v_C - v_A - v_B)} K_c(T) = K_p(T) \quad (23.43)$$

Thus, if $v_C - v_A - v_B = 0$, $K_c(T) = K_p(T)$, where $K_p(T)$ is the equilibrium constant at constant pressure. In Equation (4.20), $\mu(p,T)$ is expressed as:

$$\mu(p,T) = -k_B T \ln[q'(T)k_B T] + k_B T \ln p = \mu^0(T) + k_B T \ln p \tag{23.44}$$

or, $p = \exp[(\mu - \mu^0)/k_B T]$, which substituted in Equation (23.43), leads to:

$$\frac{p_C^{v_C}}{p_A^{v_A} p_B^{v_B}} = \exp-[(v_C\mu_C^0 - v_A\mu_A^0 - v_B\mu_B^0)/k_B T] = \exp[-\Delta G^0 / k_B T] = K_p(T) \tag{23.45}$$

23.8 Protein-Ligand Binding

Thus far, we have discussed an ideal gas system, where the equilibrium constant can be calculated analytically [e.g., Equation (23.39)]. This is not the case in realistic systems, where K_c can only be obtained numerically. We are interested in a fundamental problem in biology—calculating the *absolute* free energy of binding of a ligand (typically a small molecule) to the active site of a protein. More specifically, the system consists of a ligand, $(L)_w$, and a protein, $(P)_w$, separated in water and their complex, $(PL)_w$, in water. In equilibrium, the reaction is:

$$(L)_w + (P)_w \rightleftharpoons (PL)_w \tag{23.46}$$

Because ΔA^0 is of central interest in structural biology and drug design, the related literature is huge including methodological developments and interesting applications (see [4–55] and references cited therein). A seminal theoretical study on this subject (within the *NpT* ensemble) is due to Gilson et al. [7]. However, since we treat this problem within the framework of the canonical ensemble, we shall rely on the theory developed later by Boresch et al. [8]. Thus, in the present binding case at equilibrium, Equation (23.28) leads to:

$$v_{PL}\mu_{PL} - v_L\mu_L - v_P\mu_P = 0, \tag{23.47}$$

where $\Delta v = 1 - 1 - 1 = -1$ [Equation (23.33)]. The chemical potential of species i is [Equation (23.32)]:

$$\mu_{i,w} = \mu_{i,w}^0 + k_B T \ln \frac{\gamma_i c_i}{c^0} \tag{23.48}$$

where $\mu_{i,w}^0$, c_i, and c^0 are the standard chemical potential, the concentration of i, and the standard concentration, respectively. Our derivation is based on the assumption of low concentration of the three species, meaning that their activity coefficients, γ_i, are close to unity ($\gamma_i \approx 1$). Thus, we use $\gamma_i = 1$, which, is justified at low c (as said above) and at atmospheric pressure, but not necessarily in the cell. For the present case, one also obtains [Equation (23.34)],

$$\Delta A^0 = \mu_{PL,w}^0 - \mu_{L,w}^0 - \mu_{P,w}^0 = -k_B T \ln K_c, \tag{23.49}$$

where ΔA^0 is the standard free energy and K_c is the equilibrium constant.

The theory presented thus far is based on thermodynamic grounds, which do not provide the standard chemical potentials, μ^0, and thus the desired ΔA^0; these values will be obtained based on statistical mechanics considerations applied to a dilute system: We treat a single protein, P, and a single ligand, L, immersed in a large "box" of water molecules of volume V, where P and L are in equilibrium with their complex, PL (Figure 23.1).

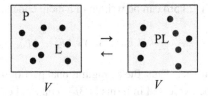

FIGURE 23.1 Assuming high dilution, one considers a protein molecule (P) and a ligand (L) in a large box with water molecules, where the separated P and L are in equilibrium with their complex, PL. This situation is emphasized by depicting two boxes rather than one, for each of the two states.

When separated, it is assumed that P and L can scan the entire volume *independently*, therefore $c(L) = c(P) = 1/V$; the standard concentration, c^0, can be expressed as $c^0 = 1/V^0$, where the standard volume, V^0 is,

$$V^0 = (1 \text{ liter}) / (\text{Avogadro number}) = 10^{27} \text{Å}^3 / (6.02 \times 10^{23}) = 1660 \text{ Å}^3 \tag{23.50}$$

The chemical potential is the difference between the Helmholtz free energies of systems differing by a single particle [Equation (4.15)]; thus, denoting L, P, and PL by $i = 1, 2, 3$, one obtains,

$$\mu_{i,w} = A_{i,N} - A_{0,N} = -k_B T \ln \frac{Q_{i,N}}{Q_{0,N}} \tag{23.51}$$

where i, N denotes molecule i surrounded by N water molecules; 0, N is a box containing N waters only, and $Q_{0,N}$ is the partition function of pure water. Equations (23.48–23.51) lead to:

$$\mu_{i,w}^0 = \mu_{i,w} - k_B T \ln \frac{c_i}{c^0} = -k_B T \ln \frac{V^0 Q_{i,N}}{V Q_{0,N}}, \tag{23.52}$$

and inserting Equation (23.52) in Equation (23.49) leads to the absolute free energy of binding, ΔA^0:

$$\Delta A^0 = \mu_{PL,w}^0 - \mu_{L,w}^0 - \mu_{P,w}^0 = -k_B T \ln \frac{V Q_{PL,N} Q_{0,N}}{V^0 Q_{P,N} Q_{L,N}} \tag{23.53}$$

Equation (23.53) can further be simplified as Q is a product of two independent terms—the configurational part (Z) and the velocities part. The velocities part is the product $\Pi_i (\Lambda_i)^{-3}$, where Λ_i is the de Broglie wavelength of atom i. Because the partial expression of Equation (23.53),

$$\frac{Q_{PL,N} Q_{0,N}}{Q_{P,N} Q_{L,N}} \tag{23.54}$$

has the same number of atoms i (protein, ligand, and water) in the numerator and denominator, the Λ_i products are cancelled and ΔA^0 becomes dependent of the configurational partition functions alone,

$$\Delta A^0 = -k_B T \ln \frac{V Z_{PL,N} Z_{0,N}}{V^0 Z_{P,N} Z_{L,N}} \tag{23.55}$$

It proves convenient to multiply and divide the term under the logarithm by $Z_{L,0}$—the partition function of the ligand in vacuum; one obtains:

$$\Delta A^0 = -k_B T \ln \frac{V}{V^0} + [F'_{PL,N} - F'_{P,N} - F'_{L,0}] - [F'_{L,N} - F'_{0,N} - F'_{L,0}] \tag{23.56}$$

where $F_i' = -k_B T \ln Z_i$. Equation (23.56) can be written in a more concise way as:

$$\Delta A^0 = -k_B T \ln(V / V^0) + \Delta F_P' - \Delta F_{sol}' \tag{23.57}$$

However, one can make a step further. Thus, the configurational partition function, $Z(x_1,...,x_{3K})$ of a molecule of K atoms in water can be expressed in terms of $3K-6$ internal coordinates $(x_1',...,x_{3K-6}')$ and six external coordinates. The external coordinates are the three Cartesians (X, Y, Z) of an arbitrary chosen atom (center of mass, etc.) and three orientation coordinates (e.g., Euler angles ξ_1, ξ_2, ξ_3). Integration over the external coordinates leads to:

$$
\begin{aligned}
Z &= \int dx_1 ... dx_{3K} \exp[-W(x_1...x_{3K}) / k_B T] \\
&= \int dX dY dZ d\xi_1 d\xi_2 d\xi_3 dx_1' ... dx_{3K-6}' |J| \exp[-W(x_1'...x_{3K-6}')/k_B T] \\
&= 8\pi^2 V \int dx_1' ... dx_{3K-6}' |J| \exp[-W(x_1'...x_{3K-6}') / k_B T] \\
&= 8\pi^2 V \bar{Z}(x_1',...x_{3K-6}')
\end{aligned}
\tag{23.58}
$$

W is the potential of mean force resulting from the (formal) integration over the solvent degrees of freedom for each set of $3K-6$ coordinates of the molecule. $|J|$ is the Jacobian of the transformation from Cartesian to internal coordinates. One obtains:

$$\Delta A^0 = -k_B T \ln \frac{V Z_{PL,N} Z_{0,N}}{V^0 Z_{P,N} Z_{L,N}} = -k_B T \ln \frac{\bar{Z}_{PL,N} Z_{0,N}}{8\pi^2 V^0 \bar{Z}_{P,N} \bar{Z}_{L,N}} \tag{23.59}$$

where the bar over Z means that Z depends only on internal coordinates. Equation (23.59) shows that ΔA^0 depends on V^0, but not explicitly on V. One obtains,

$$\Delta A^0 = k_B T \ln(8\pi^2 V^0) + [F_{PL,N} - F_{P,N}] - [F_{L,N} - F_{0,N}] \tag{23.60}$$

where $F_i = -k_B T \ln \bar{Z}_i$. Equation (23.60) can be written in a more concise way as:

$$\Delta A^0 = k_B T \ln(8\pi^2 V^0) + \Delta F_P - \Delta F_{sol}. \tag{23.61}$$

ΔF_P and ΔF_{sol} and $\Delta F_P'$ and $\Delta F_{sol}'$ are free energy differences defined for the protein and the solvent environments, respectively. The absolute Gibbs free energy $\Delta G^0 \sim \Delta A^0$ since $\Delta G^0 = \Delta A^0 + P^0 \Delta \bar{V}_{PL}$, where $P^0 \Delta \bar{V}_{PL}$ is small and can be neglected [7,8].

23.8.1 Standard Methods for Calculating ΔA^0

As has been stated earlier, a great deal of work has been devoted in the last 30 years to the issue of absolute free energy of binding, with respect to methodology enhancements and testing against the experiment (e.g., [4–55] and references cited therein). The methods used can be divided into two categories—methods consisting of detailed modeling and rigorous statistical mechanics and those that do not meet strictly this condition (e.g., [40–52]). Thus, while in most studies the systems are modeled in detail by the conventional force fields, long-range electrostatics has been treated on various levels of approximation. The standard procedure, which is generally considered the most satisfactory, is immersing the protein in a large container with explicit solvent and applying periodic boundary conditions with particle mesh Ewald [56], which, however, becomes a time consuming process for a large system. To save computer time, more approximate models have been developed, where only part of the protein close to the active site is considered in detail and is thus covered by a relatively small sphere of explicit water; the effect of the more distant parts is treated by a "mean field" where long-range electrostatics (reaction field) is taken into account. Such modeling has been developed, for example, by Warshel's group (implemented within the program MOLARIS [57–59]) and by Roux's group who developed the spherical solvent boundary potential [30] and the generalized solvent boundary potential [32]. MOLARIS is a highly functional simulation program standing by itself, while spherical solvent boundary potential/generalized solvent boundary potential has been incorporated within the program CHARMM (see Section 16.1).

We shall concentrate here on the most rigorous techniques and modeling. Typically, these techniques consist of thermodynamic cycles, where the interactions between the ligand and its environment are decreased to zero (annihilated) in both the active site and the bulk solution. While this annihilation is carried out with methods that have already been introduced (e.g., TI, FEP, and Jarzynski's identity), their implementation is not trivial and should be discussed. Excellent reviews on the absolute free energy of binding are available [9,10,12–14,24,25]. The theory presented here is based on the systematic work of Gilson et al. [7] and the following study of Boresch et al. [8], outlined in the previous sections.

From the statistical mechanics point of view, a *rigorous* approach for calculating ΔA^0 is based on Equations (23.56) and (23.57), where TI is usually applied to evaluate independently $\Delta F_{\mathrm{P}}'$ and $\Delta F_{\mathrm{sol}}'$. The ligand in the complex is treated by slowly annihilating its interactions with the environment (protein and water), while the intra-ligand interactions are kept intact. At the end of the process, the system will consist of a protein in water (P,N) and a ligand (L,0), moving freely within the volume, V, without "feeling" the protein and water; thus, this TI process leads exactly to $\Delta F_{\mathrm{P}}'$ [Equations (23.56) and (23.57)]. As explained in Section 17.5, the interactions are eliminated with the help of a parameter λ, $0 \leq \lambda \leq 1$, which multiplies only the ligand-environment interactions of the force field. Thus, for $\lambda = 1$, these interactions are intact and they decrease to zero as λ gradually vanishes. The adequate TI (commonly based on MD simulations) is,

$$\Delta F^{\mathrm{TI}} = \int_{\lambda=1}^{\lambda=0} \frac{\partial E(\lambda)}{\partial \lambda} d\lambda. \tag{23.62}$$

A similar process is applied to the ligand in water, which is separated from the protein. Thus, the water-ligand interactions are gradually eliminated by decreasing λ, and the final system consists of a box of pure water (N,0) and a ligand (L,0) moving freely within the entire volume, V; this integration leads to $\Delta F_{\mathrm{sol}}'$ [Equations (23.56) and (23.57)].

This process, which leads to ΔA^0 [Equations (23.56) and (23.57)], can be described within the framework of the thermodynamic cycle (Section 17.5) depicted in Figure 23.2. Thus, defining $F_{\mathrm{A}} \to F_{\mathrm{B}} \equiv F_{\mathrm{B}} - F_{\mathrm{A}} = \Delta F^{\mathrm{TI}}$ or F^{TI}, the sum of the free energy components accumulated in a clockwise manner is zero, which leads to ΔA^0 up to an additive constant:

$$-\Delta F_{\mathrm{P}}' + 0 + \Delta F_{\mathrm{sol}}' + \Delta A^0 = 0 \quad \to \quad \Delta A^0 = \Delta F_{\mathrm{P}}' - \Delta F_{\mathrm{sol}}' \tag{23.63}$$

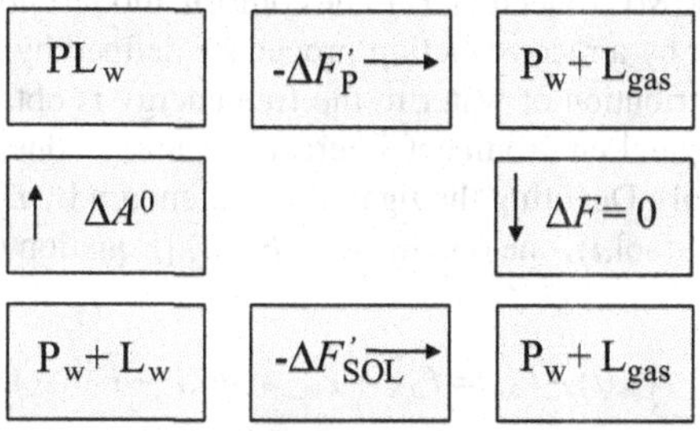

FIGURE 23.2 A diagram illustrating the double annihilation method (DAM). According to our definition $F_{\mathrm{A}} \to F_{\mathrm{B}}$ means $F_{\mathrm{B}} - F_{\mathrm{A}}$. We are interested in ΔA^0, the (standard) free energy for transferring the ligand from water to the active site of the protein. This is achieved from the clockwise thermodynamic cycle: $-\Delta F_{\mathrm{P}}' + 0 + \Delta F_{\mathrm{sol}}' + \Delta A^0 = 0$ or $\Delta A^0 = \Delta F_{\mathrm{P}}' - \Delta F_{\mathrm{sol}}'$ [up to the additive term $-k_{\mathrm{B}}T\ln(V/V^0)$, see Equation (23.57)]. (For the calculation of these free energy components, see text).

This (rigorous) approach is called the double annihilation method (DAM) [4–8,22,26,27]. It can be performed quite reliably in the solvent (water) environment for calculating $\Delta F'_{\text{sol}}$, as long as the total charge of the ligand is zero. On the other hand, it has been argued that calculating $\Delta F'_{\text{p}}$ in the protein environment is not straightforward. Thus, in the final stages of TI, where the ligand-protein interactions are weak (small λ), the ligand might leave the active site, starting to "wander" within the inhomogeneous volume (due to the protein), which would make it difficult to obtain converged results; this effect is sometimes called "the end-point problem" [8]. However, DAM has been studied extensively by Pande's group [22,26] and Fujitani et al. [27] as applied to a set of ligands bound to the protein FKBP12, where a reasonable agreement with the experiment has been achieved.

The end-point problem has been rigorously solved by adding harmonic restraints, which hold the ligand in the active site during the gradual elimination of the ligand-environment interactions. The effect (bias) of these artificial restraints (applied by a suitable TI procedure) on $\Delta F'_{\text{p}}$ is removed later by releasing them analytically or computationally. Because of the additional integration steps involved in this procedure, it is called the "double decoupling method" (DDM) [7,8]. DDM has been developed systematically during the years [4,11,29,31,33,34,55], where various implementation issues have been improved, in particular, by Roux's group (e.g., optimizing the force constants of the harmonic restraints), and a large number of complexes have been successfully studied (see the review [11]). Historically, translational and angular restraining potentials were introduced rigorously first by Hermans' group [4,29], and these ideas were extended later to more complex systems by Roux's group and others [11]. Still, treating the various restraints requires extra work and might need some expertise [13].

With another approach, first suggested by Jorgensen [54], ΔA^0 is obtained by calculating the potential of mean force of a ligand, which is initially placed far from the protein and is gradually moved to the active site. This approach has been developed further by Woo and Roux by establishing a rigorous procedure based on restraints and calculation of the potential of mean force by umbrella sampling [60]. Woo and Roux have demonstrated the advantage of this approach over DAM and DDM for handling a highly charged ligand.

23.8.2 Calculating ΔA^0 by HSMD-TI

Another approach for calculating ΔA^0 is based on HSMD-TI described earlier in Sections 19.6 and 19.7 for a peptide in two different microstates; here, instead, the peptide (ligand) is studied in solvent and in the protein environment. HSMD-TI was introduced first for calculating competing structures of mobile loops in proteins [60–62] and, subsequently, for calculating the absolute free energy of binding of ligands to proteins [63–66]. The implementation of the method starts (in accordance with Equations (23.58) and (23.59)) by defining three atoms and three related Euler angles for the ligand in solvent and for the bound protein; these coordinates are kept constant during the MD simulations. Then, each system (that is, the ligand-solvent and the complex) is equilibrated and two production MD trajectories are carried out. Subsequently, a three-stage HSMD-TI process is applied to both trajectories.

For the ligand-solvent, the stages are: (1) a small set of n ligand-water conformations (frames), $i = 1,\ldots,n$ are extracted from the MD trajectory. (2) The conformational entropy of the ligand, $S^A_{\text{ligand}}(i,\text{sol})$ is calculated for each frame, i, by a reconstruction procedure defined by HSMD (see Section 19.6). (3) Using a parameter, λ, the contribution of water to the free energy is obtained by TI, where for a *fixed* ligand conformation, i, the ligand-environment interactions are gradually *decreased* from their full value to zero, leading to $F_i^{\text{TI}}(\text{sol})$. Denoting the ligand's intra-energy by $E_{\text{intra-ligand}}$ and its free energy by $F_{\text{ligand}}(\text{sol},i) = E_{\text{intra-ligand}} - TS^A_{\text{ligand}}(\text{sol},i)$, one obtains for $\Delta F_{\text{sol}}(i)$ [Equation (23.60)]:

$$\Delta F_{\text{sol}}(i) = F_{L,N}(i) - F_{0,N} = F_{0,N} + F_{\text{ligand}}(\text{sol},i) - F^{\text{TI}}(\text{sol},i) - F_{0,N}$$

$$= -F^{\text{TI}}(\text{sol},i) + F_{\text{ligand}}(\text{sol},i) \qquad (23.64)$$

Thus, HSMD-TI provides the *absolute* $F_{L,N}$ consisting of three components: the free energy of the ligand, $F_{\text{ligand}}(\text{sol}, i)$, the free energy due to the ligand-water interaction, $F^{\text{TI}}(\text{sol}, i)$, and the free energy of bulk water, $F_{0,N}$, which luckily is cancelled in $\Delta F_{sol}(i)$ [Equation (23.64)]. In these calculations, the ligand's structure is held fixed, thus contributing to stability. On the other hand, with DAM, $\mu_{i,w} = A_{i,N} - A_{0,N} = -k_B T \ln \frac{Q_{i,N}}{Q_{0,N}}$ is obtained from a *single* λ-based TI, where both the ligand and water are moved during simulation, and the entropy is not provided.

In the protein environment, the purpose is to calculate the absolute value of $F_{PL,N}$, which is obtained: (1) by reconstructing the ligand's entropy in the active site—$S_{ligand}^A(i,p)$, (2) by eliminating the ligand-water and ligand-protein interactions with TI (using λ), which leads to $F_i^{TI}(p)$. (3) Calculation of $F_{P,N}$, is unnecessary since it gets cancelled in ΔF_P. However, the reconstruction of the ligand's frames, i, is more complicated than in solvent because the ligand is not bonded to the protein and the position and orientation of its three reference (external) atoms can change within the active site; however, their contribution to the entropy cannot be calculated analytically (as for the ligand in solvent). Therefore, this "external entropy," $S_{external}$ should be calculated numerically, in addition to the conformational entropy of the ligand, $S_{ligand}^A(i,p)$.

Calculation of the external entropy [64,66] is done by reconstructing the transition probability densities (TPs) of the three reference atoms (defined for the ligand in solvent) with positions $x_1(i)$, $x_2(i)$, and $x_3(i)$ in the laboratory frame. For that, one also determines three fixed coordinates in the laboratory frame, y_1, y_2, and y_3, where the relative position of x_1 is defined by a "dihedral angle," α_1, a "bond" angle, α_2, and the distance $r = |x_1 - y_3|$ (Figure 2, [28]). The first transition probability, TP(1), is obtained by defining three bins around these variables (at i), where their product is denoted by $V_{cube}(i)$ and the Jacobian is $J = r^2 \sin(\alpha_2)$. Then, n_f future conformations of the *entire* ligand are generated, and one calculates the number of times, n_{visit}, the three bins have been visited *simultaneously*. The contribution of atom 1 and frame i to $S_{external}$ (1) is $-k_B \ln[TP_i(1)]$, where $TP_i(1)$ is based on Equation (19.43):

$$TP_i(1) = n_{visit} / [n_f V_{cube}(i) J] \tag{23.65}$$

$S_{external}$ (1) is obtained by averaging $-k_B \ln[TP_i(1)]$ over the sample of n frames i. After atom 1 has been reconstructed, it becomes fixed at its position in conformation i and the contribution of atom 2 is calculated in the same manner. However, TP(2) consists only of a dihedral and a bond angle (based on y_2, y_3, x_1, and x_2, where $|x_1 - x_2|$ is assumed to be constant). TP(3) is based on a single dihedral angle; thus, one obtains the three averages (over n frames), $S_{external}(j) = -k_B TP_i(j)$ $j = 1, 2,$ and 3, where:

$$S_{external} = S_{external}(1) + S_{external}(2) + S_{external}(3). \tag{23.66}$$

$S_{external}$ constitutes a measure of the global movement (that is, the conformational freedom) of the bound ligand in the active site, providing, thereby, some estimation for the size of this site. Summing up all the free energy-entropy components calculated for snapshot i in the protein environment leads to ΔF_p (i) [Equations (23.60) and (23.61)],

$$\Delta F_p(i) = F_{PL,N}(i) - F_{P,N}(i) = [F_{P,N} + F_{ligand}(p,i) - F^{TI}(p,i) - TS_{external}(i)] - F_{P,N}. \tag{23.67}$$

$$= -F^{TI}(p,i) + F_{ligand}(p,i) - TS_{external}(i)$$

Both Equations (23.64) and (23.67) are defined for snapshot i. However, as discussed in Section 19.2, if the entropy, $S(i)$, obtained by reconstructing frame i is correct, the free energy of frame i, $F(i) = E(i) - TS(i)$, is equal to the correct F (that is, F is a property with zero fluctuation). Therefore, for a very long integration, large n_f and small enough bins $\Delta F_{sol}(i)$ and $\Delta F_p(i)$ would lead to the corresponding correct free energies [see Equations (19.4)–(19.9)]; this applies to any i within the sub-spaces of interest. However, in practice, these free energies are always approximate, and their accuracy improves as the above three conditions are better satisfied. Therefore, ΔF_{sol} and ΔF_p are not obtained from a single frame, but by averaging $\Delta F_{sol}(i)$ and $\Delta F_p(i)$ over a small sample of frames of size n, where n can be decreased with

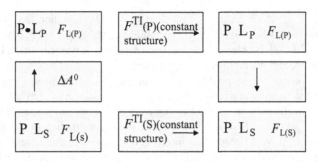

FIGURE 23.3 Illustration of the thermodynamic cycle for HSMD-TI. We start at the upper left corner with the protein-ligand complex, where the intra-energy and the HSMD entropy of the structurally *fixed* ligand are calculated leading, respectively, to $F_{L(P)} = E_{L(P)} - TS_{L(P)}$. Then, the ligand-environment interactions are gradually reduced by TI, leading to the contribution, $F^{TI}_{(P)}$; as a result, the ligand in the active site does not feel the protein and water around. The same is carried out on the solvent side, leading to $F_{L(S)}$ and $F^{TI}_{(S)}$. The contributions to the clockwise cycle are thus, $F^{TI}_{(P)} - F_{L(P)} + F_{L(S)} - F^{TI}_{(S)} + \Delta A^0 = 0$ or $\Delta A^0 = F^{TI}_{(S)} - F^{TI}_{(P)} + F_{L(P)} - F_{L(S)}$ or shorter, $\Delta A^0 = \Delta F^{TI} - \Delta F_L$, where the external entropy is included in $F_{L(P)}$. This equation is identical to Equation (23.68) up to an additive constant.

increasing n_f, the integration time, etc.; in practice, $5 \leq n \leq 20$. The total absolute free energy of binding, ΔA^0, is based on the *averaged* parameters (over n):

$$\Delta A^0 = \Delta F_p - \Delta F_{sol} + k_B T \ln(8\pi^2 V^0)$$

$$= [E_{lintra\text{-}ligand}(p) - TS_{ligand}(p) - TS_{external} - F^{TI}(p)]$$

$$- [E_{lintra\text{-}ligand}(sol) - TS_{ligand}(sol) - F^{TI}(sol)] + k_B T \ln(8\pi^2 V^0) \tag{23.68}$$

$$= \Delta E_{intra\text{-}ligand} - T\Delta S_{ligand} - TS_{external} - \Delta F^{TI} + k_B T \ln(8\pi^2 V^0).$$

This equation can be described by the thermodynamic cycle of Figure 23.3.

It should also be pointed out that because with HSMD, the whole future is scanned, the reconstruction procedure can, in principle, be started from any atom, that is, from the "first" atom of a chain-like ligand, from the "last" one (in an opposite direction), or from any middle atom, where the order in which the two branches are reconstructed is arbitrary. In practice, a large enough n_f is expected to lead to close results for $S^A_{ligand} + S_{external}$ for different reconstructing points. However, $S_{external}$ based on a suitable set of atoms might increase the information about the size and shape of the active site.

Finally, we emphasize again that in both environments, the ligand's structure is kept *fixed* during TI. This is of a technical importance, in particular, on the protein side since the end-point problem (encountered with DAM) is eliminated and the need to apply restraints (as with DDM) is avoided! Also, unlike other methods, HSMD-TI provides the decrease in the ligand's entropy as it is transferred from the solvent environment to the active site—an information of academic interest, which is expected to be helpful in rational drug design as well.

23.8.3 HSMD-TI Applied to the FKBP12-FK506 Complex: Equilibration

To demonstrate the performance of HSMD-TI, we summarize here results obtained [66] for the relatively large ligand FK506 (126 atoms) complexed with the protein FKBP12—where $\Delta A^0 = -12.8$ kcal/mol is known experimentally, as well as the crystal structure of the complex. The calculations were made with the AMBER 10 package (see Section 16.3), where the protein was modeled by the AMBER99 force field [67] and the ligand by GAFF [68] (with the AM1-BCC partial charges [69,70]). The system was immersed in a large container of TIP3P water [71] (see Section 16.5) and long-range electrostatics were taken into account by particle mesh Ewald (Section 10.9.1), with a cut-off of 10 Å [56]. The unit cell of the periodic system (based on Protein Data Bank (PDB) code 1fkf, 1991 atoms [72]) was defined by

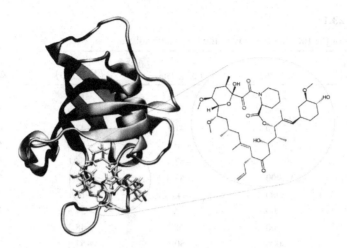

FIGURE 23.4 3 dimensional representation of the FKBP12-FK506 complex, with a 2 dimensional image of the ligand.

constructing a truncated octahedron around the protein, which was filled with 3592 water molecules. The minimum distance between the protein and the walls of the cell was 8 Å (This cell size is comparable to that used in previous studies of FK506; also, to be compatible with these studies, the three histidine residues have a positive charge.). To neutralize the system, 19 randomly selected water molecules were replaced by 10 chloride and 9 sodium ions (Figure 23.4).

The system was equilibrated by a standard procedure based on a set of energy minimizations and MD simulations, with a time step of 2 femtoseconds (fs); the SHAKE algorithm [73] (Section 10.6) was applied to all hydrogens. Finally, a 100 nanosecond (ns) constant volume MD production run at 300K [time step of 2 picoseconds (ps)] was carried out, where the first 0.2 ns trajectory was used for an initial equilibration. Three intervals along the trajectory were defined ([0.2–2], [2–5], and [5–100] ns (from now on, the word "ns" will be omitted in most cases)). For each of these three intervals, a set of up to $n = 10$ equally separated frames were extracted and used for the HSMD-TI analysis. For each of the extracted frames, and for the rest of the analysis, three successive atoms of the protein were fixed at their positions in structure i, to remove the center-of-mass translation and rotation. A similar procedure was applied to the ligand, FK506, in the solvent system (without the protein), which was solvated by 754 water molecules in a truncated octahedron. No counter-ions were added to this neutral system, and $n = 10$ frames were extracted for its analysis.

23.8.4 The Internal and External Entropies

To make a distinction between the *external* entropy and $\bar{S}^{A}_{\text{ligand}}$, we call the latter the *internal* entropy. The reconstruction of the internal entropy (in both the solvent and the protein) is applied to $n = 10$ FK506 frames (denoted i), each consisting of 57 atoms ($k' = 1,...,57$) and 114 angles α_k ($1 \leq k \leq K = 114$). A reconstruction step (out of 57) starts from conformation i with a 240 ps production run, where a future FK506 conformation is stored for a later analysis every 20 fs, thus, each step consists of $n_f = 12,000$ future conformations, where the first 200 are dropped for equilibration. The number of counts, n_{visit}, for each pair of bins is calculated leading to $\text{TP}_{k'}$ [Equation (19.43)], where the product of the 57 TPs is the distribution, ρ^{HS} [Equation (19.44)], which leads to an upper bound, S^{A}_{ligand} [Equation (19.45)] for the internal entropy. An estimation, $\bar{S}^{A}_{\text{ligand}}$, based on averaging the contribution to the entropy of the n frames is:

$$\bar{S}^{A}_{\text{ligand}} = -(k_B / n) \sum_{t=1}^{n} \ln \rho^{\text{HS}}(t) \tag{23.69}$$

TABLE 23.1

Results for the Internal Entropy of the Ligand in Solvent

Bin Size, $\delta = \Delta\alpha_k/l$	n_f	$TS_{\text{ligand}}^{A}(\text{sol})$	$TS_{\text{ligand}}^{A}(\text{p})$	$T\Delta S_{\text{ligand}}^{A}$
$\Delta\alpha_k/90$	2000	−4.4	−11.4	7.0
	4000	−27.9	−35.0	7.1
	6000	−41.1	−47.7	6.6
	8000	−50.1	−57.0	6.9
	10,000	−57.4	−64.1	6.7
	12,000	−63.0	−70.1	7.1
$\Delta\alpha_k/120$	2000	−4.8	−11.1	6.3
	4000	−27.5	−34.8	7.3
	6000	−41.2	−48.3	7.1
	8000	−50.3	−57.5	7.2
	10,000	−57.5	−64.6	7.1
	12,000	−63.1	−70.2	7.1
$\Delta\alpha_k/150$	2000	−4.8	−11.5	6.7
	4000	−27.9	−34.8	6.9
	6000	−41.4	−48.3	6.9
	8000	−50.5	−57.5	7.0
	10,000	−57.8	−64.7	6.9
	12,000	−63.4	−70.5	7.1
Converged				7.1 ± 1.2

Source: Reprinted with permission from General, I.J. and Meirovitch, H., *J. Chem. TheoryComput.*, 9, 4609–4619, 2013. Copyright 2019 American Chemical Society [66].

Notes: $S_{\text{ligand}}^{A}(\text{sol})$ and $S_{\text{ligand}}^{A}(\text{p})$ are the internal entropy of the ligand in solvent and in the active site of the protein: their difference is $\Delta S_{\text{ligand}}^{A} = S_{\text{ligand}}^{A}(\text{sol}) - S_{\text{ligand}}^{A}(\text{p})$; the complex is locally equilibrated in the [0.2–2] region. The results are based on reconstructing $n = 10$ structures selected homogeneously from MD samples of 1.8 ns for the solvent and the protein environments. The results are calculated as functions of $\delta = \Delta\alpha_k/l$ and n_f – the bin size and the sample size of the future chains, respectively [see Equation (19.43)]. S_{ligand}^{A} is defined up to an additive constant that is expected to be the same for both environments. The (best) results for $n_f = 12,000$ are bold-faced. The statistical errors for $TS_{\text{ligand}}^{A}(\text{sol})$ and $TS_{\text{ligand}}^{A}(\text{p})$ are ±1.6 and ±1.2 kcal/mol, respectively. Using $n = 5$ for all regions has led to an increase of the errors to: ±2.2 for $TS_{\text{ligand}}^{A}(\text{p})$ and ±1.5 kcal/mol for $T\Delta S_{\text{ligand}}^{A}$ (see result in Table 23.3).

The results for TS_{ligand}^{A} ([0.2–2] region) appear in Table 23.1 as functions of n_f and the bin size $\delta = \Delta\alpha_k/l$; $\Delta\alpha_k = \alpha_k(\text{max}) - \alpha_k(\text{min})$, where $\alpha_k(\text{max})$ and $\alpha_k(\text{min})$ are the maximum and minimum values of α_k found in each sample. As expected by theory, the results for TS_{ligand}^{A} decrease systematically as the approximation improves, that is, with increasing n_f, however, these results have not converged even for $n_f = 12,000$. On the other hand, for each bin size, $\delta = \Delta\alpha_k/l$, studied ($l = 90$–150), the differences, $T\Delta S_{\text{ligand}}^{A}(n_f)$, show convergence to the same value. Thus, we consider $T\Delta S_{\text{ligand}}(l = 90) = 7.1 \pm 1.2$ kcal/mol to be the exact result within the error bars. This is a relatively large difference between the internal conformational entropies of the two environments, which also reflects the relatively large size of FK506; one should compare this difference to the much smaller value $\Delta S_{\text{ligand}} = 1.7 \pm 0.2$ kcal/mol obtained for the smaller ligand, L8 [64]. The results in the other regions for TS_{ligand}^{A} and $T\Delta S_{\text{ligand}}^{A}$ behave in a similar way to those in the [0.2–2], where the converged values – $T\Delta S_{\text{ligand}}$ are comparable (see Table 23.3). The stable results obtained for $T\Delta S_{\text{ligand}}$ demonstrate that the errors in both $S_{\text{ligand}}^{A}(\text{p})$ and $S_{\text{ligand}}^{A}(\text{sol})$ for the same parameters are approximately equal.

To check the effect of the sample size, n, on the internal entropy, calculations with $n = 5$ were also carried out in the protein environment. It was found that in the [0.2–2] region, both samples ($n = 5$ and

10) lead to the same value of ΔA^0 within the error bars (see Table 23.3). Therefore, $n = 5$ was also used in the [2–5] and [5–100] regions. Notice that the results in the solvent environment are always based on $n = 10$ since they have been found to be stable along a long trajectory. All these entropy calculations are summarized in Table 23.3 and will be discussed later.

For the external entropy, we discuss below results for $TS_{external}(i)$ (for the [0.2–2] region) as functions of three increasing n_f values (2000, 4000, and 12,000), while the bin size, $\delta = \Delta\alpha_k/90$, is constant (as similar results were obtained for the other bin sizes [Equation (23.66)]). As expected, for each of the three atoms, the entropy decreases as the approximation improves; thus, the result, $TS_{external}(1) = 2.3$ kcal/mol is actually constant for all n_f values, while $TS_{external}(2) = -1.0$ kcal/mol is converged only from $n_f = 8000$. On the other hand, the results for $TS_{external}(3)$ are not converged, where $TS_{external}(3) = -0.5$ kcal/mol (for $n_f = 12,000$) is thus an upper bound, making the sum of these three entropies, $TS_{external} = 0.8$ kcal/mol an upper bound as well [Equation (23.66)]. However, since the decrease of the results for $TS_{external}(3)$ is slight, one can extrapolate the converged value of $TS_{external}(3)$ to be -0.8, that is, lower by 0.3 kcal/mol from the lowest simulated value, leading thereby to $TS_{external} = 0.5$ kcal/mol (rather than 0.8).

23.8.5 TI Results for FKBP12-FK506

To each of the $n = 10$ frames of the solvent sample, a TI procedure was applied where the ligand-water interactions were turned off gradually for a *fixed* FK506 structure; for each of the 10 frames in the protein environment, both the ligand-solvent and ligand-protein interactions were decoupled. Using a parameter λ, $0 \le \lambda \le 1$, the electrostatic interactions were decoupled first (by decreasing the ligand's charge to zero), followed by decoupling the Lennard-Jones (LJ) potentials (in the presence of zero electrostatic interactions). In all, 30 λ values (windows) were used, 13 for the electrostatic interactions and 17 for LJ. In the LJ integration, a soft-core potential [74] was used with $\delta = 3$ Å^2 (see Equation (17.24)]. For a given frame (i), each integration step (window) always started from the (initial) structure of i using the corresponding step potential energy [$E(\lambda)$], followed by a 2 ns production run, where the initial 20 ps were discarded for equilibrating the system with respect to λ. This has led to a relatively large production run for calculating F^{TI} (LJ).

Results for the differences (protein minus solvent), ΔF^{TI}(ch), ΔF^{TI}(LJ), and their sum, ΔF^{TI} appear in Table 23.2, where "ch" stands for charge. The table reveals that the results for $n = 5$ and 10 (for the [0.2–2] region) are very close with a maximal deviation of 0.7 kcal/mol for ΔF^{TI}; the deviations between the results of [0.2–2] ($n = 10$) and [2–5] are 1.9, 0.1, and 2 kcal/mol for ΔF^{TI} (ch), ΔF^{TI}(LJ), and ΔF^{TI}, and they further increase to 2.8, 2, and 4.8 kcal/mol for [5–100], respectively. The significance of these deviations will be discussed in the next section.

23.8.6 ΔA^0 Results for FKBP12-FK506

Table 23.3 presents the various energetic and free energy components whose sum leads to ΔA^0 [Equation (23.68)]. We have already pointed out that the two sets of results in Table 23.2 for the [0.2–2] region are slightly different with a lower $\Delta F^{TI}(n = 5)$ value. However, this value is compensated by the higher values for $T\Delta S_{ligand}(n = 5)$ and $-TS_{external}(n = 5)$, and thus the results for ΔA^0 for $n = 5$ and $n = 10$ become equal within the error bars, which justifies the use of $n = 5$ for most of the calculations in the protein environment. The value $\Delta A^0(n = 10) = -13.6 \pm 1.1$ is equal within the error bars to the experimental value -12.8 kcal/mol [75]. A somewhat more complex compensation occurs also for the [2–5] and [5–100] regions, which leads actually to the same result, $\Delta A^0 = -16.7$ kcal/mol, – a value which, however, is lower by 3.9 kcal/mol from the experimental result.

In Section 23.8.2, we have argued that for a given subspace, ΔF_{sol} and ΔF_p can, in principle, be obtained from any *single* configuration i, where the required *extensive* reconstruction and integration are carried out only over this space (for less extensive simulations, averaging over small samples should be performed.) Thus, the fact that the [2–5] and [5–100] lead to the same ΔA^0 suggests that these regions belong to the same structural subspace (microstates) of the protein-ligand complex. On the other hand, the significantly different ΔA^0 values obtained for [0.2–2] and [2–100] suggest that these regions belong to different conformational subspaces. Theoretically, the complex is expected to better equilibrate in [5–100] than in [0.2–2], and thus to lead to the more reliable results for ΔA^0; the fact that this has not been

TABLE 23.2

Results Obtained by Thermodynamic Integration (TI)

Region	ΔF^{TI} (ch)	ΔF^{TI} (LJ)	ΔF^{TI}
[0.2–2] ns ($n = 10$)	−4.9	−17.8	−22.7
[0.2–2] ns ($n = 5$)	−5.2	−18.2	−23.4
[2–5] ns ($n = 5$)	−6.8	−17.9	−24.7
[5–100] ns ($n = 5$)	−7.7	−19.8	−27.5

Source: Reprinted with permission from General, I.J. and Meirovitch, H., *J. Chem. Theory Comput.*, 9, 4609–4619, 2013. Copyright 2019 American Chemical Society.

Notes: The table presents the free energy components (in kcal/mol) obtained by TI for different regions within the [0.2–100] ns trajectory.[0.2–2], [2–5], and [5–100] ns stand for regions along the 100 ns MD trajectory (with no restraints). The results in the solvent environment are based on a sample of $n = 10$ structures, while in most cases, $n = 5$ is used for the complex; only for the [0.2–2] region the complex results for $n = 10$ are also provided. Δ is the difference between the complex and solvent results. ΔF^{TI} (ch) and ΔF^{TI} (LJ) were obtained by thermodynamic integration for the electrostatic (ch) and Lennard-Jones interactions, respectively, ΔF^{TI} is their sum. $T = 300$K is the absolute temperature. The statistical errors for [0.2–2] and $n = 10$ are ±1.0 and ±2.2.

TABLE 23.3

Results (in kcal/mol) for the Various Components Leading to ΔA^0 – the Absolute Free Energy of Binding

Region	$\Delta E_{\text{intra-ligand}}$	ΔF^{TI}	$-T\Delta S_{\text{ligand}}$	$-TS_{\text{external}}$	$k_B T\ln(8\pi^2 V^0)$	Total ≡ ΔA^0	ΔA^0 (exp.)
[0.2–2] ns ($n = 10$)	−4.5	−22.7	7.1	−0.5	7	−13.6	−12.8
[0.2–2] ns ($n = 5$)	−4.5	−23.4	7.4	−0.3	7	−13.8	−12.8
[2–5] ns	−3.4	−24.7	5.3	−0.8	7	−16.6	−12.8
[5–100] ns	−4.0	−27.5	8.1	−0.3	7	−16.7	−12.8

Source: Reprinted with permission from General, I.J. and Meirovitch, H., *J. Chem. Theory Comput.*, 9, 4609–4619, 2013. Copyright 2019 American Chemical Society.

Notes: Results in the solvent environment are based on a sample of $n = 10$ structures (frames), while in most cases $n = 5$ is used for the complex; only for the [0.2–2] region the complex results for $n = 10$ are provided. Δ is the difference between the complex and the solvent results. $\Delta E_{\text{intra-ligand}}$ is the intra-ligand energy, ΔF^{TI} was obtained by thermodynamic integration (see Table 23.2), and $T = 300$K is the absolute temperature. $\Delta S_{\text{internal}}$ is the internal entropy of the ligand. ΔA^0 is defined in Equation (23.69). ΔA^0 (exp.) is the experimental value [75]. The statistical errors for [0.2–2], $n = 10$ are 0.7 and 1.1 kcal/mol for $\Delta E_{\text{intra-ligand}}$ and ΔA^0, respectively. The corresponding errors for the other cases ($n = 5$) are up to 0.3 kcal/mol larger.

materialized stems probably from deficiencies in the force field, which, however, appears to be ineffective in [0.2–2] close to the crystal structure, where a very good result for ΔA^0 is obtained. This poses a practical dilemma for a researcher in the field – whether to perform a local equilibration (close to the crystal structure) or prefer a global (time consuming) one.

Both approaches have been adopted in the literature. For example, Fujitani et al. [27], who calculated the absolute binding free energy (ΔA^0) of FKBP12 complexed with various ligands, used 20 ns initial MD equilibrations for the complex, obtaining good results. Thus, for the present ligand described by two different force fields, Fujitani et al. obtained $\Delta A^0 = -10.1$ and -12.1 kcal/mol (see Table 23.4). This suggests that the sensitivity of the results for the initial equilibration depends mainly on the specific modeling and the system in hand. Also, as was pointed out in previous studies, long

TABLE 23.4

Results in kcal/mol for ΔA^0 Obtained by Different Methods and Force Fields

Group	ΔA^0 (kcal/mol)	D_{exp}	Comments
Experiment	−12.8	0	
HSMD_TI	−13.6 ± 1.1	−0.8	
Fujitany et al. [27]	−12.1	0.7	Force field 2
Fujitany et al. [27]	−10.1	2.7	Force field 1
Pande's group [26]	−9.5	3.2	Early work
Pande's group [22]	−11.8 ± 1.14	1.0	More recent work
Roux's group [33]	−10.1 ± 1.2	2.7	Initial structure: X-ray
Roux's group [34]	−10.8 ± 3.0	2.0	Optimal initial structure

Source: Reprinted with permission from General, I.J. and Meirovitch, H., *J. Chem. Theory Comput.*, 9, 4609–4619, 2013. Copyright 2019 American Chemical Society.

Note: D_{exp} is the deviation from the experimental value.

"production" simulations might be needed for a flexible system to adequately sample the different microstates visited by the system [24,37,38].

On the other hand, in calculations of *relative* free energy of complexes of FKBP12, Lamb and Jorgensen [49] used an MC sampling procedure based on "variations of all bond angles and dihedrals of the ligand and protein side chains as well as overall rotation and translation of the ligand and water molecules. *The protein backbone atoms were held fixed in their crystallographic positions.*" In a following paper from this group [50], the *absolute* binding free energy (ΔA^0) of several FKBP12 complexes was studied using the same methodology and modeling, with a very good agreement to the experiment. Very good results for the absolute free energy of binding were obtained also by Singh and Warshel [41], who applied weak (distance) harmonic restraints [$k = 0.01$ kcal/(mol·Å²)] to the crystallographic positions of all atoms within a spherical radius of 18 Å around the center of the active site; a metal ion was restrained there stronger with $k = 10$ kcal/(mol·Å²).

To investigate further this dilemma, HSMD-TI was also applied to two models of FKBP12-FK506, where the backbone atoms of FKBP12 were restrained harmonically to their crystallographic positions using a weak and a stronger force constant, $k = 1$ and $k = 10$ kcal/(mol·Å²), respectively. However, the corresponding results obtained, $\Delta A^0 = -25.1$ and −5.7 kcal/mol, strongly deviate from the experimental value, suggesting that this approach should be adopted with caution since the required strength of the restraints is a priori unknown.

The sensitivity of the results to the force field, equilibration, boundary conditions, and the method used is reflected in the large range of results for FKBP12-FK506 obtained in the literature for ΔA^0, which are summarized in Table 23.4.

The table reveals that only three out of seven results are within 1 kcal/mol from the experimental value (which is the target accuracy accepted in the field), while some of the results deviate from it by close to 3 kcal/mol. The fact that only rigorous statistical mechanics methods are involved suggests that an essential culprit for the above spread of results for ΔA^0 is inaccurate modeling (even though the lack of adequate sampling is always a factor that should not be ignored). Therefore, significant efforts should be devoted for improving the existing force fields.

Notice that the HSMD-TI result in the table is one of the best. However, this conclusion is not unbiased since the result, $\Delta A^0 = -13.6$, was chosen to appear in the table because it is closer to the *known* experimental value than the highly equilibrated result $\Delta A^0 = -16.8$ kcal/mol obtained in the [5–100] region! Clearly, such biases might infect other studies, and there has been a wide recognition in the field that performance comparisons of methods should be based on blind tests, where the experimental ΔA^0 is unknown to the competitors. This has led to the initiative SAMPL (statistical assessment of modeling of proteins and ligands), where several groups of targets have already been announced and blindly tested by different methods in recent years (see [76–79] and references cited therein).

23.9 Summary

We have described above the HSMD-TI method for calculating the absolute free energy of binding (ΔA^0) applied to the relatively large ligand, FK506, in the complex, FKBP12-FK506. HSMD-TI is an exact method in the sense that, in principle, its accuracy depends only on the amount of computer time invested. Also, HSMD-TI eliminates the end-point problem (encountered with DAM) and thus also the need for applying restraints (as with DDM). With HSMD-TI, ΔA^0 is obtained as a sum of several components, among them, ΔS_{ligand} – the difference in the internal conformational entropy of the ligand as it is transferred from the bulk solvent to the active site; calculation of ΔS_{ligand} (based on a reconstruction procedure) is not shared by other techniques, being unique to HSMD-TI. ΔS_{ligand} can provide microscopic insights into the binding mechanism and is thus important in rational drug design. The result $T\Delta S_{ligand} = 7.1 \pm 1.2$ kcal/mol obtained in the [0.2–2] region is relatively large due to the relatively large size of the ligand. HSMD-TI also provides the external entropy, $S_{external}$, which constitutes a measure for the global move of the ligand in the active site. However, the result for $S_{external}$ and $\Delta S_{ligand} = 7.1$ kcal/mol cannot be corroborated by the experiment. HSMD-TI is applicable to other problems, such as the calculation of solvation free energies or host-guest affinities. The different HSMD-TI results obtained in the three regions along the trajectory, and the spread of results for ΔA^0 obtained by other methods, reflect deficiencies in the current force fields. This problem, which has led to the SAMPL initiative, constitutes a major stumbling block for the theoretical prediction of ΔA^0 – a "hot" area in structural biology, mainly due to its relation to drug design. To make progress, in addition to the improvement of modeling, methods should be developed for handling large flexible ligands and ligands which carry charge.

REFERENCES

1. W. J. Moore. *Physical Chemistry.* (Prentice-Hall, London, 1962). Chapters 5 and 6.
2. T. L. Hill. *An Introduction to Statistical Thermodynamics.* (Dover, New York, 1986). Chapter 10.
3. D. A. McQuarrie. *Statistical Mechanics.* (Harper and Collins, New York, 1976). Chapter 9.
4. J. Hermans and S. Shankar. The free energy of xenon binding to myoglobin from molecular dynamics simulation. *Isr. J. Chem.* **27**, 225–227 (1986).
5. W. L. Jorgensen, J. K. Buckner, S. Boudon and J. Tirado-Rives. Efficient computation of absolute free energies of binding by computer simulations: Application to methane dimer in water. *J. Chem. Phys.* **89**, 3742–3746 (1988).
6. S. Miyamoto and P. A. Kollman. Absolute and relative binding free energy calculations of the interaction of biotin and its analogs with streptavidin using molecular dynamics/free energy perturbation approaches. *Proteins* **16**, 226–245 (1993).
7. M. K. Gilson, J. A. Given, B. L. Bush and J. A. McCammon. The statistical thermodynamic basis for computing of binding affinities: A critical review. *Biophys. J.* **72**, 1047–1069 (1997).
8. S. Boresch, F. Tttinger, M. Leitgeb and M. Karplus. Absolute binding free energies: A qualitative approach for their calculation. *J. Phys. Chem. B* **107**, 9535–9551 (2003).
9. M. K. Gilson and H.-X. Zhou. Calculation of protein-ligand binding affinities. *Ann. Rev. Biophys. Biomol. Struct.* **36**, 21–42 (2007).
10. N. Foloppe and R. Hubbard. Towards predictive ligand design with free-energy based computational methods? *Curr. Med. Chem.* **13**, 3583–3608 (2007).
11. Y. Deng and B. Roux. Computations of standard binding free energies with molecular dynamics simulations. *J. Phys. Chem. B* **113**, 2234–2246 (2009).
12. H.-X. Zhou and M. K. Gilson. Theory of free energy and entropy in noncovalent binding. *Chem. Rev.* **109**, 4092–4107 (2009).
13. A. Pohorille, C. Jarzynski and C. Chipot. Good practices in free-energy calculations. *J. Phys. Chem. B* **114**, 10235–10253 (2010).
14. C. Chipot and A. Pohorille (Eds.). *Free Energy Calculations, Theory and Applications in Chemistry and Biology.* (Springer, Berlin, 2007).
15. S. Miyamoto and P. A. Kollman. What determines the strength of noncovalent association of ligands to proteins in aqueous solution. *Proc. Natl. Acad. Sci. USA* **90**, 8402–8406 (1993).

16. L. Wang, B. J. Berne and R. A. Friesner. On achieving high accuracy and reliability in the calculation of relative protein–ligand binding affinities. *Proc. Natl. Acad. Sci. USA* **109**, 1937–1942 (2012).

17. I. V. Khavrutski and A. Wallqvist. Improved binding free energy predictions from single reference thermodynamic integration augmented with Hamiltonian replica exchange. *J. Chem. Theory Comput.* **7**, 3001–3011 (2011).

18. Y. Meng, D. S. Dashti and A. E. Roitberg. Computing alchemical free energy differences with Hamiltonian replica exchange molecular dynamics (H-REMD) simulations. *J. Chem. Theory Comput.* **7**, 2721–2727 (2011).

19. S. Riniker, C. D. Christ, H. S. Hansen, P. H. Hünenberger, C. Oostenbrink, D. Steiner and W. F. van Gunsteren. Calculation of relative free energies for ligand-protein binding, solvation, and conformational transitions using the GROMOS software. *J. Phys. Chem.* **115**, 13570–13577 (2001).

20. J. L. Knight and C. L. Brooks, III. Applying efficient implicit nongeometric constraints in alchemical free energy simulation. *J. Chem. Phys.* **32**, 3423–3432 (2011).

21. C. Oostenbrink, M. M. H. van Lipzig and W. F. van Gunsteren. Applications of molecular dynamics simulations in drug design. In *"Comprehensive Medicinal Chemistry II" Vol. 4, Computer-Assisted Drug Design*, J. B. Taylor and D. J. Triggle (Eds.), 651–668 (Elsevier, Amsterdam, 2007).

22. G. Jayachandran, M. R. Shirts, S. Park and V. J. Pande. Parallelized-over-parts computation of absolute binding free energy with docking and molecular dynamics. *J. Chem. Phys.* **125**, 084901-12 (2006).

23. D. Zhang, J. Gullingsrud and J. A. McCammon. Potential of mean force for a acetylcholine unbinding from the alpha7 nicotinic acetylcholine receptor ligand-binding domain. *J. Am. Chem. Soc.* **128**, 3019–3026 (2006).

24. D. L. Mobley and K. A. Dill. Binding of small-molecule ligands to proteins: "What you see" is not always "what you get". *Structure* **17**, 489–498 (2009).

25. J. Åqvist, V. B. Luzhkov and B. O. Brandsdal. Ligand binding affinities from MD simulations. *Acc. Chem. Res.* **35**, 358–365 (2002).

26. H. Fujitani, Y. Tanida, M. Ito, G. Jayachandran, C. D. Snow, M. R. Shirts, E. J. Sorin and V. J. Pande. Direct calculation of the binding free energies of FKBP ligands. *J. Chem. Phys.* **123**, 084108-5 (2005).

27. H. Fujitani, Y. Tanida and A. Matsuura. Massively parallel computation of absolute binding free energy with well equilibrated states. *Phys. Rev. E* **79**, 021914-12 (2009).

28. S. B. Dixit and C. Chipot. Can absolute free energies of association be estimated from molecular mechanical simulations? The biotin-streptavidin system revisited. *J. Phys. Chem. A* **105**, 9795–9799 (2001).

29. J. Hermans and L. Wang. Inclusion of loss of translational and rotational freedom in theoretical estimates of free energies of binding. Application to a complex of benzene and mutant T4 lysozyme. *J. Am. Chem. Soc.* **119**, 2707–2714 (1997).

30. D. Beglov and B. Roux. Finite representation of an infinite bulk system: Solvent boundary potential for computer simulations. *J. Chem. Phys.* **100**, 9050–9063 (1994).

31. B. Roux, M. Nina, R. Pomés and J. C. Smith. Thermodynamic stability of water molecules in the bacteriorhodopsin proton channel: A molecular dynamics free energy perturbation study. *Biophys. J.* **71**, 670–681 (1996).

32. W. Im, S. Bernèche and B. Roux. Generalized solvent boundary potential for computer simulations. *J. Chem. Phys.* **114**, 2924–2937 (2001).

33. J. Wang, Y. Deng and B. Roux. Absolute binding free energy calculations using molecular dynamics simulations with restraint potentials. *Biophys. J.* **91**, 2798–2814 (2006).

34. W. Jiang and B. Roux. Free energy perturbation Hamiltonian replica-exchange molecular dynamics (FEP/H-REMD) for absolute ligand binding free energy calculations. *J. Chem. Theory Comput.* **6**, 2559–2565 (2010).

35. D. Hamelberg and J. A. McCammon. Standard free energy of releasing a localized water molecule from the binding pockets of proteins: Double-decoupling method. *J. Am. Chem. Soc.* **126**, 7683–7689 (2004).

36. D. L. Mobley, J. D. Chodera and K. A. Dill. On the use of orientational restraint and symmetry corrections in alchemical free energy calculations. *J. Chem. Phys.* **125**, 084902-16 (2006).

37. D. L. Mobley, J. D. Chodera and K. A. Dill. Confine-and-release method: Obtaining correct binding free energies in the presence of protein conformational change. *J. Chem. Theory Comput.* **3**, 1231–1235 (2007).

38. D. L. Mobley, A. P. Graves, J. D. Chodera, A. C. McReynolds, B. K. Shoichet and K. A. Dill. Predicting absolute ligand binding free energies to a simple model site. *J. Mol. Biol.* **371**, 1118–1134 (2007).

39. M. R. Shirts, D. L. Mobley, J. D. Chodera and V. S. Pande. Accurate and efficient corrections for missing dispersion interactions in molecular simulations. *J. Phys. Chem. B* **111**, 13052–13063 (2007).

40. Y. Y. Sham, Z. T. Chu, H. Tao, and A. Warshel. Examining methods for calculations of binding free energies: LRA, LIE, PDLD-LRA and PDLD/S-LRA calculations of ligands binding to an HIV Protease. *Proteins* **39**, 393–407 (2000).

41. N. Singh and A. Warshel. A comprehensive examination of the contributions to the binding entropy of protein–ligand complexes. *Proteins* **78**, 1724–1735 (2010).

42. N. Singh and A. Warshel. Absolute binding free energy calculations: On the accuracy of computational scoring of protein-ligand interactions. *Proteins* **78**, 1705–1723 (2010).

43. J. Wang, R. Dixon and P. A. Kollman. Ranking ligand binding affinities with avidin: A molecular dynamic-based interaction energy study. *Proteins* **34**, 69–81 (1999).

44. B. Kuhn and P. A. Kollman. A ligand that is predicted to bind better to avidin than biotin: Insight from computational fluorine scanning. *J. Am. Chem. Soc.* **122**, 3909–3916 (2000).

45. Y. Tong, Y. Mei, Y. L. Li, Y. C. G. Ji and J. Z. H. Zhang. Electrostatic polarization makes a substantial contribution to the free energy of avidin-biotin binding. *J. Am. Chem. Soc.* **132**, 5137–5142 (2010).

46. J. M. J. Swanson, R. H. Henchman and J. A. McCammon. Revisiting free energy calculations: A theoretical connection to MM/PBSA and direct calculation of the association free energy. *Biophys. J.* **86**, 67–74 (2004).

47. J. Kongsted and U. Ryde. An improved method to predict the entropy term with MM/PBSA approach.). *J. Comput. Aided Mol. Des.* **23**, 53–71 (2009).

48. S. Genheden, T. Luchko, S. Gusarov, A. Kovalenko and U. Ryde. An MM/3D-RISM approach for ligand binding affinities. *J. Phys. Chem. B* **114**, 8505–8516 (2010).

49. M. L. Lamb and W. L. Jorgensen. Investigations of neurotropic inhibitors of FK506 binding protein via Monte Carlo simulations. *J. Med. Chem.* **41**, 3928–3939 (1998).

50. M. L. Lamb, J. Tirado-Rives and W. L. Jorgensen. Estimation of the binding affinity of FKBP12 inhibitors using a linear response method. *Bioorg. Med. Chem.* **7**, 851–860 (1999).

51. T. Lazaridis, A. Masunov and F. Gandolfo. Contributions to the binding free energy of ligands to avidin and streptavidin. *Proteins* **47**, 194–208 (2002).

52. M. S. Lee and M. A. Olson. Calculation of absolute protein-ligand binding affinity using path and end-point approaches. *Biophys. J.* **90**, 864–877 (2006).

53. W. L. Jorgensen. Free energy calculations: A breakthrough for modeling organic chemistry in solution. *Acc. Chem. Res.* **22**, 184–189 (1989).

54. W. L. Jorgensen. Interactions between amides in solution and the thermodynamics of weak binding. *J. Am. Chem. Soc.* **111**, 3770–3772 (1989).

55. Y. Deng and B. Roux. Calculation of standard binding free energies: Aromatic molecules in the T4 lysozyme L99A mutant. *J. Chem. Theory Comput.* **2**, 1255–1273 (2006).

56. T. A. Darden, D. M. York and L. G. Pedersen. Particle mesh Ewald: An $N \cdot \log(N)$ method for Ewald sums in large systems. *J. Chem. Phys.* **98**, 10089–10092 (1993).

57. A. Warshel, P. K. Sharma, M. Kato and W. W. Parson. Modeling electrostatic effects in proteins. *Biochim. Biophys. Acta* **1764**, 1647–1676 (2006).

58. E. King and A. Warshel. Surface constrained all-atom solvent model for effective simulations of polar solutions. *J. Chem. Phys.* **91**, 3647–3661 (1989).

59. F. S. Lee, Z. T. Chu and A. Warshel. Microscopic and semimicroscopic calculations of electrostatic energies in proteins by the polaris and enzymix programs. *J. Comput. Chem.* **14**, 161–185 (1993).

60. S. Cheluvaraja, M. Mihailescu and H. Meirovitch. Entropy and free energy of a mobile loop in explicit water. *J. Phys. Chem. B* **112**, 9512–9522 (2008).

61. M. Mihailescu and H. Meirovitch. Absolute free energy and entropy of a mobile loop of the enzyme AcetylCholineEsterase. *J. Phys. Chem. B* **113**, 7950–7964 (2009).

62. I. J. General and H. Meirovitch. Relative stability of the open and closed conformations of the active site loop of streptavidin. *J. Chem. Phys.* **134**, 025104-17 (2011).

63. I. J. General, R. Dragomirova and H. Meirovitch. A new method for calculating the absolute free energy of binding: the effect of a mobile loop on the avidin/biotin complex. *J. Phys. Chem. B.* **115**, 168–175 (2011).

64. I. J. General, R. Dragomirova and H. Meirovitch. Calculation of the absolute free energy of binding and related entropies with the HSMD-TI method: The FKBP12-L8 complex. *J. Chem. Theory Comput.* **7**, 4196–4207 (2011).

65. I. J. General, R. Dragomirova and H. Meirovitch. Absolute free energy of binding of avidin/biotin, revisited. *J. Phys. Chem. B* **116**, 6628–6636 (2012).
66. I. J. General and H. Meirovitch. Absolute free energy of binding and entropy of the FKBP12-FK506 complex: Effects of the force field. *J. Chem. Theory Comput.* **9**, 4609–4619 (2013).
67. W. D. Cornell, P. Cieplak, C. L. Bayly, I. R. Gould, K. M. Merz, Jr., D. M. Ferguson, D. C. Spellmeyer, T. Fox, J. W. Caldwell and P. A. Kollman. A second generation force field for the simulation of proteins, nucleic acids, and organic molecules. *J. Am. Chem. Soc.* **117**, 5179–5197 (1995).
68. J. Wang, R. M. Wolf, J. W. Calldwell, P. A. Kollman and D. A. Case. Development and testing of a general Amber force field. *J. Comput. Chem.* **25**, 1157–1174 (2004).
69. A. Jakalian, B. L. Bush, D. B. Jack and C. I. Bayly. Fast, efficient generation of high-quality atomic charges. AM1-BCC model: I. Method. *J. Comput. Chem.* **21**, 79–157 (2000).
70. A. Jakalian, D. B. Jack and C. I. Bayly. Fast, efficient generation of high-quality atomic charges. AM1-BCC model: II. Parameterization and validation. *J. Comput. Chem.* **23**, 1623–1641 (2002).
71. W. L. Jorgensen, J. Chandrasekhar, J. D. Madura, R. W. Impey and M. L. Klein. Comparison of simple potential functions for simulating liquid water. *J. Chem. Phys.* **79**, 926–935 (1983).
72. G. D. Van Duyne, R. F. Standaert, P. A. Karplus, S. L. Schreiber and J. Clardy. Atomic structure of FKBP-FK506, an immunophilin-immunosuppressant complex. *Science* **252**, 839–842 (1991).
73. M. P. Allen and D. J. Tildesley. *Computer Simulation of Liquids.* (Clarenden Press, Oxford, 1987).
74. M. Zacharias, T. P. Straatsma and J. A. McCammon. Separation-shifted scaling, a new scaling method for Lennard-Jones interactions in thermodynamic integration *J. Chem. Phys.* **100**, 9025–9031 (1994).
75. D. A. Holt, J. I. Luengo, D. S. Yamashita, et al. Design, synthesis, and kinetic evaluation of high-affinity FKBP ligands and the x-ray crystal structures of their complexes with FKBP12. *J. Am. Chem. Soc.* **115**, 9925–9938 (1993).
76. A. G. Skillman. SAMPL3. Blinded prediction of host-guest binding affinities, hydration free energies and trypsin inhibitors. *J. Comput. Aided Mol. Des.* **26**, 473–474 (2012).
77. H. S. Muddana, C. D. Varnado, C. W. Bielawski, A. R Urbach, L. Isaacs, M. T. Geballe and M. K. Gilson. Blind prediction of host-guest binding affinities: A new SAMPL3 challenge. *J. Comput. Aided Mol. Des.* **26**, 475–487 (2012).
78. T. Sulea, H. Hogues and E. O. Purisima. Exhaustive search and solvated interaction energy (SIE) for virtual screening and affinity prediction. *J. Comput. Aided Mol. Des.* **26**, 617–633 (2012).
79. C. Bannan, C. Caitlin, K. H. Burley, M. Chiu, M. R. Shirts, M. K. Gilson and D. L. Mobley. Blind prediction of cyclohexane–water distribution coefficients from the SAMPL5 challenge. *J. Comput. Aided Mol. Des.* **30**, 927–944 (2016).

Appendix

A1 Some Useful Integrals

$$\int_{-\infty}^{\infty} \exp- ax^2 dx = \sqrt{\frac{\pi}{a}} \tag{A1.1}$$

$$\int_{-\infty}^{\infty} x^2 \exp- ax^2 dx = \frac{\sqrt{\pi}}{2a^{3/2}} \tag{A1.2}$$

$$\int_{-\infty}^{\infty} x^4 \exp- ax^2 dx = \frac{3\sqrt{\pi}}{4a^{5/2}} \tag{A1.3}$$

$$\int_{-\infty}^{\infty} \exp(-ax^2 - 2bx)dx = \sqrt{\frac{\pi}{a}} \exp\left(\frac{b^2}{a}\right) \tag{A1.4}$$

A2 Exact Differential

Let

$$dz(x,y) = M(x,y)dx + N(x,y)dy \tag{A2.1}$$

where:

$$M(x,y) = \frac{\partial z(x,y)}{\partial x} \quad \text{and} \quad N(x,y) = \frac{\partial z(x,y)}{\partial y} \tag{A2.2}$$

The condition for an exact differential is:

$$\frac{\partial M(x,y)}{\partial y} = \frac{\partial N(x,y)}{\partial x} \quad \rightarrow \quad \frac{\partial z^2(x,y)}{\partial x \partial y} = \frac{\partial z^2(x,y)}{\partial y \partial x} \tag{A2.3}$$

Then the integral over $dz(x, y)$ is independent of the path.

A3 Random Number Generator

Suppose a probability space consisting of two elementary events, i and j, with probabilities 0.2 and 0.8, respectively. We want to carry out n independent experiments (product space) using a computer. We use a program called "*random number generator*" (RMG) that provides numbers within (0,1] distributed *uniformly*. For each experiment, a new random number, r, is generated: If $r \leq 0.2$, event i is chosen; if $r > 0.2$, j is chosen. To demonstrate how the RMG works, we provide below a simple RMG (which probably is not the best), where a random number required in the main program is provided by an external function.

Main Program

```
Initialization
real*8 seed
seed=8957321.d0

.

x=random(seed)          random(seed) is 0 < x≤1

.

end

Function random(seed)

seed=mod(69069.d0*|seed|+1.d0,2³²)
random=seed/2³²

return
end
comments:
2³²=4,294,967,296.d0
mod(a,b) is the remainder of a/b, e.g., mod(17,5)=2
```

A4 Jensen's Inequality [1]

The Jensen inequality states that if f is a convex function and P_i is a probability distribution, where $\Sigma \, P_i = 1$, then:

$$\sum_i P_i f(x_i) \leq f\left[\sum_i P_i x_i\right] \tag{A4.1}$$

Defining, $f = -x\ln x$, and the distribution, P_i', where $\Sigma P_i' = 1$, and $x_i = P_i' / P_i$ leads to:

$$-\sum_i P_i \frac{P_i'}{P_i} \ln(P_i' / P_i) \leq -\left[\sum_i P_i \frac{P_i'}{P_i}\right] \ln \sum_i P_i \frac{P_i'}{P_i} \tag{A4.2}$$

$$-\sum_i P_i' \ln(P_i' / P_i) \leq -\left[\sum_i P_i'\right] \ln \sum_i P_i' = 0 \tag{A4.3}$$

$$-\sum_i P_i' \ln(P_i' / P_i) \leq 0 \rightarrow \quad -\sum_i P_i \ln(P_i / P_i') \leq 0 \tag{A4.4}$$

Or

$$-\sum_i P_i \ln P_i \leq -\sum_i P_i \ln P_i' \tag{A4.5}$$

1. E. Parzen. *Modern Probability Theory and its Application.* (John Wiley & Sons, New York, 1960).

Index

Printed in the United States
By Bookmasters